Barron's Review Course Series

Let's Review:

Math B

Lawrence S. Leff
Former Assistant Principal, Mathematics Supervision
Franklin D. Roosevelt High School
Brooklyn, New York

Dedication

To Rhona . . .
For the understanding,
for the love,
and with love.

All inquiries should be addressed to:
Barron's Educational Series, Inc.
250 Wireless Boulevard
Hauppauge, New York 11788
http://www.barronseduc.com

Library of Congress Catalog Card No.: 2001052431

International Standard Book No.: 0-7641-1656-8

Library of Congress Cataloging-in-Publication Data

Leff, Lawrence S.
Let's review. Math B / Lawrence S. Leff.
p. cm.—(Barron's review course series)
Includes index.
ISBN 0-7641-1656-8
1. Mathematics. 2. Mathematics—Study and teaching (Secondary)
—New York (State) I. Title: Math B. II. Title. III. Series.
QA39.3.L44 2002
510—dc21

2001052431

PRINTED IN CANADA
9 8 7 6 5 4 3 2

TABLE OF CONTENTS

PREFACE

Let's Review: Math B offers in a *single book* complete coverage of the topics, concepts, and graphing calculator skills that are described in the New York State Core Curriculum for Mathematics B and are tested on the *new* Mathematic B Regents Examination.

For Which Course Can This Book Be Used?

Students entering high school in September 2001 or thereafter must complete 3 years of mathematics. To qualify for an *Advanced* Regents Diploma, these students must pass either the Mathematics B Regents Examination or, for as long as it continues to be offered, the Course III Regents Examination. *Let's Review: Math B* is designed to be used in any mathematics course in which students are being prepared to take the New York State Mathematics B Regents Exam. Not only does this book teach to the new Mathematics B Regents Exam, but also it presents the topics and skills needed to bridge the gap between a Math A-level course and more advanced college preparatory courses in mathematics.

What Special Features Does This Book Have?

- *A Compact Format Designed for Self-Study and Rapid Learning*
  The clear writing style quickly identifies essential ideas while avoiding unnecessary details. Each section begins with a brief "Key Ideas" section that promotes understanding by motivating or summarizing the material that follows. Helpful diagrams, convenient summaries, and numerous demonstration examples with worked-out solutions will be appreciated by students who need an easy-to-read book that provides complete and systematic preparation for the Mathematics B Regents Examination.
- *Flexible Topic Organization*
  For easy reference, major topics are grouped by their branches of mathematics. This feature helps make this book compatible with the various ways in which school systems and teachers may choose to arrange the Math B course topics.
- *Math B Skills and Regents Question Types*
  Ongoing preparation for the Mathematics B Regents Examination is provided through demonstration and practice Regents types of examples. Special consideration is given to new curriculum topics such as

constructing linear and nonlinear functions from data, finding maximum or minimum values of quadratic functions that arise in contextualized settings, and graphing calculator approaches to traditional algebraic topics. Actual Mathematics B Regents Examinations with answers are included at the end of the book.

- *Graphing Calculator Instruction*
  Topics have been arranged to permit natural integration of the graphing calculator into the development of the subject. The main features of the Texas Instruments TI-83 family of graphing calculators are introduced as they are needed to solve problems, rather than relegated to an appendix. Step-by-step calculator solutions are enhanced by actual screen shots. Use of the calculator to help make connections between graphical and algebraic representations of problems results in a deeper understanding of the underlying concepts.
- *Data Analysis*
  Linear and nonlinear regression models, standard deviation, and normal curves are discussed with the help of the special statistics features of the graphing calculator.
- *Answers and Solution Hints to Practice Exercises*
  The practice exercises at the end of each section, headed "Check Your Understanding," include Regents types of questions at different levels of difficulty, as well as questions designed to build understanding, skill, and confidence. Answers and solution hints to the practice exercises are provided at the end of the book.

Who Should Use This Book?

- *Students*
  Students enrolled in Regents-level mathematics classes will find this book helpful if they need either additional explanation and practice on a troublesome topic being studied in class or review on specific topics when they are preparing for the Mathematics B Regents Examination.
- *Teachers*
  Since this book is designed to be compatible with all styles of instruction and curriculum organization, classroom teachers will want to include *Let's Review: Math B* in their personal, departmental, and school libraries. Teachers will find it a valuable lesson-planning aid as well as a helpful source of classroom exercises, homework problems, and test questions.
- *School Districts and Mathematics Departments*
  School districts and Mathematics Departments may wish to adopt this book for their Math B classes as a cost-effective way of supplementing existing materials with up-to-date coverage of Math B topics, detailed TI-83 graphing calculator instruction at the keystroke level, and

thorough Regents examination preparation in a *single* volume. As illustrated in the accompanying table, the units and chapter topics can be used as the foundation of a comprehensive $1\frac{1}{2}$-year course that follows Mathematics A and prepares students for the Mathematics B Regents Examination.

Term	Major Topics	Chapters
1	UNIT I Geometric proofs; coordinates; circles and angle measurement	1–4
2	UNITS II and III Algebraic methods; functions, graphing, and calculator solutions; transformations	5–10
3	UNITS IV and V Regression analysis; probability; standard deviation and normal curve; trigonometric functions and laws; solving triangles	11–15

LAWRENCE S. LEFF
December 2001

Unit One

GEOMETRIC PROOFS AND CIRCLES

CHAPTER 1

PROOFS WITH TRIANGLES

1.1 BASIC GEOMETRIC TERMS AND FACTS

KEY IDEAS

Geometry has its own special vocabulary. Before you can investigate and prove geometric relationships, you must know some basic geometric terms and facts.

Congruent Figures

Figures that have exactly the same size and shape are **congruent**. The symbol $\cong$ means "is congruent to."

- Two *line segments* are congruent if they have the same length. If $AB = 6$ inches and $CD = 6$ inches, then $AB = CD$ and, as a result, $\overline{AB} \cong \overline{CD}$.
- Two *angles* are congruent if they have the same degree measure. If $m\angle 1 = 40$ and $m\angle 2 = 40$, then $m\angle 1 = m\angle 2$ and, as a result, $\angle 1 \cong \angle 2$.
- Two *triangles* are congruent if they agree in all six pairs of corresponding parts, as shown in Figure 1.1. Corresponding sides lie opposite corresponding angles.

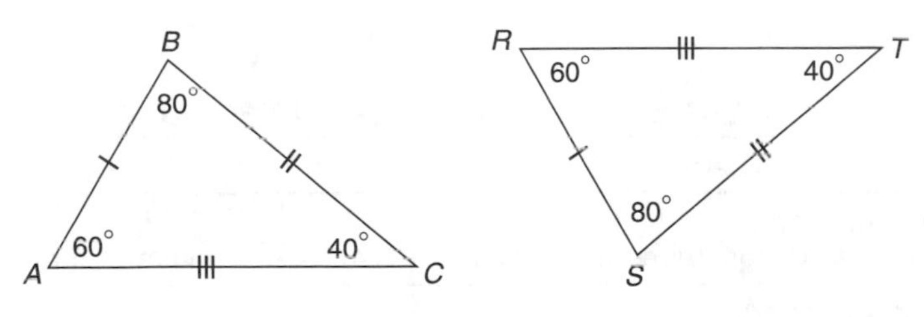

Figure 1.1 $\triangle ABC \cong \triangle RST$

Midpoints, Bisectors, and Medians

- In Figure 1.2, M is the *midpoint* of $\overline{AB}$ since $\overline{AM} \cong \overline{BM}$ or, equivalently, $AM = BM = \frac{1}{2}AB$. A **midpoint** divides a segment into two congruent segments.

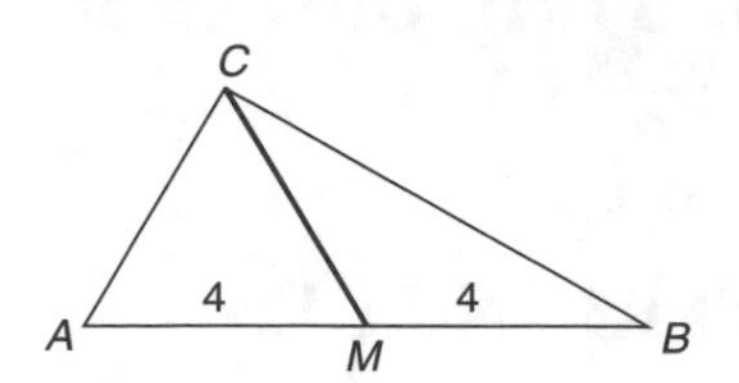

Figure 1.2 Midpoint and Median of a Triangle

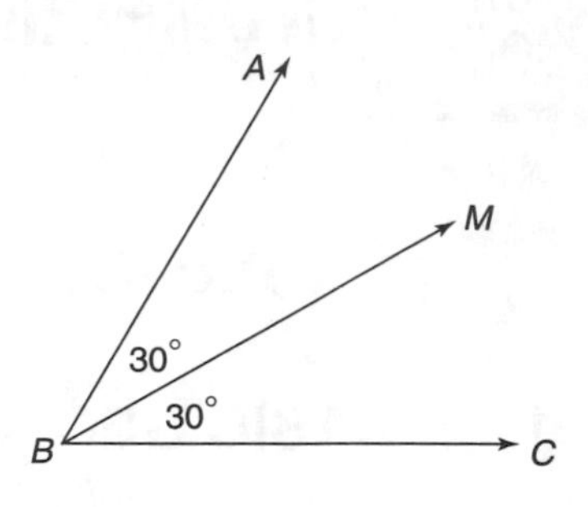

Figure 1.3 Angle Bisector

- In Figure 1.2, $\overline{CM}$ *bisects* $\overline{AB}$ since it intersects $\overline{AB}$ at its midpoint. A **bisector of a segment** is a line or ray that divides the segment into two congruent segments. Similarly, a **bisector of an angle** is a line or ray that divides the angle into two congruent angles, as shown in Figure 1.3. Here, $\overrightarrow{BM}$ bisects $\angle ABC$, so $\angle ABM \cong \angle MBC$.
- In Figure 1.2, in $\triangle ABC$, $\overline{CM}$ is the median to side $\overline{AB}$. A **median** of a triangle is a segment whose endpoints are a vertex of the triangle and the midpoint of the opposite side.

Congruence and Equality Relations

When working with line segments or angles, the properties of congruence (or equality) summarized in the accompanying table may be needed.

PROPERTIES OF CONGRUENCE AND EQUALITY

Property	Example
REFLEXIVE PROPERTY: A quantity is congruent (or equal) to itself.	$\overline{AB} \cong \overline{AB}$, and $m\angle 1 = m\angle 1$.
SYMMETRIC PROPERTY: The quantities on either side of a congruence (equal) sign can be interchanged.	If $\angle 1 \cong \angle 2$, then $\angle 2 \cong \angle 1$. If $AB = CD$, then $CD = AB$.
TRANSITIVE PROPERTY: If two quantities are congruent (equal) to the same quantity, they are congruent (equal) to each other.	If $\angle 1 \cong \angle 2$ and $\angle 2 \cong \angle 3$, then $\angle 1 \cong \angle 3$.
SUBSTITUTION PROPERTY: A quantity may be substituted for its equal.	If $m\angle 1 + m\angle 2 = 90$ and $\angle 2 \cong \angle 3$, then $m\angle 1 + m\angle 3 = 90$.

Postulate Versus Theorem

Some beginning facts in geometry, called *postulates*, are so basic that they cannot be arrived at using simpler facts. A **postulate** is a statement that is accepted without proof. For instance, the observation "Two points determine a line" is a postulate.

A general statement that can be demonstrated to be true is called a **theorem**. The statement "The sum of the measures of the three angles of a triangle is 180" is a fact that can be proved, so it is a theorem. Theorems are often stated in the conditional form "*If statement 1, then statement 2*," where statement 1 is the hypothesis or "given," and statement 2 is the conclusion or the part that needs to be proved.

Perpendicular Lines and Altitudes

Perpendicular lines ($\perp$) are two lines that intersect at right (90°) angles. In Figure 1.4, right angle 1 is congruent to right angle 2 since both angles measure 90°.

THEOREM: *Right angles are congruent.*

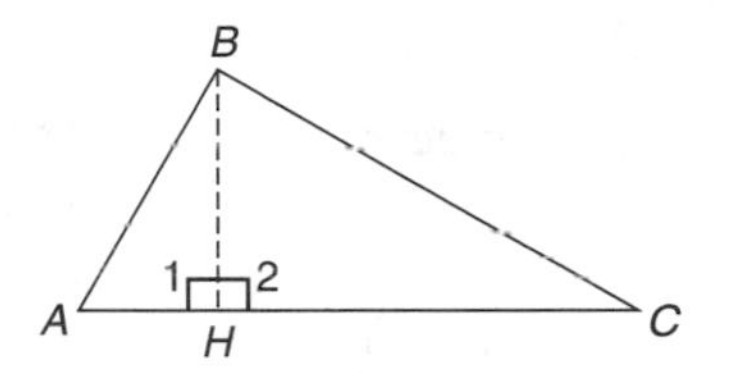

Figure 1.4 Altitude $\overline{BH}$ Is Perpendicular to $\overline{AC}$

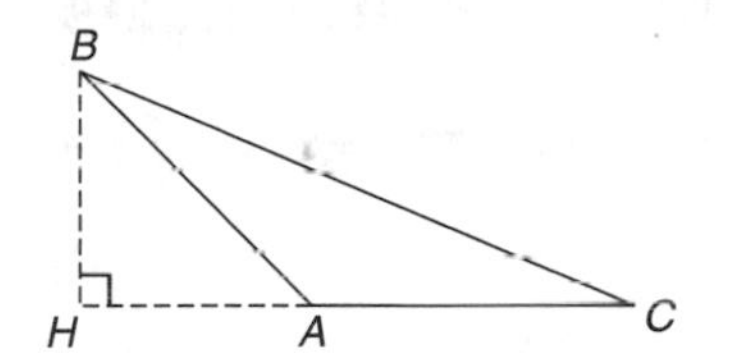

Figure 1.5 Altitude $\overline{BH}$ Intersects the Opposite Side When Extended

An **altitude of a triangle** is a segment drawn from one vertex perpendicular to the opposite side (see Figure 1.4) or perpendicular to the line that contains the opposite side, as in Figure 1.5.

Knowing When Lines Are Perpendicular

Suppose that, in Figure 1.4, it is not given that $\overline{BH} \perp \overline{AC}$ and $\angle 1$ and $\angle 2$ are not marked as right angles. If $m\angle 1 = 85$ and $m\angle 2 = 95$, then $\overline{BH} \not\perp \overline{AC}$. On the other hand, if adjacent angles 1 and 2 are equal in measure, then $\overline{BH} \perp \overline{AC}$ since $m\angle 1 = m\angle 2 = 90$.

THEOREM: *If two lines intersect to form congruent adjacent angles, then the lines are perpendicular.*

Complementary Angles

Complementary angles are two angles whose degree measures add up to 90, as shown in Figure 1.6.

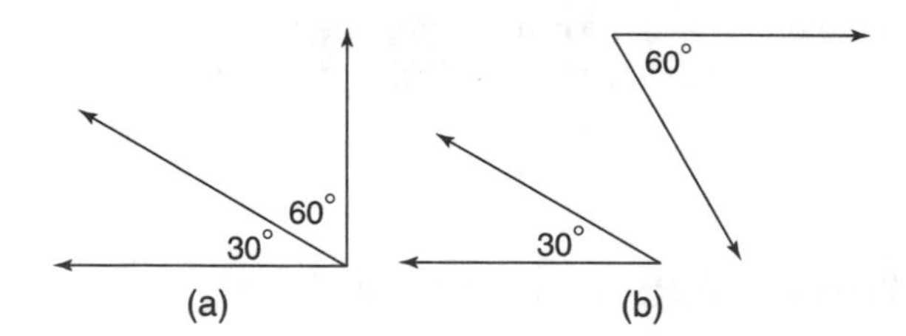

Figure 1.6 Pairs of Complementary Angles

THEOREM: *If the noncommon sides of two adjacent angles are perpendicular, as in Figure 1.6(a), the angles are complementary.*

Supplementary Angles

Supplementary angles are two angles whose degree measures add up to 180.

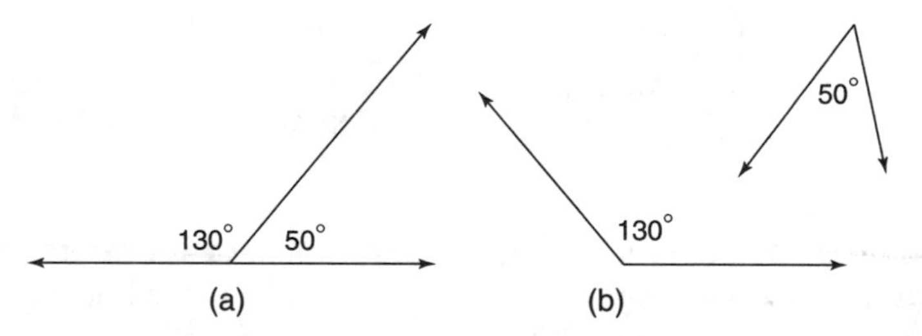

Figure 1.7 Pairs of Supplementary Angles

THEOREM: *If the exterior sides of two adjacent angles are opposite rays, as in Figure 1.7(a), then the angles are supplementary.*

THEOREM: *Supplements (or complements) of the same or congruent angles are congruent.* For example, if $\angle X$ and $\angle Y$ are both supplementary to $\angle A$ ($m\angle X + m\angle A = 180$ and $m\angle Y + m\angle A = 180$), then $\angle X \cong \angle Y$.

Vertical Angles

Vertical angles are opposite angles formed when two lines intersect, as shown in Figure 1.8.

THEOREM: *Vertical angles are congruent.*

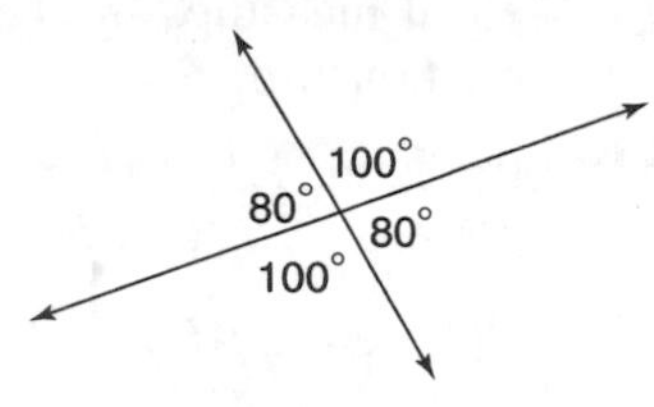

Figure 1.8 Vertical Angles

Arithmetic Properties of Equality

Drawing conclusions about whether pairs of segments or pairs of angles are congruent may depend on adding, subtracting, multiplying, or dividing the lengths of the segments or the measures of the angles.

- **ADDITION PROPERTY:** If equals are added to equals, the sums are equal. In Figure 1.9, if it is given that $AD = FC$, then adding CD to AC and FC makes the sums equal:

$$\underbrace{AD + CD}_{AC} = \underbrace{FC + CD}_{FD}$$

or $\overline{AC} \cong \overline{FD}$

Figure 1.9 Addition Property

- **SUBTRACTION PROPERTY:** If equals are subtracted from equals, the differences are equal.

In Figure 1.10, if $m\angle 1 = m\angle 2$, then subtracting $m\angle MOB$ from the degree measures of angles 1 and 2 makes the differences equal:

$$\underbrace{m\angle 1 - m\angle MOB}_{m\angle TOM} = \underbrace{m\angle 2 - m\angle MOB}_{m\angle WOB}$$

or $\angle TOM \cong \angle WOB$

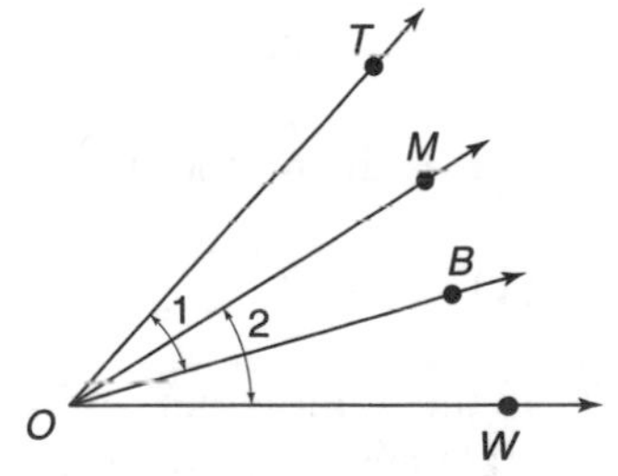

Figure 1.10 Subtraction Property

- **MULTIPLICATION AND DIVISION PROPERTIES:** Multiplying or dividing equal measures by the same nonzero quantity produces an equivalent relationship.

  In Figure 1.11, $AB = BC$, E is the midpoint of $\overline{AB}$, and F is the midpoint of $\overline{BC}$. Hence, $AE = \frac{1}{2}AB$ and $CF = \frac{1}{2}BC$. Since $\frac{1}{2}AB$ and $\frac{1}{2}BC$ are halves of equals, they are equal to each other, so $AE = CF$.

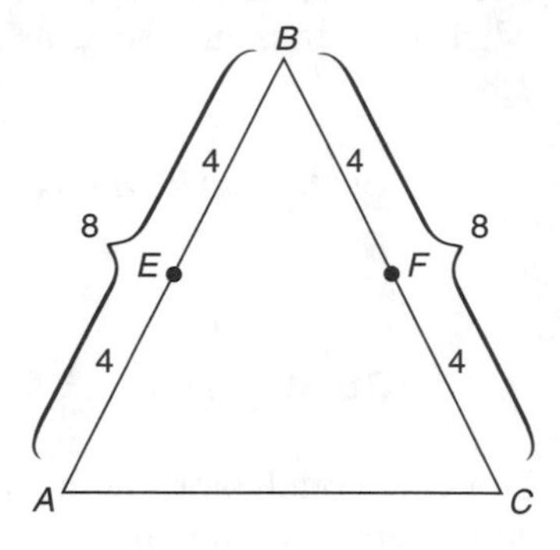

Figure 1.11 Multiplication Property

Check Your Understanding of Section 1.1

Multiple Choice

1. If M is the midpoint of $\overline{AB}$, which statement is false?
(1) $\frac{AB}{2} = MB$
(2) $AB - MB = AM$
(3) $AM + AB = MB$
(4) $AM = MB$

2. If C is the midpoint of $\overline{AB}$ and D is the midpoint of $\overline{AC}$, which statement is true?
(1) $AC > BC$ (2) $AD < CD$ (3) $DB = AC$ (4) $DB = 3CD$

3. If the complement of $\angle A$ is greater than the supplement of $\angle B$, which statement must be true?
(1) $m\angle A + m\angle B = 180$ (2) $m\angle A + m\angle B = 90$ (3) $m\angle A < m\angle B$
(4) $m\angle A > m\angle B$

4. The degree measure of one angle exceeds two times the degree measure of its complement by 21. What is the degree measure of the smaller angle?
(1) 23 (2) 53 (3) 67 (4) 127

5. In the accompanying figure, what is the value of y?
(1) 20 (2) 30 (3) 45 (4) 60

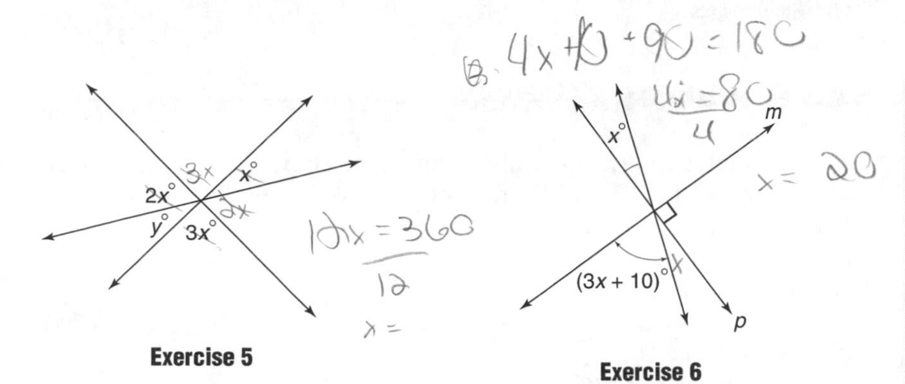

6. In the accompanying figure, line $m \perp$ line p. What is the value of x?
(1) 15 (2) 20 (3) 24 (4) 30

1.2 PROVING TRIANGLES CONGRUENT

KEY IDEAS

To prove that two triangles are congruent, show that a particular set of three pairs of corresponding parts are congruent: $SAS \cong SAS$, $ASA \cong ASA$, or $SSS \cong SSS$. For example, $SAS \cong SAS$ means that, if two sides, S, and the included angle, A, of one triangle are congruent to the corresponding parts of another triangle, as shown in the accompanying figure, then the triangles are congruent.

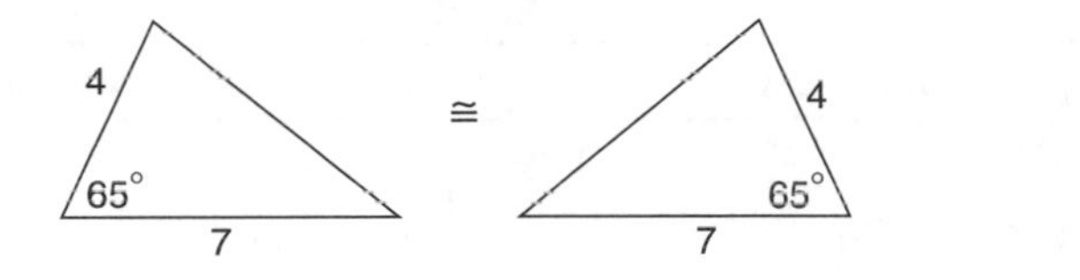

General Strategy for Proving Triangles Congruent

The setup of a geometric proof usually includes a diagram, the facts that are "Given" about that diagram, and the geometric relationship that you need to "Prove" is true. To prove that two triangles are congruent:

- Mark the diagram with the "Given." Then mark off any other pairs of parts that are congruent, such as common angles, common sides, vertical angles, and congruent parts formed by midpoints or bisectors.
- Decide on the congruence method to use.
- Write the formal proof. Show your reasoning as a set of numbered statements in the left column of a table with the corresponding reasons in the right column, as illustrated in Example 1.

SAS Postulate: $SAS \cong SAS$

To prove that two triangles are congruent, show that two sides and the included angle of one triangle are congruent to the corresponding parts of the other triangle.

Example 1

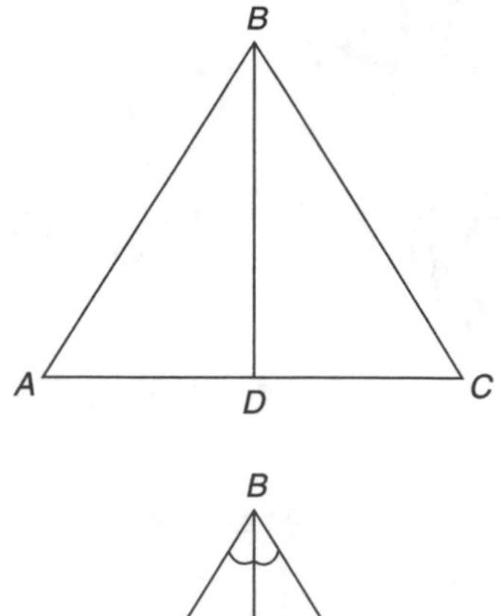

Given: $\overline{AB} \cong \overline{BC}$,
$\overline{BD}$ bisects $\angle ABC$.
Prove: $\triangle ADB \cong \triangle CDB$.

Solution: See the proof.

PLAN:

- Mark the diagram with the "Given." Indicate that $\overline{BD}$ is a side of both triangles by marking it with a cross, X.
- Since $\overline{AB} \cong \overline{BC}$ (***Side***), $\angle ABD \cong \angle CBD$ (***Angle***), and $\overline{BD} \cong \overline{BD}$ (***Side***), use the SAS postulate to prove the two triangles congruent.
- Write the proof.

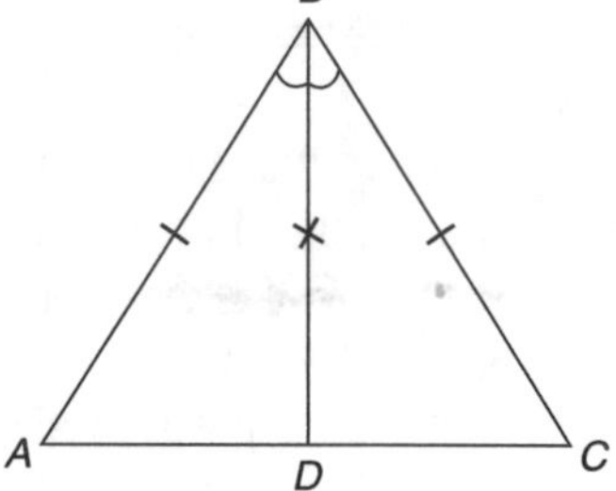

PROOF

Statement		Reason
1. $\overline{AB} \cong \overline{BC}$	*Side*	**1.** Given.
2. $\overline{BD}$ bisects $\angle ABC$.		**2.** Given.
3. $\angle ABD \cong \angle CBD$	*Angle*	**3.** A bisector divides an angle into two congruent angles.
4. $\overline{BD} \cong \overline{BD}$	*Side*	**4.** Reflexive property of congruence.
5. $\triangle ADB \cong \triangle CDB$		**5.** SAS postulate.

TIP

The "Reason" column of a proof may include the following types of statements:

- A fact contained in the "Given."
- A property of equality or a property of congruence.
- A definition, postulate, or theorem.

ASA Postulate: $ASA \cong ASA$

To prove that two triangles are congruent, show that two angles and the included side of one triangle are congruent to the corresponding parts of the other triangle.

Example 2

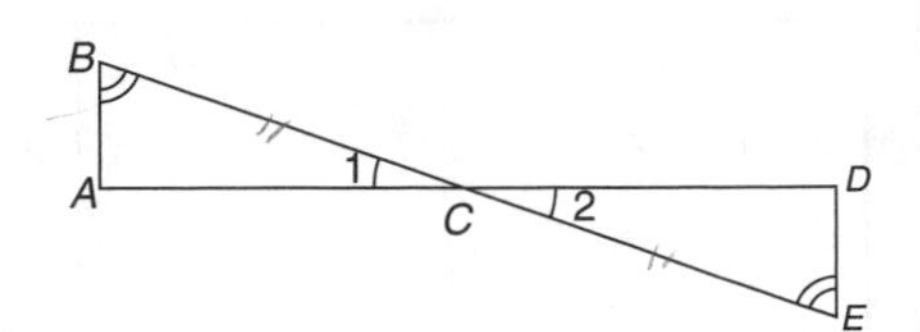

Given: $\overline{AD}$ bisects $\overline{BE}$,
$\angle B \cong \angle E$.
Prove: $\triangle ABC \cong \triangle DEC$.

Solution: See the proof.

PLAN: Since $\angle B \cong \angle E$, $\overline{BC} \cong \overline{EC}$, and $\angle 1 \cong \angle 2$ (vertical angles), prove the triangles congruent by using the ASA postulate.

PROOF

Statement		Reason
1. $\angle B \cong \angle E$	*Angle*	**1.** Given.
2. $\overline{AD}$ bisects $\overline{BE}$		**2.** Given.
3. $\overline{BC} \cong \overline{EC}$	*Side*	**3.** A bisector divides a segment into two congruent segments.
4. $\angle 1 \cong \angle 2$	*Angle*	**4.** Vertical angles are congruent.
5. $\triangle ABC \cong \triangle DEC$		**5.** ASA postulate.

SSS Postulate: $SSS \cong SSS$

To prove that two triangles are congruent, show that the three sides of one triangle are congruent to the corresponding parts of the other triangle.

Example 3

Given: $\overline{AB} \cong \overline{DE}$,
$\overline{BC} \cong \overline{EF}$,
$\overline{AF} \cong \overline{DC}$.
Prove: $\triangle ABC \cong \triangle DEF$.

Solution: See the proof.

PLAN: It is given that $\overline{AF} \cong \overline{DC}$, and you know that $\overline{FC} \cong \overline{FC}$. Thus:

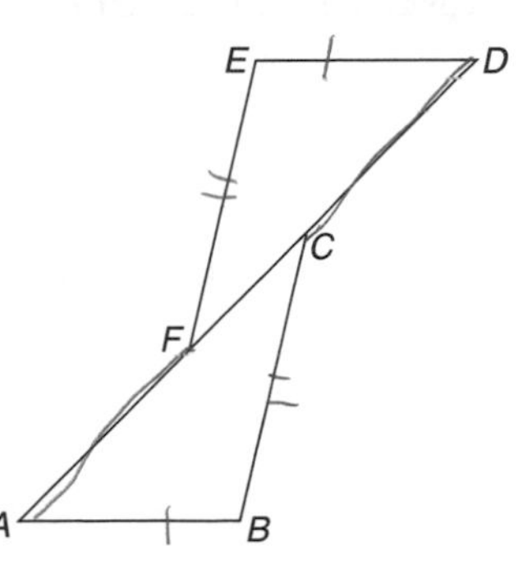

$$\underbrace{AF + FC}_{AC} = \underbrace{DC + FC}_{DF},$$

so $\overline{AC} \cong \overline{DF}$, and the two triangles are congruent by the SSS postulate.

PROOF

Statement		Reason
1. $\overline{AB} \cong \overline{DE}$	*Side*	**1.** Given.
2. $\overline{BC} \cong \overline{EF}$	*Side*	**2.** Given.
3. $AF = DC$		**3.** Given.
4. $FC = FC$		**4.** Reflexive property of equality.
5. $\underbrace{AF + FC}_{\downarrow} = \underbrace{DC + FC}_{\downarrow}$		**5.** Addition property of equality.
6. $AC = DF$		**6.** Substitution property of equality.
7. $\overline{AC} \cong \overline{DF}$	*Side*	**7.** Segments that have the same length are congruent.
8. $\triangle ABC \cong \triangle DEF$		**8.** SSS postulate.

Proving Overlapping Triangles Congruent

When trying to prove that overlapping triangles are congruent, look for a side or an angle that is shared by the two triangles. After copying the diagram in your notebook, you should outline the sides of the two triangles using two pencils of different colors. If you don't have colored pencils handy, outline the sides of one of the triangles with a thick, heavy line.

Example 4

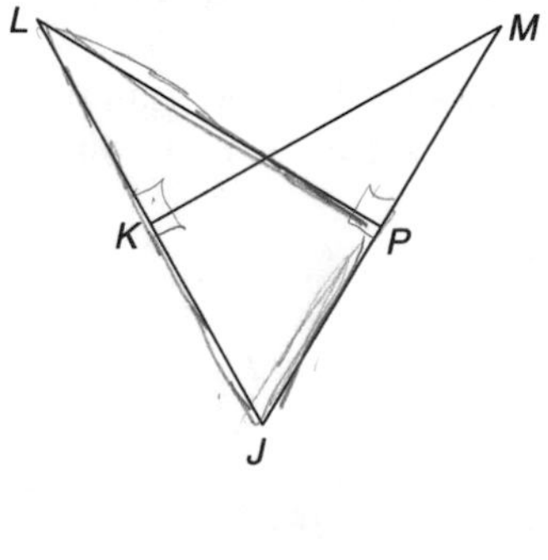

Given: $\overline{MK} \perp \overline{JL}$, $\overline{LP} \perp \overline{JM}$,
$\overline{JK} \cong \overline{JP}$.
Prove: $\triangle JMK \cong \triangle JLP$.

L M K P J

Solution: See the proof.

PLAN: Since $\angle J \cong \angle J$ (*Angle*), $JK \cong JP$ (*Side*), and $\angle 1 \cong \angle 2$ (*Angle*), prove the triangles congruent by using the ASA postulate.

PROOF

Statement		Reason
1. $\overline{MK} \perp \overline{JL}$, $\overline{LP} \perp \overline{JM}$		1. Given.
2. Angles 1 and 2 are right angles.		2. Perpendicular lines intersect to form right angles.
3. $\angle 1 \cong \angle 2$	*Angle*	3. All right angles are congruent.
4. $\overline{JK} \cong \overline{JP}$	*Side*	4. Given.
5. $\angle J \cong \angle J$	*Angle*	5. Reflexive property of congruence.
6. $\triangle JMK \cong \triangle JLP$		6. ASA postulate.

Check Your Understanding of Section 1.2

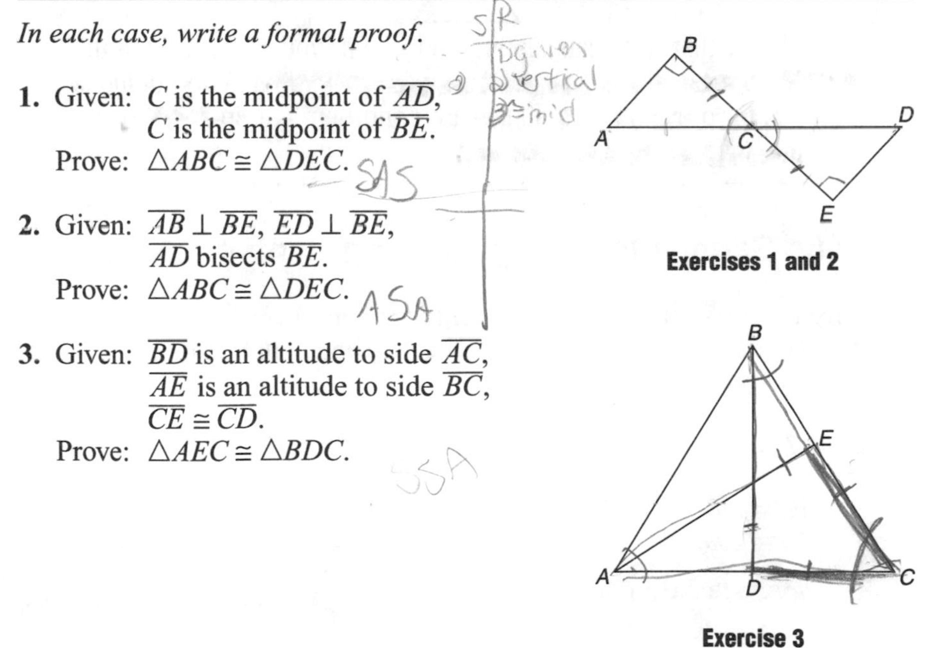

In each case, write a formal proof.

1. Given: C is the midpoint of $\overline{AD}$,
   C is the midpoint of $\overline{BE}$.
   Prove: $\triangle ABC \cong \triangle DEC$.

2. Given: $\overline{AB} \perp \overline{BE}$, $\overline{ED} \perp \overline{BE}$,
   $\overline{AD}$ bisects $\overline{BE}$.
   Prove: $\triangle ABC \cong \triangle DEC$.

3. Given: $\overline{BD}$ is an altitude to side $\overline{AC}$,
   $\overline{AE}$ is an altitude to side $\overline{BC}$,
   $\overline{CE} \cong \overline{CD}$.
   Prove: $\triangle AEC \cong \triangle BDC$.

Exercises 1 and 2

Exercise 3

4. Given: $\overline{AB} \cong \overline{BC}$, $\angle A \cong \angle C$.
Prove: $\triangle AEB \cong \triangle CDB$.

5. Given: $\overline{AE} \perp \overline{BC}$, $\overline{CD} \perp \overline{AB}$,
$\overline{BD} \cong \overline{BE}$.
Prove: $\triangle ABE \cong \triangle CBD$.

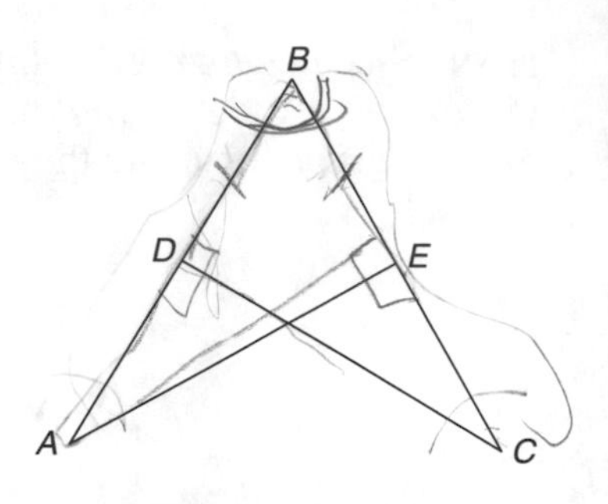

Exercises 4 and 5

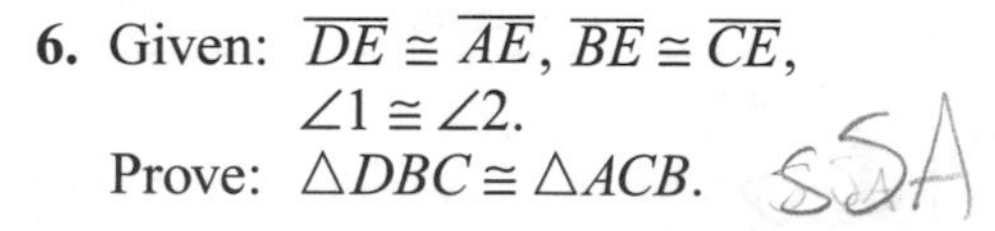

6. Given: $\overline{DE} \cong \overline{AE}$, $\overline{BE} \cong \overline{CE}$,
$\angle 1 \cong \angle 2$.
Prove: $\triangle DBC \cong \triangle ACB$.

7. Given: $\overline{AB} \perp \overline{BE}$, $\overline{DC} \perp \overline{CE}$,
$\angle 1 \cong \angle 2$,
$\overline{AB} \cong \overline{DC}$.
Prove: $\triangle ABC \cong \triangle DCB$.

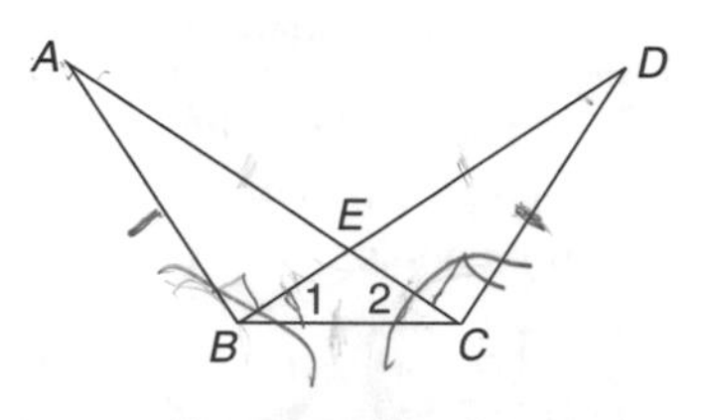

Exercises 6 and 7

1.3 APPLYING CONGRUENT TRIANGLES

KEY IDEAS

If you know that two triangles are congruent, you can conclude that any pair of matching parts must also be congruent. Two additional methods for proving two triangles are congruent are *AAS* ≅ *AAS* and, in a right triangle, *Hy-Leg* ≅ *Hy-Leg*.

Proving Segments or Angles Congruent

To prove that two angles or two segments are congruent, look for two triangles that contain these parts. Then use the "Given" to help prove these triangles are congruent.

Example 1

Given: $\overline{AC} \cong \overline{DB}$, $\overline{AB} \cong \overline{DC}$.
Prove: $\angle A \cong \angle D$.

Solution: See the proof.

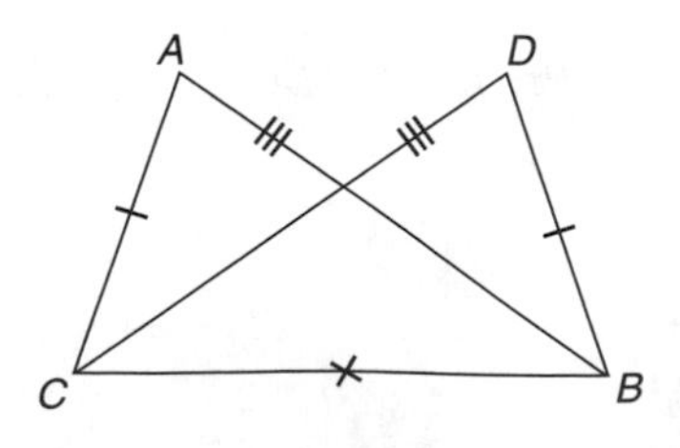

PLAN: *Work backwards* from the "Prove" by answering two questions:

- Which triangles contain angles A and D as parts? Since $\angle A$ is an angle of $\triangle ABC$ and $\angle D$ is an angle of $\triangle DBC$, first prove $\triangle ABC \cong \triangle DCB$.
- How can these triangles be proved congruent? Prove $\triangle ABC \cong \triangle DCB$ by the SSS postulate. Since "*C*orresponding *P*arts of *C*ongruent *T*riangles are *C*ongruent," $\angle A \cong \angle D$. The expression enclosed by quotation marks is sometimes abbreviated as CPCTC.

PROOF

Statement		Reason
1. $\overline{AC} \cong \overline{DB}$, $\overline{AB} \cong \overline{DC}$	*Side* *Side*	**1.** Given.
2. $\overline{BC} \cong \overline{BC}$	*Side*	**2.** Reflexive property of congruence.
3. $\triangle ABC \cong \triangle DCB$		**3.** SSS postulate.
4. $\angle A \cong \angle D$		**4.** CPCTC.

Proving Statements

To prove a conditional statement that has the form "If . . . then . . . ," set up a diagram with the information in the *if* clause as the "Given" and the information in the *then* clause as the "Prove."

Example 2

Prove: If a point lies on the perpendicular bisector of a line segment, then it is equidistant from the endpoints of the line segment.

Solution: The "Given" is taken from the *if* clause of the statement to be proved, while the "Prove" is taken from the *then* clause.

Given: Line $\ell \perp \overline{AB}$ at M,
$\overline{AM} \cong \overline{BM}$,
point P is any point on ℓ,
$\overline{AP}$ and $\overline{BP}$ are drawn.
Prove: $AP = BP$.

PLAN: Show that $\triangle AMP \cong \triangle BMP$ by SAS. By CPCTC, $\overline{AP} \cong \overline{BP}$. Therefore, $AP = BP$.

The formal proof is left for you to complete.

Angle-Angle-Side Theorem ($AAS \cong AAS$)

AAS THEOREM: *Two triangles are congruent when two angles and the side opposite one of these angles in one triangle are congruent to the corresponding parts of the other triangle.*

Example 3

Given: $\angle B \cong \angle C$,
$\overline{AH}$ is the altitude to side $\overline{BC}$.
Prove: $\overline{AB} \cong \overline{AC}$.

Solution:

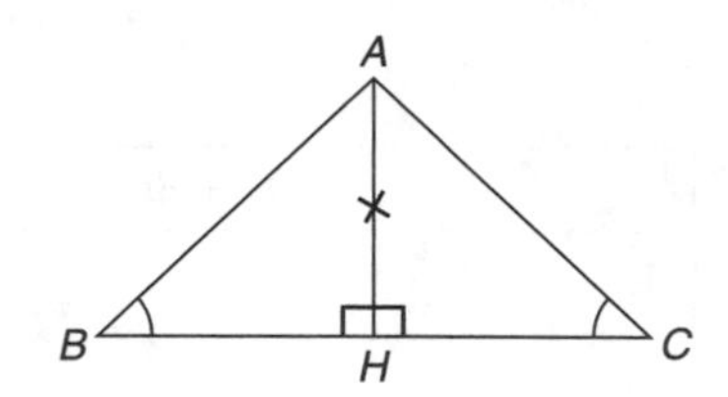

PLAN: Prove that $\triangle AHB \cong \triangle AHC$ by the AAS Theorem since $\angle B \cong \angle C$ (Given), $\angle AHB \cong \angle AHC$ (An altitude forms right angles with the side to which it is drawn), and $\overline{AH} \cong \overline{AH}$ (Reflexive property). The formal proof is left for you to complete.

Hypotenuse-Leg Theorem ($Hy\text{-}Leg \cong Hy\text{-}Leg$)

HY-LEG THEOREM: *Two* right *triangles are congruent when the hypotenuse and a leg of one right triangle are congruent to the corresponding parts of the other right triangle.*

Example 4

Given: $\overline{BA} \perp \overline{AD}$, $\overline{BC} \perp \overline{CD}$,
$\overline{AD} \cong \overline{CD}$.
Prove: $\overline{BD}$ bisects $\angle ABC$.

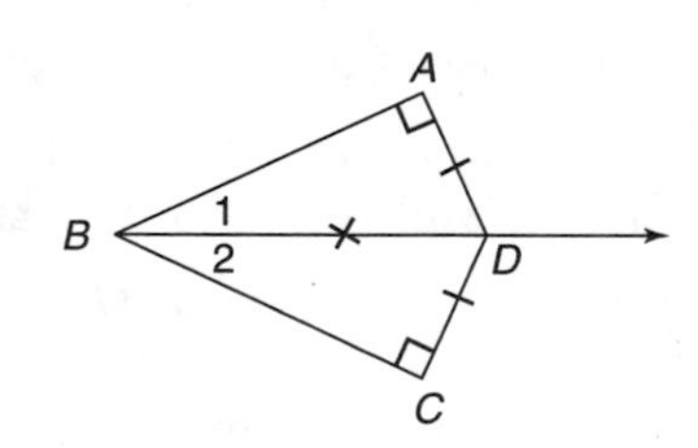

Solution: See the proof.

PLAN: Show that $\angle 1 \cong \angle 2$ by proving that $\triangle BAD \cong \triangle BCD$, using the Hy-Leg Theorem.

PROOF

Statement		Reason
1. $\overline{BA} \perp \overline{AD}$, $\overline{BC} \perp \overline{CD}$		1. Given.
2. Triangles *BAD* and *BCD* are right triangles.		2. A triangle two of whose sides are perpendicular is a right triangle.
3. $\overline{BD} \cong \overline{BD}$	*Hyp*	3. Reflexive property of congruence.
4. $\overline{AD} \cong \overline{CD}$	*Leg*	4. Given.
5. $\triangle BAD \cong \triangle BCD$		5. Hy-Leg Theorem.
6. $\angle 1 \cong \angle 2$		6. CPCTC.
7. $\overrightarrow{BD}$ bisects $\angle ABC$.		7. A ray that divides an angle into two congruent angles bisects the angle.

Proving a Segment Has a Special Property

To prove that a segment is:

- A *median*, show that it divides another segment into two congruent parts. In Figure 1.12, to prove $\overline{BX}$ is a median, first prove $\overline{AX} \cong \overline{CX}$.
- An *angle bisector*, show that it divides an angle into two congruent parts. In Figure 1.12, to prove $\overline{BX}$ bisects $\angle ABC$, first prove $\angle ABX \cong \angle CBX$.
- *Perpendicular* to the side to which it is drawn, show that it forms congruent adjacent angles with that side. In Figure 1.12, to prove $\overline{BX} \perp \overline{AC}$, first prove $\angle 1 \cong \angle 2$.

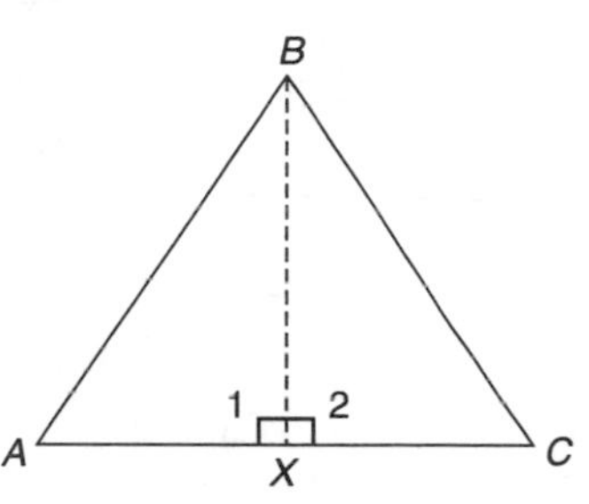

Figure 1.12 Proving $\overline{BX}$ Is a Special Segment

Proving a Geometric Construction

Congruent triangles can be used to help prove that a geometric construction is valid. For example, Figure 1.13 shows the construction of line ℓ perpendicular to $\overleftrightarrow{AB}$ at point *P*. The arcs are constructed so that $\overline{AP} \cong \overline{BP}$ and $\overline{AQ} \cong \overline{BP}$. Since $\overline{PQ} \cong \overline{PQ}$, $\triangle APQ \cong \triangle BPQ$ by SSS. Angles APQ and BPQ are congruent and adjacent, so each must be a right angle, making line $\ell \perp \overleftrightarrow{AB}$.

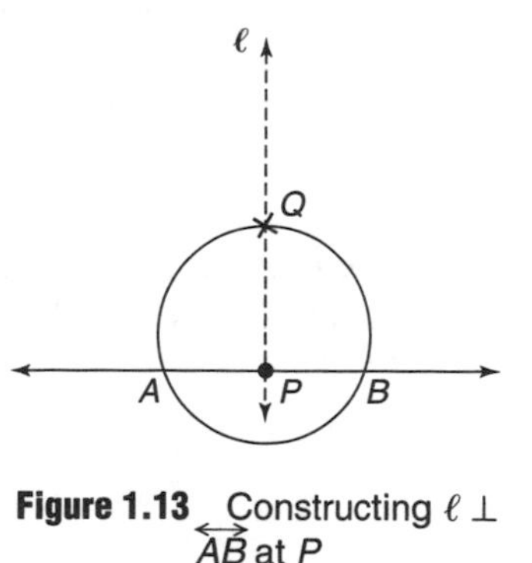

Figure 1.13 Constructing $\ell \perp \overleftrightarrow{AB}$ at *P*

Check Your Understanding of Section 1.3

A. Multiple Choice

1. Which set of statements can be used to prove that $\triangle A'B'C'$ is congruent to $\triangle ABC$?

(1) $\angle A \cong A'$, $\angle B \cong \angle B'$, and $\angle C \cong \angle C'$

(2) $\overline{AB} \cong \overline{A'B'}$, $\overline{BC} \cong \overline{B'C'}$, and $\angle A' \cong \angle A$

(3) $\overline{AB} \cong \overline{A'B'}$, $\angle A' \cong \angle A$, and $\angle C \cong \angle C'$

(4) $\overline{AC} \cong \overline{A'C'}$, $\overline{BC} \cong \overline{B'C'}$, and $\angle A' \cong \angle A$

2. In the accompanying diagram, $\overline{MN} \perp \overline{NP}$, $\overline{QP} \perp \overline{PN}$, O is the midpoint of $\overline{NP}$, and $\overline{MN} \cong \overline{QP}$. Which reason would be *least* likely to be used to prove $\triangle MNO \cong \triangle QPO$?

(1) *Hy-Leg* $\cong$ *Hy-Leg* (3) $SAS \cong SAS$

(2) $AAS \cong AAS$ (4) $ASA \cong ASA$

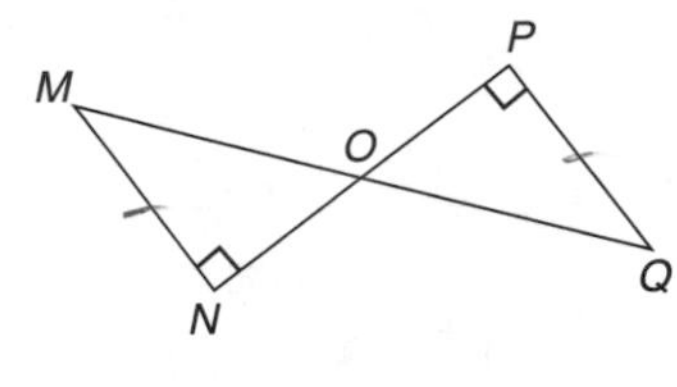

Exercise 2

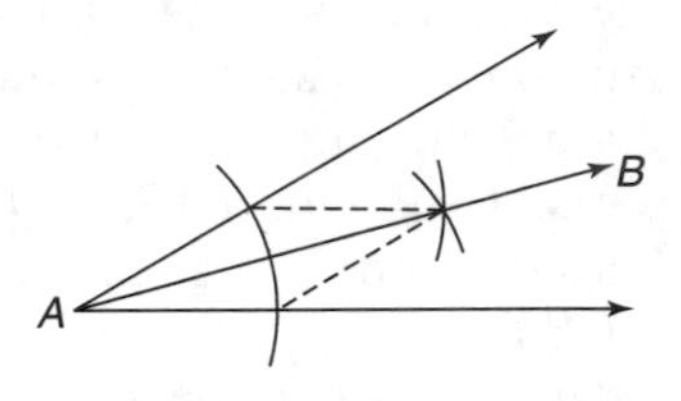

Exercise 3

3. In the accompanying diagram, the bisector of an angle has been constructed. In proving this construction, which reason is used for the congruence involved?

(1) ASA (2) SSS (3) AAS (4) SAS

B. In each case, write a formal proof.

4. Given: $\overline{AD} \cong \overline{CD}$, $\overline{DB} \perp \overline{AC}$.
Prove: $\angle 1 \cong \angle 2$.

5. Given: $\angle 1 \cong \angle 2$, $\overline{DB} \perp \overline{AC}$.
Prove: $\overline{DB}$ is a median.

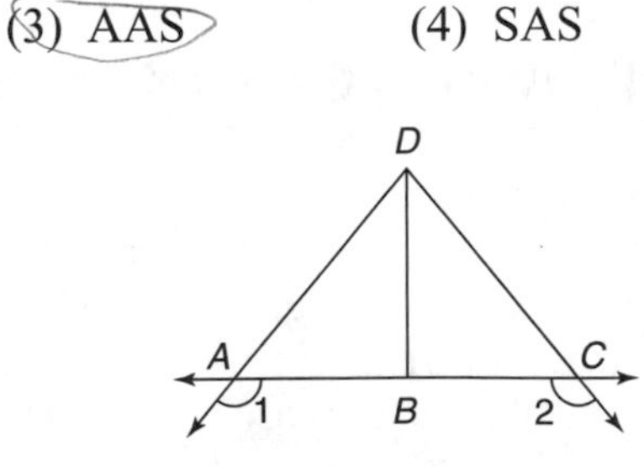

Exercises 4 and 5

6. Given: $\overline{BD}$ bisects $\angle ABC$,
$\angle BAD \cong \angle BCD$.
Prove: $\overline{AD} \cong \overline{CD}$.

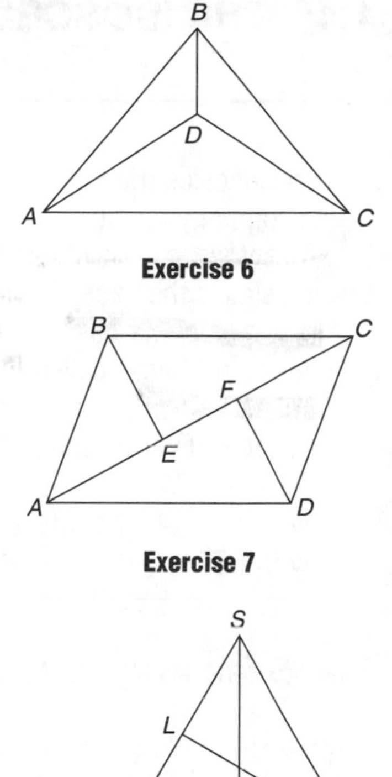

Exercise 6

7. Given: $\overline{BE} \perp \overline{AC}$, $\overline{DF} \perp \overline{AC}$,
$\overline{AF} \cong \overline{CE}$, $\overline{AB} \cong \overline{CD}$.
Prove: $\overline{BE} \cong \overline{DF}$.

Exercise 7

8. Given: $\overline{TL} \perp \overline{RS}$, $\overline{SW} \perp \overline{RT}$,
$\overline{TL} \cong \overline{SW}$.
Prove: $\overline{SL} \cong \overline{TW}$.

9. Given: $\overline{TL}$ is the altitude to $\overline{RS}$,
$\overline{SW}$ is the altitude to $\overline{RT}$,
$\overline{RS} \cong \overline{RT}$.
Prove: $\overline{RW} \cong \overline{RL}$.

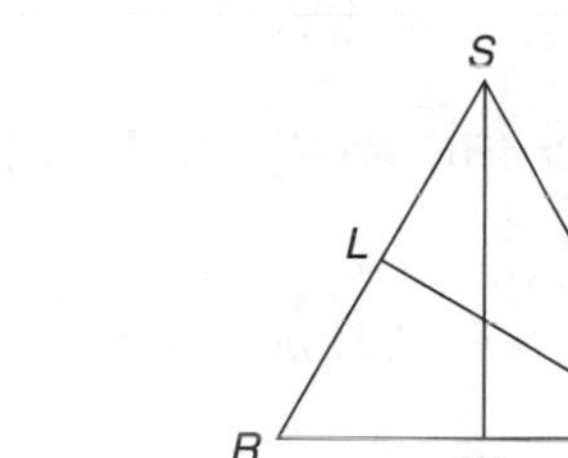

Exercises 8 and 9

10. Prove that the median drawn to the base of an isosceles triangle is perpendicular to the base.

11. Given: $\overline{FD} \perp \overline{AB}$, $\overline{FE} \perp \overline{BC}$,
F is the midpoint of $\overline{AC}$,
$\overline{AD} \cong \overline{CE}$.
Prove: a. $\triangle ADF \cong \triangle CEF$.
b. $\overline{BF}$ bisects $\angle DFE$.

Exercise 11

1.4 THE ISOSCELES TRIANGLE

KEY IDEAS

An isosceles triangle, as shown in the accompanying figure, has two congruent sides called **legs**. The remaining side is the **base**. The **vertex angle** is opposite the base. The angles opposite the legs, called **base angles**, are congruent since drawing the bisector of vertex angle B in isosceles triangle ABC makes $\triangle ABD \cong \triangle CBD$ by $SAS \cong SAS$. As a result, $\angle A \cong \angle C$ by CPCTC.

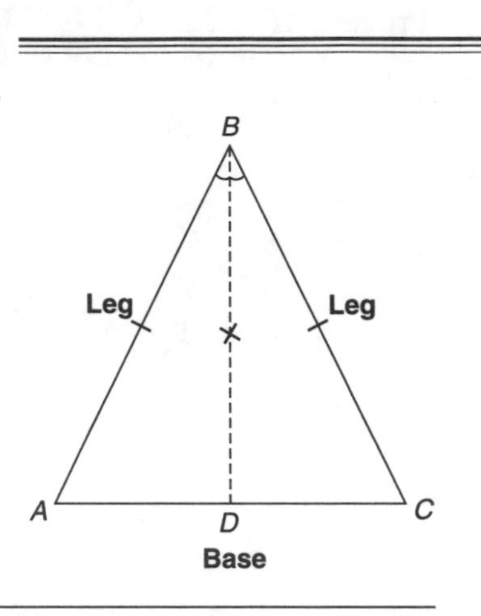

The Base Angles Theorem

The BASE ANGLES THEOREM states that *if two sides of a triangle are congruent, then the angles opposite them are congruent.*

Example 1

Given: $\overline{SR} \cong \overline{ST}$,
$\overline{MP} \perp \overline{RS}$, $\overline{MQ} \perp \overline{ST}$,
M is the midpoint of $\overline{RT}$.
Prove: $\overline{MP} \cong \overline{MQ}$.

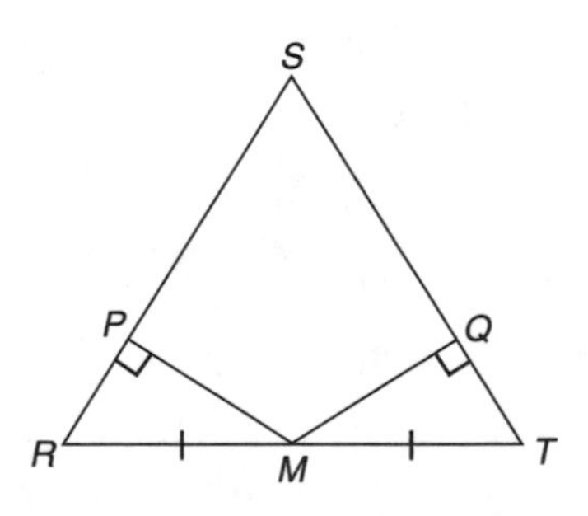

Solution: See the proof.

PLAN: By the application of the Base Angles Theorem, $\angle R \cong \angle T$. Marking off the diagram suggests that triangles MPR and MQT may be proved congruent by using the AAS Theorem.

PROOF

Statement		Reason
1. $\overline{SR} \cong \overline{ST}$		**1.** Given.
2. $\angle R \cong \angle T$	*Angle*	**2.** If two sides of a triangle are congruent, then the angles opposite them are congruent.
3. $\overline{MP} \perp \overline{RS}$, $\overline{MQ} \perp \overline{ST}$		**3.** Given.
4. Angles *MPR* and *MQT* are right angles.		**4.** Perpendicular lines intersect to form right angles.
5. $\angle MPR \cong \angle MQT$	*Angle*	**5.** All right angles are congruent.
6. *M* is the midpoint of $\overline{RT}$.		**6.** Given.
7. $\overline{RM} \cong \overline{TM}$	*Side*	**7.** A midpoint divides a segment into two congruent segments.
8. $\triangle MPR \cong \triangle MQT$		**8.** AAS Theorem.
9. $\overline{MP} \cong \overline{MQ}$		**9.** CPCTC.

Using the Converse of the Base Angles Theorem

The CONVERSE OF THE BASE ANGLES THEOREM states that *if two angles of a triangle are congruent, then the sides opposite them are congruent.*

Example 2

Given: $\overline{CEA} \cong \overline{CDB}$,
$\angle PAB \cong \angle PBA$.
Prove: $\overline{PE} \cong \overline{PD}$.

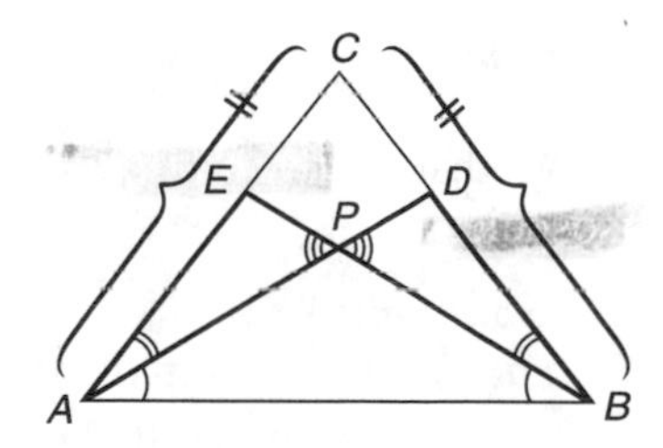

Solution: See the proof.

PLAN: Show $\triangle EPA \cong \triangle DPB$ by the ASA postulate. Angles *EPA* and *DPB* are congruent since they are vertical angles. In $\triangle APB$, apply the converse of the Base Angles Theorem; then $\overline{AP} \cong \overline{BP}$. Use the subtraction property to obtain the other pair of corresponding congruent angles:

$$\begin{array}{l} m\angle CAB = m\angle CBA \text{ (Base Angles Theorem)} \\ -m\angle PAB = m\angle PBA \text{ (Given)} \\ \hline m\angle EAP = m\angle DBP \end{array}$$

PROOF

Statement		Reason
1. $\angle EPA \cong \angle DPB$	*Angle*	**1.** Vertical angles are congruent.
2. $\angle PAB \cong \angle PBA$		**2.** Given.
3. $\overline{AP} \cong \overline{BP}$	*Side*	**3.** If two angles of a triangle are congruent, then the sides opposite them are congruent.
4. $\overline{CEA} \cong \overline{CDB}$		**4.** Given.
5. $\angle CAB \cong \angle CBA$		**5.** If two sides of a triangle are congruent, then the angles opposite them are congruent.
6. $m\angle CAB - m\angle PAB = m\angle CBA - m\angle PBA$		**6.** Subtraction property of equality.
7. $\angle EAP \cong \angle DBP$	*Angle*	**7.** Substitution principle.
8. $\triangle EPA \cong \triangle DPB$		**8.** ASA postulate.
9. $\overline{PE} \cong \overline{PD}$		**9.** CPCTC.

Proving a Triangle Isosceles

To prove that a triangle is isosceles, show that one of the following statements is true:

- The triangle has a pair of congruent sides.
- The triangle has a pair of congruent angles.

Example 3

Prove that, if two altitudes of a triangle are congruent, then the triangle is isosceles.

Solution:

Given: $\overline{CD}$ is the altitude to $\overline{AB}$,
$\overline{AE}$ is the altitude to $\overline{BC}$,
$\overline{CD} \cong \overline{AE}$.

Prove: $\triangle ABC$ is isosceles.

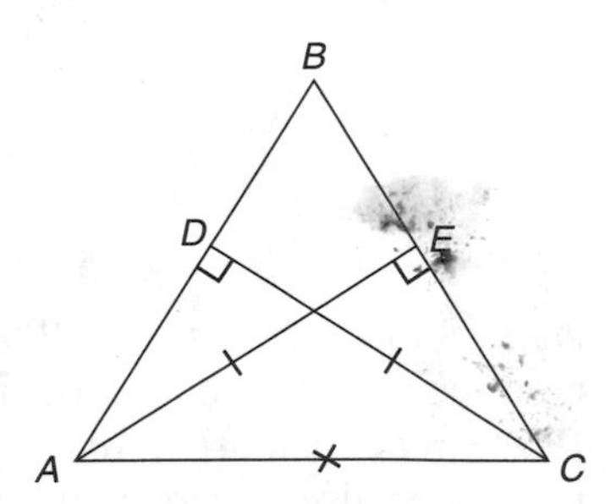

PLAN: Show that $\angle BAC \cong \angle BCA$ by proving $\triangle ADC \cong \triangle CEA$. Marking off the diagram suggests that the Hy-Leg Theorem be used:

$$\overline{AC} \cong \overline{AC} \text{ (Hy)} \qquad \text{and} \qquad \overline{CD} \cong \overline{AE} \text{ (Leg).}$$

PROOF

Statement		Reason
1. $\overline{CD}$ is the altitude to $\overline{AB}$, $\overline{AE}$ is the altitude to $\overline{BC}$.		1. Given.
2. Triangles *ADC* and *CEA* are right triangles.		2. A triangle that contains a right angle is a right triangle. (***Note:*** This step consolidates several obvious steps.)
3. $\overline{CD} \cong \overline{AE}$	*Leg*	3. Given.
4. $\overline{AC} \cong \overline{AC}$	*Hy*	4. Reflexive property of congruence.
5. $\triangle ADC \cong \triangle CEA$		5. Hy-Leg Theorem.
6. $\angle BAC \cong \angle BCA$		6. CPCTC.
7. $\triangle ABC$ is isosceles.		7. A triangle that has a pair of congruent angles is isosceles.

Check Your Understanding of Section 1.4

In each case, write a formal proof.

1. Given: $\angle 2 \cong \angle 4$, $\angle BDA \cong \angle BDC$.
   Prove: $\triangle ABC$ is isosceles.

2. Given: $\angle 1 \cong \angle 3$, $\overline{AB} \cong \overline{BC}$.
   Prove: $\triangle ADC$ is isosceles.

3. Given: $\angle 1 \cong \angle 2$, $\overline{AB} \cong \overline{BC}$, F is the midpoint of $\overline{AB}$, G is the midpoint of $\overline{BC}$.
   Prove: $\overline{FD} \cong \overline{GE}$.

4. Given: $\angle 1 \cong \angle 2$, $\overline{FD} \perp \overline{AC}$, $\overline{GE} \perp \overline{AC}$, $\overline{AE} \cong \overline{CD}$.
   Prove: $\triangle ABC$ is isosceles.

5. Given: $\overline{AB} \cong \overline{BC}$, $\overline{AE} \cong \overline{CD}$.
   Prove: $\triangle DBE$ is isosceles.

6. Given: $\angle BDE \cong \angle BED$, $\angle ABE \cong \angle CBD$.
   Prove: $\triangle ABC$ is isosceles.

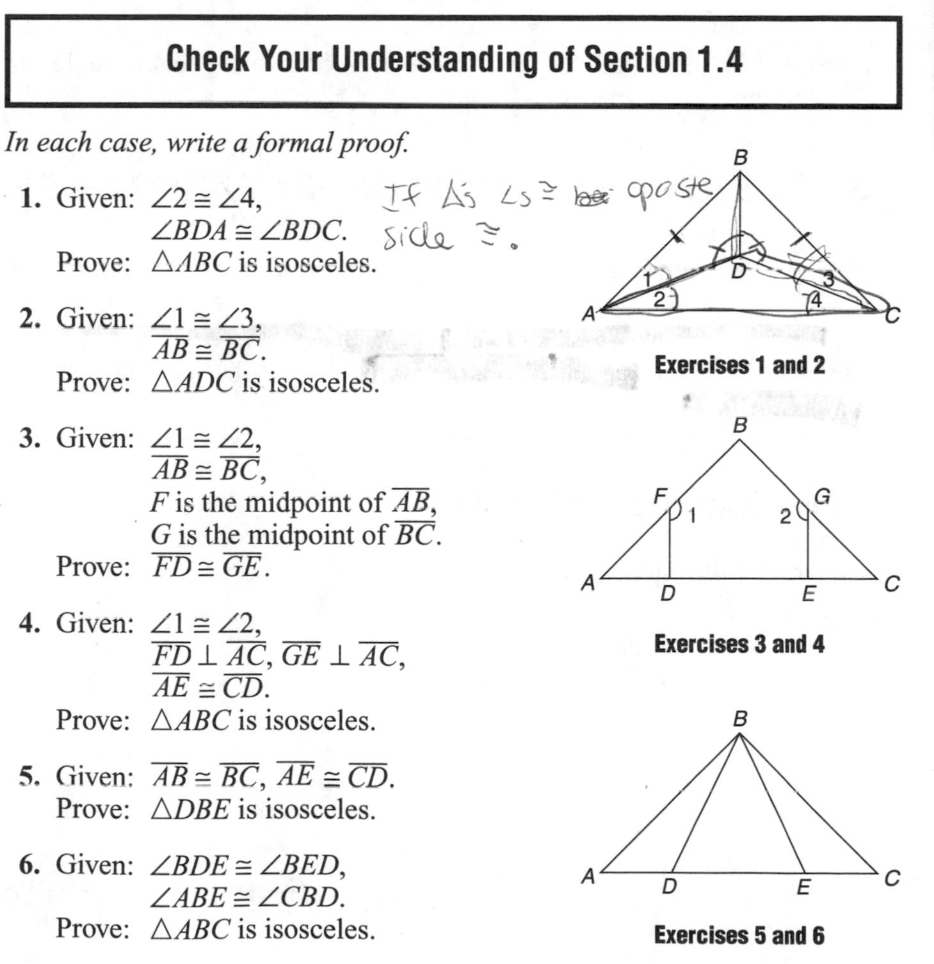

Exercises 1 and 2

Exercises 3 and 4

Exercises 5 and 6

7. Given: $\overline{DF} \cong \overline{CF}$, $\overline{AD} \cong \overline{FC}$,
$\overline{BC} \cong \overline{ED}$.
Prove: $\angle B \cong \angle E$.

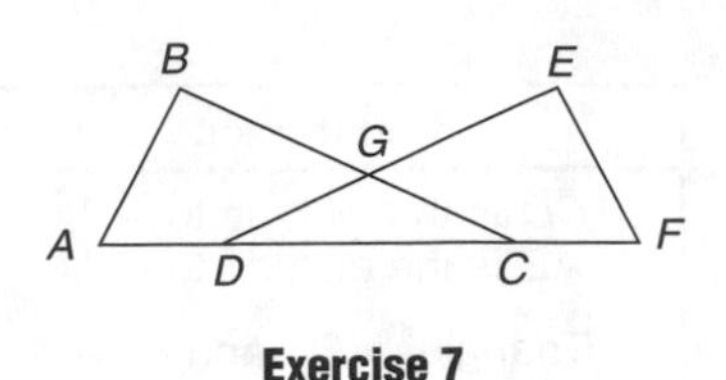

Exercise 7

8. Prove that the medians drawn to the legs of an isosceles triangle are congruent.

9. Given: $\overline{BE} \cong \overline{FC}$, $\overline{AF} \cong \overline{DE}$,
$\overline{AF}$ and $\overline{DE}$ bisect each other at G.
Prove: $\overline{AB} \cong \overline{DC}$.

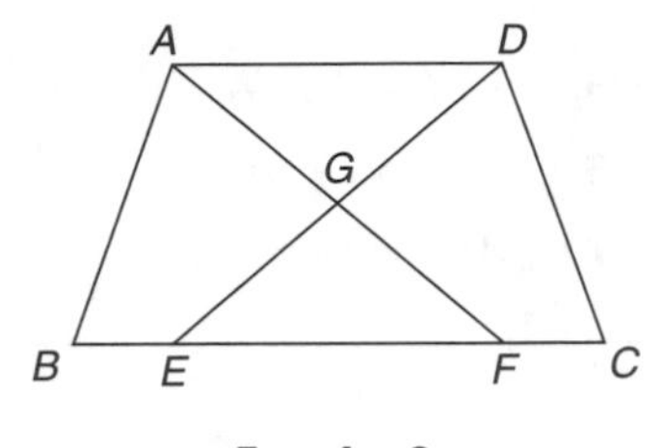

Exercise 9

10. Prove that the altitudes drawn to the legs of an isosceles triangle are congruent.

1.5 USING PAIRS OF CONGRUENT TRIANGLES

KEY IDEAS

It may be necessary to prove that one pair of triangles is congruent in order to obtain a pair of congruent parts that can be used to help prove a second pair of triangles congruent.

Double-Congruence Proofs

Consider the following example:

Given: $\overline{AB} \cong \overline{CB}$,
E is the midpoint of $\overline{AC}$.
Prove: $\triangle AED \cong \triangle CED$.

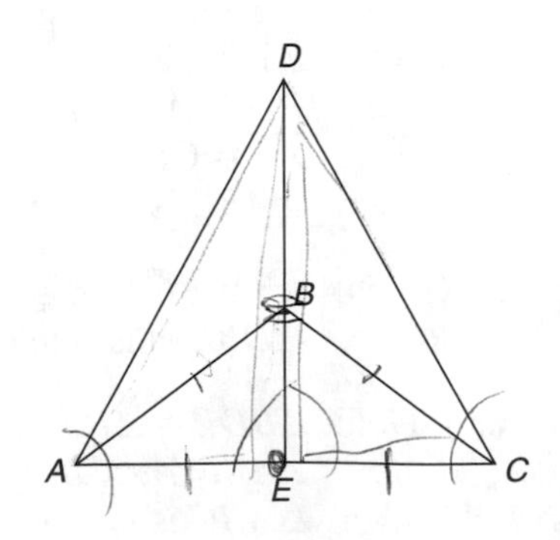

Since there does not appear to be enough information in the "Given" to prove that triangles AED and CED are congruent, try to prove a *different* pair of triangles (triangles AEB and CEB) congruent in order to obtain an additional pair of congruent parts ($\angle AED \cong \angle CED$) that can be used to prove the original pair of triangles congruent.

Here is the formal proof:

PROOF

Statement		Reason
Part I. To prove $\triangle AEB \cong \triangle CEB$:		
1. $\overline{AB} \cong \overline{CB}$	*Side*	**1.** Given.
2. E is the midpoint of $\overline{AC}$.		**2.** Given.
3. $\overline{AE} \cong \overline{CE}$	*Side*	**3.** A midpoint divides a segment into two congruent segments.
4. $\overline{BE} \cong \overline{BE}$		**4.** Reflexive property of congruence.
5. $\triangle AEB \cong \triangle CEB$		**5.** SSS postulate.
Part II. To prove $\triangle AED \cong \triangle CED$:		
6. $\angle AED \cong \angle CED$	*Angle*	**6.** CPCTC.
7. $\overline{DE} \cong \overline{DE}$	*Side*	**7.** Reflexive property of congruence.
8. $\triangle AED \cong \triangle CED$		**8.** SAS postulate.

Proofs in Paragraph Form

Rather than use a two-column format, it is sometimes convenient to summarize the key steps of a mathematical proof in one or more easy-to-read paragraphs. Regardless of which format is chosen, the statements should be logically organized and lead to the desired conclusion by using valid mathematical reasoning.

Example

Given: $\overline{AB} \cong \overline{AC}$, $\overline{BD} \cong \overline{CE}$, $\overline{BF} \perp \overline{AD}$, $\overline{CG} \perp \overline{AE}$.
Prove: $\overline{DE} \cong \overline{AE}$.

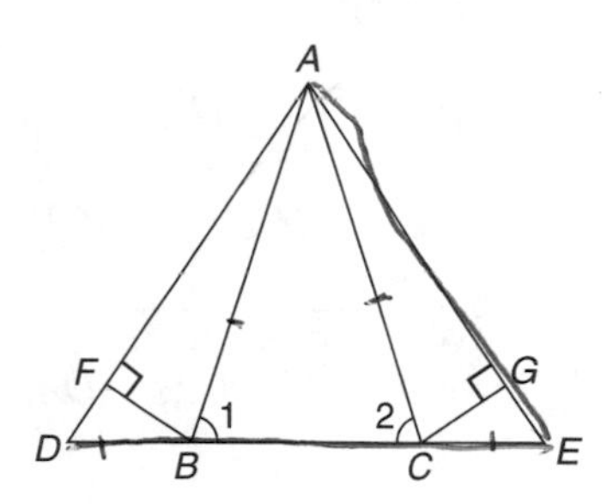

Solution: See the proof.

PARAGRAPH PROOF

To prove $\overline{DF} \cong \overline{EG}$, I must prove $\triangle DFB \cong \triangle EGC$. Before I can prove these triangles congruent, I need to show that $\triangle ABD \cong \triangle ACE$. These triangles are congruent by $SAS \cong SAS$ since $\overline{AB} \cong \overline{AC}$ (Given); $\angle ABD \cong \angle ACE$ (Supplements of congruent angles 1 and 2 are congruent); $\overline{BD} \cong \overline{CE}$ (Given). Because $\triangle ABD \cong \triangle ACE$, $\angle D \cong \angle E$. I can use these congruent angles to prove that $\triangle DFB \cong \triangle EGC$. I know that $\angle D \cong \angle E$ (CPCTC), $\angle DFB \cong \angle EGC$ (Right angles are congruent), and $\overline{BD} \cong \overline{CE}$. Therefore, $\triangle DFB \cong \triangle EGC$ by $AAS \cong AAS$. Hence, $\overline{DF} \cong \overline{EG}$ by CPCTC.

Check Your Understanding of Section 1.5

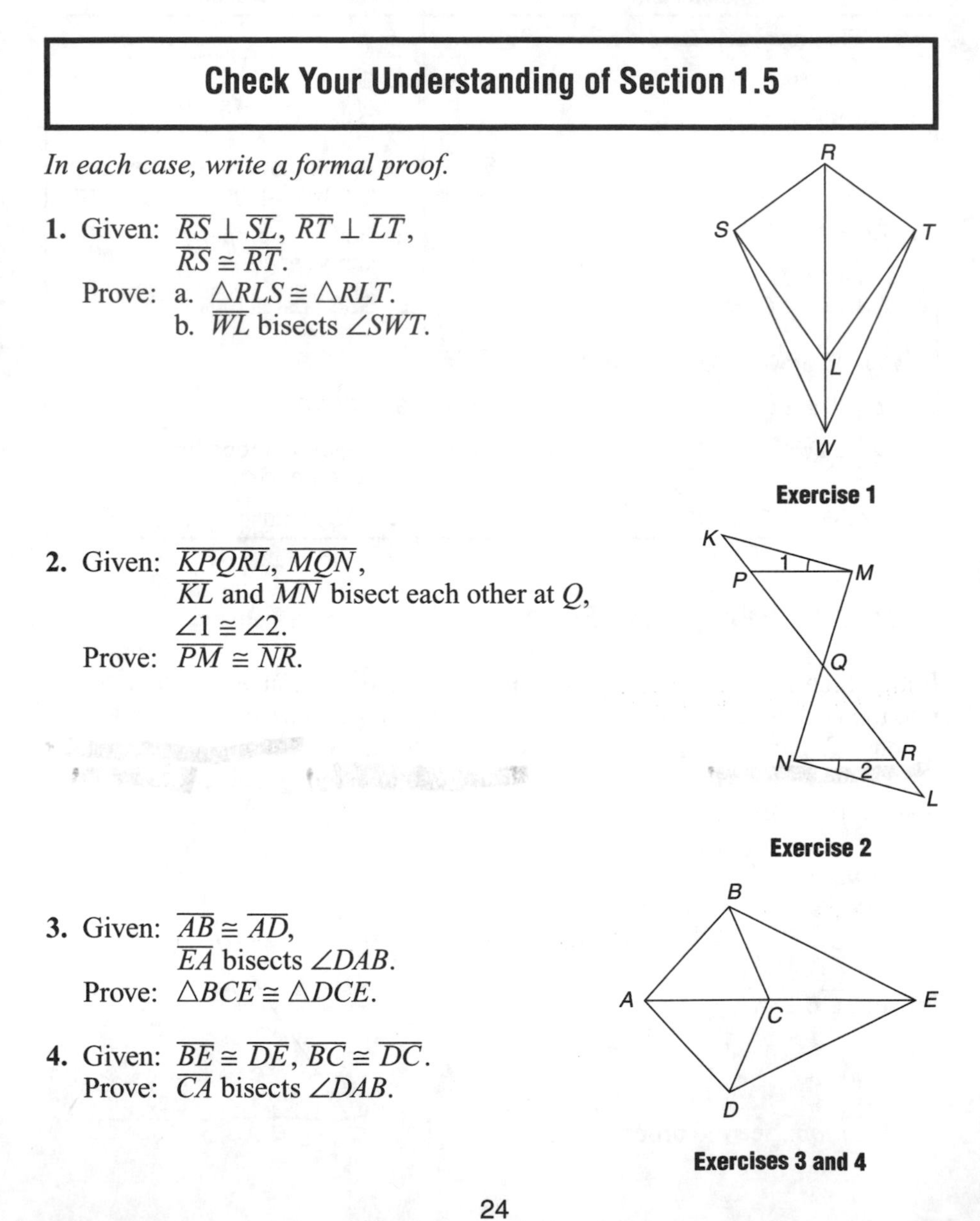

In each case, write a formal proof.

1. Given: $\overline{RS} \perp \overline{SL}$, $\overline{RT} \perp \overline{LT}$,
$\overline{RS} \cong \overline{RT}$.
Prove: a. $\triangle RLS \cong \triangle RLT$.
b. $\overline{WL}$ bisects $\angle SWT$.

Exercise 1

2. Given: $\overline{KPQRL}$, $\overline{MQN}$,
$\overline{KL}$ and $\overline{MN}$ bisect each other at Q,
$\angle 1 \cong \angle 2$.
Prove: $\overline{PM} \cong \overline{NR}$.

Exercise 2

3. Given: $\overline{AB} \cong \overline{AD}$,
$\overline{EA}$ bisects $\angle DAB$.
Prove: $\triangle BCE \cong \triangle DCE$.

4. Given: $\overline{BE} \cong \overline{DE}$, $\overline{BC} \cong \overline{DC}$.
Prove: $\overline{CA}$ bisects $\angle DAB$.

Exercises 3 and 4

5. Given: $\angle FAC\ \angle FCA$,
$\overline{FD} \perp \overline{AB}$, $\overline{FE} \perp \overline{BC}$.
Prove: $\overline{BF}$ bisects $\angle DBE$.

6. Given: $\overline{BD} \cong \overline{BE}$,
$\overline{FD} \cong \overline{FE}$.
Prove: $\triangle AFC$ is isosceles.

7. In the accompanying diagram, the perpendicular bisector of $\overline{AB}$, $\overleftrightarrow{PQ}$, has been constructed by drawing arcs with the same radius setting, using points A and B as centers.

Prove that this construction of $\overleftrightarrow{PQ}$ is valid.

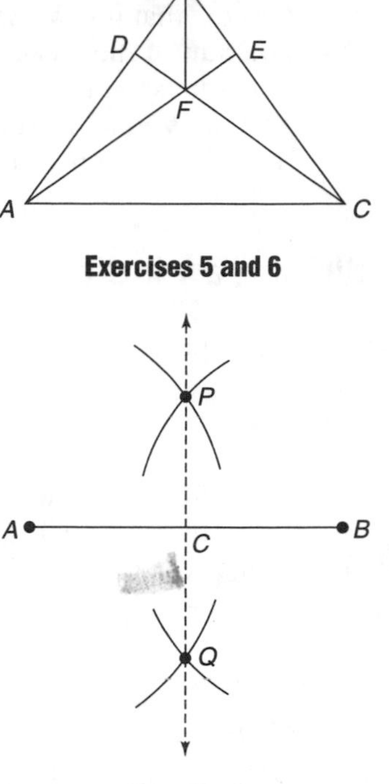

Exercises 5 and 6

Exercise 7

1.6 INDIRECT PROOFS

Sometimes it is necessary to prove a statement *indirectly* by showing that its opposite cannot be true.

Strategy for Writing an Indirect Proof

To prove a statement *indirectly*:

- Assume that the opposite of what you want to prove is true.
- Show that this assumption leads to a contradiction of a known fact.
- State that what you want to prove is true.

Example 1

Alex tells his friend that it is *not* possible to construct a triangle with two obtuse angles. Prove that Alex is correct.

Solution: Use an indirect proof:

- Assume that it *is* possible to draw a triangle with two obtuse angles.
- The sum of the measures of two obtuse angles is more than 180°. This contradicts the fact that the sum of the measures of the three angles of a triangle is equal to 180°.
- Hence, it is *not* possible to construct a triangle with two obtuse angles.

Writing an Indirect Proof

An indirect proof can be organized using a two-column format.

Example 2

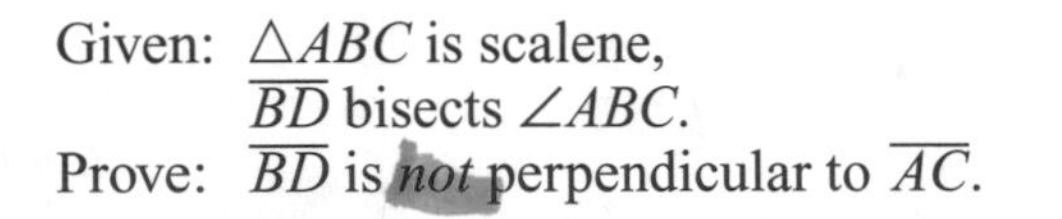

Given: $\triangle ABC$ is scalene,
$\overline{BD}$ bisects $\angle ABC$.
Prove: $\overline{BD}$ is *not* perpendicular to $\overline{AC}$.

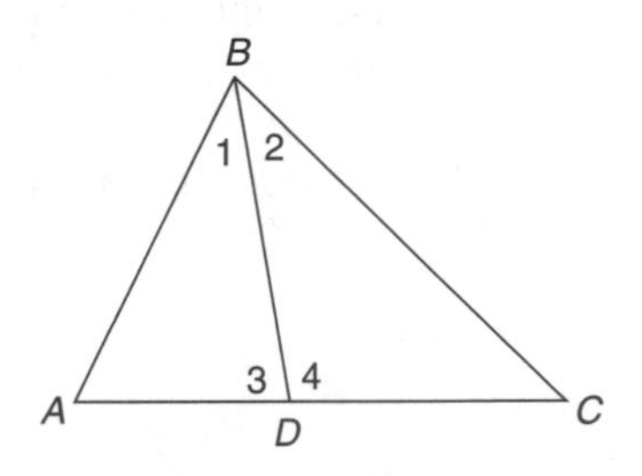

Solution: See the proof.

PLAN: Assume $\overline{BD} \perp \overline{AC}$, and show that this means $\triangle ADB \cong \triangle CDB$. Therefore, $\overline{AB} \cong \overline{BC}$, which contradicts the "Given" that $\triangle ABC$ is scalene.

PROOF

Statement		Reason
1. Assume $\overline{BD} \perp \overline{AC}$.		**1.** Assume that the opposite of what needs to be proved is true.
2. $\angle 3 \cong \angle 4$	*Angle*	**2.** Perpendicular lines intersect to form congruent adjacent angles.
3. $\overline{BD} \cong \overline{BD}$	*Side*	**3.** Reflexive property of congruence.
4. $\angle 1 \cong \angle 2$	*Angle*	**4.** It is given that $\overline{BD}$ bisects $\angle ABC$.
5. $\triangle ADB \cong \triangle CDB$		**5.** ASA postulate.
6. $\overline{AB} \cong \overline{BC}$		**6.** CPCTC.
7. $\triangle ABC$ is scalene.		**7.** Given.
8. Statement 7 contradicts statement 6.		**8.** In a scalene triangle, no two sides have the same length.
9. $\overline{BD}$ is *not* perpendicular to $\overline{AC}$.		**9.** Because statement 1 leads to a contradiction, its opposite is true.

Indirect Proofs in Paragraph Form

Another way of presenting the proof in Example 2 is by summarizing the key steps of the proof in a paragraph.

PARAGRAPH PROOF

Assume $\overline{BD} \perp \overline{AC}$. Thus, $\angle 1 \cong \angle 2$, $\overline{BD} \cong \overline{BD}$, and $\angle 3 \cong \angle 4$, so $\triangle ADB \cong \triangle CDB$ by $ASA \cong ASA$. By CPCTC, $\overline{AB} \cong \overline{BC}$, which contradicts the "Given" that $\triangle ABC$ is scalene. Since the assumption that $\overline{BD} \perp \overline{AC}$ is false, $\overline{BD}$ is *not* perpendicular to $\overline{AC}$.

Check Your Understanding of Section 1.6

In each case, write a formal proof.

1. Given: $\angle 1 \cong \angle 2$,
$\overline{BD}$ does *not* bisect $\angle ABC$.
Prove: $\overline{AB} \ncong \overline{BC}$.

2. Given: Scalene triangle JAM with altitudes $\overline{JK}$ and $\overline{ML}$ drawn to $\overline{AM}$ and $\overline{AJ}$, respectively.
Prove: $\overline{JK} \ncong \overline{ML}$.

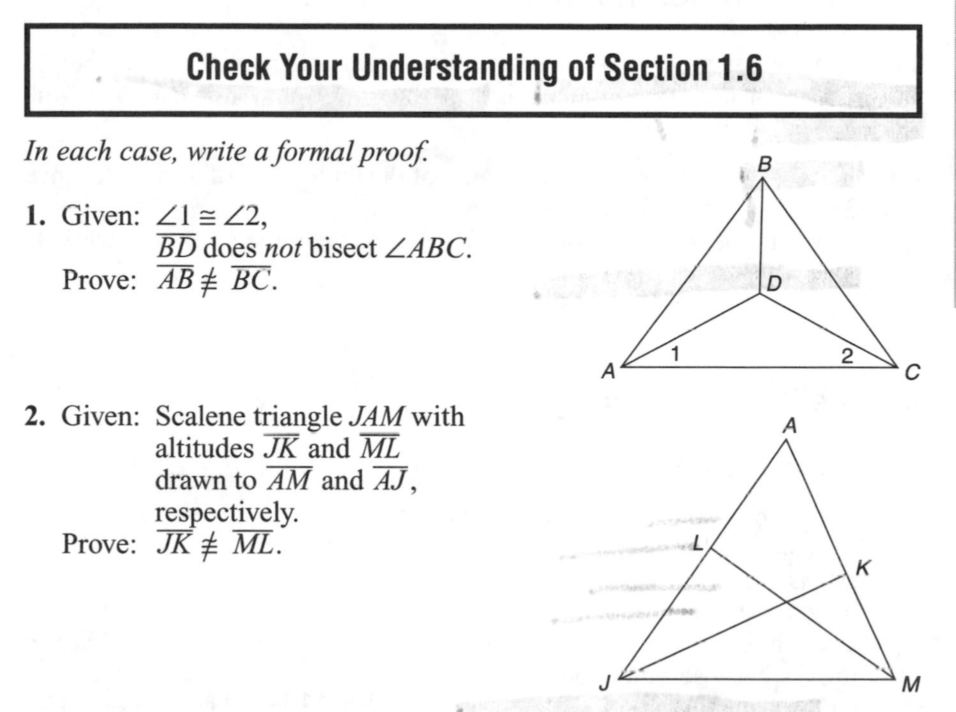

3. Given scalene triangle ABC with M the midpoint of side $\overline{AB}$. Prove that the two segments drawn from M perpendicular to the other two sides of the triangle are *not* congruent.

4. Given: $\overline{AB} \cong \overline{AC}$,
$\overline{BD} \ncong \overline{CD}$.
Prove: $\overline{BE} \ncong \overline{EC}$.

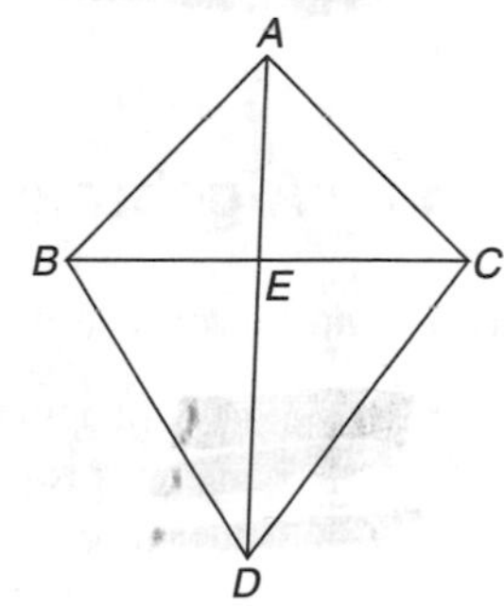

1.7 GEOMETRIC INEQUALITIES

KEY IDEAS

General triangle-inequality relationships, together with algebraic properties of inequalities, can be used to help prove inequality relationships in particular triangles.

Side-Length Restrictions in a Triangle

Three positive numbers can represent the lengths of the sides of a triangle only if each of the three numbers is less than the sum of the lengths of the other two sides. For example:

- 3, 5, and 7 can represent the lengths of the three sides of a triangle since $3 < 5 + 7$, $5 < 3 + 7$, *and* $7 < 3 + 5$.
- 1, 5, and 7 *cannot* represent the lengths of the three sides of a triangle since 7 is *not* less than $1 + 5$.

Inequalities in a Triangle

In $\triangle ABC$, shown in Figure 1.14:

- $\overline{AB}$ is the longest side since it is opposite the largest angle.
- $\overline{BC}$ is the shortest side since it is opposite the smallest angle.
- Then $m\angle 1 > m\angle A$ and $m\angle 1 > m\angle C$ because the measure of an exterior angle of a triangle is greater than the measure of either nonadjacent interior angle.

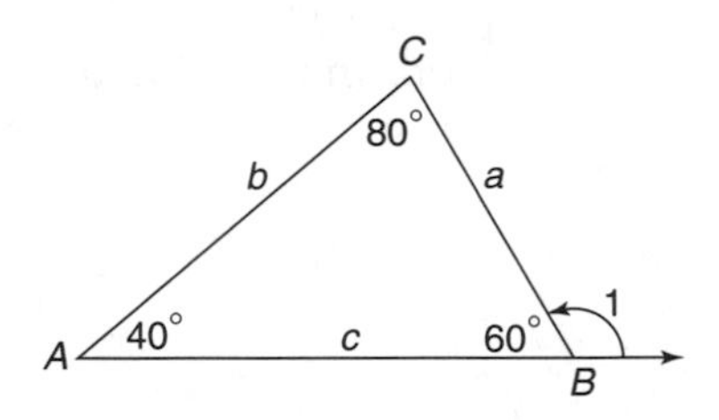

Figure 1.14 Triangle Inequalities

Some Algebraic Inequality Relationships

If a, b and c represent real numbers, then these inequality properties are true:

- Transitive property: If $a > b$ and $b > c$, then $a > c$.
- Substitution property: If $a > b$ and $b = c$, then $a > c$.
- Comparison property: If $a = b + c$ and $c > 0$, then $a > b$.

Applying Triangle Inequalities

In any triangle, the largest angle lies opposite the longest side and the smallest angle lies opposite the shortest side.

Example 1

Multiple Choice:

In $\triangle PQR$, $m\angle P = 51$ and $m\angle Q = 57$. Which expression is true?

(1) $QR > PQ$ (2) $PR > PQ$ (3) $PQ = QR$ (4) $PQ > QR$

Solution: **(4)**

Find the degree measure of the remaining angle of the triangle. Since $m\angle P + m\angle Q + m\angle R = 180$:

$$\begin{aligned} m\angle R &= 180 - (51 + 57) \\ &= 180 - 108 \\ &= 72 \end{aligned}$$

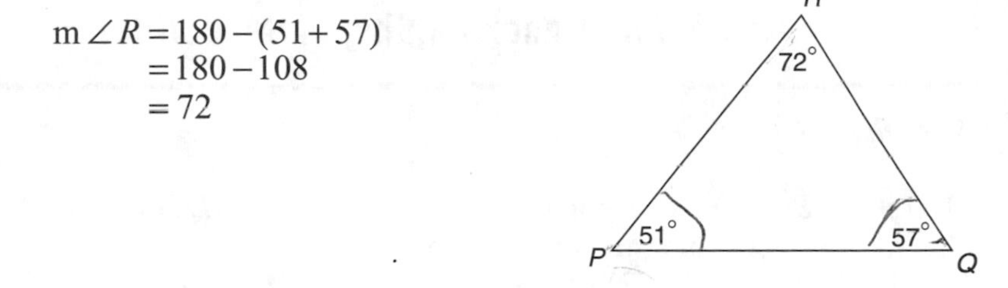

Since $\angle R$ is the largest angle of $\triangle PQR$, the side opposite, $\overline{PQ}$, is the longest side of the triangle. Hence, $PQ > QR$.

Example 2

Given: $\overline{BD}$ bisects $\angle ABC$.
Prove: $AB > AD$.

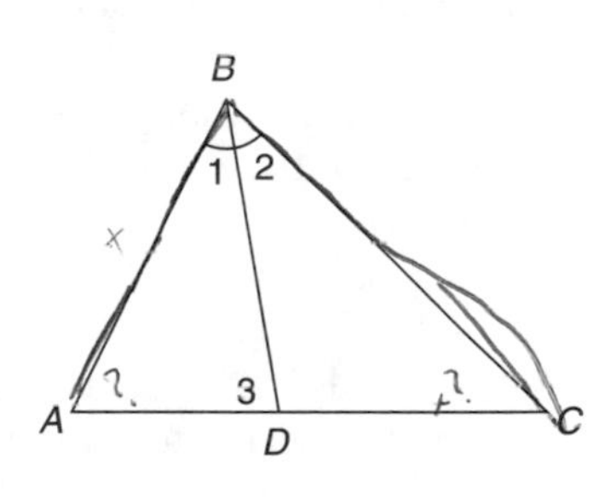

Solution: See proof.

PLAN: To prove $AB > AD$, first establish that $m\angle 3$ (the angle opposite $\overline{AB}$ is greater than $m\angle 1$ (the angle opposite $\overline{AD}$).

PROOF

Statement	Reason
1. $m\angle 3 > m\angle 2$	1. The degree measure of an exterior angle of a triangle is greater than the degree measure of either nonadjacent interior angle.
2. $\overline{BD}$ bisects $\angle ABC$.	2. Given.
3. $m\angle 1 = m\angle 2$	3. A bisector divides an angle into two angles having the same degree measure.
4. $m\angle 3 > m\angle 1$	4. Substitution property of inequalities.
5. $AB > AD$	5. If two angles of a triangle are not equal in degree measure, then the sides opposite are not equal and the longer side is opposite the larger angle.

Check Your Understanding of Section 1.7

A. Multiple Choice

1. In $\triangle ABC$, D is a point on $\overline{AC}$ such that $\overline{BD}$ bisects $\angle ABC$. If $m\angle ABC = 60$ and $m\angle C = 70$, then
   (1) $AC > AB$ (2) $BD > BC$ (3) $AD > BD$ (4) $AB > AD$

2. In $\triangle ABC$, D is a point on $\overline{AC}$, $m\angle A > m\angle B$, and E is a point on $\overline{BC}$ such that $\overline{CE} \cong \overline{CD}$. Which statement is *always* true?
   (1) $m\angle CDE > m\angle A$
   (2) $m\angle B > m\angle CED$
   (3) $EB > AD$
   (4) $EB = AD$

3. In isosceles triangle ABC, $\overline{AC} \cong \overline{BC}$ and D is a point lying between A and B on base $\overline{AB}$. If $\overline{CD}$ is drawn, which statement is *always* true?
   (1) $AB > CD$
   (2) $CD < AC$
   (3) $m\angle A > m\angle ADC$
   (4) $m\angle B > m\angle BDC$

4. If $\frac{1}{4}$ and $\frac{1}{2}$ represents the lengths of two sides of an isosceles triangle, then the perimeter of the triangle is

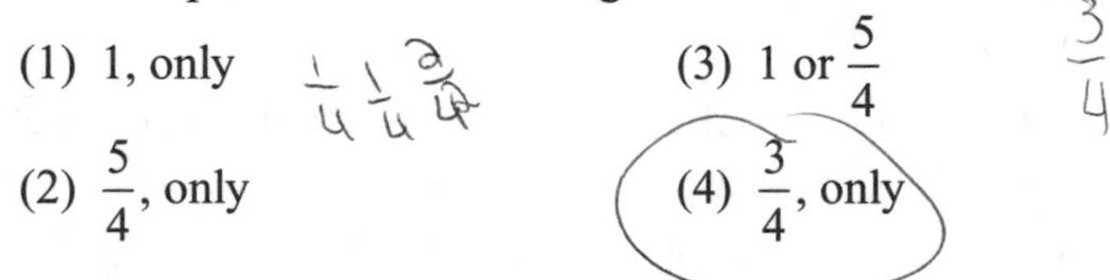

   (1) 1, only
   (2) $\dfrac{5}{4}$, only
   (3) 1 or $\dfrac{5}{4}$
   (4) $\dfrac{3}{4}$, only

5. If D is any point on side $\overline{AB}$ of equilateral triangle ABC, which statement is *always* true?
   (1) $CD > DB$
   (2) $AB < CD$
   (3) $m\angle A > m\angle ADC$
   (4) $m\angle B > m\angle BDC$

6. A box contains one 2-inch rod, one 3-inch rod, one 4-inch rod, and one 5-inch rod. What is the maximum number of different triangles that can be made using these rods as sides?

(1) 1 (2) 2 (3) 3 (4) 4

B. In each case, write a formal proof.

7. Given: $m\angle 1 = m\angle 2$.
Prove: $AD > ED$.

8. Given: $\overline{AC} \cong \overline{BC}$, $\overline{AD} \cong \overline{BD}$.
Prove: $AD > DE$.

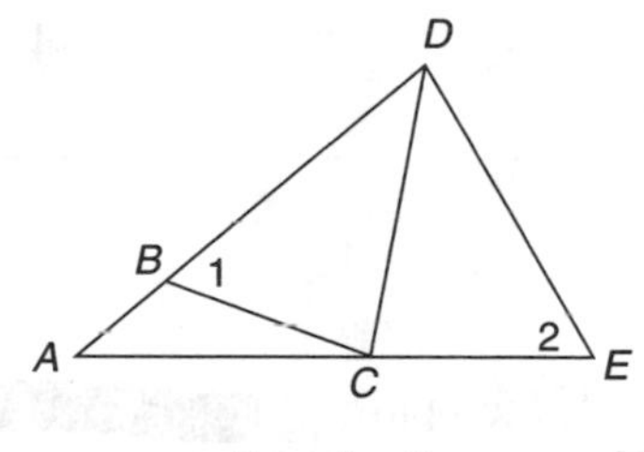

Exercise 7

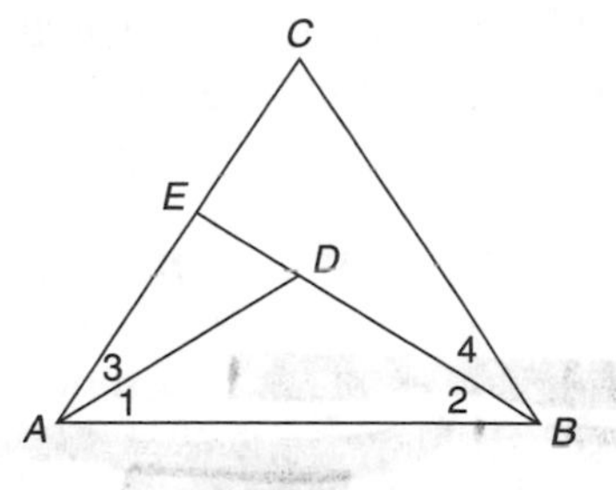

Exercise 8

9. Given: $\overline{CB}$ is drawn to side $\overline{AE}$ of $\triangle AEC$,
$\overline{AD}$ is drawn to side $\overline{BC}$ of $\triangle ABC$.
Prove: $m\angle 4 > m\angle AEC$.

10. Given: $\overline{CB}$ is drawn to side $\overline{AE}$ of $\triangle AEC$,
$\overline{AD}$ bisects $\angle CAB$,
$AD > BD$.
Prove: $AC > DC$.

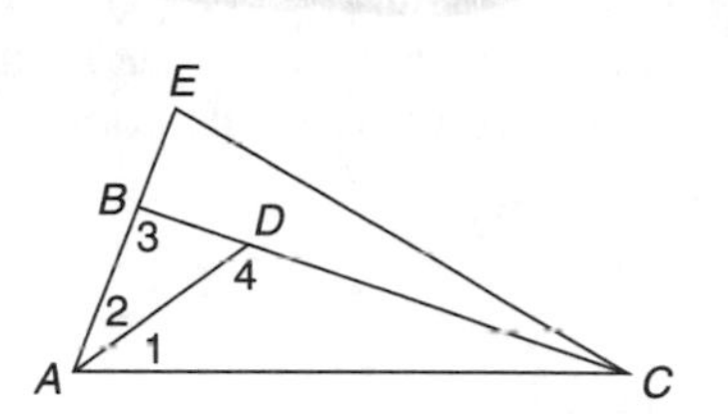

Exercises 9 and 10

CHAPTER 2
QUADRILATERALS AND COORDINATES

2.1 PARALLEL LINES

KEY IDEAS

Proofs involving congruent triangles may also make use of the properties of parallel lines.

Properties of Parallel Lines

If two lines are parallel, as shown in Figure 2.1, then pairs of acute angles are congruent, pairs of obtuse angles are congruent, and any acute angle and any obtuse angle are supplementary. When one of these facts is needed as a reason in a formal proof, you need to write the appropriate theorem.

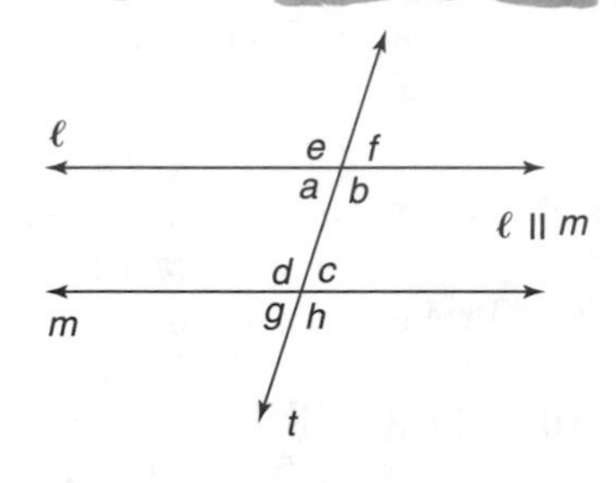

Figure 2.1 Angles Formed by Parallel Lines Intersected by Transversal *t*

- THEOREM: *If two lines are parallel, then alternate interior angles are congruent.*
  In Figure 2.1: $\angle a \cong \angle c$ and $\angle b \cong \angle d$.
- THEOREM: *If two lines are parallel, then corresponding angles are congruent.*
  In Figure 2.1: $\angle e \cong \angle d$ and $\angle a \cong \angle g$; $\angle f \cong \angle c$ and $\angle b \cong \angle h$.
- THEOREM: *If two lines are parallel, then interior angles on the same side of the transversal are supplementary.*
  In Figure 2.1: $m\angle a + m\angle d = 180$ and $m\angle b + m\angle c = 180$.

Example 1

Given: $\overline{DM} \parallel \overline{BL}, \overline{AD} \parallel \overline{BC}, \overline{AL} \cong \overline{CM}$.
Prove: $\overline{AD} \cong \overline{BC}$.

Solution: See the proof.

PLAN: Mark the diagram with the "Given." Identify pairs of angles that are congruent as a result of the parallel lines. Show that $\triangle ADM \cong \triangle CBL$ by $ASA \cong ASA$.

Statement		Reason
1. $\overline{DM} \parallel \overline{BC}$		**1.** Given.
2. $\angle 1 \cong \angle 2$	*Angle*	**2.** If two lines are parallel, alternate interior angles are congruent.
3. $\overline{AL} \cong \overline{CM}$		**3.** Given.
4. $\overline{LM} \cong \overline{LM}$		**4.** Reflexive property of congruence.
5. $AL + LM = CM + LM$		**5.** Addition property of equality.
6. $\overline{AM} \cong \overline{CL}$	*Side*	**6.** Substitution property.
7. $\overline{AD} \parallel \overline{BC}$		**7.** Given.
8. $\angle 3 \cong \angle 4$	*Angle*	**8.** If two lines are parallel, alternate interior angles are congruent.
9. $\triangle ADM \cong \triangle CBL$		**9.** $ASA \cong ASA$.
10. $\overline{AD} \cong \overline{BC}$		**10.** CPCTC.

Proving Lines Are Parallel

To prove two lines are parallel, show that any one of the following is true:

- A pair of alternate interior angles are congruent.
- A pair of corresponding angles are congruent.
- A pair of interior angles on the same side of the transversal are supplementary.
- The two lines are perpendicular to the same line.

Example 2

Given: $\overline{MP} \cong \overline{ST}, \overline{PL} \cong \overline{RT}, \overline{MP} \parallel \overline{ST}$.
Prove: $\overline{RS} \parallel \overline{LM}$.

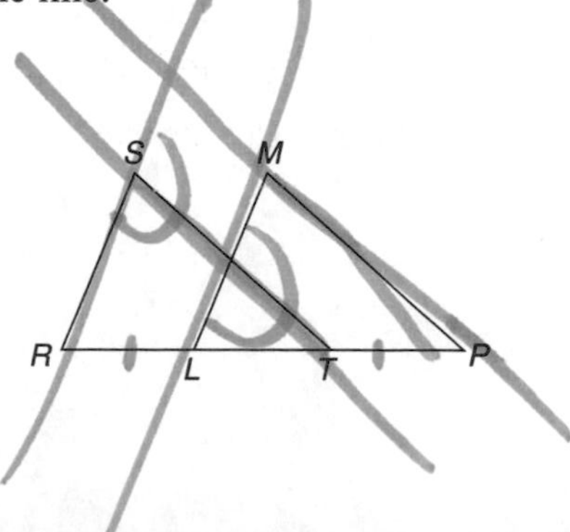

Solution: See the proof.

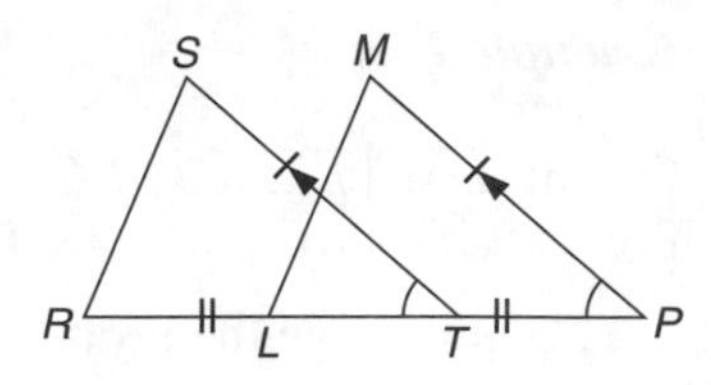

PLAN: Prove $\triangle RST \cong \triangle LMP$ by SAS to get $\angle STR \cong \angle MPL$. Since corresponding angles are congruent, $\overline{RS} \parallel \overline{LM}$.

PROOF

Statement		Reason
1. $\overline{MP} \cong \overline{ST}$	*Side*	1. Given.
2. $\overline{MP} \parallel \overline{ST}$		2. Given.
3. $\angle MPL \cong \angle STR$	*Angle*	3. If two lines are parallel, then their corresponding angles are congruent.
4. $\overline{PL} \cong \overline{RT}$	*Side*	4. Given.
5. $\triangle RST \cong \triangle LMP$		5. SAS postulate.
6. $\angle SRT \cong \angle MLP$		6. CPCTC.
7. $\overline{RS} \parallel \overline{LM}$		7. Two lines are parallel if a pair of corresponding angles are congruent.

Check Your Understanding of Section 2.1

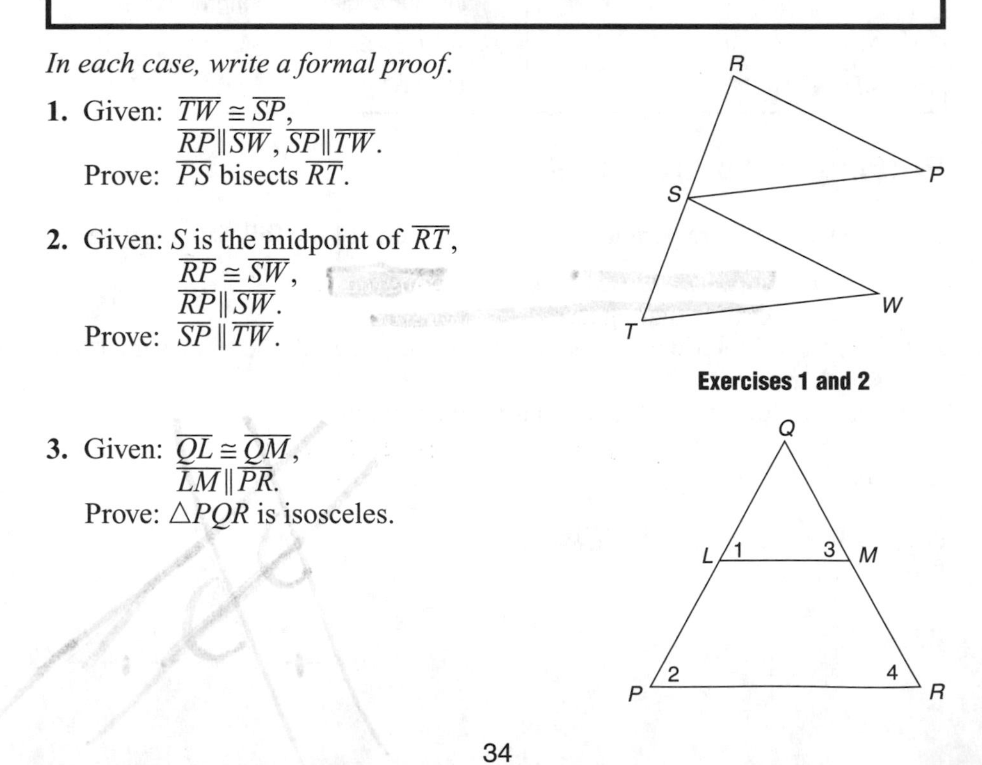

In each case, write a formal proof.

1. Given: $\overline{TW} \cong \overline{SP}$,
   $\overline{RP} \parallel \overline{SW}$, $\overline{SP} \parallel \overline{TW}$.
   Prove: $\overline{PS}$ bisects $\overline{RT}$.

2. Given: S is the midpoint of $\overline{RT}$,
   $\overline{RP} \cong \overline{SW}$,
   $\overline{RP} \parallel \overline{SW}$.
   Prove: $\overline{SP} \parallel \overline{TW}$.

Exercises 1 and 2

3. Given: $\overline{QL} \cong \overline{QM}$,
   $\overline{LM} \parallel \overline{PR}$.
   Prove: $\triangle PQR$ is isosceles.

4. Given: $\overline{RS}$ intersects $\overleftrightarrow{ARB}$ and $\overleftrightarrow{CST}$,
$\overleftrightarrow{ARB} \parallel \overleftrightarrow{CST}$,
$\overline{RT}$ bisects $\angle BRS$,
M is the midpoint of $\overline{RT}$,
$\overline{SM}$ is drawn.
Prove: a. $\overline{RS} \cong \overline{ST}$.
b. $\overline{SM}$ bisects $\angle RST$.

5. Given: $\triangle ABC$, $\overline{CM}$ is the median to $\overline{AB}$,
$\overline{CM}$ is extended to point P
so that $\overline{CM} \cong \overline{MP}$,
$\overline{AP}$ is drawn.
Prove: $\overline{AP} \parallel \overline{CB}$.

6. Given: $AC > AB$, $\overline{DE} \cong \overline{CE}$.
Prove: $\overline{AB}$ is *not* parallel to $\overline{DE}$.

7. Given: $\overline{BC} \parallel \overline{AD}$,
$\triangle ABC$ is *not* isosceles.
Prove: $\overline{AC}$ does *not* bisect $\angle BAD$.

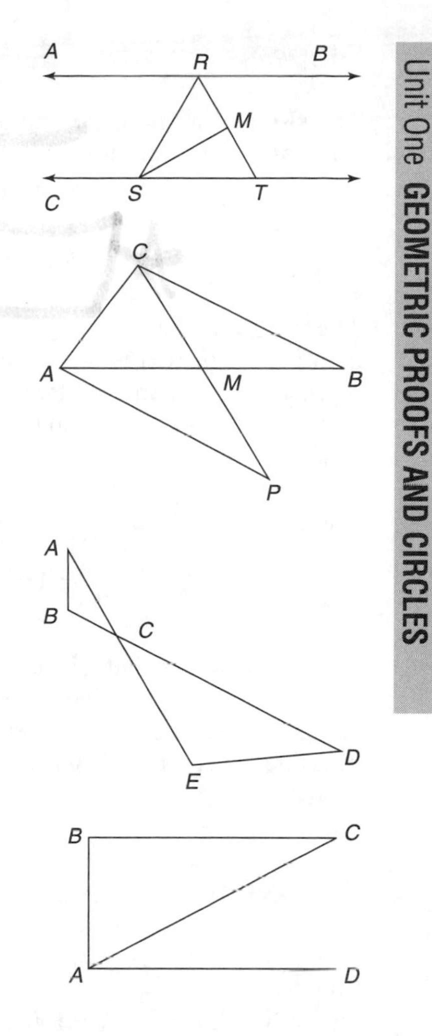

2.2 WRITING EQUATIONS OF LINES

KEY IDEAS

If you know the slope of a line and one point on the line, you can determine an equation for the line. If you are given two points on a line, you can determine an equation of the line by first calculating its slope.

Slope

The **slope** of a line is a number that measures its steepness. If a nonvertical line passes through points $A(x_A, y_A)$ and $B(x_B, y_B)$, as shown in Figure 2.2, then the slope of $\overleftrightarrow{AB}$ can be calculated using this formula:

$$\text{Slope of } \overleftrightarrow{AB} = \frac{\Delta y}{\Delta x} = \frac{y_B - y_A}{y_B - x_A}.$$

The notation Δy is read as "delta y" and means the change in the y-coordinates of two points on the line. Similarly, Δx is read as "delta x" and represents the change in the x-coordinates of the same two points taken in the same order. For example, the slope of the line that contains points $A(0, -4)$ and $B(3, 5)$ is calculated as follows:

$$\text{Slope of } \overleftrightarrow{AB} = \frac{\Delta y}{\Delta x} = \frac{5-(-4)}{3-0} = \frac{9}{3} = 3.$$

Figure 2.2 shows that, since the slope is a positive number, the line joining $A(0, -4)$ and $B(3, 5)$ rises as x increases. If the slope of a line is negative, then the line falls as x increases.

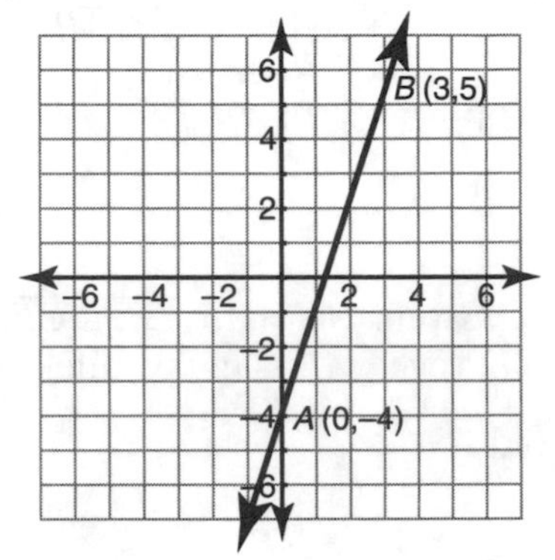

Figure 2.2 Calculating the Slope of a Line

Slopes of Horizontal and Vertical Lines

A *horizontal line* is parallel to the x-axis so, between any two points on the line, $\Delta y = 0$. A *vertical line* is parallel to the y-axis so, between any two points on the line, $\Delta x = 0$. Therefore:

- The slope of a horizontal line is 0.
- The slope of a vertical line is undefined.

Slopes of Parallel and Perpendicular Lines

Special slope relationships exist between pairs of parallel and perpendicular lines.

- *Parallel lines* have the same slope. To prove that two nonvertical lines are parallel, show that the lines have the same slope.
- *Perpendicular lines* have slopes that are negative reciprocals. To show that a nonvertical line is perpendicular to another line, show that the product of their slopes is -1.

Equation of a Line: Slope-Intercept Form

If an equation of a line has the form $y = mx + b$, then m, the coefficient of x, is the slope of the line and the constant b is the y-intercept. If an equation of line ℓ is $y + 4x = 3$, then $y = -4x + 3$, so the slope of line ℓ is -4 and the y-intercept is 3. Also:

- If line p is *parallel* to line ℓ, then the slope of line p is also -4 since parallel lines have equal slopes.
- If line q is *perpendicular* to line ℓ, then the slope of line q is $\frac{1}{4}$ since the product of the slopes of perpendicular lines is -1.

Example 1

The equations for two lines are $3x + y - 8 = 0$ and $-2x + by + 9 = 0$. For what value of b will the two lines be perpendicular?

Solution: $\boldsymbol{b = 6}$

- If $3x + y - 8 = 0$, then $y = -3x + 8$, so $m_1 = -3$.
- If $-2x + by + 9 = 0$, then $by = 2x - 9$, so $y = \frac{2}{b}x - \frac{9}{b}$ and $m_2 = \frac{2}{b}$.
- Since the product of the slopes of perpendicular lines is -1:

$$m_1 \cdot m_2 = -1$$
$$(-3)\left(\frac{2}{b}\right) = -1$$
$$\frac{-6}{b} = -1$$
$$6 = b$$

Example 2

Write an equation of the line whose y-intercept is -3 and that is perpendicular to the line $y = -\frac{1}{2}x + 4$.

Solution: $\boldsymbol{y = 2x - 3}$

The slope of the given line is $-\frac{1}{2}$. The lines are perpendicular, so the slope, m, of the desired line is 2 since the product of $-\frac{1}{2}$ and 2 is -1. Since $m = 2$ and $b = -3$, an equation of the desired line is $y = 2x - 3$.

Example 3

Write an equation of the line that is parallel to the line $y + 3x = 5$ and passes through point (1, 4).

Solution: $\mathbf{y = -3x + 7}$

The slope of the desired line is -3, so $y = -3x + b$. Since (1,4) is a point on the line, find b by letting $x = 1$ and $y = 4$:

$$4 = -3(1) + b$$
$$7 = b$$

An equation of the line is $y = -3x + 7$.

Writing an Equation of a Line Given Two Points

Two points determine a line. To write an equation of a line that contains points (1, −7) and (3, 5) in $y = mx + b$ form, find m and then b:

- Determine the slope: $m = \frac{5-(-7)}{3-1} = \frac{12}{2} = 6.$
- Let $m = 6$: $y = 6x + b.$
- Use $(x, y) = (3, 5)$ to find b: $5 = 6(3) + b$, so $b = -13$.

Hence, an equation of the line is $y = 6x - 13$.

Example 4

What is an equation of the line that is the perpendicular bisector of the segment whose endpoints are $A(0, 7)$ and $B(-5, 2)$?

Solution: $\mathbf{y = -x + 2}$

- The coordinates of the midpoint $(\overline{x}, \overline{y})$ of a line segment whose endpoints are $A(x_A, y_A)$ and $B(x_B, y_B)$ are

  $$\overline{x} = \frac{x_A + x_B}{2} \quad \text{and} \quad \overline{y} = \frac{y_A + y_B}{2}.$$

  Thus, the coordinates of the midpoint of the segment with endpoints $A(0, 7)$ and $B(-5, 2)$ are

  $$\overline{x} = \frac{0-5}{2} = \frac{-5}{2} \quad \text{and} \quad \overline{y} = \frac{7+2}{2} = \frac{9}{2}.$$

- The slope of the line segment that contains points (0, 7) and (−5, 2) is $\frac{2-7}{-5-(0)} = \frac{-5}{-5} = 1$. Hence, the slope of the perpendicular line is -1.

- Since the perpendicular bisector of $\overline{AB}$ contains $(\frac{-5}{2}, \frac{9}{2})$ and has a slope of -1, its equation is $y = -1 \cdot x + b$. To find b, let $x = -\frac{5}{2}$ and $y = \frac{9}{2}$; then

$$\frac{9}{2} = -\left(-\frac{5}{2}\right) + b$$

$$b = \frac{9}{2} - \frac{5}{2} = \frac{4}{2} = 2$$

Hence, an equation of this line is $y = -x + 2$.

TIP

You can use your graphing calculator to find an equation of the line that contains two given points by entering the *x*-values in list L1 and the corresponding *y*-values in list L2 and then calculating a regression line by selecting the **LinReg** instruction from the STAT CALC menu. The procedure is explained in Chapter 11.

Check Your Understanding of Section 2.2

A. Multiple Choice

1. If line ℓ is perpendicular to line m and the slope of line ℓ is undefined, what is the slope of line m?
(1) 1 (2) undefined (3) 0 (4) -1

2. The graph of $x - 3y = 6$ is parallel to the graph of
(1) $y = -3x + 7$ (2) $y = -\frac{1}{3} + 5$ (3) $y = 3x - 8$ (4) $y = \frac{1}{3}x + 8$

3. The graph of which equation is perpendicular to the graph of $y - 3 = \frac{1}{2}x$?
(1) $y = -\frac{1}{2}x + 5$ (2) $2y = x + 3$ (3) $y = 2x + 5$ (4) $y = -2x + 3$

4. Which is an equation of the line that is parallel to $y = 2x - 8$ and passes through point $(0, -3)$?
(1) $y = 2x + 3$ (2) $y = 2x - 3$ (3) $y = -\frac{1}{2}x + 3$ (4) $y = -\frac{1}{2}x - 3$

5. Which is an equation of the line that is parallel to $y - 3x + 5 = 0$ and has the same y-intercept as $y = -2x + 7$?
(1) $y = 3x - 2$ (2) $y = -2x - 5$ (3) $y = 3x + 7$ (4) $y = -2x - 7$

6. Which is an equation of a line that has an x-intercept of -3 and a y-intercept of 4?
(1) $3x + 4y = 12$ (2) $4y - 3x = 12$
(3) $4x + 3y = 12$ (4) $3y - 4x = 12$

B. *In each case, show how you arrived at your answer by clearly indicating all of the necessary steps, formula substitutions, diagrams, graphs, charts, etc.*

7. Kevin knows that $-40°C = -40°F$, $20°C = 68°F$, and that the conversion relationship between Celsius and Fahrenheit temperatures is linear. What is an equation of the line that contains all paired conversion temperatures of the form (°C, °F)?

8. The equations for two lines are $2x + y + 6 = 0$ and $kx + 4y - 8 = 0$. For what value of k will the two lines be perpendicular?

9. The equations for two lines are $3y - 2x = 6$ and $3x + ky = -7$. For what value of k will the two lines be parallel?

10. Determine an equation of the line that is the perpendicular bisector of the segment whose endpoints are $(-1, 8)$ and $(-5, 2)$.

11. Determine an equation of the line that contains the point of intersection of the lines $x + y = 1$ and $2x - 3y = 7$, and is perpendicular to the line $x - 2y = 0$.

12. If the line $2y + 5x = k$ is the perpendicular bisector of the segments whose endpoints are $A(-8, t)$ and $B(2, -3)$, find the values of k and t.

13. a. Prove that $A(1, 2)$, $B(7, 0)$, and $C(3, -2)$ are the vertices of a right triangle.
b. Write an equation of the line that contains the median to the hypotenuse.

14. a. Prove that $A(1, -2)$, $B(2, 5)$, and $C(-2, 1)$ are the vertices of a right triangle.
b. Write an equation of the line that is parallel to the hypotenuse and passes through the vertex of the right angle.

15. The vertices of $\triangle ABC$ are $A(0, 6)$, $B(-8, 0)$, and $C(0, 0)$.
a. Write an equation of the altitude from point C to $\overline{AB}$. Name this line p.
b. Write an equation of the line that contains the midpoints of $\overline{AC}$ and $\overline{BC}$. Name this line q.
c. Do lines p and q intersect at right angles? Give a reason for your answer.

2.3 PARALLELOGRAMS

KEY IDEAS

A **parallelogram** is a quadrilateral, denoted by the symbol ▱, in which both pairs of opposite sides are parallel. Proofs involving congruent triangles may also make use of the properties of parallelograms.

Facts About Parallelograms

In a parallelogram:

- Opposite angles and opposite sides are congruent. In Figure 2.3:

  $\angle A \cong \angle C$ and $\angle B \cong \angle D$,
  $\overline{AB} \cong \overline{CD}$ and $\overline{AD} \cong \overline{BC}$.

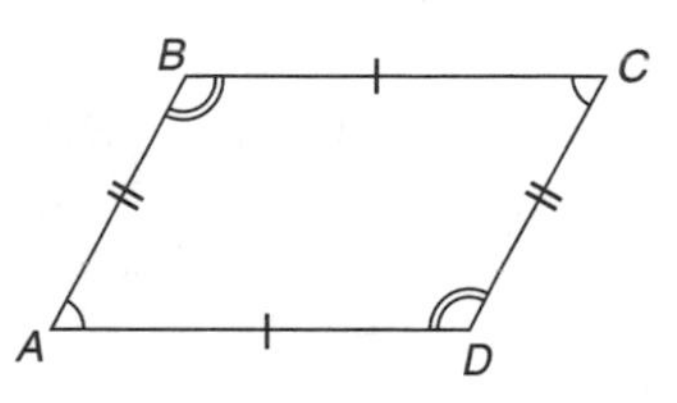

Figure 2.3 Angles and Sides of a Parallelogram

- The diagonals bisect each other. In Figure 2.4:

  $\overline{AE} \cong \overline{CE}$ and $\overline{BE} \cong \overline{DE}$.

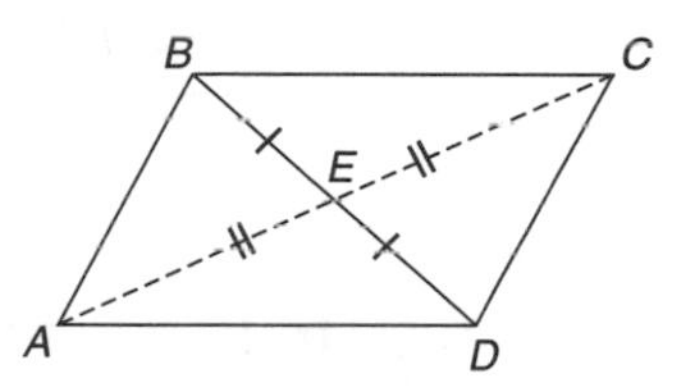

Figure 2.4 Diagonals of a Parallelogram

Example 1

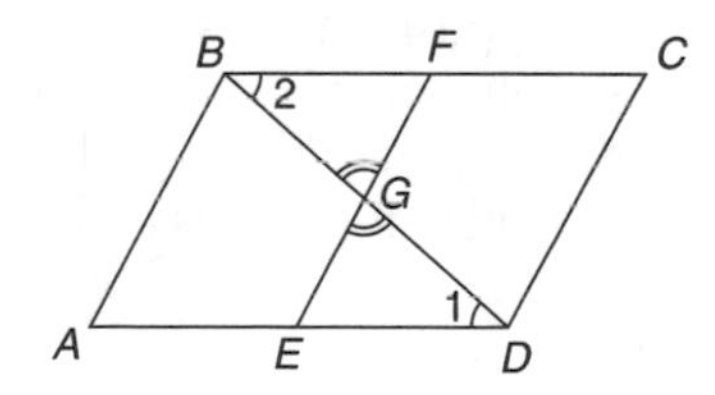

Given: ▱$ABCD$,
E is the midpoint of $\overline{AD}$,
F is the midpoint of $\overline{BC}$.
Prove: G is the midpoint of $\overline{EF}$.

Solution: See the proof.

PLAN: Show that $\triangle DGE \cong \triangle BGF$ by the AAS Theorem.

PROOF

Statement		Reason
1. $\square ABCD$		**1.** Given.
2. $\angle DGE \cong \angle BGF$	*Angle*	**2.** Vertical angles are congruent.
3. $\overline{AD} \parallel \overline{BC}$		**3.** Opposite sides of a parallelogram are parallel.
4. $\angle 1 \cong \angle 2$	*Angle*	**4.** If two lines are parallel, then alternate interior angles are congruent.
5. $AD = BC$		**5.** Opposite sides of a parallelogram are equal in length.
6. E is the midpoint of $\overline{AD}$, F is the midpoint of $\overline{BC}$.		**6.** Given.
7. $DE = \frac{1}{2} AD, BF = \frac{1}{2} BC$		**7.** Definition of midpoint.
8. $DE = BF$		**8.** Halves of equals are equal.
9. $\overline{DE} \cong \overline{BF}$	*Side*	**9.** Segments having equal lengths are congruent.
10. $\triangle DGE \cong \triangle BGF$		**10.** AAS Theorem.
11. $\overline{EG} \cong \overline{FG}$		**11.** CPCTC.
12. G is the midpoint of $\overline{EF}$.		**12.** Definition of midpoint.

Proving That a Quadrilateral Is a Parallelogram

To prove that a quadrilateral is a parallelogram, show that any one of the following statements is true:

- Opposite sides are parallel.
- Opposite angles are congruent.
- Opposite sides are congruent.
- Diagonals bisect each other.
- One pair of sides are both parallel and congruent.

Example 2

Given: $\square ABCD$,
$\overline{BE} \perp \overline{AC}$, $\overline{DF} \perp \overline{AC}$.
Prove: $BEDF$ is a parallelogram.

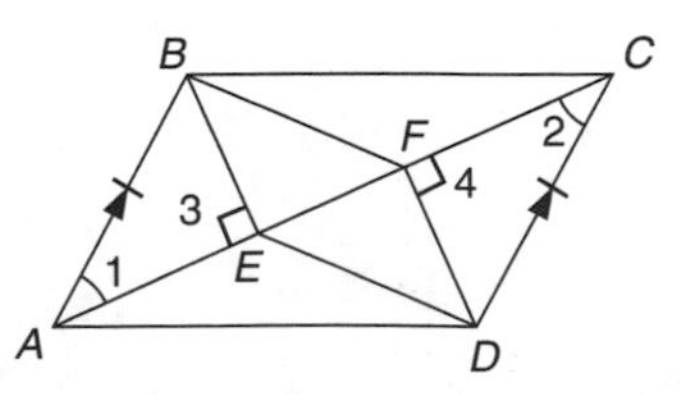

Solution: See the proof.

PLAN: Prove that $\overline{BE}$ and $\overline{DF}$ are both parallel and congruent, so that quadrilateral *BEDF* is a parallelogram.

PROOF

Statement		Reason
1. $\overline{BE} \perp \overline{AC}, \overline{DF} \perp \overline{AC}$		1. Given.
2. $\overline{BE} \parallel \overline{DF}$		2. If two lines are perpendicular to the same line, they are parallel.
3. $\square ABCD$		3. Given.
4. $\overline{AB} \parallel \overline{DC}$		4. Opposite sides of a parallelogram are parallel.
5. $\angle 1 \cong \angle 2$	*Angle*	5. If two lines are parallel, then alternate interior angles are congruent.
6. Angles 3 and 4 are right angles.		6. Perpendicular lines intersect to form right angles.
7. $\angle 3 \cong \angle 4$	*Angle*	7. All right angles are congruent.
8. $\overline{AB} \cong \overline{DC}$	*Side*	8. Opposite sides of a parallelogram are congruent.
9. $\triangle AEB \cong \triangle CFD$		9. AAS Theorem.
10. $\overline{BE} \cong \overline{DF}$		10. CPCTC.
11. Quadrilateral *BEDF* is a parallelogram.		11. If one pair of sides of a quadrilateral are both parallel and congruent, then the quadrilateral is a parallelogram.

Example 3

The vertices of a quadrilateral are $A(-1, -2)$, $B(4, 0)$, $C(5, 3)$, and $D(0, 1)$. Prove that quadrilateral *ABCD* is a parallelogram.

Solution:

- The midpoint of diagonal $\overline{AC}$ is $\left(\frac{-1+5}{2}, \frac{-2+3}{2}\right) = \left(2, \frac{1}{2}\right)$.
- The midpoint of diagonal $\overline{BD}$ is $\left(\frac{4+0}{2}, \frac{0+1}{2}\right) = \left(2, \frac{1}{2}\right)$.
- Since $\overline{AC}$ and $\overline{BD}$ have the same midpoint, the diagonals of quadrilateral *ABCD* bisect each other and, as a result, *ABCD* is a parallelogram.

TIP

You can prove that *ABCD* is a parallelogram in other ways. For example, using the slope formula, show that opposite sides have equal slopes and, as a result, are parallel.

Check Your Understanding of Section 2.3

In each case, write a formal proof.

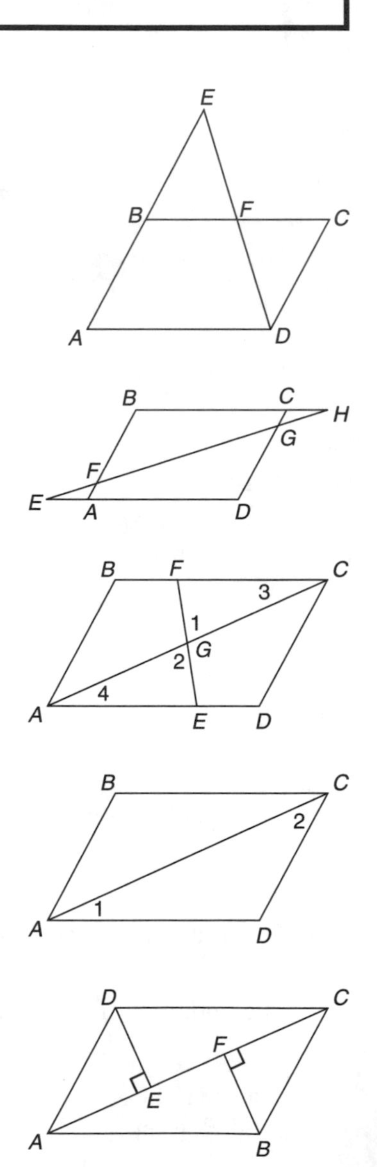

1. Given: $\square ABCD$,
   B is the midpoint of $\overline{AE}$.
   Prove: $\overline{EF} \cong \overline{FD}$.

2. Given: $\square ABCD$, $\overline{EF} \cong \overline{HG}$.
   Prove: $\overline{AF} \cong \overline{CG}$.

3. Given: Quadrilateral $ABCD$,
   $\overline{FG} \cong \overline{EG}, \overline{AG} \cong \overline{CG}$,
   $\angle B \cong \angle D$.
   Prove: a. $\overline{BC} \cong \overline{DA}$.
   b. $ABCD$ is a parallelogram.

4. Given: $ABCD$ is a parallelogram,
   $AD > DC$.
   Prove: $m\angle BAC > m\angle DAC$.

5. Given: Quadrilateral $ABCD$,
   diagonal $\overline{AEFC}$,
   $\overline{DE} \perp \overline{AC}, \overline{BF} \perp \overline{AC}$,
   $\overline{AE} \cong \overline{CF}, \overline{DE} \cong \overline{BF}$.
   Prove: $ABCD$ is a parallelogram.

6. Given: $\square BMDL$,
$\overline{AL} \cong \overline{CM}$.
Prove: $ABCD$ is a parallelogram.

7. Given: $\square ABCD$,
$\angle ABL \cong \angle CDM$.
Prove: $BLDM$ is a parallelogram.

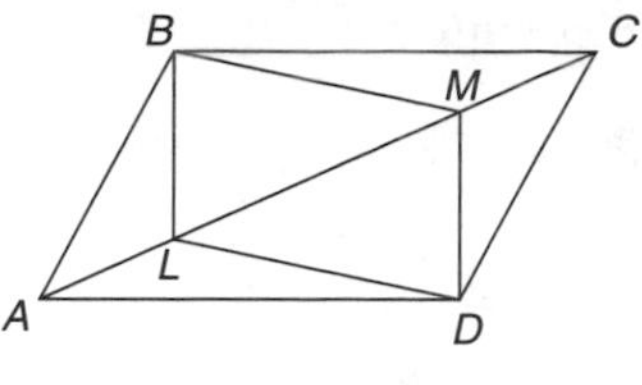

Exercises 6–7

8. Prove that $A(-2, -1)$, $B(6, 2)$, $C(7, 7)$, and $D(-1, 4)$ are the vertices of a parallelogram.
9. a. Prove that $A(-2, 8)$, $B(6, 6)$, $C(0, 2)$, and $D(-4, 14)$ are *not* the vertices of a parallelogram.
   b. Prove that the quadrilateral formed by joining the midpoints of the sides of quadrilateral $ABCD$ is a parallelogram.
10. The vertices of parallelogram $STAR$ are $S(1, 1)$, $T(-2, 3)$, $A(0, k)$, and $R(3, -5)$.
   a. What is the value of k?
   b. Write an equation of diagonal $\overline{AS}$.

2.4 SPECIAL PARALLELOGRAMS

KEY IDEAS

A parallelogram may have four equal angles, four equal sides, or both of these properties.

Rectangle

A **rectangle** (see Figure 2.5) is a parallelogram that has these special properties:

- All the properties of a parallelogram.
- Four right angles.
- Congruent diagonals.

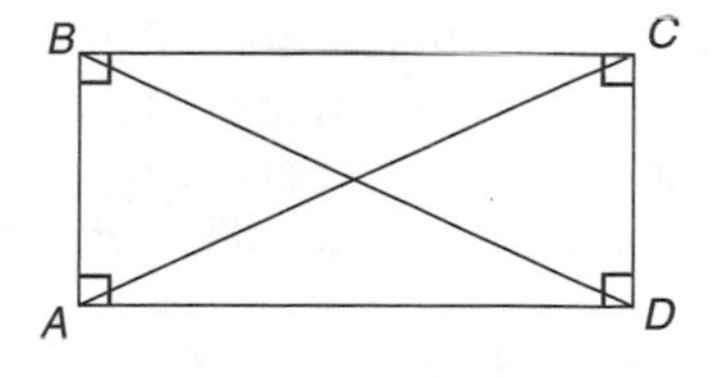

Figure 2.5 Rectangle $ABCD$ with $\overline{AC} \cong \overline{BD}$

Example 1

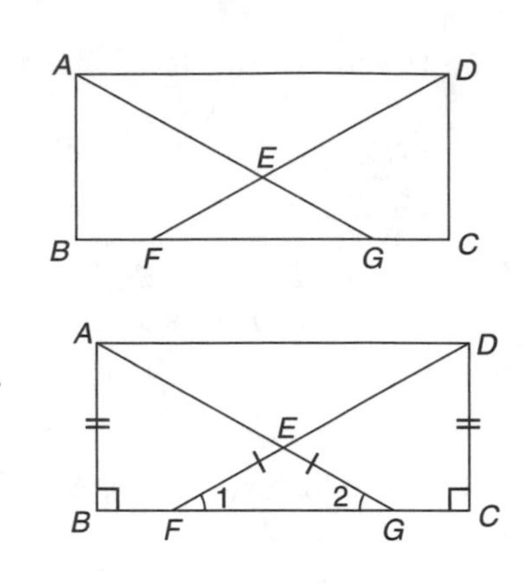

Given: Rectangle $ABCD$,
$\overline{EF} \cong \overline{EG}$.
Prove: $\overline{AG} \cong \overline{DF}$.

Solution: See the proof.

PLAN: Prove $\triangle ABG \cong \triangle DCF$ by $AAS \cong AAS$.

PROOF

Statement		Reason
1. Rectangle $ABCD$		1. Given.
2. Angles B and C are right angles.		2. A rectangle contains four right angles.
3. $\angle B \cong \angle C$	*Angle*	3. All right angles are congruent.
4. $\overline{EF} \cong \overline{EG}$		4. Given.
5. $\angle 1 \cong \angle 2$	*Angle*	5. If two sides of a triangle are congruent, the angles opposite these sides are congruent.
6. $\overline{AB} \cong \overline{CD}$	*Side*	6. Opposite sides of a rectangle are congruent.
7. $\triangle ABG \cong \triangle DCF$		7. $AAS \cong AAS$.
8. $\overline{AG} \cong \overline{DF}$		8. Corresponding sides of congruent triangles are congruent.

Proving That a Quadrilateral Is a Rectangle

To show that a quadrilateral is a rectangle, show that the quadrilateral is a parallelogram that contains a right angle or is a parallelogram with congruent diagonals. If a parallelogram does *not* contain a right angle or does *not* have congruent diagonals, then it is *not* a rectangle.

Example 2

Quadrilateral $GAME$ has vertices $G(-2, 2)$, $A(8, -1)$, $M(9, 3)$, and $E(3, 2)$.
a. Prove that $GAME$ is a parallelogram.
b. Prove that $GAME$ is *not* a rectangle.

Solutions: See the coordinate proof

a. PLAN: A quadrilateral is a parallelogram if its diagonals bisect each other. To show that the diagonals of *GAME* bisect each other, show that the diagonals have the same midpoint.

PROOF:

- The midpoint of diagonal $\overline{GM}$ is $\left(\frac{2+9}{2}, \frac{-2+3}{2}\right) = \left(\frac{11}{2}, \frac{1}{2}\right)$.
- The midpoint of diagonal $\overline{AE}$ is $\left(\frac{8+3}{2}, \frac{-1+2}{2}\right) = \left(\frac{11}{2}, \frac{1}{2}\right)$.
- The midpoint of each diagonal is $(\frac{11}{2}, \frac{1}{2})$. Since the diagonals have the same midpoint, they bisect each other and, as a result, *GAME* is a parallelogram.

b. PLAN: A parallelogram is *not* a rectangle if its diagonals are *not* congruent. To show that the diagonals of *GAME* are not congruent, show that they have unequal lengths.

PROOF:

- The length of a line segment is

$$\sqrt{(\Delta x)^2 + (\Delta y)^2},$$

where Δx and Δy represent the differences in the x- and y-coordinates, respectively, of the endpoints of the line segment. Hence:

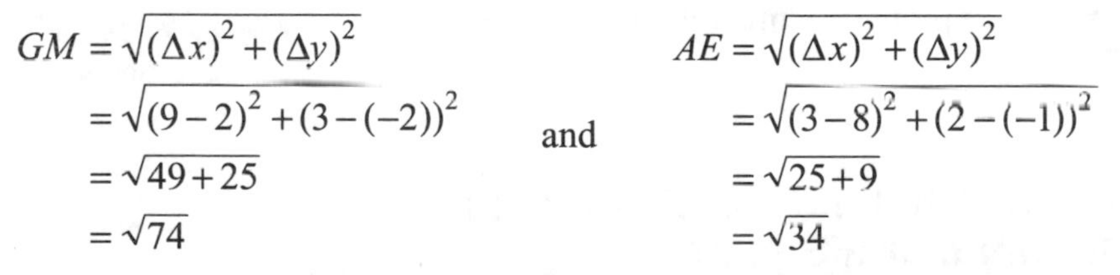

$$GM = \sqrt{(\Delta x)^2 + (\Delta y)^2} = \sqrt{(9-2)^2 + (3-(-2))^2} = \sqrt{49+25} = \sqrt{74}$$

and

$$AE = \sqrt{(\Delta x)^2 + (\Delta y)^2} = \sqrt{(3-8)^2 + (2-(-1))^2} = \sqrt{25+9} = \sqrt{34}$$

- Since $GM = \sqrt{74}$ and $AE = \sqrt{34}$, diagonals $\overline{GM}$ and $\overline{AE}$ are *not* congruent and, as a result, parallelogram *GAME* is *not* a rectangle.

You can also prove that *GAME* is not a rectangle by showing that the slopes of any pair of adjacent sides are *not* negative reciprocals and, as a result, do not form a right angle.

Rhombus

A **rhombus** is a parallelogram that has these special properties:

- All the properties of a parallelogram.
- Four congruent sides. In Figure 2.6:

$$AB = BC = CD = AD.$$

- Diagonals that bisect opposite pairs of angles. In Figure 2.6:

$$\angle 1 \cong \angle 2 \text{ and } \angle 3 \cong \angle 4,$$

$$\angle 5 \cong \angle 6 \text{ and } \angle 7 \cong \angle 8.$$

- Perpendicular diagonals. In Figure 2.6, $\overline{AC} \perp \overline{BD}$. Because a rhombus is a parallelogram, $AE = EC$ and $BE = ED$.

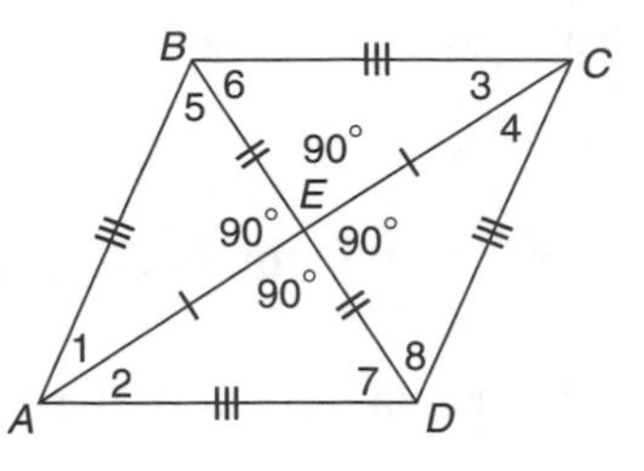

Figure 2.6 Properties of a Rhombus

Square

A **square** is a rectangle with four congruent sides. In addition, a square has these special properties (see Figure 2.7):

- All the properties of a parallelogram.
- All the properties of a rectangle.
- All the properties of a rhombus.

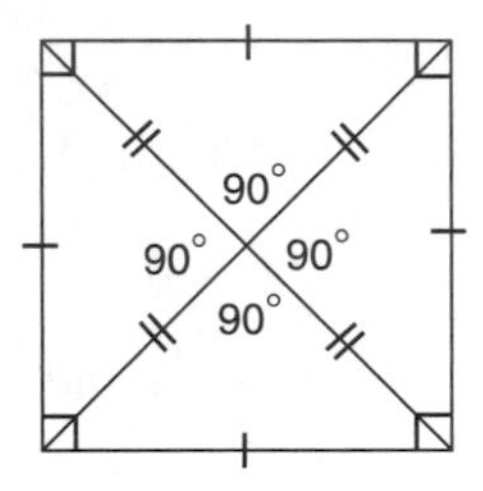

Figure 2.7 Properties of a Square

Proving That a Quadrilateral Is a Rhombus or a Square

To prove that a quadrilateral is a *rhombus*, show that the quadrilateral has four congruent sides *or* show that it is a parallelogram for which one of the following statements is true:

- A pair of adjacent sides are congruent.
- The diagonals intersect at right angles.
- The opposite angles are bisected by the diagonals.

To prove that a quadrilateral is a *square*, show that one of the following statements is true:

- The quadrilateral is a rectangle with a pair of congruent adjacent sides.
- The quadrilateral is a rhombus that contains a right angle.

Check Your Understanding of Section 2.4

A. Multiple Choice

1. A parallelogram must be a rectangle if its diagonals
(1) bisect each other
(2) bisect the angles to which they are drawn
(3) are perpendicular to each other
(4) are congruent

2. The diagonals of a rhombus do *not*
(1) bisect each other
(2) bisect the angles to which they are drawn
(3) intersect at right angles
(4) have the same length

3. If the diagonals of a parallelogram are perpendicular but not congruent, then the parallelogram is
(1) a rectangle
(2) a rhombus
(3) a square
(4) an isosceles trapezoid

4. In rhombus *PQRS*, diagonals $\overline{PR}$ and $\overline{QS}$ intersect at *T*. Which statement is *always* true?
(1) Quadrilateral *PQRS* is a square.
(2) Triangle *RTQ* is a right triangle.
(3) Triangle *PQS* is equilateral.
(4) Diagonals $\overline{PR}$ and $\overline{QS}$ are congruent.

5. Which statements describe the properties of the diagonals of a rectangle?
I. The diagonals are congruent.
II. The diagonals are perpendicular.
III. The diagonals bisect each other.
(1) II and III, only
(2) I and II, only
(3) I and III, only
(4) I, II, and III

B. *In each case, show how you arrived at your answer by clearly indicating all of the necessary steps, formula substitutions, diagrams, graphs, charts, etc.*

6. Prove that $M(-2, -1)$, $A(1, 6)$, $T(8, 3)$, and $H(5, -4)$ are the vertices of a square.

7. a. Prove that $A(-3, 6)$, $B(6, 0)$, $C(9, -9)$, and $D(0, -3)$ are the vertices of a parallelogram.
b. Prove that parallelogram *ABCD* is *not* a rhombus.

8. a. Prove that $R(2, 1)$, $E(10, 7)$, $C(7, 11)$ and $T(-1, 5)$ are the vertices of a rectangle.

b. Prove that $\square RECT$ is *not* a square.

9. a. Prove that $R(-5, 0)$, $H(2, -1)$, $O(7, 4)$, and $M(0, 5)$ are the vertices of a rhombus.

b. Prove that $\square RHOM$ is *not* a square.

10. The vertices of $\triangle ABC$ are $A(0, 0)$, $B(2, 6)$, and $C(4, 2)$. If $\overline{LM}$ connects the midpoints of sides $\overline{AB}$ and $\overline{AC}$, prove that:

a. $LM = \frac{1}{2}BC$ b. $\overline{LM} \parallel \overline{BC}$

11. Given: Rhombus $ABCD$, $\overline{BL} \cong \overline{CM}$, $\overline{AL} \cong \overline{BM}$.

Prove: $ABCD$ is a square.

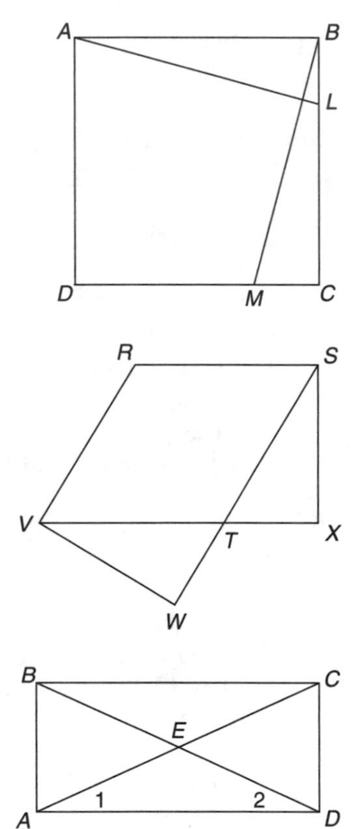

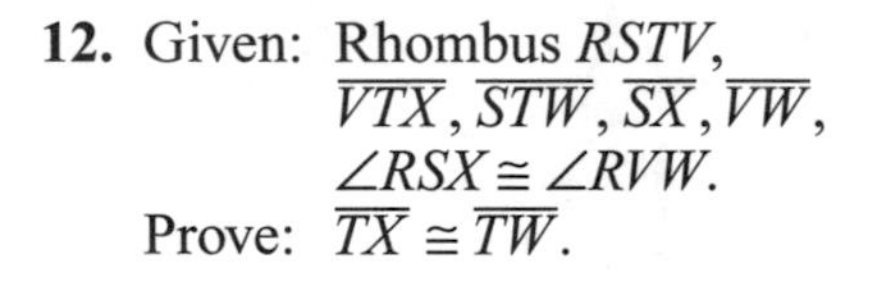

12. Given: Rhombus $RSTV$, $\overline{VTX}$, $\overline{STW}$, $\overline{SX}$, $\overline{VW}$, $\angle RSX \cong \angle RVW$.

Prove: $\overline{TX} \cong \overline{TW}$.

13. Given: $\square ABCD$, $m\angle 2 > m\angle 1$.

Prove: $\square ABCD$ is *not* a rectangle.

14. Given: Square $ABCD$,
$\angle 1 \cong \angle 2$.
Prove: $\overline{BE} \cong \overline{DF}$.

15. Given: Rectangle $ABCD$,
$\angle 1 \cong \angle 2$, $\angle BEF \cong \angle DFE$.
Prove: $ABCD$ is a square.

16. Given: Rectangle $ABCD$,
$\overline{BE} \cong \overline{DF}$, $\angle CEF \cong \angle CFE$.
Prove: a. $ABCD$ is a square.
b. $\triangle EAF$ is isosceles.

Exercises 15 and 16

2.5 TRAPEZOIDS

KEY IDEAS

A **trapezoid**, shown in the accompanying figure, is a quadrilateral that has exactly one pair of parallel sides, called **bases**. The nonparallel sides of a trapezoid are the **legs**.

Properties of an Isosceles Trapezoid

In an **isosceles trapezoid** (Figure 2.8):

- The legs are congruent:

$$\overline{AB} \cong \overline{DC}.$$

- The upper and lower base angles are congruent:

$$\angle A \cong \angle D \quad \text{and} \quad \angle B \cong \angle C.$$

- The diagonals are congruent:

$$\overline{AC} \cong \overline{DB}.$$

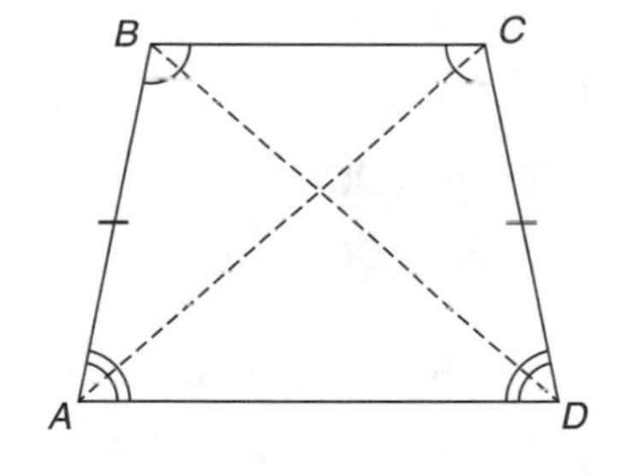

Figure 2.8 Isosceles Trapezoid

Example 1

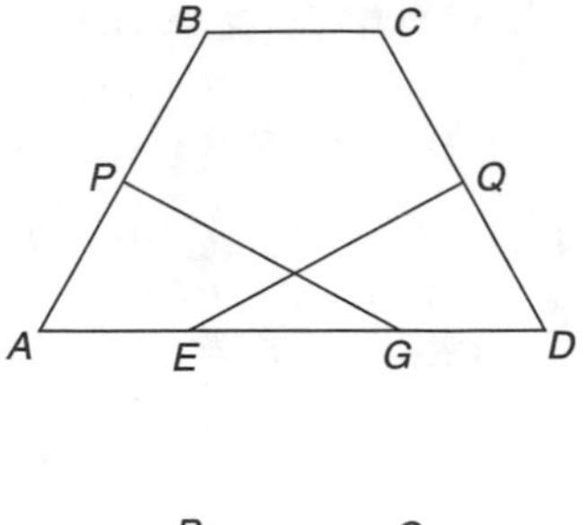

Given: Isosceles trapezoid with $\overline{BC} \parallel \overline{AD}$, $\overline{GP} \perp \overline{AB}$, $\overline{EQ} \perp \overline{CD}$; P and Q are midpoints of $\overline{AB}$ and $\overline{CD}$, respectively.
Prove: $\overline{GP} \cong \overline{EQ}$.

Solution: See the proof.

PLAN: Prove $\triangle APG \cong \triangle EQD$ by $ASA \cong ASA$.

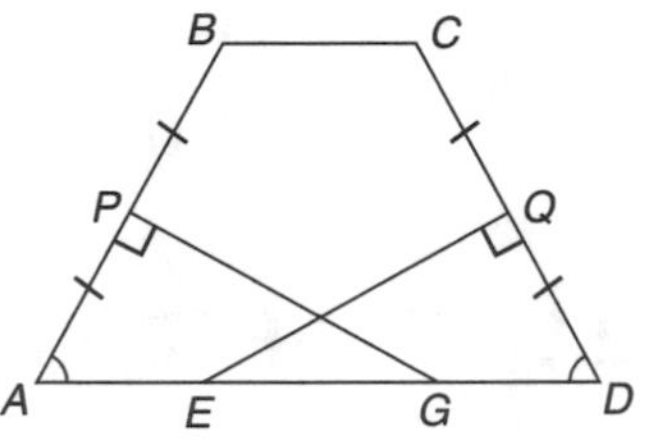

PROOF

Statement		Reason
1. Isosceles trapezoid with $\overline{BC} \parallel \overline{AD}$.		1. Given.
2. $\angle A \cong \angle D$	*Angle*	2. Base angles of an isosceles trapezoid are congruent.
3. P and Q are midpoints of $\overline{AB}$ and $\overline{CD}$, respectively.		3. Given.
4. $AP = \frac{1}{2}AB$ and $DQ = \frac{1}{2}DC$		4. Definition of midpoint.
5. $AB = DC$		5. The legs of an isosceles trapezoid are congruent and, as a result, have the same length.
6. $AP = DQ$		6. Halves of equals are equal.
7. $\overline{AP} \cong \overline{DQ}$	*Side*	7. Line segments that are equal in length are congruent.
8. $\overline{GP} \perp \overline{AB}$ and $\overline{EQ} \perp \overline{CD}$		8. Given.
9. $\angle APG$ and $\angle DQE$ are right angles.		9. Perpendicular lines intersect to form right angles.
10. $\angle APG \cong \angle DQE$	*Angle*	10. All right angles are congruent.
11. $\triangle APG \cong \triangle EQD$		11. $ASA \cong ASA$.
12. $\overline{GP} \cong \overline{EQ}$		12. Corresponding sides of congruent triangles are congruent.

Proving That a Trapezoid Is Isosceles

To prove that a trapezoid is an *isosceles trapezoid*, show that one of the following statements is true:

- The legs are congruent.
- The lower (or upper) base angles are congruent.
- The diagonals are congruent.

Example 2

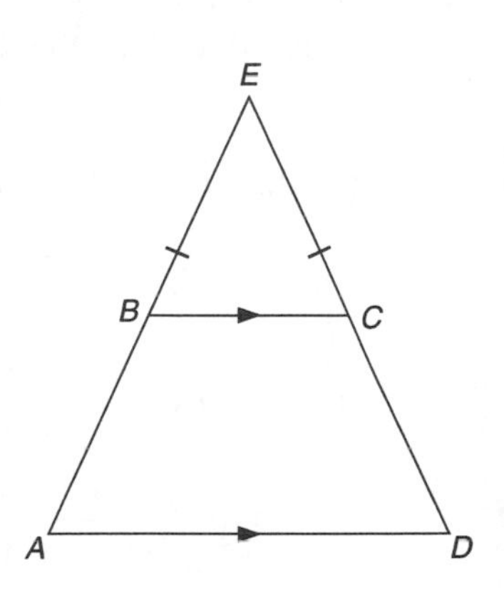

Given: Trapezoid $ABCD$,
$\overline{BC} \parallel \overline{AD}$, $\overline{EB} \cong \overline{EC}$.
Prove: $ABCD$ is an isosceles trapezoid.

Solution: See the proof.

PLAN: show that $\angle A \cong \angle D$.

PROOF

Statement	Reason
1. Trapezoid $ABCD$ with $\overline{BC} \parallel \overline{AD}$, $\overline{EB} \cong \overline{EC}$	**1.** Given.
2. $\angle EBC \cong \angle ECB$	**2.** If two sides of a triangle ($\triangle EBC$) are congruent, then the angles opposite these sides are congruent.
3. $\angle A \cong \angle EBC$, $\angle D \cong \angle ECB$	**3.** If two lines are parallel, then corresponding angles are congruent.
4. $\angle A \cong \angle D$	**4.** Transitive property.
5. $ABCD$ is an isosceles trapezoid.	**5.** If a trapezoid has a pair of congruent base angles, then it is an isosceles trapezoid.

Proving That a Quadrilateral Is or Is *Not* a Trapezoid

Unlike a parallelogram, a trapezoid has exactly *one* pair of parallel sides.

- To prove that a quadrilateral is a trapezoid, show that two sides are parallel *and* that the other two sides are *not* parallel.
- To prove that a quadrilateral is *not* a trapezoid, show that both pairs of opposite sides are parallel *or* that both pairs of opposite sides are *not* parallel.

Example 3

The coordinates of the vertices of quadrilateral *ABCD* are $A(-2, 0)$, $B(10, 3)$, $C(5, 7)$, and $D(1, 6)$. Prove that quadrilateral *ABCD* is a trapezoid.

Solution: See the proof.

PLAN: To show that a quadrilateral is a trapezoid, show that one pair of sides is parallel and that the other pair of sides is *not* parallel. To show that two lines are parallel, show that these lines have the same slope.

PROOF:

- Find the slopes of the four sides of $\square ABCD$:

$$\text{slope of } \overline{AB} = \frac{y_B - y_A}{x_B - x_A} = \frac{3-0}{10-(-2)} = \frac{3}{12} = \frac{1}{4}$$

$$\text{slope of } \overline{BC} = \frac{y_C - y_B}{x_C - x_B} = \frac{7-3}{5-10} = \frac{4}{-5} = -\frac{4}{5}$$

$$\text{slope of } \overline{CD} = \frac{y_D - y_C}{x_D - x_C} = \frac{6-7}{1-5} = \frac{-1}{-4} = \frac{1}{4}$$

$$\text{slope of } \overline{AD} = \frac{y_D - y_A}{x_D - x_A} = \frac{6-0}{1-(-2)} = \frac{6}{3} = 2$$

$\overline{AB} \parallel \overline{CD}$

- Compare the slopes of opposite sides. Side $\overline{AB}$ is parallel to side $\overline{CD}$ since the sides have the same slope. Side $\overline{BC}$ is *not* parallel to side $\overline{AD}$ since the slopes of the sides are *not* equal.
- Quadrilateral *ABCD* is a trapezoid since it contains exactly one pair of parallel sides.

Example 4

Given: Trapezoid *ROSE* with diagonals $\overline{RS}$ and $\overline{EO}$ intersecting at point *M*.
Prove: Diagonals $\overline{RS}$ and $\overline{EO}$ do *not* bisect each other.

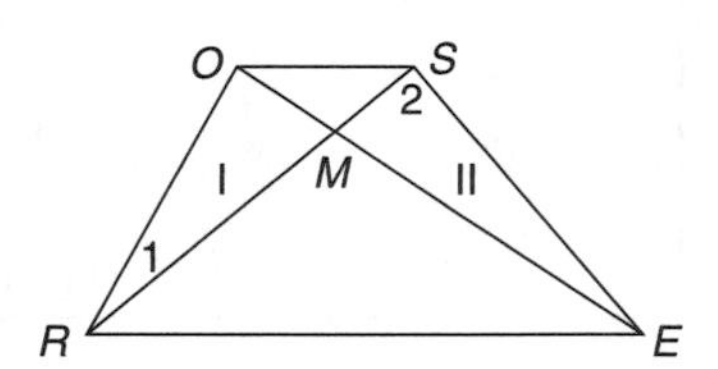

Solution: See the proof.

PLAN: Use an indirect proof. Assume $\overline{RS}$ and $\overline{EO}$ bisect each other, and show that, as a result, $\triangle I \cong \triangle II$. Then $\angle 1 \cong \angle 2$, so $\overline{OR} \parallel \overline{SE}$, which contradicts the "Given" that the quadrilateral is a trapezoid.

PROOF

Statement	Reason
1. Trapezoid *ROSE* with $\overline{OS} \parallel \overline{RE}$, and diagonals $\overline{RS}$ and $\overline{EO}$ intersecting at point *M*.	1. Given.
2. Diagonals $\overline{RS}$ and $\overline{EO}$ bisect each other.	2. Assume this is true.
3. $\overline{OM} \cong \overline{EM}$ *Side*	3. A bisector divides a segment into two congruent segments.
4. $\angle OMR \cong \angle SMR$ *Angle*	4. Vertical angles are congruent.
5. $\overline{RM} \cong \overline{SM}$ *Side*	5. A bisector divides a segment into two congruent segments.
6. $\triangle \text{I} \cong \triangle \text{II}$	6. *SAS* ≅ *SAS*.
7. $\angle ORM \cong \angle ESM$	7. CPCTC.
8. $\overline{OR} \parallel \overline{SE}$.	8. If two lines are cut by a transversal and alternate interior angles are congruent, then the lines are parallel.
9. Statement 8 contradicts the given.	9. A trapezoid has exactly one pair of parallel sides.
10. Statement 2 is false.	10. A statement that leads to a contradiction is false.
11. $\overline{RS}$ and $\overline{EO}$ do *not* bisect each other.	11. The opposite of a false statement is true.

Check Your Understanding of Section 2.5

In each case, show how you arrived at your answer by clearly indicating all of the necessary steps, formula substitutions, diagrams, graphs, charts, etc.

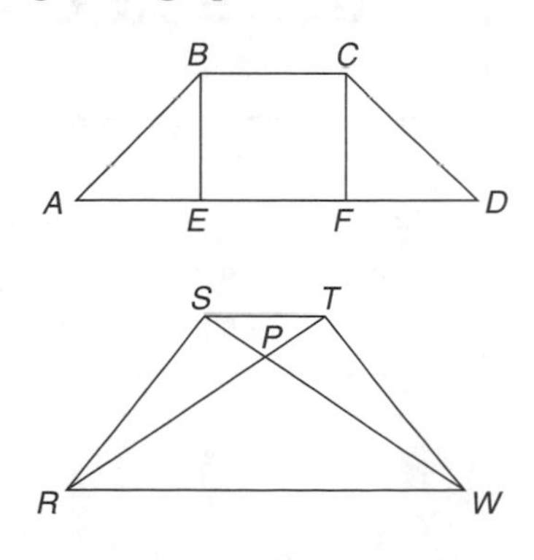

1. Given: Trapezoid *ABCD*, $\overline{AD} \parallel \overline{BC}$, $\overline{BE}$ and $\overline{CF}$ are altitudes drawn to $\overline{AD}$, $\overline{AE} \cong \overline{DF}$.
   Prove: Trapezoid *ABCD* is isosceles.

2. Given: Isosceles trapezoid *RSTW*.
   Prove: $\triangle RPW$ is isosceles.

3. Given: Trapezoid $ABCD$,
$\overline{EF} \cong \overline{EG}$, $\overline{AF} \cong \overline{DG}$, $\overline{BG} \cong \overline{CF}$.
Prove: Trapezoid $ABCD$ is isosceles.

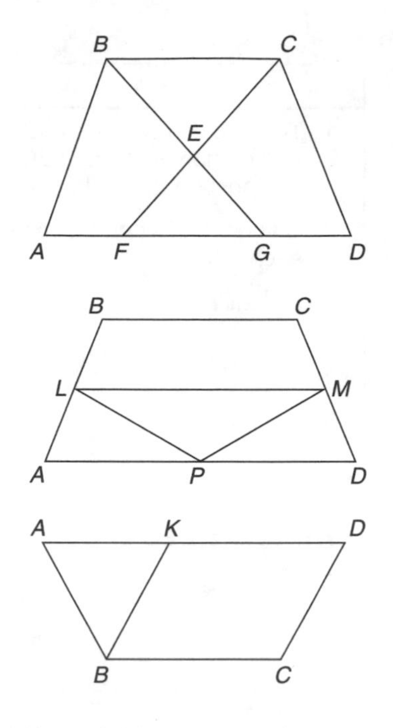

4. Given: Trapezoid $ABCD$ with median $\overline{LM}$, P is the midpoint of $\overline{AD}$,
$\overline{LP} \cong \overline{MP}$.
Prove: Trapezoid $ABCD$ is isosceles.

5. Given: Isosceles trapezoid $ABCD$,
$\angle BAK \cong \angle BKA$.
Prove: $BKDC$ is a parallelogram.

6. Prove that points $P(-3, -4)$, $Q(9, 5)$, $R(-1, 10)$, and $S(-5, 7)$ are the vertices of an isosceles trapezoid.

7. Given points $A(-2, 3)$, $B(1, 0)$, $C(7, 6)$, and $D(0, 5)$.
a. Prove that quadrilateral $ABCD$ is a trapezoid.
b. If points B, $E(h, k)$, and C are collinear, find the values of h and k such that points A, B, E, and D are the vertices of a parallelogram.
c. Prove that $ABED$ is a rectangle.

8. The coordinates of the vertices of quadrilateral $JAKE$ are $J(0, 3a)$, $A(3a, 3a)$, $K(4a, 0)$, and $E(-a, 0)$. Prove that $JAKE$ is an isosceles trapezoid.

2.6 GENERAL COORDINATE PROOFS

KEY IDEAS

Generalizations about a figure may be easier to prove if the figure is placed in the coordinate plane, using variables and zeros as coordinates of the vertices of the figure.

Proofs Using General Coordinates

To prove that a figure has a special property, place the figure in the coordinate plane with general coordinates. Then use the slope, midpoint, or distance formula, depending on the relationship you want to prove.

Example 1

The vertices of right triangle ABC are $A(2a, 0)$, $B(0, 2b)$, and $C(0, 0)$. Prove that the median drawn to the hypotenuse of right triangle ABC is one-half of the length of the hypotenuse.

Solution: See the proof.

PROOF:

- The midpoint of hypotenuse $\overline{AB}$ is $\left(\dfrac{0+2a}{2}, \dfrac{2b+0}{2}\right) = (a, b)$.
- Find the length of median $\overline{CM}$ by using the distance formula:

$$CM = \sqrt{(\Delta x)^2 + (\Delta y)^2}$$
$$= \sqrt{(a-0)^2 + (b-0)^2}$$
$$= \sqrt{a^2 + b^2}$$

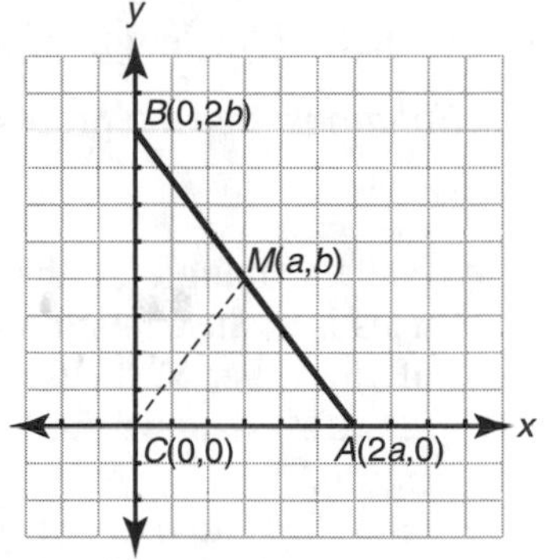

- Find the length of hypotenuse $\overline{AB}$ by using the distance formula:

$$AB = \sqrt{(\Delta x)^2 + (\Delta y)^2}$$
$$= \sqrt{(2a-0)^2 + (0-2b)^2}$$
$$= \sqrt{4a^2 + 4b^2}$$
$$= 2\sqrt{a^2 + b^2}$$

- Since median $CM = \sqrt{a^2 + b^2}$ and hypotenuse $AB = 2\sqrt{a^2 + b^2}$, the length of the median drawn to the hypotenuse of a right triangle is one-half of the length of the hypotenuse.

Because general coordinates were used for the vertices of the right triangle, the property proved in this example is true for any right triangle.

Positioning Figures Using General Coordinates

When you need to determine the general coordinates of the vertices of a figure being placed in the coordinate plane, follow these guidelines:

- Try to place one of the vertices at the origin.
- Place at least one of the sides on a coordinate axis.
- Make use of the defining properties of the figure. For example, place square *ABCD* so that *A* is at the origin, adjacent sides $\overline{AB}$ and $\overline{AD}$ coincide with the *x*- or *y*-axis, and all four sides have the same length.

Example 2

Prove that the line segment joining the midpoints of two sides of a triangle is:

a. parallel to the third side of the triangle
b. one-half of the length of the third side

Solutions: See the proofs.

a. PROOF:

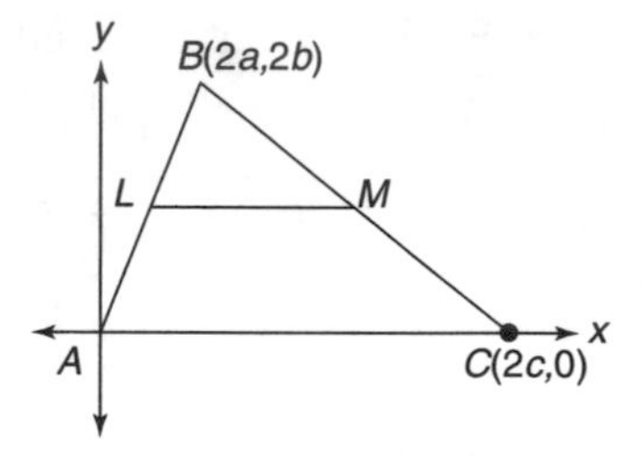

- Place a triangle so that one of the vertices is at the origin and the coordinates of the other vertices are (2*a*, 2*b*) and (2*c*, 0), as shown in the accompanying figure.
- Find the midpoints of $\overline{AB}$ and $\overline{BC}$:

$$L(\bar{x}, \bar{y}) = L\left(\frac{0+2a}{2}, \frac{0+2b}{2}\right) = L(a, b)$$

$$M(\bar{x}, \bar{y}) = M\left(\frac{2a+2c}{2}, \frac{2b+0}{2}\right) = M(a+c, b)$$

- Points *L* and *M* have the same *y*-coordinate, so $\overline{LM}$ is a horizontal segment. Since $\overline{AC}$ lies on the *x*-axis, it is also a horizontal segment. Since two horizontal segments are parallel, $\overline{LM} \parallel \overline{AC}$.

b. PROOF:

- $LM = (a + c) - a = c$
- $AC = 2c - 0 = 2c$
- Since $LM = c$ and $AC = 2c$, $LM = \frac{1}{2}AC$.

TIP

In Example 2, choosing $2a$, $2b$, and $2c$, rather than a, b, and c, eliminates the need to work with fractions when calculating the midpoints of $\overline{AB}$ and $\overline{BC}$.

Check Your Understanding of Section 2.6

In each case, show how you arrived at your answer by clearly indicating all of the necessary steps, formula substitutions, diagrams, graphs, charts, etc.

1. The coordinates of the vertices of $\triangle ABC$ are $A(-2a, 0)$, $B(0, 2b)$ and $C(2a, 0)$. Using coordinates, prove that medians $\overline{AH}$ and $\overline{CK}$ drawn to the legs of isosceles triangle ABC are congruent.

2. The coordinates of the vertices of square $ABCD$ are $A(0, 0)$, $B(0, t)$, $C(x, y)$, and $D(t, 0)$.
   a. Express the coordinates of C in terms of t.
   b. Use the distance formula to prove that the diagonals of square $ABCD$ are congruent.
   c. Use the slope formula to prove that the diagonals of square $ABCD$ are perpendicular.

3. Using variables as coordinates, place a rectangle in the coordinate plane and prove that the diagonals of the rectangle are congruent.

4. The coordinates of the vertices of quadrilateral $MATH$ are $M(0, 0)$, $A(r, t)$, $T(s, t)$, and $H(s - r, 0)$. Using coordinates, prove that:
   a. Quadrilateral $MATH$ is a parallelogram.
   b. The diagonals of quadrilateral $MATH$ are not necessarily congruent.

5. The coordinates of the vertices of quadrilateral $ABCD$ are $A(0, 0)$, $B(a, b)$, $C(c, b)$, and $D(a + c, 0)$. Using coordinates, prove that $ABCD$ is an isosceles trapezoid.

6. The coordinates of the vertices of parallelogram $ABCD$ are $A(0, 0)$, $B(b, y)$, $C(a + b, y)$, and $D(a, 0)$.
   a. If $ABCD$ is a rhombus, express y in terms of a and b.
   b. Using coordinates, prove that the diagonals of rhombus $ABCD$ are perpendicular.

7. The vertices of isosceles trapezoid $ABCD$ with bases $\overline{BC}$ and $\overline{AD}$ are $A(0, 0)$, $B(b, c)$, $C(h, k)$, and $D(a, 0)$.

a. Express h and k in terms of a, b, or c.
b. Using coordinates, prove that the diagonals of $ABCD$ are congruent.

8. The median of a trapezoid is the segment whose endpoints are the midpoints of the nonparallel sides. If $A(0, 0)$, $B(0, 2b)$, $C(2a, 2b)$, and $D(2d, 0)$ are the vertices of a trapezoid, prove that the length of the median of trapezoid $ABCD$ is one-half the sum of the lengths of the two bases.

9 and 10. Use the theorem proved in Example 2 on page 58 to prove each of the following *without* using coordinates.

9. Given: $RSTW$ is a parallelogram,
B is the midpoint of $\overline{SW}$,
C is the midpoint of $\overline{ST}$.
Prove: $WACT$ is a parallelogram.

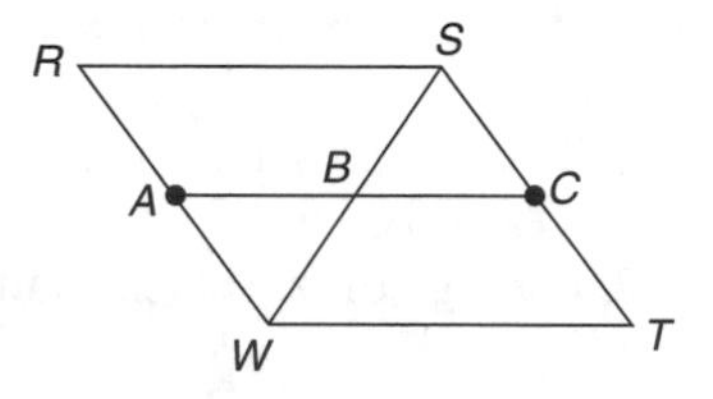

10. If the midpoints of the sides of a rectangle are joined consecutively, the resulting quadrilateral is a rhombus.

CHAPTER 3
SIMILAR AND RIGHT TRIANGLES

3.1 RATIO AND PROPORTION

KEY IDEAS

A *ratio* compares two numbers that have the *same* units of measurement. The ratio of x and y can be written in any one of the following three ways:

$$\frac{x}{y}, \qquad x{:}y, \qquad x \text{ to } y.$$

An equation that states that two ratios are equal is called a **proportion**. To find an unknown member of a proportion, cross-multiply and solve for the unknown term.

Ratio Versus Rate

A *ratio* and a *rate* each compare two numbers.

- A **ratio** is a comparison by division of two quantities that have the *same* units of measurement. For example, if John is 24 years old and Glen is 8 years old, then the *ratio* of John's age to Glen's age is 3 to 1:

$$\frac{\text{John's age}}{\text{Glen's age}} = \frac{24 \cancel{\text{ years}}}{8 \cancel{\text{ years}}} = \frac{3}{1} \text{ or } 3 \text{ to } 1.$$

  Notice that a ratio has no units of measurement.
- A **rate** is a comparison by division of two quantities that have *different* units of measurement. For example, if Jaime earns \$300 for working 16 hours, then Jaime's *rate* of pay per hour is \$18.75:

$$\frac{\$300}{16 \text{ hours}} = \$18.75 \text{ per hour.}$$

Example 1

If Tamika runs $\frac{3}{5}$ of a mile in 2 minutes 30 seconds, what is her average rate, in miles per minute?

Solution: $\mathbf{\dfrac{6}{25}}$

Since $2\,\text{min}\ 30\,\text{sec} = 2\frac{1}{2}\ \text{min} = \frac{5}{2}\ \text{min}$:

$$\begin{aligned}\text{Rate} = \frac{\text{number of miles}}{\text{number of minutes}} &= \frac{3}{5}\ \text{mi} \div \frac{5}{2}\ \text{min}\\ &= \frac{3}{5} \times \frac{2}{5}\ \text{mi/min}\\ &= \frac{6}{25}\ \text{mi/min}\end{aligned}$$

Example 2

On a trip, a student drove 40 miles per hour for 2 hours and then drove 30 miles per hour for 3 hours. What is the student's average rate of speed, in miles per hour, for the whole trip?

Solution: **34**

On the trip, the student drove 170 mi in 5 hr since:

- $40\frac{\text{mi}}{\text{hr}} \times 2\,\text{hr} = 80\,\text{mi}$ in the first 2 hr
- $30\frac{\text{mi}}{\text{hr}} \times 3\,\text{hr} = 90\,\text{mi}$ for the next 3 hr.

Hence:

$$\begin{aligned}\text{Average rate of speed (mph)} &= \frac{\text{total distance in miles}}{\text{total time in hours}}\\ &= \frac{170\ \text{mi}}{5\ \text{hr}}\\ &= 34\ \text{mph}\end{aligned}$$

Solving Problems with Ratios

If two whole numbers are in the ratio of 2 to 1, the pair of numbers can be 2 and 1, 4 and 2, 6 and 3, and so forth. Since, in each of these possibilities after the first, 2 and 1 are multiplied by the same nonzero number, say x, we can represent the set of all numbers whose ratio is 2 to 1 by $2x$ and x.

Example 3

The perimeter of a rectangle is 56 centimeters. If the length and width of the rectangle are in the ratio of 3 to 1, what is the length of the rectangle?

Solution: **21 cm**

Let x = width of the rectangle. Then $3x$ = length of the rectangle. Since the perimeter of a rectangle is the sum of the lengths of the four sides:

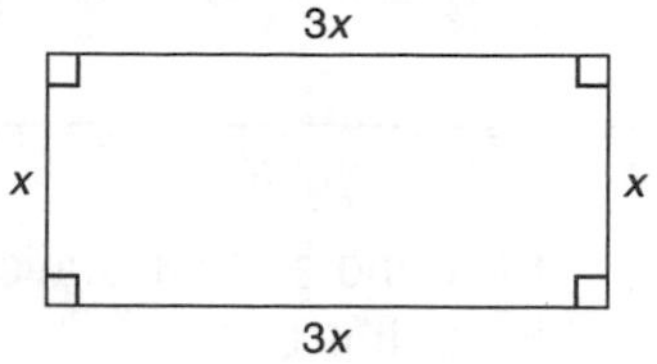

3x
x
x
3x

$$x + 3x + x + 3x = 56$$
$$8x = 56$$
$$x = \frac{56}{8} = 7$$

Hence, length $= 3x = 3(7\text{ cm}) = 21\text{ cm}$.

Cross-Multiplication Rule

In a proportion, the cross-products are equal. For example,

$$\frac{2}{6} = \frac{4}{12}$$
$$6 \times 4 = 2 \times 12$$
$$24 = 24$$

MATH FACT

If $\frac{a}{b} = \frac{c}{d}$, then $b \times c = a \times d$, provided that denominators b and d are not 0. The terms b and c of the proportion are called the **means**, and the terms a and d are the **extremes**. Thus, *in a proportion the product of the means is equal to the product of the extremes.*

Example 4

Find the positive value of x: $\frac{x+3}{x} = \frac{x-1}{6}$.

Solution: $\boldsymbol{x = 9}$

After cross multiplying:

$$x(x-1) = 6(x+3)$$
$$x^2 - x = 6x + 18$$
$$x^2 - 7x - 18 = 0$$

Thus, $(x - 9)(x + 2) = 0$, so $x = 9$ or $x = -2$. Since the problem asks for the positive value of x, the answer is $x = 9$.

Factoring and quadratic equations are reviewed in Sections 6.4 and 8.3 of this book.

Percentage Mixture Problems

If a mixture contains n% of an ingredient, then

$$\frac{\text{Amount of ingredient}}{\text{Total amount of mixture}} = \frac{n}{100}.$$

Example 5

How many liters of water must be evaporated from 84 liters of a 20% salt solution to raise it to a 35% salt solution?

Solution: **36**

Let x represent the number of liters of water that must be evaporated ("subtracted") from the original mixture. Then:

$$\frac{\text{Amount of salt}}{\text{Amount of new mixture}} = \frac{0.20 \times 84}{84 - x} = \frac{35}{100}$$

$$\frac{16.8}{84 - x} = \frac{7}{20}$$

$$(16.8)(20) = 7(84 - x)$$

$$336 = 588 - 7x$$

$$7x = 252$$

$$x = \frac{252}{7} = 36$$

The number of liters of water that must be evaporated is 36.

Geometric Mean

In $\frac{8}{4} = \frac{4}{2}$, 4 is the *geometric mean* of 8 and 2. The **geometric mean** of two positive numbers a and b is the positive number x for which $\frac{a}{x} = \frac{x}{b}$.

Example 6

What is the geometric mean of 3 and 27?

Solution: **9**

Let x represent the geometric mean of 3 and 27; then $\frac{3}{x} = \frac{x}{27}$, so $x \cdot x = 3 \cdot 27 = 81$. Since $x^2 = 81$, $x = \sqrt{81} = 9$.

Writing Equivalent Proportions

Changing the positions of the terms of the proportion in certain ways creates equivalent proportions. Proportions equivalent to $\frac{1}{2} = \frac{4}{8}$ can be formed by:

- Inverting each ratio: $\frac{2}{1} = \frac{8}{4}$
- Interchanging the means: $\frac{1}{4} = \frac{2}{8}$
- Interchanging the extremes: $\frac{8}{2} = \frac{4}{1}$

Check Your Understanding of Section 3.1

A. Multiple Choice

1. On a certain map, $\frac{3}{8}$ inch represents 120 miles. How many miles does $1\frac{3}{4}$ inches represent?
(1) 400 (2) 480 (3) 520 (4) 560

2. The population of a bacteria culture doubles in number every 12 minutes. The ratio of the number of bacteria at the end of 1 hour to the number of bacteria at the beginning of that hour is
(1) 60 : 1 (2) 32 : 1 (3) 16 : 1 (4) 8 : 1

3. On her first trip, Sari biked 24 miles in T hours. The following week Sari biked 32 miles in T hours. The ratio of her average speed on her second trip to her average speed on her first trip is
(1) $\frac{3}{4}$ (2) $\frac{2}{3}$ (3) $\frac{4}{3}$ (4) $\frac{3}{2}$

4. What is the geometric mean of $\frac{1}{3}x$ and $27x^3$ when $x \neq 0$?
(1) $3x^2$ (2) $9x$ (3) $3x$ (4) $9x^2$

5. A bicyclist travels 6 miles in 20 minutes. Which expression does *not* represent the rate of speed of the bicyclist?

(1) $\frac{3}{10}$ mi/min
(2) $3\frac{1}{3}$ min/mi
(3) 18 mph
(4) $\frac{1}{120}$ mi/sec

B. *In each case, show how you arrived at your answer by clearly indicating all of the necessary steps, formula substitutions, diagrams, graphs, charts, etc.*

6–8 Solve each proportion for x.

6. $\frac{4}{11} = \frac{x+6}{2x}$ **7.** $\frac{10-x}{5} = \frac{7-x}{2}$ **8.** $\frac{x-4}{2} = \frac{3x}{x+4}$

9. At a certain intersection, the ratio of the number of seconds the traffic light is red, yellow, and green in each complete cycle is 3 to 1 to 8. If a complete cycle of the light takes exactly 1 minute, find the number of seconds during which the light is *not* red.

10. Kyle weighs 147 pounds. On planet Zeno, Kyle would weigh 98 pounds and Margie would weigh 78 pounds. How much does Margie weigh on Earth?

11. During an hour of prime-time television programming, the ratio of the number of minutes of television programs to the number of minutes of commercial interruptions is about 11 to 2. What is the approximate number of minutes of commercial interruptions in $1\frac{1}{2}$ hours of prime-time television programming?

12. A certain solution of salt and water was 15% salt. When 50 kilograms of water was evaporated from this solution, it became a 25% salt solution. Find the total number of kilograms of the original solution.

13. A ball is dropped from a height of 192 feet. On the third bounce the ball reaches a height of 24 feet. The height that the ball reaches after each bounce is the same fraction of the height that the ball reached on the previous bounce. What is the ratio of the height the ball reaches on the fifth bounce to the height the ball reached on the second bounce?

14. A tank contains 40 liters of a solution that is 25% acid. How many liters of water must be added to make a new solution that is 20% acid?

15. A chemist has one solution that is 35% pure acid and another solution that is 75% pure acid. How many cubic centimeters of each solution must be used to produce 80 cubic centimeters of solution that is 50% acid?

16. Carl runs $\frac{3}{4}$ mile in 5 minutes 30 seconds and Sonya runs $\frac{2}{3}$ mile in 5 minutes 20 seconds. At the same average rates, how many seconds longer will Sonya take to run a mile than Carl will take?

3.2 PROVING TRIANGLES SIMILAR

KEY IDEAS

Two *polygons* are similar if their corresponding angles are congruent and the lengths of corresponding sides are in proportion. To prove that two *triangles* are similar, it is necessary to prove only that two pairs of corresponding angles are congruent.

If it is known that two triangles are similar, then you may conclude that the ratios of the lengths of their corresponding sides are equal and, as a result, form a proportion.

Angle-Angle Theorem of Similarity

AA THEOREM OF SIMILARITY: *Two triangles are similar if two angles of one triangle are congruent to the corresponding angles of the other triangle.*

Example 1

Given: $\overline{CB} \perp \overline{BA}$, $\overline{CD} \perp \overline{DE}$.
Prove: $\triangle ABC \sim \triangle EDC$.

Solution: See the proof.

PLAN: Use the AA Theorem. The two triangles include right and vertical angles that yield two pairs of congruent angles.

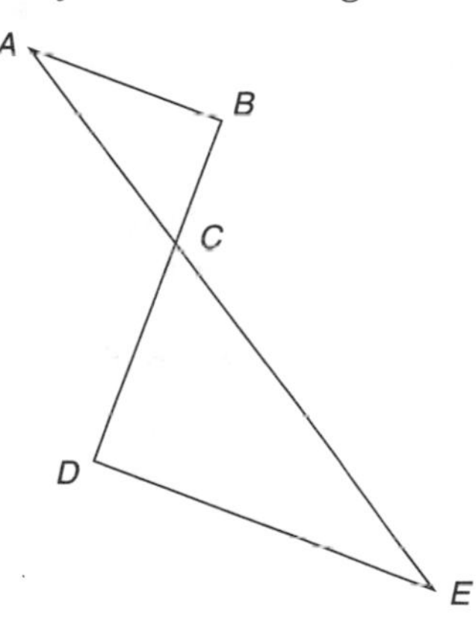

PROOF

Statement		Reason
1. $\overline{CB} \perp \overline{BA}$, $\overline{CD} \perp \overline{DE}$		1. Given.
2. Angles *ABC* and *EDC* are right angles.		2. Perpendicular lines intersect to form right angles.
3. $\angle ABC \cong \angle EDC$	*Angle*	3. All right angles are congruent.
4. $\angle ACB \cong \angle ECD$	*Angle*	4. Vertical angles are congruent.
5. $\triangle ABC \sim \triangle EDC$		5. AA Theorem.

Proving That Side Lengths Are in Proportion

To prove that the lengths of the sides of triangles I and II are in the proportion

$$\frac{\text{side of } \Delta\text{I}}{\text{corresponding side of } \Delta\text{II}} = \frac{\text{another side of } \Delta\text{I}}{\text{corresponding side of } \Delta\text{II}},$$

prove that $\Delta\text{I} \sim \Delta\text{II}$.

Example 2

Given: $\overline{AB} \parallel \overline{DE}$.

Prove: $\dfrac{EC}{BC} = \dfrac{ED}{AB}$.

Solution: See the proof.

PLAN:

- Select the triangles that contain as sides the segments named in the "Prove." Read across the proportion:

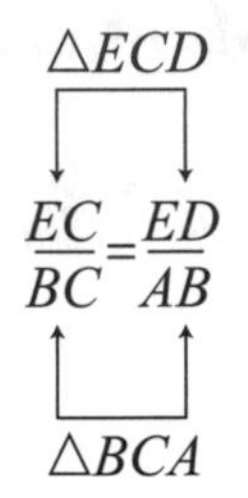

- Mark on the diagram the "Given" and all pairs of corresponding congruent angles.
- Write the proof.

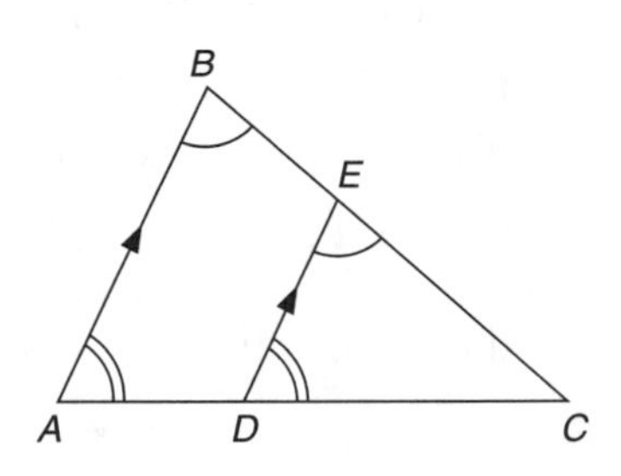

PROOF

Statement		Reason
1. $\overline{AB} \parallel \overline{DE}$		**1.** Given.
2. $\angle CED \cong \angle CBA$, $\angle CDE \cong \angle CAB$	*Angle* *Angle*	**2.** If two lines are parallel, then their corresponding angles are congruent.
3. $\triangle ECD \sim \triangle BCA$		**3.** AA Theorem.
4. $\frac{EC}{BC} = \frac{ED}{AB}$		**4.** The lengths of corresponding sides of similar triangles are in proportion.

TIP

If the original proportion in the "Prove" had been written as $\frac{EC}{ED} = \frac{BC}{AB}$, then reading *across* the proportion would *not* yield the vertices of the triangles to be proved similar. In that case, reading *down* each ratio would give the required triangles:

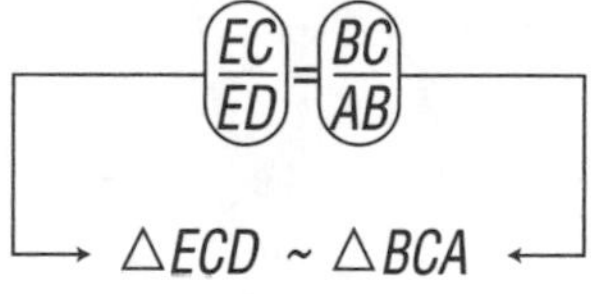

Check Your Understanding of Section 3.2

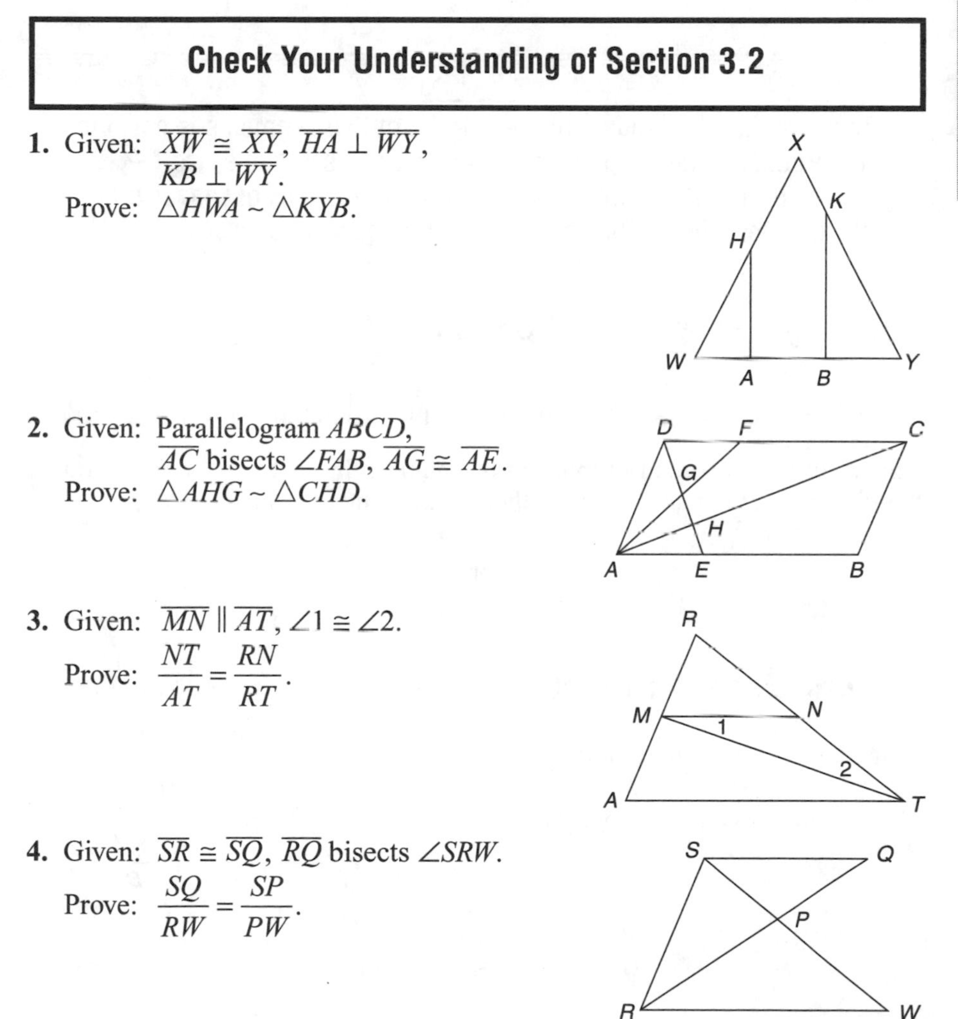

1. Given: $\overline{XW} \cong \overline{XY}$, $\overline{HA} \perp \overline{WY}$, $\overline{KB} \perp \overline{WY}$.
Prove: $\triangle HWA \sim \triangle KYB$.

2. Given: Parallelogram $ABCD$, $\overline{AC}$ bisects $\angle FAB$, $\overline{AG} \cong \overline{AE}$.
Prove: $\triangle AHG \sim \triangle CHD$.

3. Given: $\overline{MN} \parallel \overline{AT}$, $\angle 1 \cong \angle 2$.
Prove: $\frac{NT}{AT} = \frac{RN}{RT}$.

4. Given: $\overline{SR} \cong \overline{SQ}$, $\overline{RQ}$ bisects $\angle SRW$.
Prove: $\frac{SQ}{RW} = \frac{SP}{PW}$.

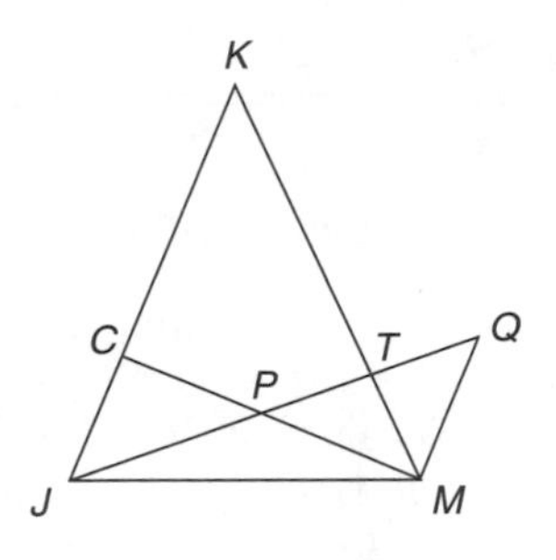

Exercises 5 and 6

5. Given: $\overline{MC} \perp \overline{JK}$, $\overline{PM} \perp \overline{MQ}$, $\overline{TP} \cong \overline{TM}$.

Prove: $\dfrac{PM}{MC} = \dfrac{PQ}{MK}$.

6. Given: T is the midpoint of $\overline{PQ}$, $\overline{MP} \cong \overline{MQ}$, $\overline{JK} \parallel \overline{MQ}$.

Prove: $\dfrac{PM}{JK} = \dfrac{TQ}{JT}$.

3.3 PROVING PRODUCTS OF SIDES EQUAL

KEY IDEAS

To prove that the product of the lengths of two segments is equal to the product of the lengths of another pair of segments, work backwards from the two products to determine the pair of triangles that contain these sides. Then prove these triangles are similar.

Forming a Proportion from a Product

Since $\frac{3}{6} = \frac{1}{2}$, $6 \times 1 = 3 \times 2$. Starting with the product $6 \times 1 = 3 \times 2$, you can figure out an equivalent proportion by making the factors of either product the means and the factors of the other product the extremes:

$$\frac{3}{6} = \frac{1}{2} \quad \text{or} \quad \frac{6}{3} = \frac{2}{1}.$$

Proving Equal Products

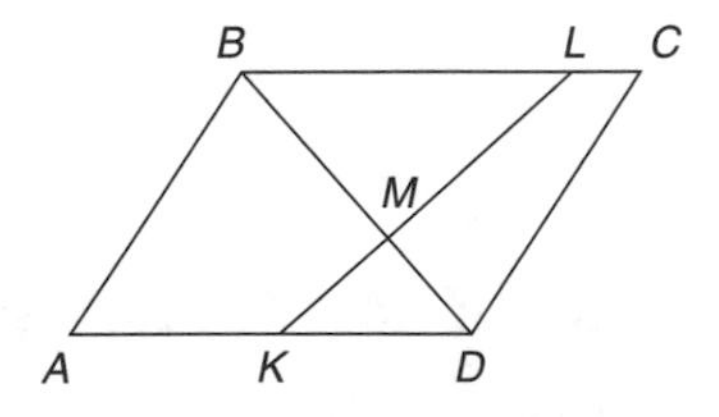

Consider the following proof:

Given: $ABCD$ is a parallelogram.
Prove: $KM \times LB = LM \times KD$.

To develop a plan, reason backwards from the "Prove" by answering three questions:

- What proportion produces the product $KM \times LB = LM \times KD$? Form an equivalent proportion by making KM and LB the extremes, and LM and KD the means:

$$\frac{KM}{LM} = \frac{KD}{LB}.$$

- Which pair of triangles must be proved similar?

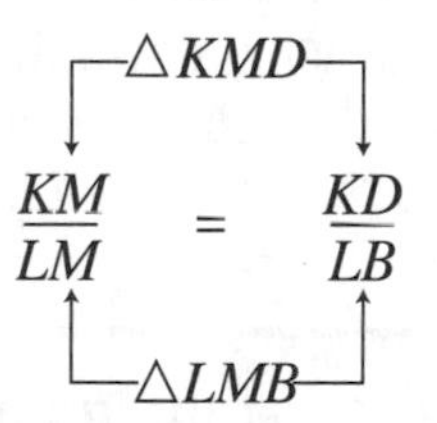

- How can I prove $\triangle KMD \sim \triangle LMB$?

Mark on the diagram the congruent pairs of angles.

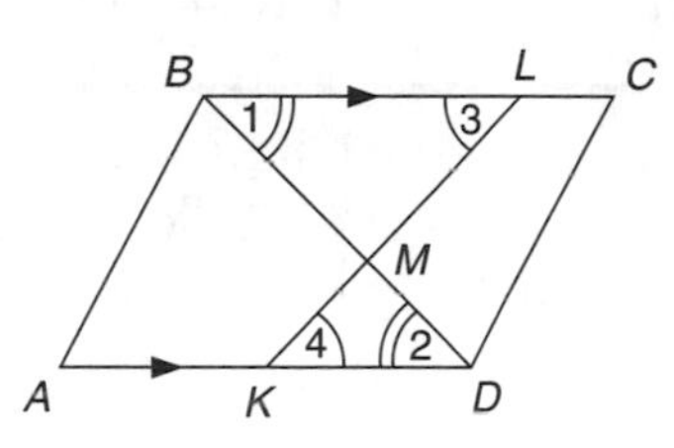

Now you are ready to write the formal proof:

PROOF

Statement	Reason
1. ▱$ABCD$	1. Given.
2. $\overline{AD} \parallel \overline{BC}$	2. Opposite sides of a parallelogram are parallel.
3. $\angle 1 \cong \angle 2$, $\angle 3 \cong \angle 4$	3. If two lines are parallel, then their alternate interior angles are congruent.
4. $\triangle KMD \sim \triangle LMB$	4. AA Theorem.
5. $\frac{KM}{LM} = \frac{KD}{LB}$	5. The lengths of corresponding sides of similar triangles are in proportion.
6. $KM \times LB = LM \times KD$	6. In a proportion, the product of the means equals the product of the extremes.

TIP

In developing the plan for the proof, suppose you formed the proportion

$$\frac{KM}{KD} = \frac{LM}{LB}.$$

Reading across the top (*K-M-L*), you do not find a set of letters that correspond to the vertices of a triangle. When this happens, read down rather than across. Reading down the first ratio (*K-M-D*) gives the vertices of one of the desired triangles, and reading down the second ratio (*L-M-B*) gives the vertices of the other triangle.

Check Your Understanding of Section 3.3

1. Given: $\overline{AF}$ bisects $\angle BAC$, $\overline{BH}$ bisects $\angle ABC$, $\overline{BC} \cong \overline{AC}$.
   Prove: $AH \times EF = BF \times EH$.

2. Given: $\overline{XY} \parallel \overline{LK}$, $\overline{XZ} \parallel \overline{JK}$.
   Prove: $JY \times ZL = XZ \times KZ$.

3. Given: $\overline{EF}$ is the median of trapezoid $ABCD$.
   Prove: $EI \times GH = IH \times EF$.

4. Given: V is a point on $\overline{ST}$ such that $\overline{RVW}$ bisects $\angle SRT$, $\overline{TW} \cong \overline{TV}$.
   Prove: $RW \times SV = RV \times TW$.

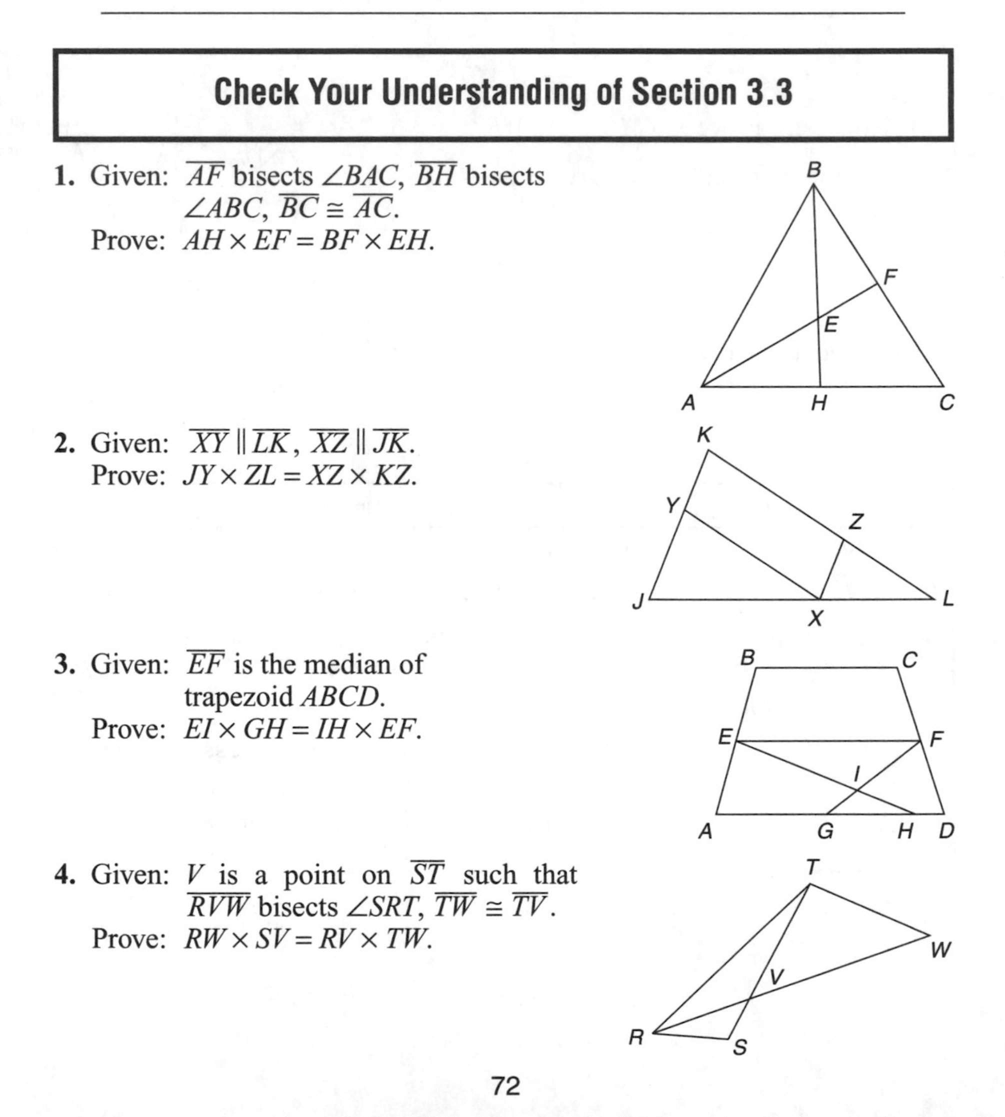

5. Given: $\triangle ABC$ with $\overline{CDA}$, $\overline{CEB}$, $\overline{AFB}$,
$\overline{DE} \parallel \overline{AB}$, $\overline{EF} \parallel \overline{AC}$,
$\overline{CF}$ intersects $\overline{DE}$ at G.
Prove: $\triangle CAF \sim \triangle FEG$.
$DG \times GF = EG \times GC$.

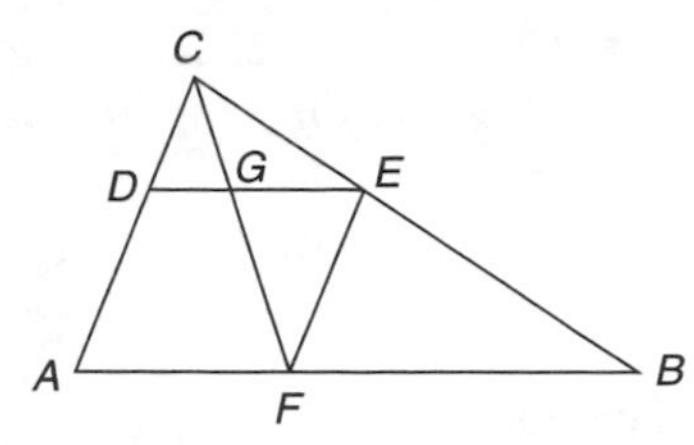

6. Rosalie thinks she discovered a new theorem. She claims that, "The product of the lengths of the legs of a right triangle is always equal to the product of the lengths of the hypotenuse and the altitude drawn to the hypotenuse." Prove or disprove Rosalie's theorem.

3.4 PROPORTIONS IN A RIGHT TRIANGLE

KEY IDEAS

In the right triangle shown in the accompanying diagram, altitude $\overline{CD}$ divides hypotenuse AB so that

$$\frac{x}{b} = \frac{b}{c}, \quad \frac{y}{a} = \frac{a}{c}, \quad \frac{x}{h} = \frac{h}{y}.$$

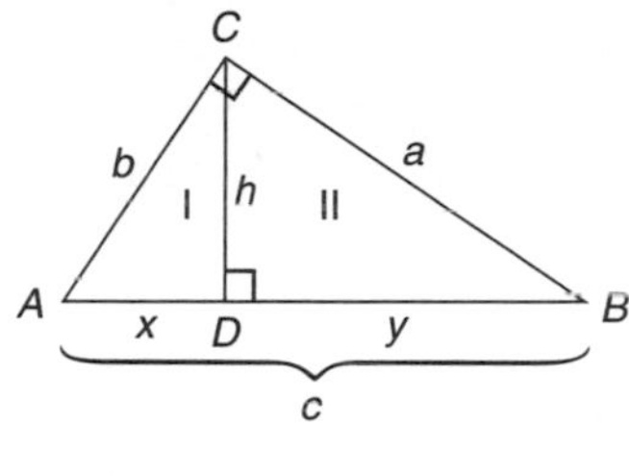

The Pythagorean theorem can be proved using these proportions.

Right Triangle Proportions

A **corollary** is a fact that follows directly from a theorem.

THEOREM: *If the altitude is drawn to the hypotenuse of a right triangle, then each right triangle formed is similar to the original triangle and similar to every other triangle.*

In right triangle ABC shown above,

$$\Delta\text{I} \sim \Delta ABC, \qquad \Delta\text{II} \sim \Delta ABC, \qquad \text{and, as a result,} \qquad \Delta\text{I} \sim \Delta\text{II}.$$

- COROLLARY 1: *The length of each leg is the geometric mean of the length of the segment that is adjacent to that leg and the length of the whole hypotenuse*. Thus:

$$\frac{AD}{AC} = \frac{AC}{AB} \quad \text{and} \quad \frac{BC}{BC} = \frac{BC}{AB}.$$

- COROLLARY 2: *The length of the altitude is the geometric mean of the lengths of the two segments of the hypotenuse.* Thus:

$$\frac{AD}{CD} = \frac{CD}{BD}.$$

Example 1

In the accompanying diagram, find the values of x and y.

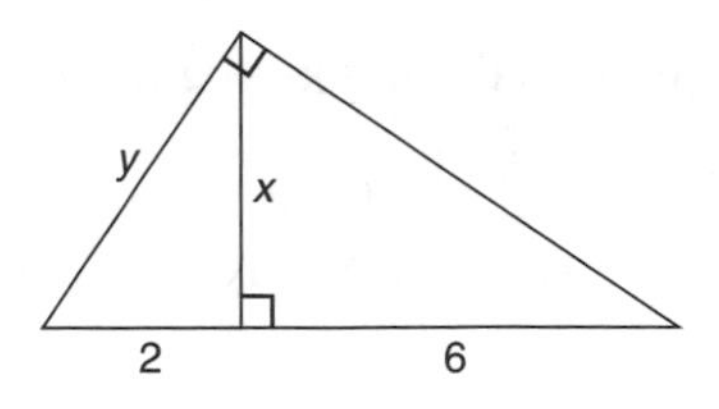

Solution: $\mathbf{x = 2\sqrt{3}, y = 4}$

- According to Corollary 2,

$$\frac{2}{x} = \frac{x}{6}$$
$$x^2 = 12$$
$$x = \sqrt{12} = 2\sqrt{3}$$

- According to Corollary 1,

$$\frac{2}{y} = \frac{y}{2+6}$$
$$y^2 = 16$$
$$y = 4$$

A Proof of the Pythagorean Theorem

PYTHAGOREAN THEOREM: *In a right triangle, $a^2 + b^2 = c^2$, where* a *and* b *represent the lengths of the two legs and* c *represents the length of the hypotenuse*. Here is an outline of a proof of this theorem that uses proportions in a right triangle.

Given: $\triangle ABC$, $\angle C$ is a right angle, c is the length of the hypotenuse, a and b are the lengths of the legs.
Prove: $a^2 + b^2 = c^2$.

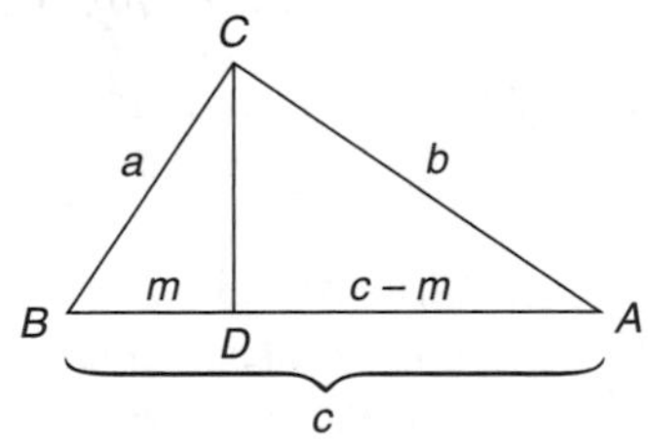

OUTLINE OF PROOF

Draw altitude $\overline{CD}$, and let $BD = m$, so $AD = c - m$, as shown in the accompanying diagram. Then apply Corollary 1:

$$\left.\begin{array}{lll} \dfrac{m}{a} = \dfrac{a}{c} & \text{so} & a^2 = \quad + mc \\ \dfrac{c-m}{b} = \dfrac{b}{c} & \text{so} & \underline{b^2 = c^2 - mc} \end{array}\right\} \text{Add the two equations.}$$

$$a^2 + b^2 = c^2 \qquad \leftarrow \text{Pythagorean theorem}$$

Pythagorean Triples

A **Pythagorean triple** is a set of three positive integers, a, b, and c, related in such a way that $a^2 + b^2 = c^2$. Here are some Pythagorean triples you should know:

- $\{3, 4, 5\}$ and multiples such as $\{6, 8, 10\}$.
- $\{5, 12, 13\}$ and multiples such as $\{10, 24, 26\}$.
- $\{8, 15, 17\}$ and multiples such as $\{16, 30, 34\}$.

Example 2

Exits A and B are on a straight highway. A motel is located at point M, as shown in the accompanying diagram, in such a way that $\angle AMB$ is a right angle. A gas station is built along the highway at point G, so that MG is the shortest distance from the gas station to the motel. If $MG = 20$ miles and $AM = 25$ miles, how far is exit A from exit B?

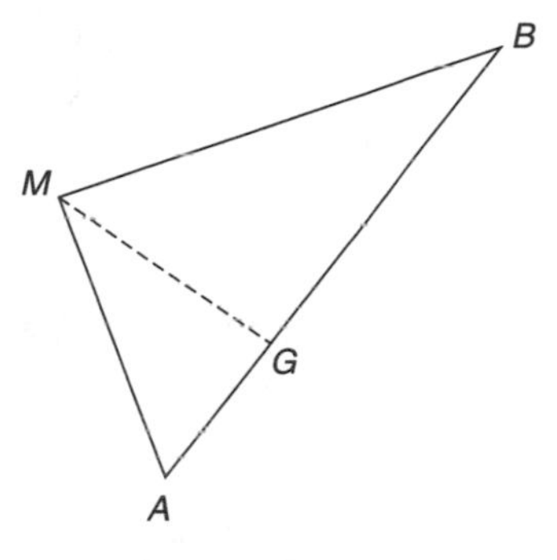

Solution: **41.7 mi**

- Since the shortest distance from a point (M) to a line (AB) is the length of the perpendicular dropped from the point to the line, $\overline{MG} \perp \overline{AB}$.
- In right triangle AGM, hypotenuse $AM = 25$ ($= 5 \times \underline{5}$) and leg $MG = 20$ ($= 5 \times \underline{4}$), so leg $AG = 15$ ($= 5 \times \underline{3}$).
- According to Corollary 1:

$$\frac{AG}{AM} = \frac{AM}{AB}$$

$$\frac{15}{25} = \frac{25}{AB}$$

$$15AB = 625$$

$$AB = \frac{625}{15} \approx 41.7$$

Exit A is 41.7 mi from exit B.

Check Your Understanding of Section 3.4

A. Multiple Choice

1. If the length of the altitude drawn to the hypotenuse of a right triangle is 10 inches, the number of inches in the lengths of the segments of the hypotenuse may be
(1) 5 and 20 (2) 2 and 5 (3) 3 and 7 (4) 50 and 50

2. At 9:00 A.M. a car starts at point A and travels north for 1 hour at an average rate of 60 miles per hour. Without stopping, the car then travels east for 2 hours at an average rate of 45 miles per hour. At noon, what is the best approximation of the distance, in miles, of the car from point A?
(1) 100 (2) 105 (3) 108 (4) 115

3. In the accompanying diagram, $\triangle FUN$ is a right triangle, $\overline{UR}$ is the altitude to hypotenuse $\overline{FN}$, $UR = 12$, and the lengths of $\overline{FR}$ and $\overline{FN}$ are in the ratio $1:10$. What is the length of $\overline{FR}$?
(1) 1 (3) 36
(2) 1.2 (4) 4

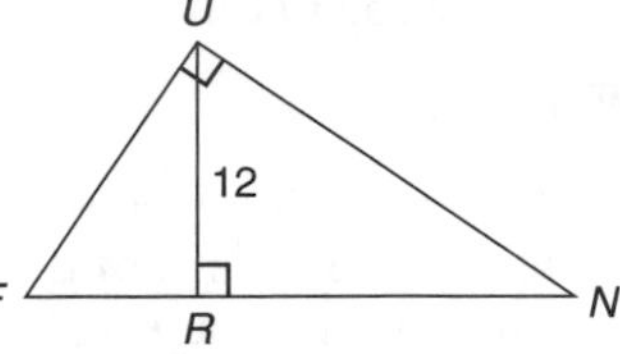

4. In right triangle ABC, $m\angle C = 90$ and altitude $\overline{CD}$ is drawn to hypotenuse $\overline{AB}$. If $AD = 4$ and $DB = 5$, what is AC?
(1) $\sqrt{20}$ (2) 6 (3) $\sqrt{45}$ (4) 9

5. In the accompanying diagram, $\triangle RST$ is a right triangle, $\overline{SU}$ is the altitude to hypotenuse $\overline{RT}$, $RS = 8$, and the ratio of RU to UT is 1 to 3. What is the length of $\overline{RT}$?
(1) 16 (3) 24
(2) 20 (4) 32

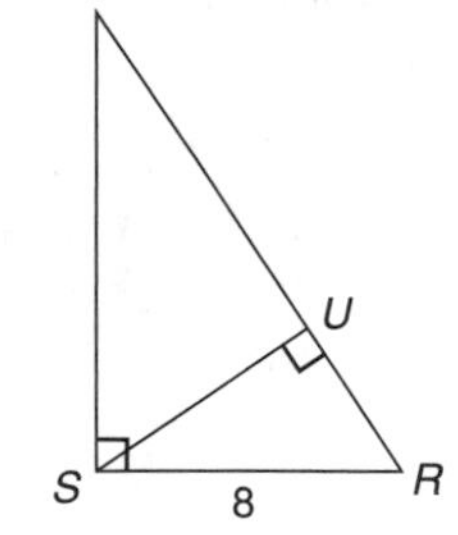

***B.** In each case, show how you arrived at your answer by clearly indicating all of the necessary steps, formula substitutions, diagrams, graphs, charts, etc.*

6. The altitude drawn to the hypotenuse of a right triangle divides the hypotenuse into two segments whose lengths are in the ratio of 1 to 4. If the length of the altitude is 8, find the length of the longer leg of the triangle.

7 and 8. In the accompanying diagrams of right triangle ABC, altitude $\overline{CD}$ is drawn to hypotenuse $\overline{AB}$.

7. Find r, s, and t.

8. Find r, s, and t.

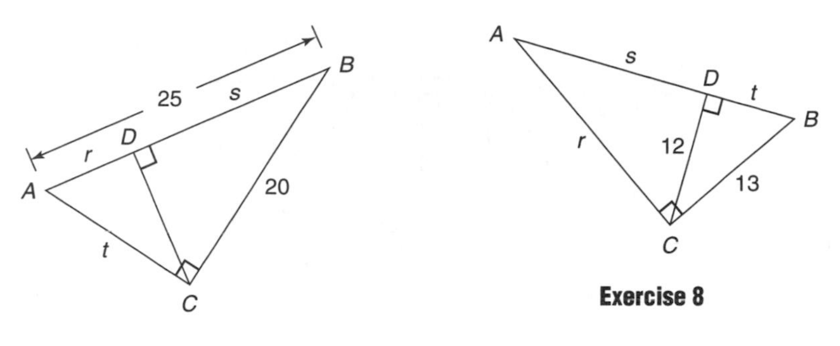

Exercise 7

Exercise 8

9. If the perimeter of a rhombus is 164 and the length of the longer diagonal is 80, find the length of the shorter diagonal.

10. In right triangle JKL, $\angle K$ is a right angle. Altitude $\overline{KH}$ intersects the hypotenuse at H in such a way that JH exceeds HL by 5. If $KH = 6$, find the length of the hypotenuse.

11. In right triangle ABC, the length of altitude $\overline{CD}$ to hypotenuse $\overline{AB}$ is 12. If the length of the longer segment of the hypotenuse exceeds the length of the shorter segment of the hypotenuse by 7, find the length of the hypotenuse.

12. In right triangle ABC, altitude $\overline{CD}$ is drawn to hypotenuse $\overline{AB}$. If AB is four times as great as AD and AC is 3 more than AD, find the length of altitude $\overline{CD}$.

13. The cross section of an attic is in the shape of an isosceles trapezoid, as shown in the accompanying figure. If the height of the attic is 9 feet, $BC = 12$ feet, and $AD = 28$ feet, find the length of $\overline{AB}$ to the *nearest foot*.

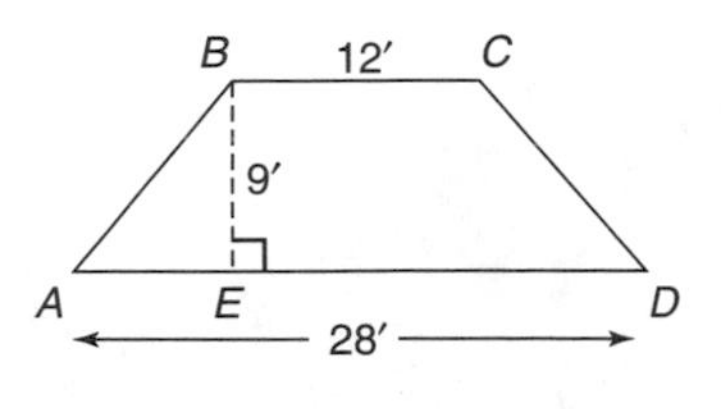

Exercise 13

14. The accompanying diagram shows a semicircular arch over a street that has a radius of 14 feet. A banner is attached to the arch at points A and B in such a way that $AE = EB = 5$ feet. How many feet above the ground are these points of attachment for the banner? Estimate to the *nearest tenth of a foot*.

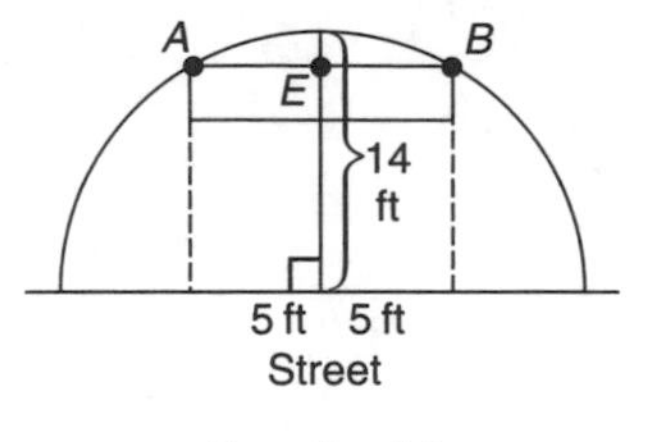

Exercise 14

15. Kathy bicycles 2 miles north, 12 miles east, and then 3 miles north again. How many miles, in a straight line, is Kathy from her starting point?

16. In right triangle ABC, altitude $\overline{CD}$ is drawn to hypotenuse $\overline{AB}$, $AD = 12$, and DB is 3 less than CD. Find, in simplest radical form, the perimeter of triangle ABC.

17. In right triangle ABC, altitude $\overline{CD}$ is drawn to hypotenuse $\overline{AB}$. If the coordinates of points A, B, and D are $(3, -1)$, $(3, 34)$, and $(3, 6)$ respectively, what are possible coordinates of point C?

18. Town A is 8 miles from town C, town B is 15 miles from town C, and angle ACB is a right angle. On the straight road that connects towns A and B, a restaurant will be built at the point that is closest to town C. To the *nearest tenth of a mile*, find:

a. the distance from town A to the restaurant
b. the distance from town C to the restaurant

3.5 SPECIAL RIGHT TRIANGLES

KEY IDEAS

In a right triangle in which the acute angles measure 30° and 60°, or 45° and 45°, special relationships exist among the lengths of the three sides of the triangle.

The 45–45 Isosceles Right Triangle

Let x represent the length of each of the congruent legs in an isosceles right triangle, as shown in Figure 3.1. According to the Pythagorean theorem, $(AB)^2 = x^2 + x^2 = 2x^2$, so $AB = \sqrt{2x^2} = x\sqrt{2}$. Hence:

$$\text{Hypotenuse} = \text{leg} \times \sqrt{2} \qquad \text{and} \qquad \text{Leg} = \frac{1}{2} \times \text{hypotenuse} \times \sqrt{2}.$$

The 30–60 Right Triangle

Suppose the length of each side of an equilateral triangle is represented by $2x$. Since drawing an altitude divides the original triangle into two congruent 30–60 right triangles, as shown in Figure 3.2, $AD = x$. Using the Pythagorean theorem, it can be shown that $BD = x\sqrt{3}$. Therefore:

- AD (opposite the 30° angle) $= \dfrac{1}{2} \times$ hypotenuse.

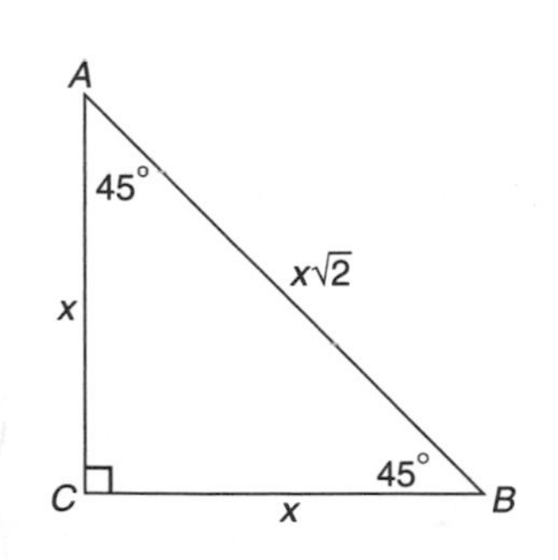

Figure 3.1 45–45 Right Triangle

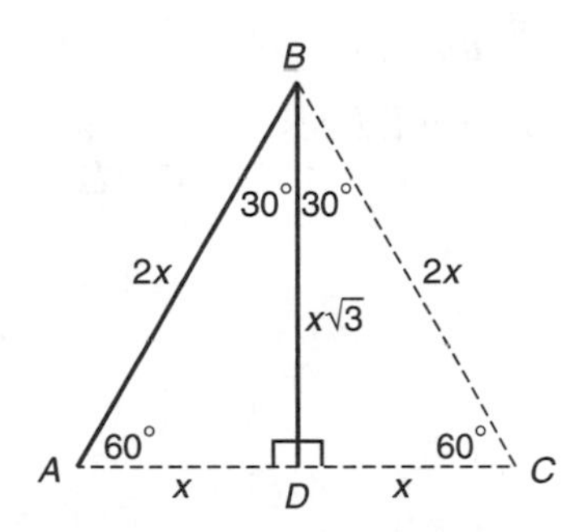

Figure 3.2 30–60 Right Triangle

- BD (opposite the 60° angle) $= \frac{1}{2} \times hypotenuse \times \sqrt{3}$.
- $BD = AD \times \sqrt{3}$

Example 1

In the accompanying diagram, find the values of x, y, and z.

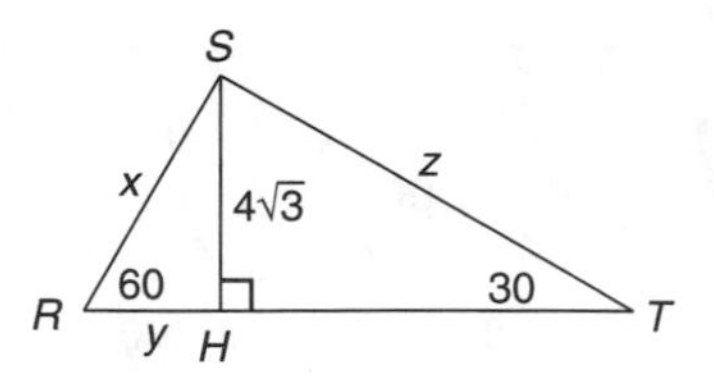

Solution: $\mathbf{x = 8, y = 4, z = 8\sqrt{3}}$

- In right triangle RHS, since $SH = \frac{1}{2} \times RS \times \sqrt{3}$,

$$x = RS = \frac{2 \times SH}{\sqrt{3}} = \frac{2 \times 4\sqrt{3}}{\sqrt{3}} = 8.$$

- In right triangle RHS, $m\angle RSH = 30$, so

$$y = \frac{1}{2} \times RS = \frac{1}{2} \times 8 = 4.$$

- In right triangle THS, since $SH = \frac{1}{2} \times ST$,

$$z = ST = 2 \times SH = 2 \times 4\sqrt{3} = 8\sqrt{3}.$$

Example 2

In isosceles trapezoid $ABCD$, a lower base angle measures 45° and the length of the shorter base is 5. If the length of an altitude is 7, find the length of the longer base.

Solution: **19**

- Drop altitudes from B and C, thereby forming two congruent right triangles, $\triangle AEB$ and $\triangle DFC$, as shown in the diagram.

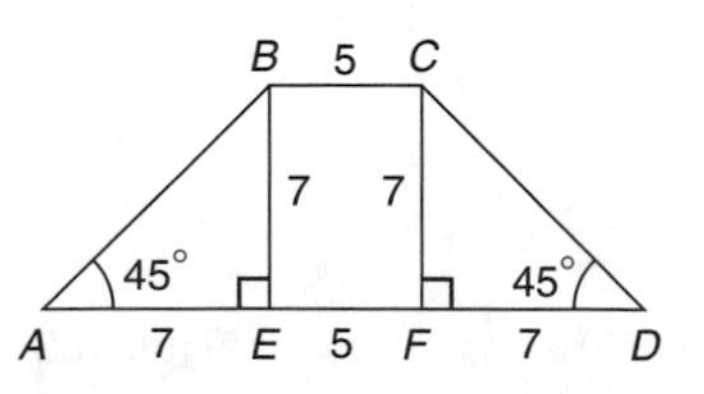

- In a 45°–45° right triangle, the legs have the same length, so

$$AE = BE = 7 \qquad \text{and} \qquad DF = CF = 7.$$

- Since quadrilateral $BEFC$ is a rectangle, $EF = BC = 5$.
- Hence, $AD = 7 + 5 + 7 = 19$.

The length of the longer base is 19.

Check Your Understanding of Section 3.5

A. Multiple Choice

1. What is the number of centimeters in the length of an altitude of an equilateral triangle whose side measures 4 centimeters?
(1) $2\sqrt{3}$ (2) 2 (3) $4\sqrt{3}$ (4) 4

2. In right triangle ABC, $\angle C$ is a right angle and $m\angle B = 30$. What is the ratio of AC to BC?
(1) $2:\sqrt{3}$ (2) $1:2$ (3) $1:\sqrt{3}$ (4) $1:1$

3. What is the perimeter of an equilateral triangle if the length of an altitude is $5\sqrt{3}$?
(1) 15 (2) 30 (3) $\frac{15}{\sqrt{3}}$ (4) 45

4. The lengths of two adjacent sides of a parallelogram are 6 and 15. If the degree measure of the included angle is 60, what is the length of the shorter diagonal of the parallelogram?
(1) $\sqrt{171}$ (2) $\sqrt{148}$ (3) $\sqrt{153}$ (4) $\sqrt{261}$

5. In the accompanying diagram of right triangles ABC and EDC, $\overline{BCD}$, $\overline{AC} \perp \overline{EC}$, $BD = 10$, $CE = 8$, and $m\angle ACB = 30$. What is the length of $\overline{AB}$?
(1) $2\sqrt{3}$ (3) $3\sqrt{3}$
(2) 3 (4) 6

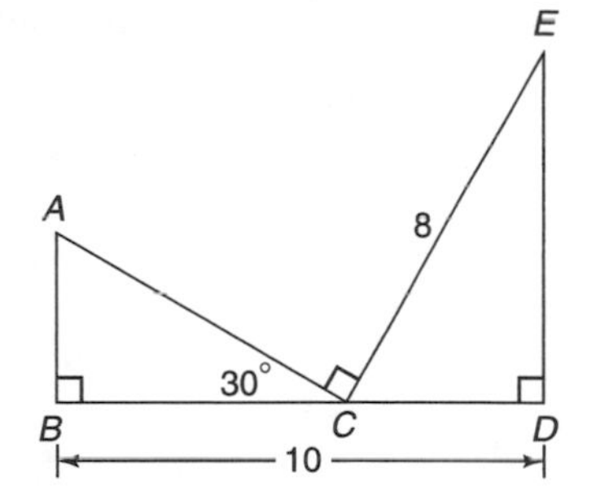

6. The perimeter of an equilateral triangle is $3s$. In terms of s, what is the area of the equilateral triangle?

(1) $4s\sqrt{3}$ (2) $\frac{s^2\sqrt{3}}{4}$ (3) $36s\sqrt{3}$ (4) $\frac{2s^2}{\sqrt{3}}$

B. *In each case, show how you arrived at your answer by clearly indicating all of the necessary steps, formula substitutions, diagrams, graphs, charts, etc.*

7. The length of each side of a rhombus is 10, and the degree measure of an angle of the rhombus is 60. What is the length of the longer diagonal of the rhombus?

8. The vertices of right triangle RAG with hypotenuse $\overline{AG}$ are $R(-2, 4)$, $A(7, 4)$, and $G(x, y)$. If $m\angle RAG = 45$, what are possible coordinates of point G?

9. At point A, a man looks up at point B. He then walks in a straight line on level ground to point D and again looks up at point B. The angles of elevation from points A and D to point B are 30° and 45°, respectively. Point B is 100 feet above the ground, as illustrated in the accompanying diagram. To the *nearest tenth of a foot*, find the distance the man walked from point A to point D.

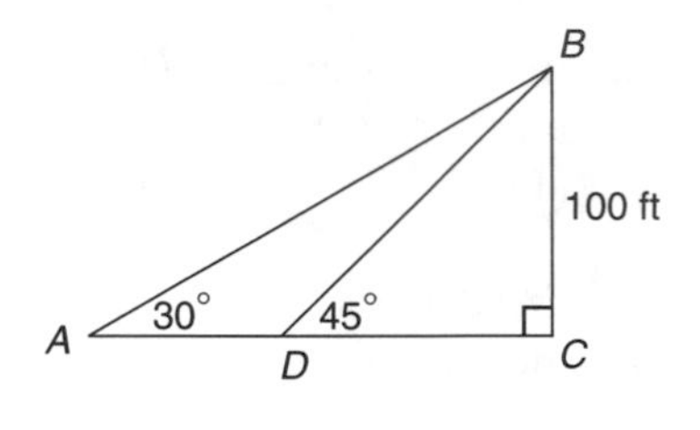

10. The perimeter of regular hexagon $ABCDEF$ is 48 inches.

a. If point O is the center of the hexagon, what is the perimeter of $\triangle AOB$?

b. What is the number of square inches in the area of $ABCDEF$?

CHAPTER 4

CIRCLES AND ANGLE MEASUREMENT

4.1 CIRCLE PARTS AND RELATIONSHIPS

KEY IDEAS

An important branch of geometry is concerned with measuring angles and the special segments that intersect a circle, shown in the accompanying diagram. A **chord** ($\overline{AB}$) is a line segment whose endpoints are any two points on the circle. A **radius** ($\overline{OR}$) is a segment whose endpoints are the center of the circle and any point on the circle. A **diameter** ($\overline{PQ}$) is a chord that passes through the center of the circle. A **secant** is a line ($\overleftrightarrow{AB}$) that intersects the circle in two points.

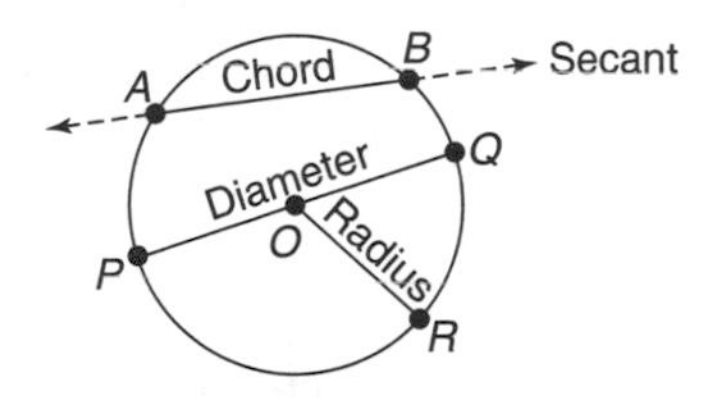

Radius Perpendicular to a Tangent

A **tangent** is a line that intersects a circle in exactly one point, called the **point of tangency**. As shown in Figure 4.1, *a radius is perpendicular to a tangent at the point of tangency.*

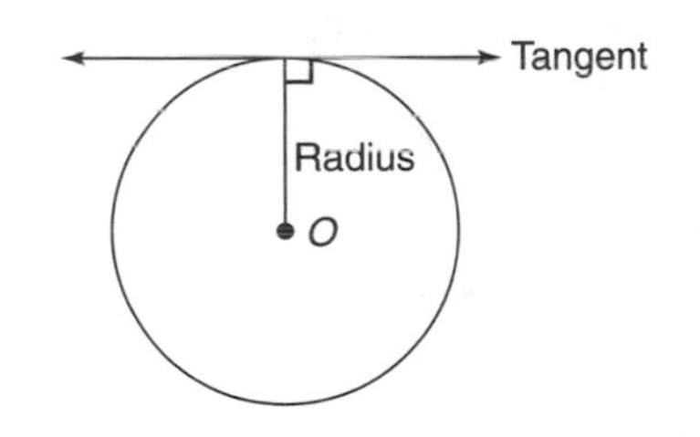

Figure 4.1 Radius Perpendicular to Tangent

Tangent Segments with the Same Endpoint

Tangents $\overline{PA}$ and $\overline{PB}$ are the legs of right triangles, as shown in Figure 4.2. Right triangles *OAP* and *OBP* are congruent by the Hy-Leg theorem since

$$\overline{OP} \cong \overline{OP} \qquad \text{(hypotenuse)}$$

$$\text{radius } \overline{OA} \cong \text{radius } \overline{OB} \quad \text{(leg).}$$

As a result, $\overline{PA} \cong \overline{PB}$ by CPCTC.

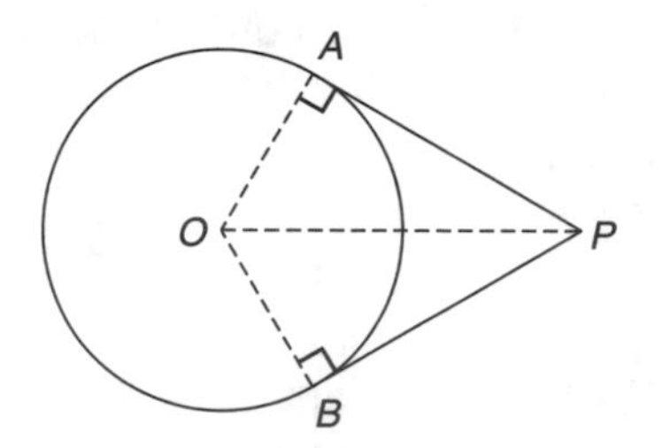

Figure 4.2 Congruent Tangent Segments

THEOREM: *If two tangents are drawn to a circle from the same point, the tangents are congruent.*

Central Angles and Arcs

A **central angle** is an angle whose vertex is the center of a circle. If, in Figure 4.3, radii $\overline{OA}$ and $\overline{OB}$ form a central angle that measures less than 180 degrees, then:

- Points *A* and *B* and the points of the circle that are in the *interior* of the central angle form a **minor arc**.
- Points *A* and *B* and the points of the circle that are in the *exterior* of the central angle form a **major arc**.

If points *A* and *B* are the endpoints of a diameter, then each arc that is formed is called a **semicircle**.

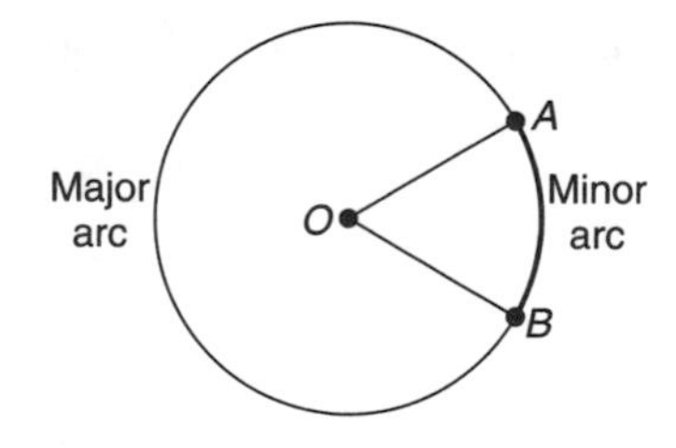

Figure 4.3 Classifying Arcs

Naming and Measuring Arcs

A minor arc is named by its two endpoints. In Figure 4.4 the minor arc intercepted by central angle *AOB* is denoted as $\overset{\frown}{AB}$, which is read as "arc *AB*." To distinguish minor arc *AB* from the major arc having the same two endpoints, a third letter is placed on the major arc. In Figure 4.4, $\overset{\frown}{APB}$ is the major arc with endpoints *A* and *B*.

- The **measure of a circle** is 360 degrees, and the **measure of a semicircle** is 180 degrees.
- The **measure of minor arc** *AB* is the degree measure of central angle *AOB* that intercepts it. Since $m\angle AOB = 50$, $m\overset{\frown}{AB} = 50$.
- The **measure of major arc** *APB* is 360 minus the degree measure of minor arc *AB*. Hence, $m\overset{\frown}{APB} = 360 - 50 = 310$.

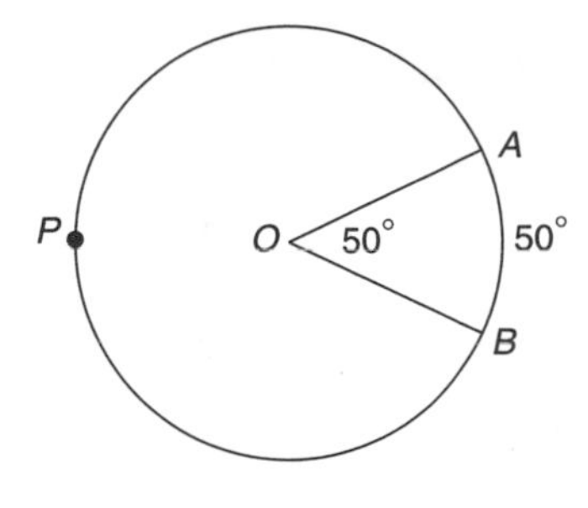

Figure 4.4 Measuring Arcs

Congruent Circles and Arcs

Congruent circles are circles with congruent radii. In the same or congruent circles, congruent central angles have **congruent arcs**, as shown in Figure 4.5. Although arcs *AB* and *CD* in Figure 4.6 have congruent central angles, the arcs are not congruent since they are in circles with unequal radii.

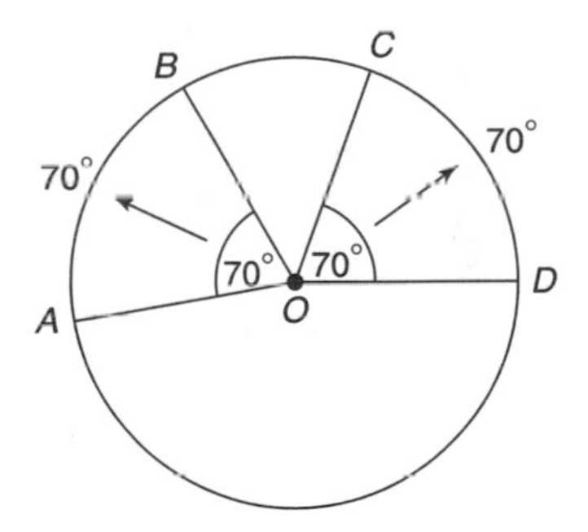

Figure 4.5 $\overset{\frown}{AB} \cong \overset{\frown}{CD}$

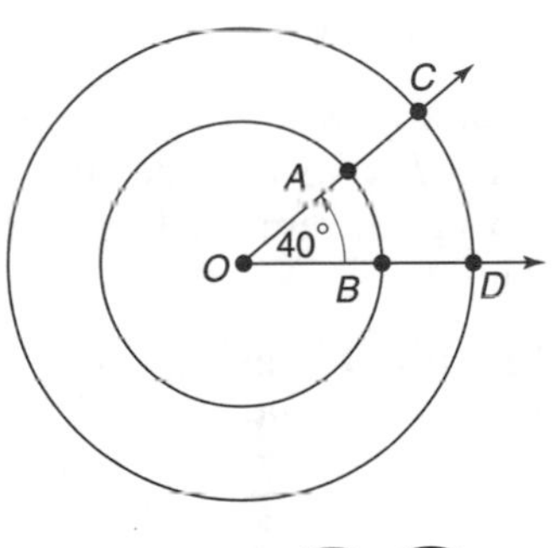

Figure 4.6 $\overset{\frown}{AB} \ncong \overset{\frown}{CD}$

Congruent Arcs and Chords

In Figure 4.7, if chords $\overline{AB}$ and $\overline{CD}$ are congruent, then $\overset{\frown}{AB} \cong \overset{\frown}{CD}$. Conversely, if arcs AB and CD are congruent, then $\overline{AB} \cong \overline{CD}$.

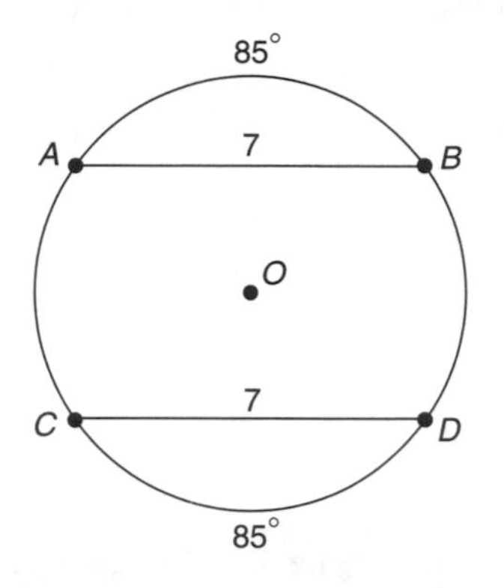

Figure 4.7 Congruent Chords and Arcs

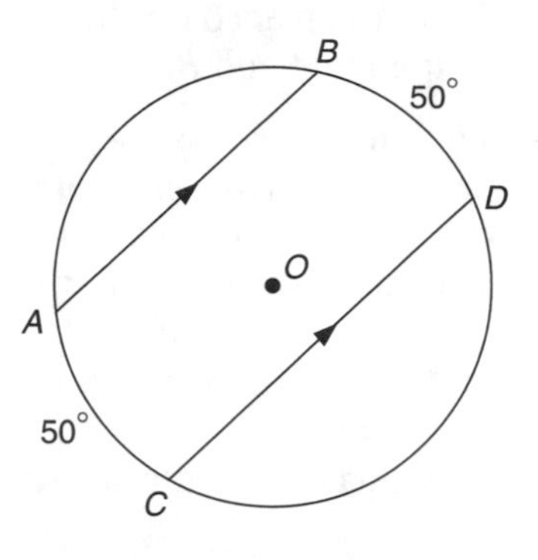

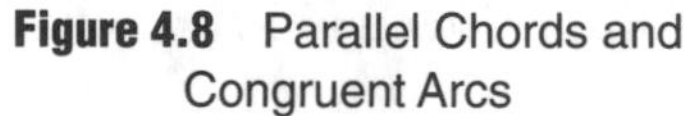
Figure 4.8 Parallel Chords and Congruent Arcs

In Figure 4.8, if chords $\overline{AB}$ and $\overline{CD}$ are parallel, then $\overset{\frown}{AC} \cong \overset{\frown}{BD}$.

Example 1

In the accompanying diagram, $\overline{AB}$ is parallel to $\overline{CD}$. If m$\overset{\frown}{AB} = 110$ and m$\overset{\frown}{CD} = 90$, what is the degree measure of central angle BOD?

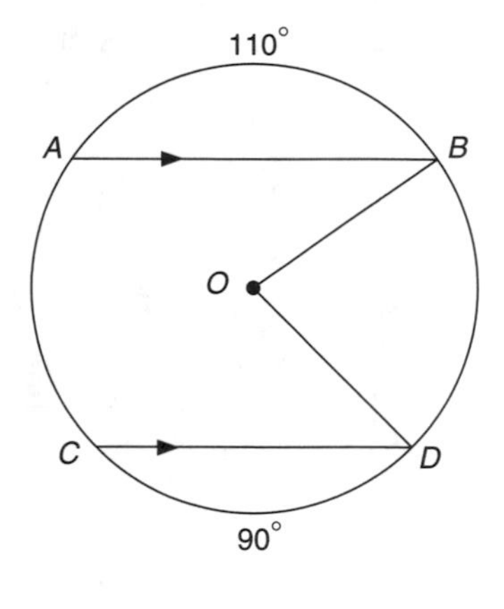

Solution: **80**

Since parallel chords intercept equal arcs, let $x = \text{m}\overset{\frown}{BD} = \text{m}\overset{\frown}{AC}$. The sum of the degree measures of the arcs that comprise a circle is 360. Thus

$$\begin{aligned} \text{m}\overset{\frown}{AB} + \text{m}\overset{\frown}{BD} + \text{m}\overset{\frown}{CD} + \text{m}\overset{\frown}{AC} &= 360 \\ 110 + x \quad + 90 \quad + x &= 360 \\ 2x + 200 &= 360 \\ x = \frac{160}{2} &= 80 \end{aligned}$$

Hence, $\text{m}\overset{\frown}{BD} = 80$. Since a central angle and its intercepted arc have the same degree measure, $\text{m}\angle BOD = 80$.

Diameter Perpendicular to a Chord

When a diameter is perpendicular to a chord, the diameter bisects the chord and its arcs, as shown in Figure 4.9.

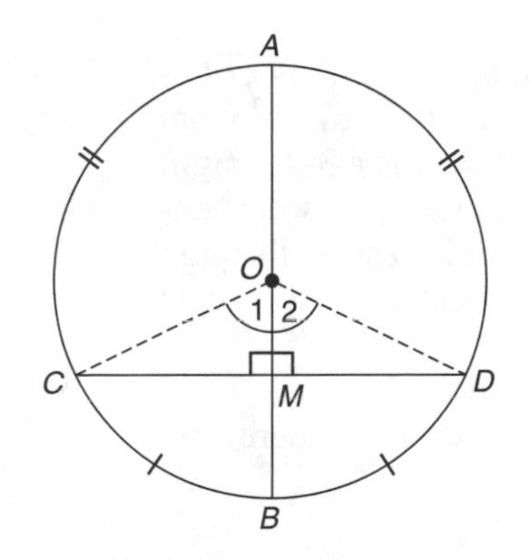

Figure 4.9 Diameter Perpendicular to Chord Makes $\overline{CM} \cong \overline{DM}$, $\overset{\frown}{CB} \cong \overset{\frown}{DB}$, and $\overset{\frown}{AC} \cong \overset{\frown}{AD}$

Example 2

Prove that a diameter that is perpendicular to a chord bisects the chord and its arcs.

Solution: See the proof.

PLAN: Refer to Figure 4.9.
Given: Diameter $\overline{AB} \perp \overline{CD}$.
Prove: a. $\overline{AB}$ bisects $\overline{CD}$.
b. $\overline{AB}$ bisects $\overset{\frown}{CD}$ and $\overset{\frown}{CAD}$.

PROOF

a. Right triangle $COM \cong$ right triangle DOM by the Hy-Leg theorem since $\overline{OC} \cong \overline{OD}$ (hypotenuse) and $\overline{OM} \cong \overline{OM}$ (leg).
   - By CPCTC, $\overline{CM} \cong \overline{DM}$, so $\overline{AB}$ bisects $\overline{CD}$.

b. Since $\angle 1 \cong \angle 2$ by CPCTC, $\overset{\frown}{CB} \cong \overset{\frown}{DB}$. Hence, $\overline{AB}$ bisects $\overset{\frown}{CD}$.
   - Supplements of congruent angles ($\angle 1 \cong \angle 2$) are congruent. As a result, $\angle AOC \cong \angle AOD$, so $\overset{\frown}{AC} \cong \overset{\frown}{AD}$. Hence, $\overline{AB}$ bisects $\overset{\frown}{CAD}$.

Example 3

A chord is 3 inches from the center of a circle whose radius is 5 inches. What is the length of the chord?

Solution: **8 in.**

In the accompanying diagram, radius $\overline{OA}$ is 5 in. and OE, the distance of chord $\overline{AB}$ from center O, is 3 in. Thus, $\triangle OEA$ is a 3-4-5 right triangle, with $AE = 4$. Since $\overline{OE}$ is perpendicular to chord $\overline{AB}$ and passes through center O, it lies on a diameter, so it bisects the chord.

Hence $AE = BE = 4$ in.

The length of the chord is, therefore, $4 + 4$ or 8 in.

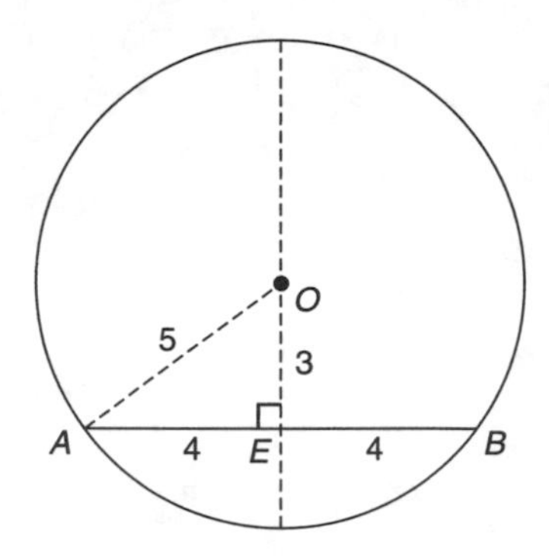

Equidistant Chords Are Congruent

In the same or congruent circles, if two chords are congruent, then the chords must be the same distance from the center. In Figure 4.10, if $\overline{AB} \cong \overline{CD}$, then $OP = OQ$.

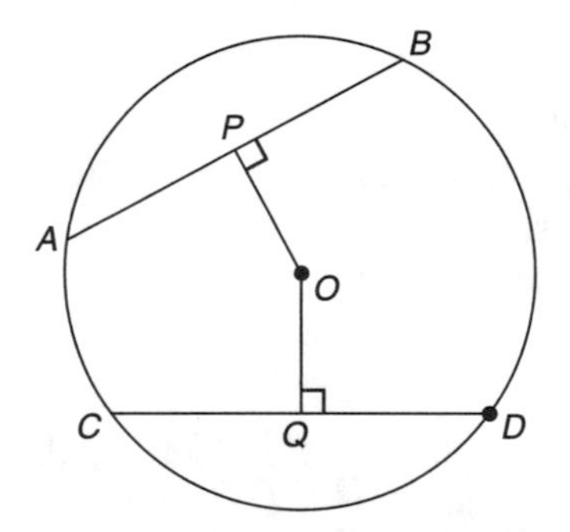

Figure 4.10 Equidistant Chords Are Congruent

Conversely, if two chords are the same distance from the center of a circle, then the two chords must be congruent. In Figure 4.10, if $OP = OQ$, then $\overline{AB} \cong \overline{CD}$.

MATH FACTS

Deciding When Two Arcs Are Congruent

In the same circle or in congruent circles, two *arcs* are congruent when any one of the following statements is true:

- The central angles that intercept the arcs are congruent.
- The chords that the arcs determine are congruent.
- The arcs are between parallel chords.
- The arcs are formed by a diameter drawn perpendicular to a chord.
- The arcs are semicircles.

Deciding When Two Chords Are Congruent

In the same circle or in congruent circles, two *chords* are congruent when either of the following statements is true:

- The arcs that the chords intercept are congruent.
- The chords are the same distance from the center.

Check Your Understanding of Section 4.1

A. Multiple Choice

1. A chord 48 centimeters in length is 7 centimeters from the center of a circle. What is the number of centimeters in the length of a radius of this circle?
(1) 25 (2) 50 (3) 55 (4) $\sqrt{2353}$

2. A chord is 5 inches from the center of a circle whose diameter is 26 inches. What is the number of centimeters in the length of the chord?
(1) 12 (2) 24 (3) 31 (4) 36

3. What is the distance, in inches, of a 30-inch chord from the center of a circle with a radius of 17 inches?
(1) 4 (2) 8 (3) 13 (4) 47

4. In the accompanying figure, $\overline{JK}$, $\overline{JL}$, and $\overline{LK}$ are tangent to circle O. What is the length of $\overline{JTL}$?
(1) 7 (3) 11
(2) 9 (4) 15

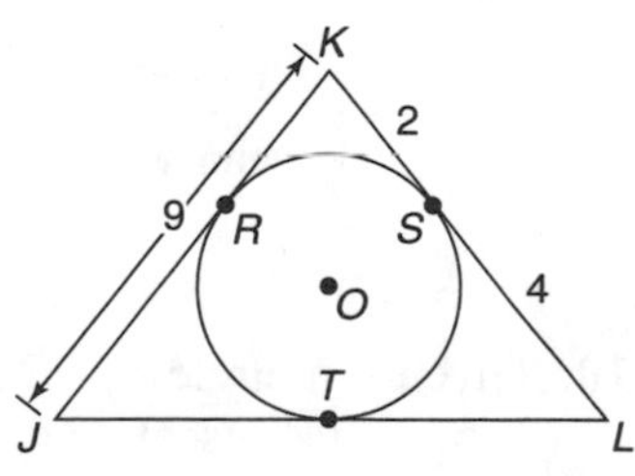

B. *In each case, show how you arrived at your answer by clearly indicating all of the necessary steps, formula substitutions, diagrams, graphs, charts, etc.*

5–7. In the accompanying diagram, $\overline{AB}$ is parallel to $\overline{CD}$ in circle O.

5. If $m\overset{\frown}{BD} = 75$ and $m\overset{\frown}{CD} = 90$, find $m\angle AOB$.

6. If $m\angle BAO = 40$ and $m\overset{\frown}{CD} = 70$, find $m\overset{\frown}{BD}$.

7. If $OC = CD$ and $m\overset{\frown}{AC} = 79$, find $m\angle AOB$.

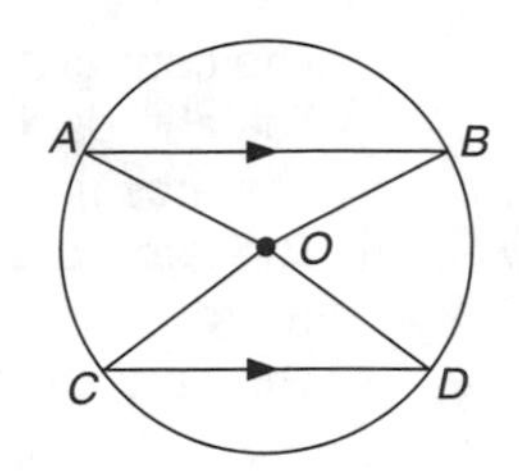

Exercises 5–7

8. Find the values of x and y.

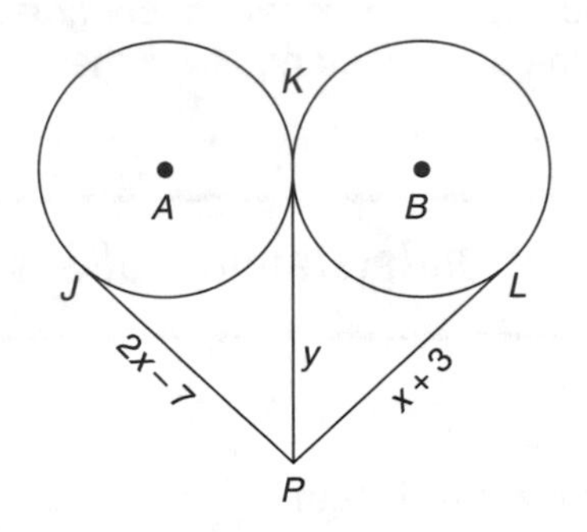

9. Given: In circle O, quadrilateral $OXEY$ is a square.
Prove: $\overset{\frown}{QP} \cong \overset{\frown}{JT}$.

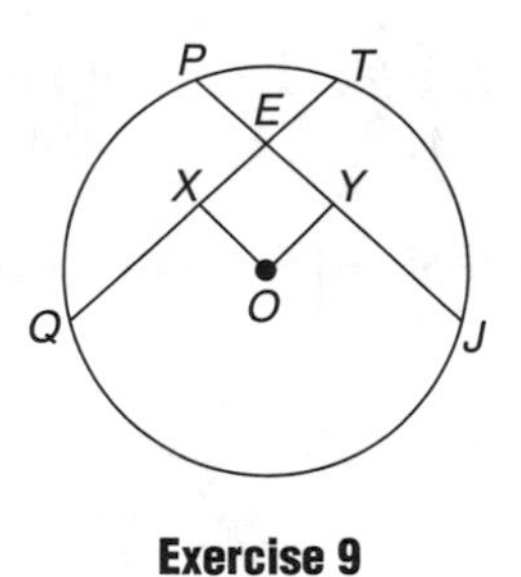

Exercise 9

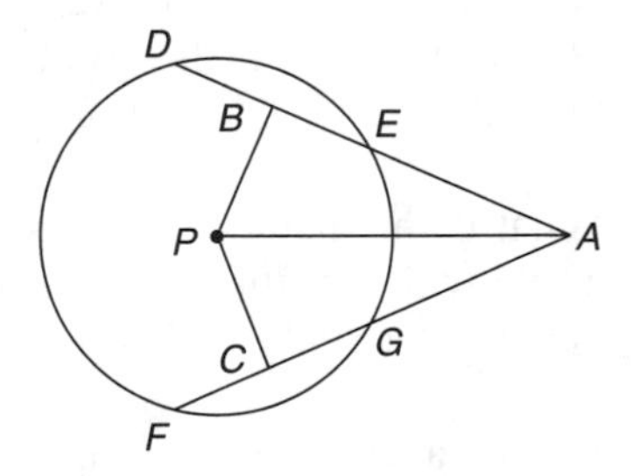

Exercise 10

10. Given: In circle P, $\overline{PB} \perp \overline{DE}$, $\overline{PC} \perp \overline{FG}$, $\overline{DE} \cong \overline{FG}$.
Prove: $\overline{PA}$ bisects $\angle FAD$.

4.2 CIRCUMFERENCE AND AREA

KEY IDEAS

The formulas for the circumference of a circle and the area of a circle can be used to find the length of an arc of the circle and the area of a "slice" or sector of the circle.

Arc Length and Sector Area

In Figure 4.11, circle O has radius length r, and a central angle of $n°$ intercepts minor arc AB. The length of minor arc AB is a fractional part of the circumference of the circle, and the area of the shaded slice, or **sector**, of the circle is the same fractional part of the area of the circle. In particular:

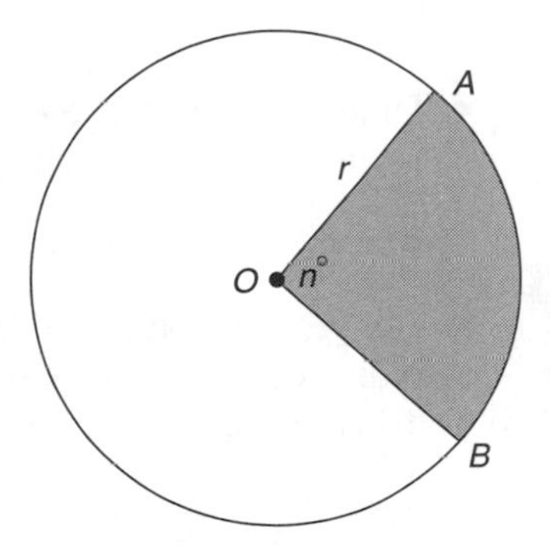

Figure 4.11 Sector *AOB*

- Circumference $= 2\pi r$
- Area πr^2
- Length of minor arc $AB = \dfrac{n}{360} \times 2\pi r$
- Area of sector $AOB = \dfrac{n}{360} \times \pi r^2$

Example 1

A regular hexagon is inscribed in a circle whose area is 144π.
a. What is the length of the minor arc intercepted by a side of the hexagon?
b. What is the area of the hexagon?

Solution: a. $\mathbf{4\pi}$

- If the area of the circle $= 144\pi$, then $r^2 = 144$, so $r = \sqrt{144} = 12$. Hence, the circumference of the circle is $2 \times \pi \times 12 = 24\pi$.
- The six congruent sides of a regular hexagon divide the circumference of a circle into six congruent arcs.
- Hence, the length of each minor intercepted arc is $\dfrac{1}{6} \times 24\pi = 4\pi$.

b. $\mathbf{36\sqrt{3}}$

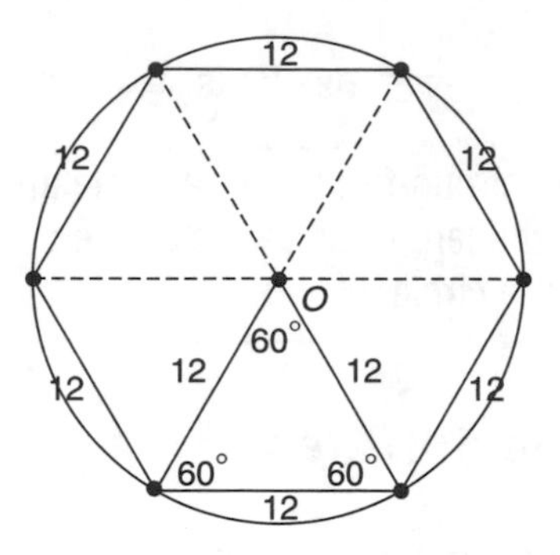

- Drawing radii to the vertices of an inscribed regular hexagon, as shown in the accompanying figure, forms six congruent equilateral triangles in which each central angle measures $\frac{360°}{6} = 60°$.
- Hence, the length of each side of the regular hexagon is equal to 12, the radius of the circle.
- To find the area of an equilateral triangle with side s, use the formula $\frac{s^2}{4} \times \sqrt{3}$:

$$\text{Area} = \frac{12^2}{4} \times \sqrt{3} = \frac{144}{4}\sqrt{3} = 36\sqrt{3}.$$

- Since a regular hexagon is comprised of six congruent triangles:

$$\text{Area of regular hexagon} = 6 \times 36\sqrt{3} = 216\sqrt{3}.$$

Example 2

Square $ABCD$ is inscribed in circle O, as shown in the accompanying figure. If the radius is 8 inches, find the area of the shaded region in terms of π.

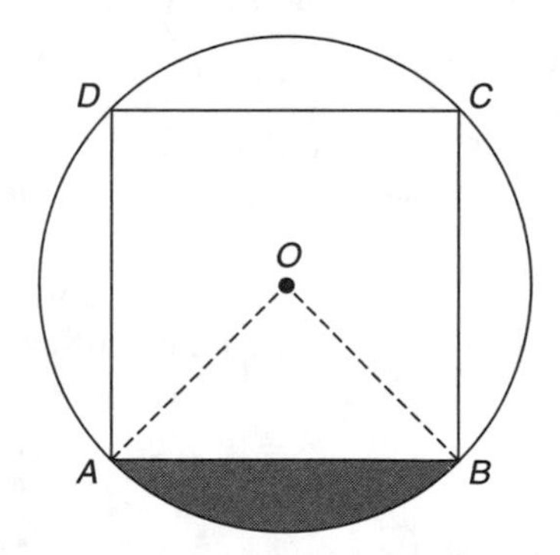

Solution: $\mathbf{16\pi - 32}$ **in.**$^{\mathbf{2}}$

Draw radii $\overline{OA}$ and $\overline{OB}$. To find the area of the shaded region, subtract the area of $\angle AOB$ from the area of sector AOB.

- Since $\widehat{AB}$ is one-fourth of the circumference of the circle, $\text{m}\widehat{AB} = \frac{1}{4} \times 360 = 90$. Therefore, $\angle AOB$ is a right angle.
- The area of circle O is $\pi \times 8^2 = 64\pi$ in.^{2} Since sector AOB is one quarter of the circle, the area of sector AOB is $\frac{1}{4} \times 64\pi = 16\pi$ in.^{2}
- The area of isosceles right triangle AOB is $\frac{1}{2} \times OA \times OB = \frac{1}{2} \times 8 \times 8 = 32$ in.^{2}
- The area of the shaded region is $16\pi - 32$ in.^{2}

Example 3

The circle $x^2 + y^2 = 100$ with center $O(0, 0)$ contains points $P(\sqrt{50}, \sqrt{50})$ and $Q(10, 0)$. What is the length of minor arc PQ in terms of π?

Solution: $\mathbf{2.5\pi}$

In general, $x^2 + y^2 = r^2$ is an equation of a circle with center at the origin and radius r.

- The radius of circle $x^2 + y^2 = 100$ is $\sqrt{100} = 10$.
- Since both the x- and the y-coordinate of point P are $\sqrt{50}$, $\text{m}\angle POQ = 45$.
- Length of $\widehat{PQ} = \frac{45}{360} \times 2\pi \times 10 = 2.5\pi$.

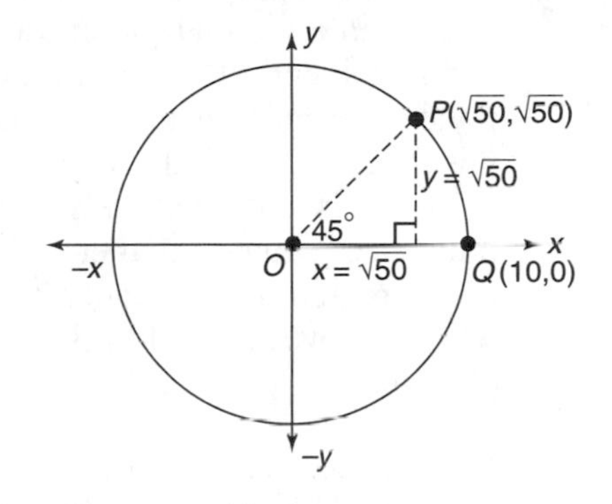

Check Your Understanding of Section 4.2

A. Multiple Choice

1. What is the area of a circle in which the coordinates of the endpoints of a diameter are (–5, 7) and (1, –1)?
(1) 10π (2) 25π (3) 50π (4) 100π

2. What is the area of the region in the first quadrant that is contained in the circle $x^2 + y^2 = 25$ but not in the circle $x^2 + y^2 = 9$?
(1) π (2) 2π (3) 3π (4) 4π

3. $P(1, \sqrt{3})$ is a point on the circle $x^2 + y^2 = 4$ whose center is at $O(0, 0)$. What is the area of the sector of circle O that is bounded by radius $\overline{OP}$ and the positive x-axis?

(1) $\frac{1}{3}\pi$ (2) $\frac{1}{2}\pi$ (3) $\frac{2}{3}\pi$ (4) π

4. $P(\sqrt{2}, \sqrt{2})$ is a point on the circle $x^2 + y^2 = 4$ whose center is at $O(0, 0)$. What is the length of the minor arc of circle O whose endpoints are point P and the point at which the circle intersects the negative x-axis?

(1) $\frac{1}{2}\pi$ (2) $\frac{2}{3}\pi$ (3) $2\sqrt{2}\pi$ (4) 2π

5. A regular hexagon is inscribed in a circle. What is the ratio of the length of a side of the hexagon to the minor arc that it intercepts?

(1) $\frac{\pi}{6}$ (2) $\frac{3}{6}$ (3) $\frac{3}{\pi}$ (4) $\frac{6}{\pi}$

6. In the accompanying figure, a radar tracking beam is centered at O and sweeps in a circular pattern an angle of 30° as the end of the beam moves from point A to point B in the counterclockwise direction. If the length of $\overset{\frown}{AB}$ of circle O is 6π kilometers, what is the number of square kilometers in the area of the region that the radar beam covers when it moves from A to B?

(1) 36π (2) 54π (3) 72π (4) 108π

7. If the circumference of a circular garden is increased by 10%, by what percent is the area of the garden increased?

(1) 10% (2) 21% (3) 31% (4) 100%

B. *In each case, clearly indicate the necessary steps, including formula substitutions, diagrams, graphs, charts, etc.*

8. The area of sector AOB of circle 0 is 60π square inches. If the radius is 12 inches, what is the number of inches in the length of minor arc AB, correct to the *nearest tenth of an inch*?

9. If an arc of 60° on circle A has the same length as an arc of 45° on circle B, what is the ratio of the area of circle B to the area of circle A?

10. From point P, tangents $\overline{PA}$ and $\overline{PB}$ are drawn to circle O. Quadrilateral $OAPB$ is formed by drawing radii $\overline{OA}$ and $\overline{OB}$. If $OP = 18$ and $m\angle APB = 60$, find the area of sector AOB.

11. The area of a sector of a circle is 20.25π square inches, and the length of the minor arc cut off by the radii that form that sector is 3π inches.
a. What is the diameter of the circle?
b. What is the degree measure of the central angle of the sector?

12. a. A medium-size circular pizza is cut into eight equal wedge-shaped pieces. If the length of the rounded edge of the crust of one of the pieces measures $4\frac{1}{2}$ inches, find the diameter of the pizza correct to the *nearest one-half inch*.
b. The area of a large circular pizza is 50% greater than the area of a medium-size circular pizza. If a large pizza is cut into eight equal wedge-shaped pieces, what is the length of the rounded edge of the crust of one of the pieces correct to the *nearest one-half inch*?

13. Equilateral triangle ABC is inscribed in circle O, as shown in the accompanying figure. If the area of circle O is 196π square inches, find in terms of π:
a. the area of sector AOB
b. the area of the shaded region

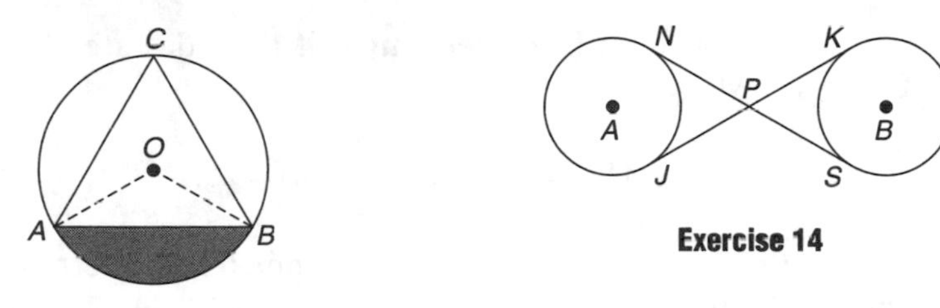

Exercise 14

Exercise 13

14. In the pulley system shown in the accompanying figure, a belt with negligible thickness is stretched around two identical wheels each having a radius of 9 inches. If $m\angle NPJ = 60$, find:
a. the degree measure of minor arc NJ
b. the length of the belt correct to the *nearest tenth of an inch*

4.3 CIRCLE ANGLE MEASUREMENT

KEY IDEAS

The vertex (pl., vertices) of an angle whose sides are chords, secants, or tangents may lie on the circle, inside the circle, or outside the circle. In each case, the location of the vertex determines the relationship between the measure of the angle and the measures of the intercepted arc or arcs.

Vertices of Angles on the Circle

If the vertex of an angle whose sides intercepts arcs is on the circle, then the degree measure of the angle is equal to one-half of the degree measure of the intercepted arc.

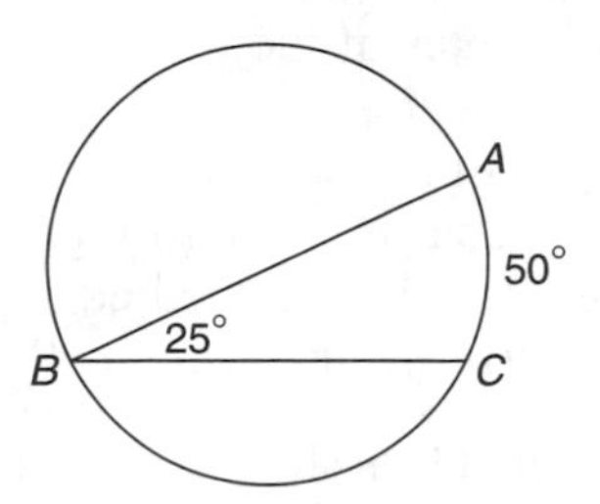

Figure 4.12 Inscribed angle

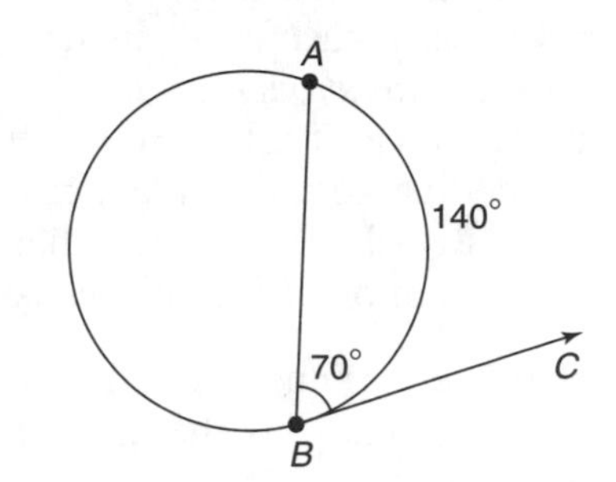

Figure 4.13 Chord-tangent angle

- The angle is called an **inscribed angle** if the sides are chords, as in Figure 4.12, where

$$m\angle ABC = \frac{1}{2} m\widehat{AC} = \frac{1}{2}(50) = 25.$$

  This angle-measurement relationship is sometimes referred to as the **Inscribed Angle Theorem**.
- The angle is called a **chord-tangent angle** if one side is a chord and the other side is a tangent, as in Figure 4.13, where

$$m\angle ABC = \frac{1}{2} m\widehat{AC} = \frac{1}{2}(140) = 70.$$

Example 1

In each case, a.–d., find the value of x.

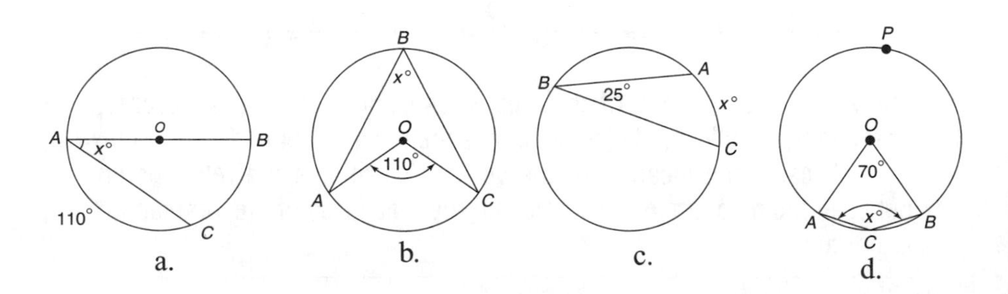

Solutions: a. **35**

$\widehat{ABC}$ is a semicircle, so m$\widehat{BC}$ = 180 – 110 = 70.

$$x = \frac{1}{2}\text{m}\widehat{BC} = \frac{1}{2}(70)$$
$$= 35$$

b. **55**

$\angle AOC$ is a central angle, so m$\widehat{AC}$ = 110.

$$\text{m}\angle ABC = \frac{1}{2}\text{m}\widehat{AC} = \frac{1}{2}(110)$$
$$= 55$$

c. **50**

The degree measure of an arc intercepted by an inscribed angle must be twice the degree measure of the inscribed angle. Hence

$$x = 2(\text{m}\angle ABC) = 2(25)$$
$$= 50$$

d. **145**

$\angle AOB$ is a central angle, so m$\widehat{ACB}$ = 70.

$$\text{m}\widehat{APB} = 360 - 70 = 290$$
$$x = \frac{1}{2}\text{m}\widehat{APB} = \frac{1}{2}(290)$$
$$= 145$$

Example 2

In the accompanying diagram, $\overrightarrow{DE}$ is tangent to circle O, at D, $\overline{AOB}$ is a diameter, and $\overline{CD}$ is parallel to $\overline{AOB}$. If m$\angle DAB$ = 21, find m$\angle CDE$.

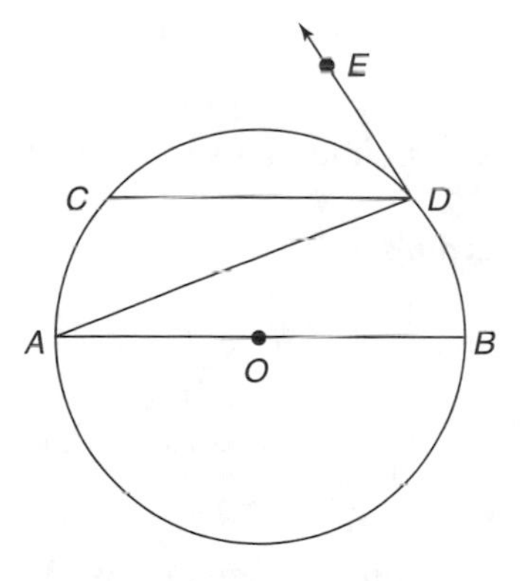

Solution: **48**

- Angle *CDE* intercepts arc *CD*, so the degree measure of this arc is needed. Since the sum of the degree measures of the arcs of a semicircle is 180:

$$m\overset{\frown}{AC} + m\overset{\frown}{CD} + m\overset{\frown}{DB} = 180.$$

- The degree measure of inscribed angle *DAB* is 21. The degree measure of its intercepted arc, $\overset{\frown}{DB}$, must be twice as great, or 42. Since parallel chords intercept equal arcs, $m\overset{\frown}{AC} = m\overset{\frown}{DB} = 42$. Hence:

$$\begin{aligned} 42 + m\overset{\frown}{CD} + 42 &= 180 \\ m\overset{\frown}{CD} &= 180 - 84 \\ &= 96 \end{aligned}$$

- Since $\angle CDE$ is formed by a chord and a tangent:

$$\begin{aligned} m\angle CDE &= \frac{1}{2} m\overset{\frown}{CD} \\ &= \frac{1}{2}(96) \\ &= 48 \end{aligned}$$

Circle Proofs

Using the Inscribed Angle Theorem, you know that:

- If two inscribed angles intercept the same arc or congruent arcs, then the inscribed angles are congruent.
- If an inscribed angle intercepts a semicircle, the angle is a right angle $\left(\frac{1}{2} \times 180° = 90°\right)$. This type of angle is sometimes described as being *inscribed in a semicircle* since the endpoints of its sides are the endpoints of a semicircle that includes the vertex of the angle.

Example 3

Given: In circle O, $\overline{AB}$ is a diameter, $\overset{\frown}{AM} \cong \overset{\frown}{DM}$.
Prove: $\triangle AMB \cong \triangle CMB$.

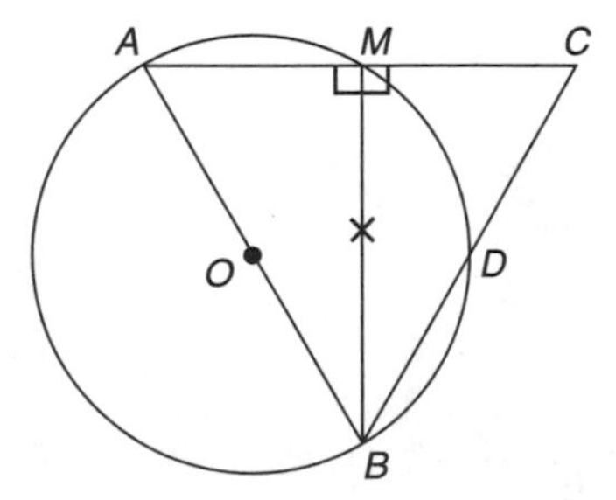

Solution: See the proof.

PLAN: The triangles can be proved congruent by using the ASA postulate.

PROOF

Statement	Reason
1. In circle O, $\overline{AB}$ is a diameter.	1. Given.
2. $\angle AMB$ is a right angle.	2. An angle inscribed in a semicircle is a right angle.
3. $\angle CMB$ is a right angle.	3. The supplement of a right angle is a right angle.
4. $\angle AMB \cong \angle CMB$ (*Angle*)	4. All right angles are congruent.
5. $\overline{MB} \cong \overline{MB}$ (*Side*)	5. Reflexive property.
6. $\widehat{AM} \cong \widehat{DM}$	6. Given.
7. $\angle ABM \cong \angle CBM$ (*Angle*)	7. Inscribed angles that intercept congruent arcs are congruent.
8. $\triangle AMB \cong \triangle CMB$	8. ASA postulate.

Vertex of Angle Inside the Circle

If two chords intersect inside a circle, as shown in Figure 4.14, then the degree measure of each angle formed is equal to one-half of the sum of the degree measures of the two arcs opposite the angle. Thus:

$$
\begin{aligned}
m\angle AEC &= \frac{1}{2}(m\widehat{AC} + m\widehat{BD}) \\
&= \frac{1}{2}(150 + 70) \\
&= 110
\end{aligned}
$$

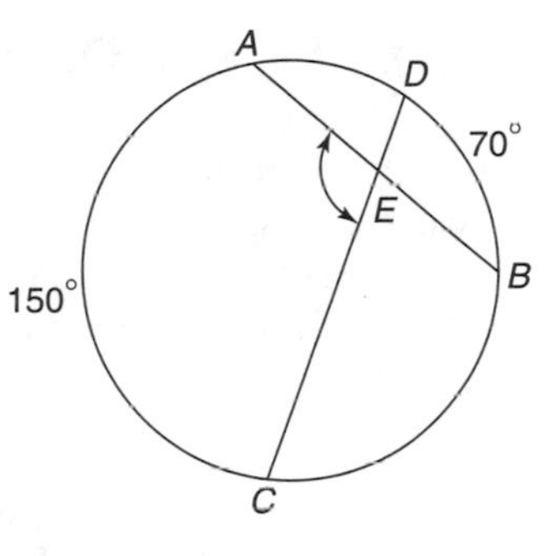

Figure 4.14 Chord-Chord Angle

Example 4

In the accompanying diagram, regular pentagon $ABCDE$ is inscribed in circle O. Chords $\overline{AD}$ and $\overline{BE}$ intersect at F, and $\overleftrightarrow{BT}$ is tangent to circle O at B. Find:

a. $m\angle ABT$ b. $m\angle AFE$

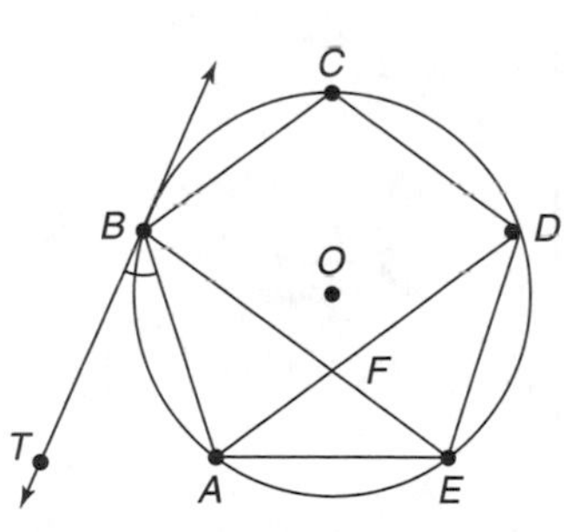

Solutions: a. **36**

In a regular pentagon all five sides have the same length, and therefore circle O is divided into five equal arcs. The degree measure of each arc is, therefore, $\frac{360}{5}$ or 72. Hence:

$$m\angle ABT = \frac{1}{2}m\widehat{AB} = \frac{1}{2}(72) = 36$$

b. **108**

Angle *AFE* intercepts arcs *BCD* and *AE*. Since

$$m\widehat{AE} = 72 \text{ and } m\widehat{BCD} = m\widehat{BC} + m\widehat{CD} = 72 + 72 = 144:$$

$$\begin{aligned} m\angle AFE &= \frac{1}{2}(m\widehat{BCD} + m\widehat{AE}) \\ &= \frac{1}{2}(144+72) \\ &= \frac{1}{2}(216) \\ &= 108 \end{aligned}$$

Vertex of Angle Outside the Circle

If the vertex of an angle lies outside the circle, the sides of the angle may be two secants, two tangents, or a secant and a tangent. In each case the degree measure of the angle is one-half of the *difference* of the degree measures of the two arcs that lie between the sides of the angle, as shown in Figure 4.15.

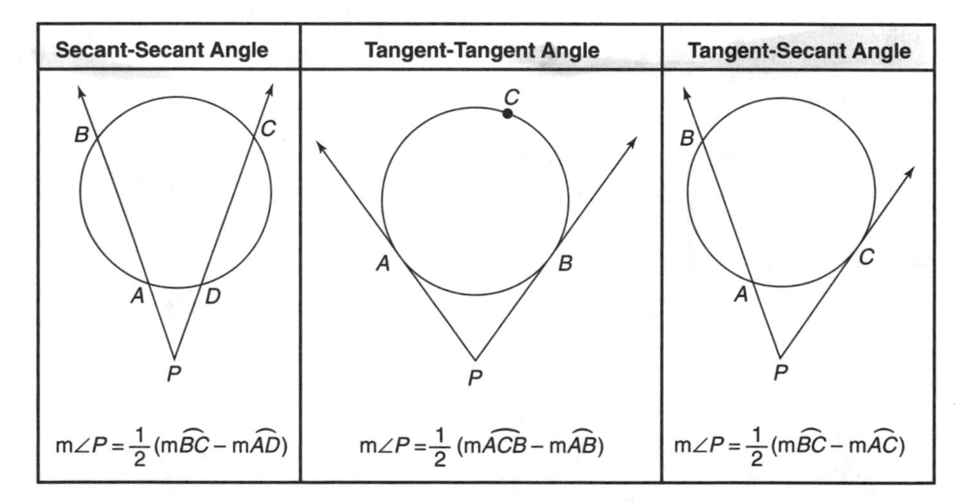

Figure 4.15 Angles Whose Vertices Are Outside the Circle. In each case the degree measure of the angle is found by taking one-half of the difference of the degree measures of the intercepted arcs

Example 5

In each case, find the value of x.

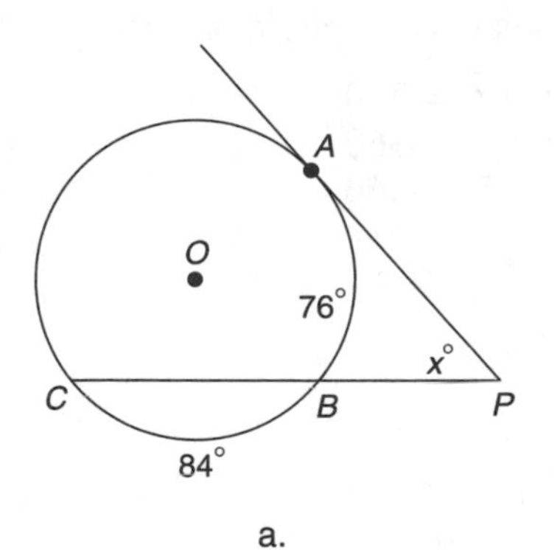

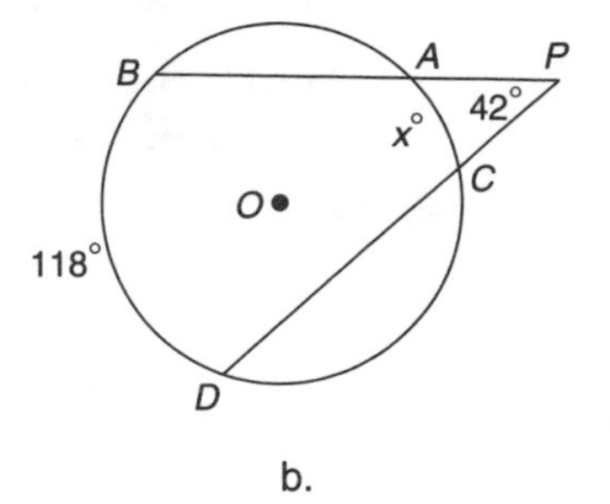

Solutions: a. **62**

$$\begin{aligned} m\widehat{AC} + 76 + 84 &= 360 \\ m\widehat{AC} &= 360 - 160 \\ &= 200 \end{aligned}$$

Hence:

$$\begin{aligned} x &= \frac{1}{2}(m\widehat{AC} - m\widehat{AB}) \\ &= \frac{1}{2}(200 - 76) \\ &= \frac{1}{2}(124) \\ &= 62 \end{aligned}$$

b. **34**

$$\begin{aligned} m\angle P &= \frac{1}{2}(m\widehat{BD} - m\widehat{AC}) \\ 42 &= \frac{1}{2}(118 - x) \\ 84 &= 118 - x \\ x &= 118 - 84 \\ &= 34 \end{aligned}$$

Example 6

In the accompanying diagram, $\overrightarrow{PA}$ is tangent to circle O at point A. Secant $\overline{PBC}$ is drawn. Chords $\overline{CA}$ and $\overline{BD}$ intersect at point E. If

$$m\widehat{AD} : m\widehat{AB} : m\widehat{DC} : m\widehat{BC} = 2:3:4:6,$$

find the degree measure of each of the numbered angles.

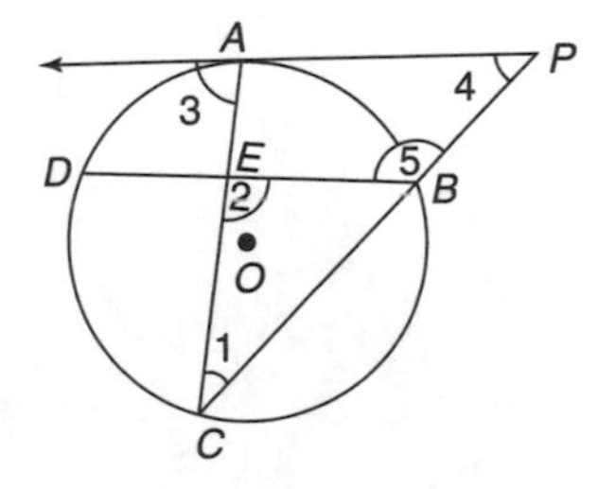

Solution: **36, 96, 72, 36, 132**

First find the degree measures of the arcs of the circle. Let $2x = \text{m}\widehat{AD}$. Then:

$$3x = \text{m}\widehat{AB}, \qquad 4x = \text{m}\widehat{CD}, \qquad 6x = \text{m}\widehat{BC}.$$

Since the sum of the degree measures of the arcs of a circle is 360:

$$\begin{aligned} \text{m}\widehat{AD} + \text{m}\widehat{AB} + \text{m}\widehat{DC} + \text{m}\widehat{BC} &= 360 \\ 2x + 3x \quad + 4x \quad + 6x \quad &= 360 \\ 15x &= 360 \\ x &= \frac{360}{15} = 24 \end{aligned}$$

Hence:

$$\begin{aligned} \text{m}\widehat{AD} &= 2x = 2(24) = 48 \\ \text{m}\widehat{AB} &= 3x = 3(24) = 72 \\ \text{m}\widehat{DC} &= 4x = 4(24) = 96 \\ \text{m}\widehat{BC} &= 6x = 6(24) = 144 \end{aligned}$$

- *∠1 is an inscribed angle:*

$$\begin{aligned} \text{m}\angle 1 &= \frac{1}{2}\text{m}\widehat{AB} \\ &= \frac{1}{2}(72) \\ &= 36 \end{aligned}$$

- *∠2 is a chord-chord angle:*

$$\begin{aligned} \text{m}\angle 2 &= \frac{1}{2}(\text{m}\widehat{AD} + \text{m}\widehat{BC}) \\ &= \frac{1}{2}(48 + 144) \\ &= 96 \end{aligned}$$

- *∠3 is a tangent-chord angle:*

$$\begin{aligned} \text{m}\angle 3 &= \frac{1}{2}\text{m}\widehat{AC} = \frac{1}{2}(\text{m}\widehat{AD} + \text{m}\widehat{DC}) \\ &= \frac{1}{2}(48 + 96) \\ &= 72 \end{aligned}$$

- *∠4 is a secant-tangent angle:*

$$\begin{aligned} \text{m}\angle 4 &= \frac{1}{2}(\text{m}\widehat{AC} - \text{m}\widehat{AB}) \\ &= \frac{1}{2}(144 - 72) \\ &= 36 \end{aligned}$$

- *∠5*: Angles 5 and *CBD* are supplementary. The measure of ∠5 may be found indirectly by first finding m∠*CBD*.

$$\begin{aligned} \text{m}\angle CBD &= \frac{1}{2}\text{m}\widehat{DC} = \frac{1}{2}(96) = 48 \\ \text{m}\angle 5 &= 180 - \text{m}\angle CBD \\ &= 180 - 48 \\ &= 132 \end{aligned}$$

MATH FACTS

The degree measure of an angle whose sides intercept arcs of a circle depends on the location of the vertex of the angle.

Position of Vertex of Angle	**Measure of Angle**
• On the circle	$\frac{1}{2}$ the degree measure of the intercepted arc
• Inside the circle	$\frac{1}{2}$ the *sum* of the degree measures of the two intercepted (opposite) arcs
• Outside the circle	$\frac{1}{2}$ the *difference* of the degree measures of the two intercepted arcs

Check Your Understanding of Section 4.3

In each case, clearly indicate the necessary steps, including formula substitutions, diagrams, graphs, charts, etc.

1–4. Find the value of x.

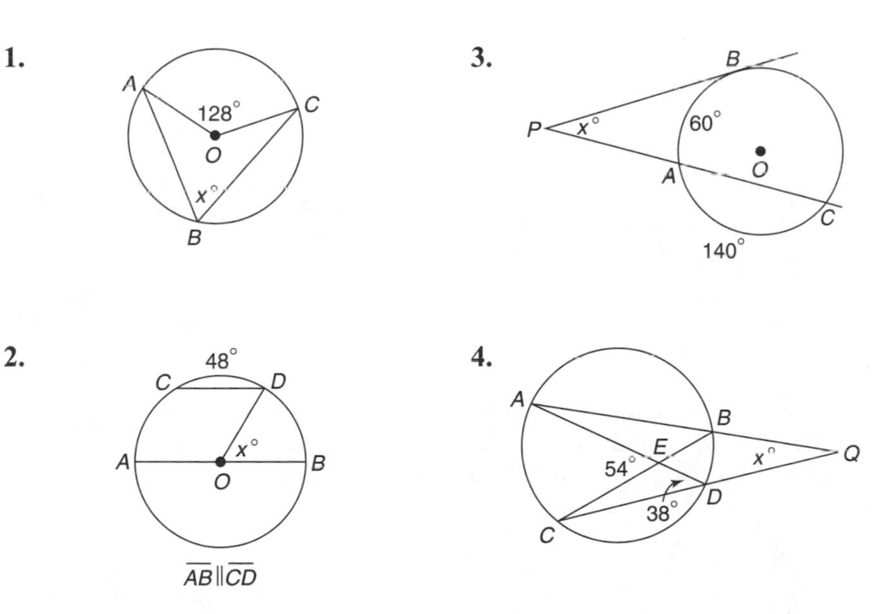

5. Point P lies outside circle O, which has a diameter of $\overline{AOC}$. The angle formed by tangent $\overline{PA}$ and secant $\overline{PBC}$ measures 30°. Find the number of degrees in the measure of minor arc CB.

6. In the accompanying diagram, chords $\overline{AB}$ and $\overline{CD}$ intersect at E. If $m\angle AEC = 4x$, $m\overset{\frown}{AC} = 120$, and $m\overset{\frown}{DB} = 2x$, what is the value of x?

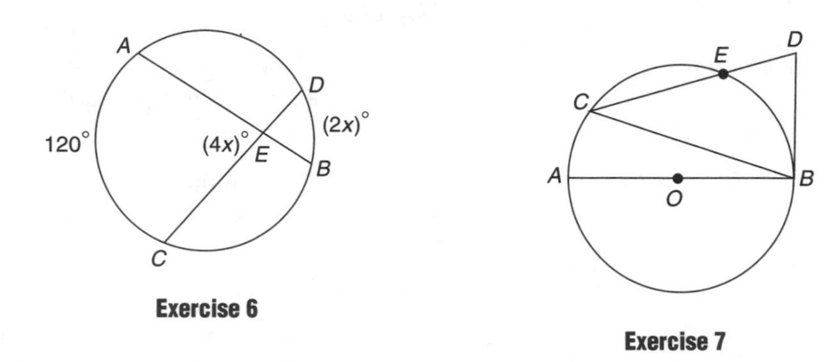

Exercise 6

Exercise 7

7. In the accompanying diagram, $\overline{CED}$ is a secant, $\overrightarrow{BD}$ is tangent to circle O at B, $\overline{BC}$ is a chord, and $\overline{BOA}$ is a diameter. If $m\overset{\frown}{AC} : m\overset{\frown}{CB} = 1:4$ and $m\overset{\frown}{CE} = 68$, find $m\angle BDC$.

8. In accompanying diagram, chord $\overline{BE}$ bisects $\angle ABC$. If $m\angle ABC = 70$ and $m\overset{\frown}{BAE} = 200$, find $m\angle AFE$.

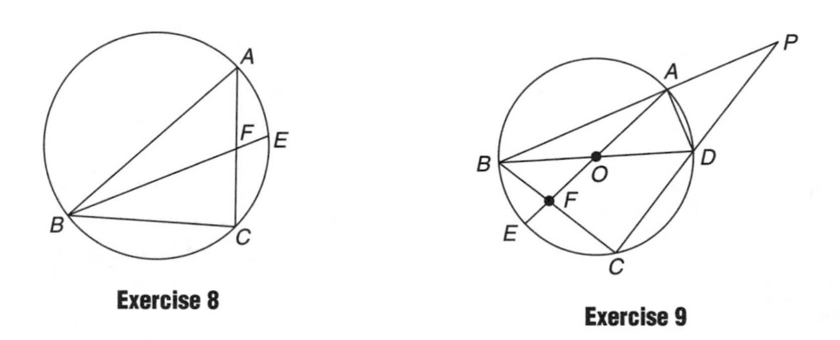

Exercise 8

Exercise 9

9. In the accompanying diagram of circle O, diameters $\overline{BD}$ and $\overline{AE}$, secants $\overline{PAB}$ and $\overline{PDC}$, and chords $\overline{AD}$ and $\overline{BC}$ are drawn. Diameter $\overline{AE}$ intersects chord $\overline{BC}$ at F. If $m\angle ABD = 20$ and $m\angle BFE = 75$, find $m\angle P$.

10. In circle O, tangent $\overrightarrow{PW}$ and $\overline{PST}$ are drawn. Chord $\overline{WA}$ is parallel to chord $\overline{ST}$. Chords $\overline{AS}$ and $\overline{WT}$ intersect at point B. If $m\widehat{WA} : m\widehat{AT} : m\widehat{ST} = 1:3:5$, find:

a. $m\angle TBS$
b. $m\angle TWP$
c. $m\angle WPT$
d. $m\angle ASP$

11. In the accompanying diagram, $\overline{PCD}$ and $\overline{PBA}$ are secants from external point P to circle O. Chords $\overline{DA}$, $\overline{DEB}$, $\overline{CEA}$, and $\overline{C}$ are drawn. $\overline{AB} \cong \overline{DC}$, $m\widehat{BC}$ is twice $m\widehat{AB}$, and $m\widehat{AD}$ is 60 more than $m\widehat{BC}$. Find:

a. $m\angle P$
b. $m\angle PCB$

12. In the accompanying diagram of circle O, $\overline{AB} \cong \overline{CD}$, $m\angle DFB = 70$, $m\widehat{AB} = 115$, and $m\widehat{DE} = 65$. Find the measure of $\angle P$.

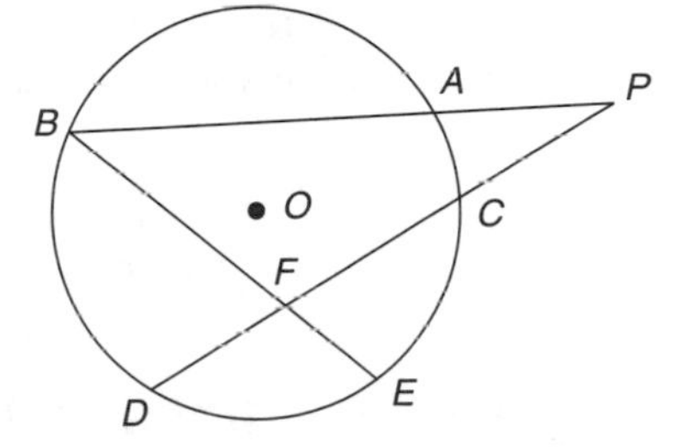

4.4 SIMILAR TRIANGLES AND CIRCLES

KEY IDEAS

Proving that triangles with chord, tangent, or secant segments as sides are similar can lead to some useful relationships involving the lengths of these segments.

Proofs with Circles

Circle angle-measurement relationships can be used to prove that triangles are similar.

Example 1

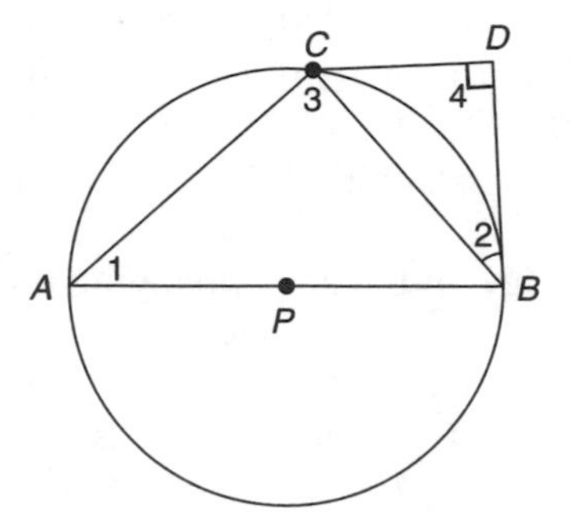

Given: $\overline{DB}$ is tangent to circle P at B, $\overline{AB}$ is a diameter, and $\overline{CD} \perp \overline{DB}$.

Prove: $\dfrac{AB}{BC} = \dfrac{BC}{CD}$.

Solution: See the proof.

PLAN: Prove that triangles ABC and CBD are similar.

PROOF

- Angle 1 is an inscribed angle, and $\angle 2$ is formed by a tangent and a chord. Since the degree measure of each of these angles is one-half of the degree measure of the same arc, $\overset{\frown}{BC}$, $\angle 1 \cong \angle 2$.
- Angle 3 is inscribed in a semicircle, and $\angle 4$ is formed by perpendicular lines. Since each angle is a right angle, $\angle 3 \cong \angle 4$.
- Hence, $\triangle ABC \sim \triangle CBD$ by the AA Theorem of Similarity. Since the lengths of corresponding sides of similar triangles are in proportion:

$$\frac{AB}{BC} = \frac{BC}{CD}.$$

Segment Relationships

In Figure 4.16, the product of the lengths of the segments of chord $\overline{AB}$ is equal to the product of the lengths of the segments of chord $\overline{CD}$ since $\triangle AED \sim \triangle CEB$. Similar triangles can also be used to prove special products involving tangent and secant segments.

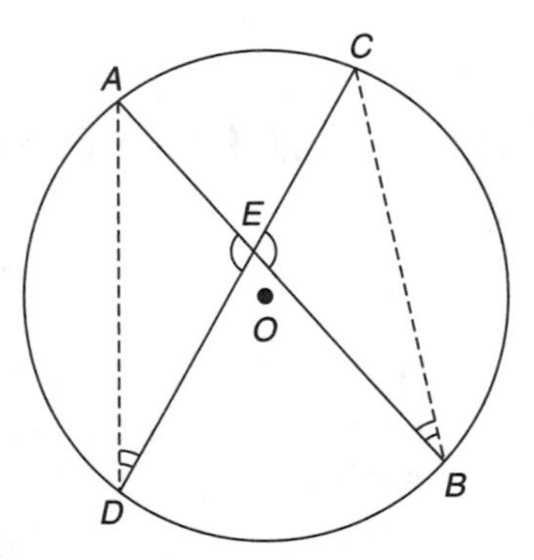

Figure 4.16 Products of Intersecting Chord Segments

MATH FACTS

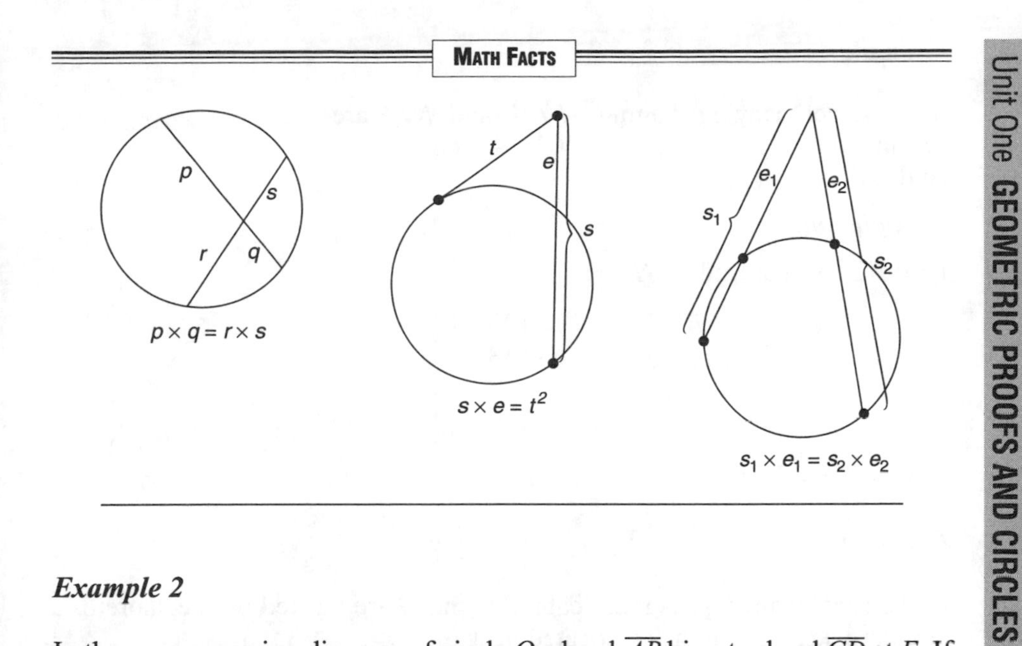

Example 2

In the accompanying diagram of circle O, chord $\overline{AB}$ bisects chord $\overline{CD}$ at E. If $AE = 4$ and $EB = 9$, what is the length of $\overline{CD}$?

Solution: **12**

If $x = CE = ED$, then:

$$\begin{aligned}(CE)(ED) &= (EB)(AE)\\ (x)(x) &= (9)(4)\\ x^2 &= 36\\ x &= \sqrt{36} = 6\end{aligned}$$

Hence, $CD = CE + ED = 6 + 6 = 12$.

Example 3

In the accompanying diagram, $\overline{PT}$ is tangent to circle O at T and $\overline{PAB}$ is a secant. If $PA = 4$ and $AB = 12$, find PT.

Solution: **8**

Since $PA = 4$ and $AB = 12$, $PAB = 4 + 12 = 16$. Hence:

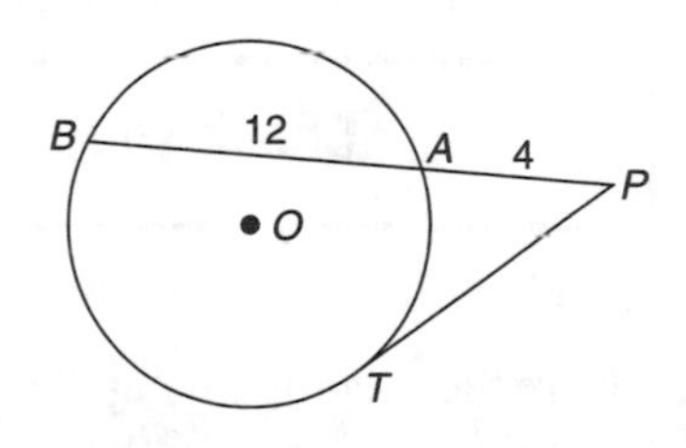

$$\begin{aligned}(PT)^2 &= (PAB)(PA)\\ &= (16)(4)\\ &= 64\\ PT &= \sqrt{64} = 8\end{aligned}$$

Example 4

In the accompanying diagram, $\overline{NEW}$ and $\overline{NTA}$ are secants to circle O. If $NE = 5$, $NT = 4$, and $TA = 6$, find WE.

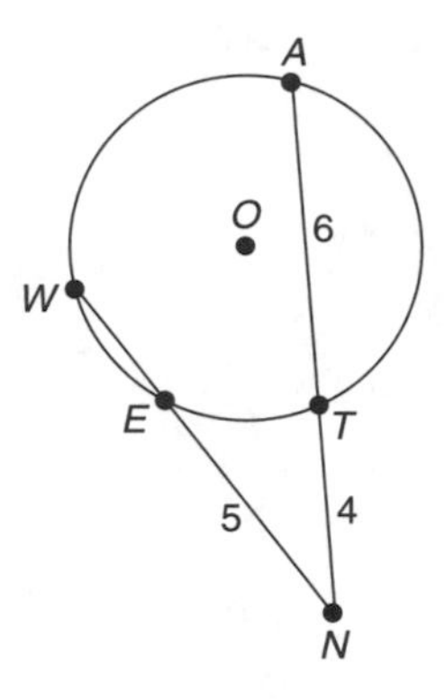

Solution: **3**

Find WE by first finding NEW:

$$\begin{aligned}(NE) \times (NEW) &= (NT) \times (NTA) \\ 5 \times (NEW) &= 4 \quad \times (4+6) \\ NEW &= \frac{40}{5} = 8\end{aligned}$$

Hence, $WE = NEW - NE = 8 - 5 = 3$.

Example 5

In the accompanying diagram, cabins B and G are located on the shore of a circular lake, and cabin L is located near the lake. Point D is a dock on the lake shore, and is collinear with cabins B and L. The road between cabins G and L is 8 miles long and is tangent to the lake. The path between cabin L and dock D is 4 miles long. What is the shortest distance, in miles, from cabin B to dock D?

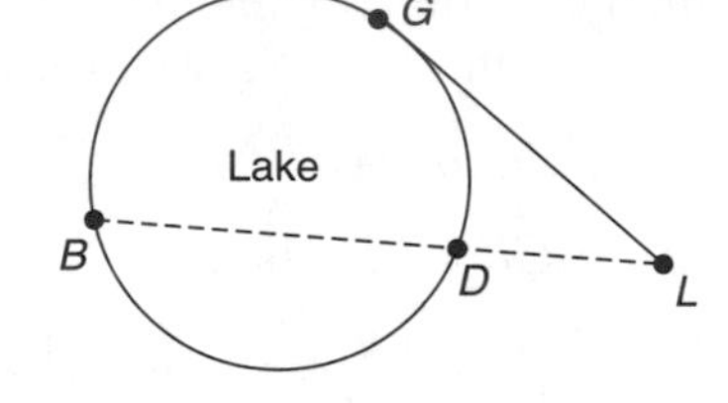

Solution: **12**

Find BD by first finding the length of secant $\overline{LDB}$:

$$\begin{aligned}(LD)(LDB) &= (LG)^2 \\ (4)(LDB) &= (8)^2 \\ LDB &= \frac{64}{4} = 16\end{aligned}$$

Hence, $BD = LDB - LD = 16 - 4 = 12$.

The shortest distance from cabin to dock is 12 mi.

Check Your Understanding of Section 4.4

A. *Multiple Choice*

1. In the accompanying diagram, $\overline{PAB}$ and $\overline{PCD}$ are secants drawn to circle O, $\overline{PA} = 8$, $\overline{PB} = 20$, and $\overline{PD} = 16$. What is the length of $\overline{PC}$?
(1) 6.4 (2) 10 (3) 12 (4) 40

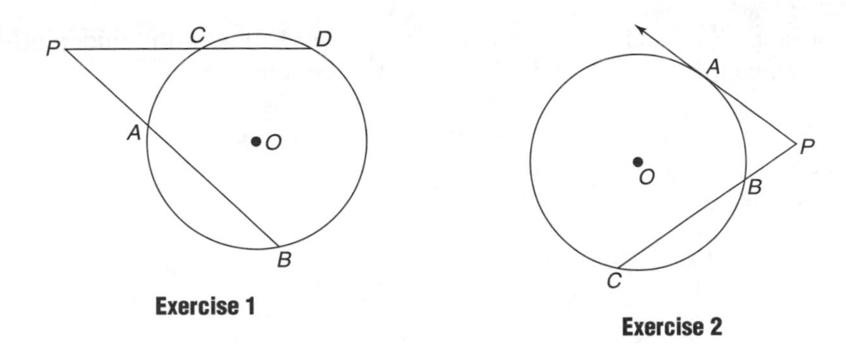

Exercise 1

Exercise 2

2. In the accompanying diagram, $\overrightarrow{PA}$ is tangent to circle O at A. If $CB = 12$ and $PB = 4$, what is the length of $\overline{PA}$?
(1) $4\sqrt{3}$ (2) 48 (3) $16\sqrt{3}$ (4) 8

3. In circle O, chords $\overline{AB}$ and $\overline{CD}$ intersect at E. If $AE = 4$, $EB = 12$, and $ED = 16$, then CE equals
(1) 19 (2) 16 (3) 3 (4) 48

B. *In each case, clearly indicate the necessary steps, including formula substitutions, diagrams, graphs, charts, etc.*

4. In circle O, chords $\overline{AB}$ and $\overline{CD}$ intersect at point E. If $AE = 2$, $CD = 9$, and $CE = 4$, find AB.

5. In circle O, diameter $\overline{AB}$ is perpendicular to chord $\overline{CD}$ at point E, $CD = 8$, and $EB = 2$. What is the length of diameter $\overline{AB}$?

6. In the accompanying diagram, secants $\overline{PAB}$ and $\overline{PCD}$ are drawn to circle O. If $PC = 5$, $CD = 7$, and $PA = 4$, find the length of a radius of circle O.

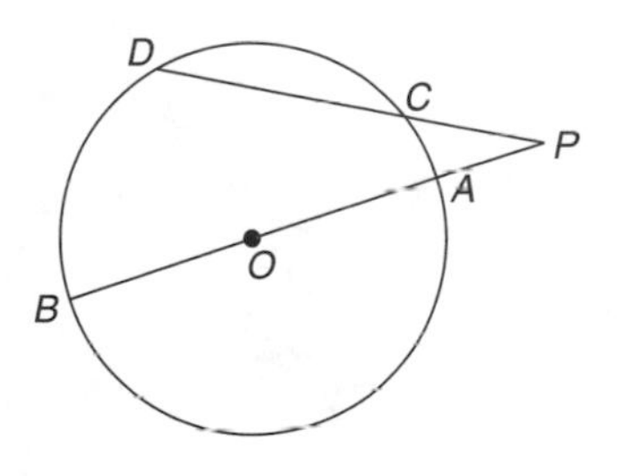

7. Chord $\overline{AB}$ bisects chord $\overline{CD}$ in the interior of circle O at point E. If $AE = 4$ and $AB = 20$, find the length of $\overline{CD}$.

8. From point P, $\overline{PA}$ is drawn tangent to circle O at A. A secant drawn from P through point O intersects the circle at points B and C. If $PA = 12$ and the length of a radius of circle O is 9, find the length of the external segment of secant $\overline{PBC}$.

9. Secants $\overline{PAB}$ and $\overline{PCD}$ are drawn to circle O. If A is the midpoint of $\overline{PAB}$, chord $CD = 14$, and $PC = 4$ what is the length of $\overline{PAB}$?

10. Tangent $\overrightarrow{PC}$ and secant $\overline{PAB}$ are drawn to circle O, as shown in the accompanying diagram. If PA is 6 less than PC and AB is 1 more than two times PC, find the length of $\overline{PC}$.

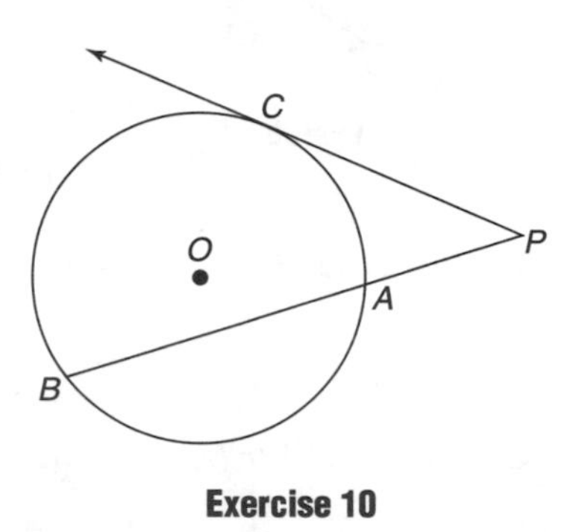

Exercise 10

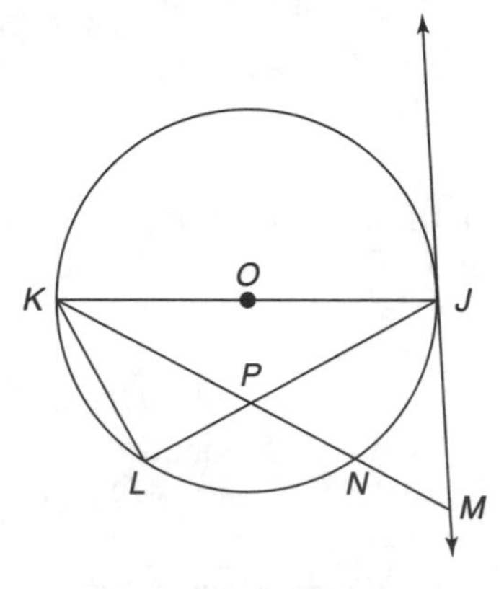

Exercises 11 and 12

11. Given: $\overline{MJ}$ is tangent to circle O at point J, $\overline{KJ}$ is a diameter, N is the midpoint of $\overline{LNJ}$.

Prove: $\dfrac{KL}{KJ} = \dfrac{KP}{KM}$.

12. Given: $\overline{MJ}$ is tangent to circle O at point J, $\overline{KJ}$ is a diameter, $\overline{JP} \cong \overline{JM}$.

Prove: $KL \times KM = JK \times LP$.

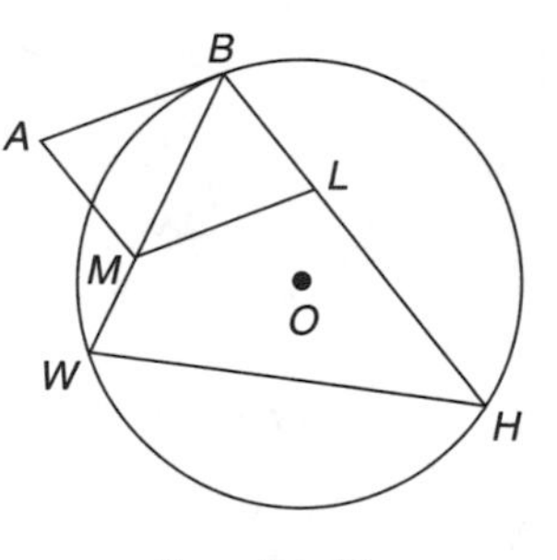

Exercise 13

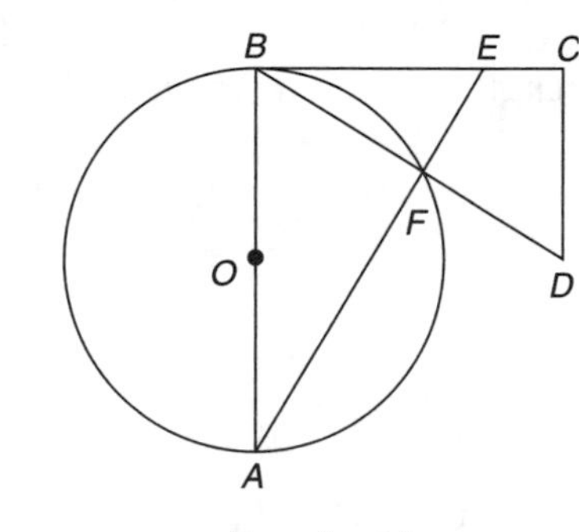

Exercise 14

13. Given: $\overline{AB}$ is tangent to circle O at point B, chords $\overline{WB}$, $\overline{BH}$, and $\overline{HW}$ are drawn; Quadrilateral $ABLM$ is a parallelogram.

Prove: $\dfrac{BL}{BW} = \dfrac{BM}{BH}$

14. Given: $\overline{CEB}$ is tangent to circle O at B, diameter $\overline{AB} \parallel \overline{CD}$. Secants $\overline{DFB}$ and $\overline{EFA}$ intersect circle O at F.

Prove: $\dfrac{BD}{AE} = \dfrac{CD}{BE}$

15. Given: $\overline{TK}$ bisects $\angle NTW$; $\overline{WK} \cong \overline{WT}$.

Prove: $\dfrac{JT}{TW} = \dfrac{TW}{TK}$

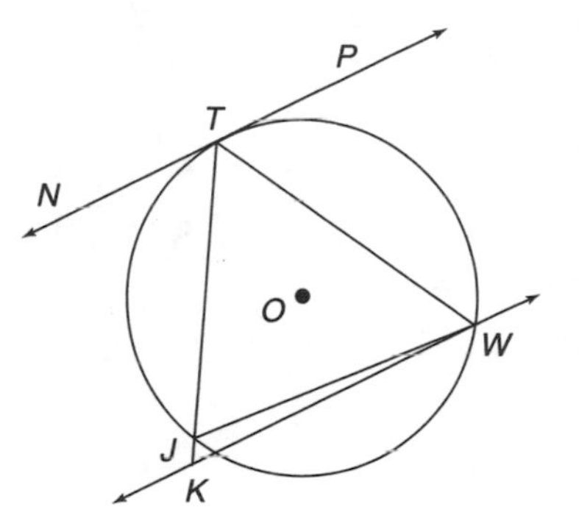

Exercise 15

Unit Two ALGEBRAIC AND GRAPHICAL REPRESENTATIONS

CHAPTER 5

ALGEBRAIC OPERATIONS

5.1 REAL NUMBERS AND EXPONENTS

KEY IDEAS

The set of **real numbers** is modeled by the continuous set of ordered points found on a number line. Real numbers include positive and negative integers, 0, fractions, decimals, and irrational numbers for which there are no exact fractional or decimal equivalents.

In the product $2 \times 2 \times 2 = 8$, 2 appears as a factor three times. Thus, $2^3 = 8$, where 3 is called an **exponent** and 2 is the **base**. Fractional exponents indicate roots of numbers.

The Set of Real Numbers and Its Subsets

The set of **real numbers** is comprised of these subsets of numbers:

- **Integers:** This subset includes positive and negative whole numbers and 0:

$$\ldots, -4, -3, -2, -1, 0, 1, 2, 3, 4, \ldots.$$

- **Rational numbers:** This subset includes numbers that can be written as the quotient of two integers, such as $\frac{5}{7}$ and $-2\left(=\frac{-2}{1}\right)$. Repeating decimal numbers, such as $0.33333\ldots\left(=\frac{1}{3}\right)$, and decimal numbers that end, such as $0.25\left(=\frac{1}{4}\right)$, are rational.
- **Irrational numbers:** This subset includes numbers that cannot be expressed as the quotient of two integers, such as $\sqrt{5}$. Nonending decimal numbers with no repeating pattern of digits, such as π $(= 3.1415926\ldots)$, are also irrational numbers.

The set of real numbers is the union of the sets of rational and irrational numbers.

Integer Exponents

The rules for working with integer exponents are summarized in the accompanying table.

RULES FOR INTEGER EXPONENTS

Exponent Law	Rule	Example
Multiplication	$a^x \times a^y = a^{x+y}$	$n^5 \times n^2 = n^7$
Quotient	$a^x \div a^y = a^{x-y}$	$n^5 \div n^2 = n^3$
Power of a Power	$(a^x)^y = a^{xy}$	$(mn^5)^2 = m^2n^{10}$
Zero Power	$a^0 = 1$ (provided that $a \neq 0$)	$4^0 = 1$
Negative Integer	$a^{-x} = \frac{1}{a^x}$ and $\frac{1}{a^{-x}} = a^x$ (provided that $a \neq 0$)	$3^{-2} = \frac{1}{3^2} = \frac{1}{9}$ and $\frac{1}{2^{-3}} = 2^3 = 8$

Square and Cube Roots

The **square root** of a number x is one of two identical numbers whose product is x. Thus, $\sqrt{9} = 3$ since $3 \times 3 = 9$; the symbol $\sqrt{}$ is called a **radical sign**, and 9, the number underneath the radical sign, is the **radicand**. Numbers whose square roots are rational are called **perfect squares**. For example, $\frac{4}{25}$ is a perfect square since $\sqrt{\frac{4}{25}} = \frac{2}{5}$, but 13 is *not* a perfect square since $\sqrt{13}$ does

not have an exact rational equivalent.

Similarly, the **cube root** of a number x is one of three identical numbers whose product is x. For example, the cube root of 64, denoted as $\sqrt[3]{64}$, is 4 since $4 \times 4 \times 4 = 64$.

In general, $\sqrt[k]{x}$ represents k equal factors of x, where k, called the **index** of the radical, is a positive whole number greater than 1. When working with square root radicals, the index 2 is omitted.

Principal Square Root

Every positive number has two square roots. For example, the two square roots of 16 are 4 and –4 since $(4)^2 = 16$ and $(-4)^2 = 16$. The positive square root of a number is called the **principal square root**. The expression $\sqrt{x}$ always names the principal square root of x. Thus, $\sqrt{16}$ stands for 4 rather than –4. The expression $\pm\sqrt{16}$ means 4 or –4.

Using Exponents to Indicate Roots

By extending the laws of exponents to include rational exponents, we know that

$$x^{\frac{1}{2}} \cdot x^{\frac{1}{2}} = x^{\frac{1}{2}+\frac{1}{2}} = x^1.$$

Since $x^{\frac{1}{2}}$ represents one of two identical factors whose product is x, then $x^{\frac{1}{2}}$ and $\sqrt{x}$ must be equivalent expressions. Similarly, $x^{\frac{1}{3}}$ and $\sqrt[3]{x}$ represent the cube root of x. In general, $x^{\frac{1}{k}}$ means $\sqrt[k]{x}$, where k indicates what root of x is to be taken.

MATH FACTS

The expression $\sqrt[k]{x}$ may or may not represent a real number:

- An even root of a negative number is *never* a real number. For example, $\sqrt{-25}$ is not a real number. Also, $\sqrt[4]{16}$ is real, but $\sqrt[4]{-16}$ is not real.
- An odd root of a real number is *always* a real number. For example, $\sqrt[3]{-8}$ is real since $(-2) \times (-2) \times (-2) = -8$.

Evaluating $x^{\frac{n}{k}}$

Since $x^{\frac{n}{k}} = \left(x^{\frac{1}{k}}\right)^n = \left(x^n\right)^{\frac{1}{k}}$, $x^{\frac{n}{k}}$ can be evaluated in either of two ways:

- Find the kth root of x and then raise the result to the nth power. For example,

$$8^{\frac{4}{3}} = \left(\sqrt[3]{8}\right)^4 = (2)^4 = 16.$$

- Raise x to the nth power and then take the kth root of the result. For example,

$$8^{\frac{4}{3}} = \sqrt[3]{8^4} = \sqrt[3]{4096} = 16.$$

MATH FACTS

$$x^{\frac{n\text{th power}}{k\text{th root}}} = \left(\sqrt[k]{x}\right)^n = \sqrt[k]{x^n},$$

where k is a whole number greater than 2. If $\frac{n}{k}$ is in lowest terms, then $x^{\frac{n}{k}}$ represents a real number only when $\sqrt[k]{x}$ is a real number.

Example

If $y = x^{\frac{2}{3}} + x^{-\frac{1}{2}} - 2x^0$, what is the value of y when $x = 64$?

Solution: $\mathbf{14\frac{1}{8}}$

Let $x = 64$. Then

$$\begin{aligned} y &= x^{\frac{2}{3}} + x^{-\frac{1}{2}} - 2x^0 \\ &= 64^{\frac{2}{3}} + 64^{-\frac{1}{2}} - 2(64^0) \\ &= \left(\sqrt[3]{64}\right)^2 + \frac{1}{\sqrt{64}} - 2(1) \\ &= 4^2 + \frac{1}{8} - 2 \\ &= 16 + \frac{1}{8} - 2 \\ &= 14\frac{1}{8} \end{aligned}$$

Using a Calculator to Evaluate Powers and Roots

If you are using the Texas Instruments TI-83 graphing calculator, quit the current screen and enter the Home Screen by pressing the [2nd] key followed by the [MODE] key, which has the QUIT command printed above it.

- To evaluate 3^7, press [3] [^] [7] [ENTER]. The answer 2187 appears on the right side of the next line.
- To evaluate $\sqrt[3]{4}$, write the cube root radical in exponential form as $4^{\frac{1}{3}}$. Then press

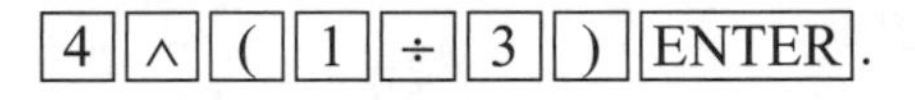

You should verify that $\sqrt[3]{4} \approx 1.587401052$, where the symbol $\approx$ is read as "is approximately equal to."

- To evaluate $8^{-\frac{2}{3}}$, use this sequence of keystrokes:

[8] [^] [(−)] [(] [2] [÷] [3] [)] [ENTER].

Verify that the answer displayed is .25.

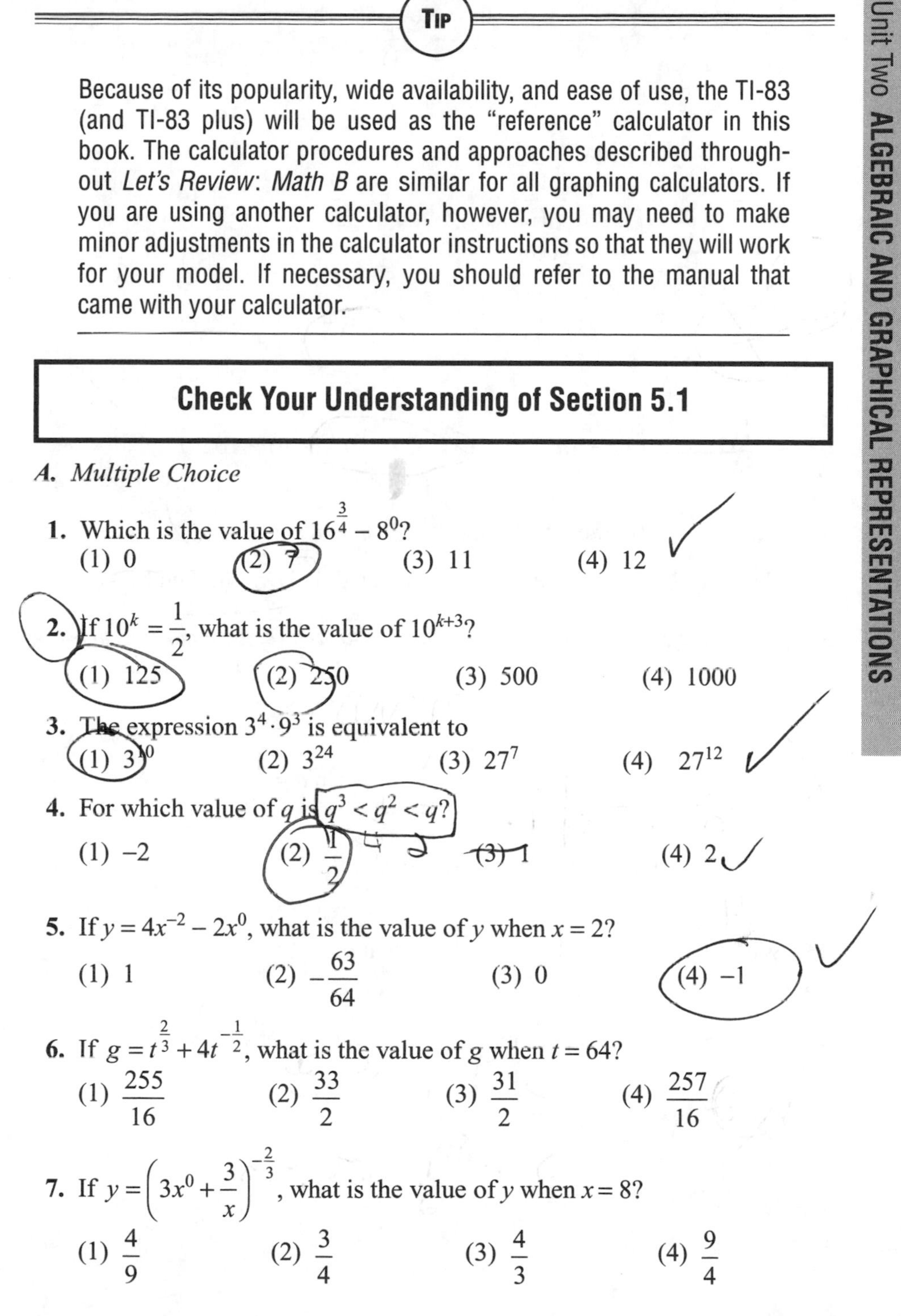

TIP

Because of its popularity, wide availability, and ease of use, the TI-83 (and TI-83 plus) will be used as the "reference" calculator in this book. The calculator procedures and approaches described throughout *Let's Review*: *Math B* are similar for all graphing calculators. If you are using another calculator, however, you may need to make minor adjustments in the calculator instructions so that they will work for your model. If necessary, you should refer to the manual that came with your calculator.

Check Your Understanding of Section 5.1

A. Multiple Choice

1. Which is the value of $16^{\frac{3}{4}} - 8^0$?
(1) 0 (2) 7 (3) 11 (4) 12

2. If $10^k = \frac{1}{2}$, what is the value of 10^{k+3}?
(1) 125 (2) 250 (3) 500 (4) 1000

3. The expression $3^4 \cdot 9^3$ is equivalent to
(1) 3^{10} (2) 3^{24} (3) 27^7 (4) 27^{12}

4. For which value of q is $q^3 < q^2 < q$?
(1) -2 (2) $\frac{1}{2}$ (3) 1 (4) 2

5. If $y = 4x^{-2} - 2x^0$, what is the value of y when $x = 2$?
(1) 1 (2) $-\frac{63}{64}$ (3) 0 (4) -1

6. If $g = t^{\frac{2}{3}} + 4t^{-\frac{1}{2}}$, what is the value of g when $t = 64$?
(1) $\frac{255}{16}$ (2) $\frac{33}{2}$ (3) $\frac{31}{2}$ (4) $\frac{257}{16}$

7. If $y = \left(3x^0 + \frac{3}{x}\right)^{-\frac{2}{3}}$, what is the value of y when $x = 8$?
(1) $\frac{4}{9}$ (2) $\frac{3}{4}$ (3) $\frac{4}{3}$ (4) $\frac{9}{4}$

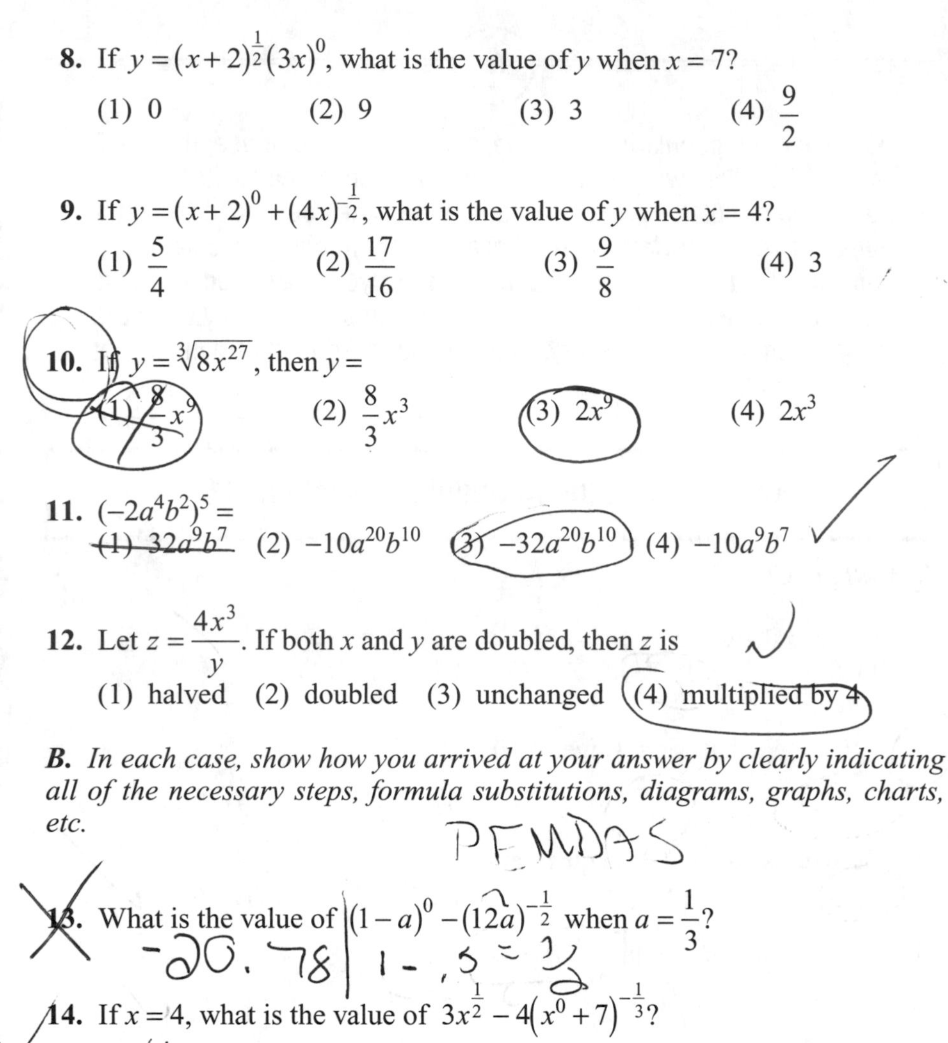

8. If $y=(x+2)^{\frac{1}{2}}(3x)^0$, what is the value of y when $x=7$?

(1) 0 (2) 9 (3) 3 (4) $\frac{9}{2}$

9. If $y=(x+2)^0+(4x)^{-\frac{1}{2}}$, what is the value of y when $x=4$?

(1) $\frac{5}{4}$ (2) $\frac{17}{16}$ (3) $\frac{9}{8}$ (4) 3

10. If $y=\sqrt[3]{8x^{27}}$, then $y=$

(1) $\frac{8}{3}x^9$ (2) $\frac{8}{3}x^3$ (3) $2x^9$ (4) $2x^3$

11. $(-2a^4b^2)^5=$

(1) $32a^9b^7$ (2) $-10a^{20}b^{10}$ (3) $-32a^{20}b^{10}$ (4) $-10a^9b^7$

12. Let $z=\frac{4x^3}{y}$. If both x and y are doubled, then z is

(1) halved (2) doubled (3) unchanged (4) multiplied by 4

B. *In each case, show how you arrived at your answer by clearly indicating all of the necessary steps, formula substitutions, diagrams, graphs, charts, etc.*

13. What is the value of $(1-a)^0-(12a)^{-\frac{1}{2}}$ when $a=\frac{1}{3}$?

14. If $x=4$, what is the value of $3x^{\frac{1}{2}}-4(x^0+7)^{-\frac{1}{3}}$?

15. What is the value of $\left(\frac{b^0-b^2}{b^0+b}\right)^{-\frac{1}{2}}$ when $b=\frac{3}{4}$?

16. Arrange the following terms from least to greatest when $0<x<1$:

$$1,\quad x,\quad \sqrt[3]{x},\quad \frac{1}{x},\quad \frac{1}{\sqrt{x}},\quad \frac{1}{\sqrt[3]{x}}$$

5.2 FACTORING METHODS

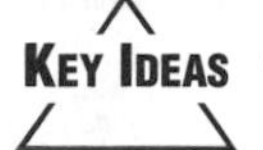

You can multiply two binomials such as (2x + 7) and (x + 3) together by writing the binomials next to each other and using FOIL:

$$\begin{aligned}(2x+7)(x+3) &= \overbrace{2x\cdot x}^{\text{First}} + \overbrace{2x\cdot 3}^{\text{Outer}}+\overbrace{(+7)(x)}^{\text{Inner}} + \overbrace{(+7)(+3)}^{\text{Last}}\\ &= 2x^2 + [6x+7x] + 21\\ &= 2x^2 + 13x + 21\end{aligned}$$

Sometimes it is useful to know which quantities were multiplied together to obtain a certain product. The process of reversing multiplication is called **factoring**.

Factoring by Reversing FOIL

Since $(x + 2)(x + 5) = x^2 + 7x + 10$, we know that $x^2 + 7x + 10 = (x + 2)(x + 5)$. The binomial factors contain 2 and 5 since these are the only two integers that, when multiplied together, give 10, the last term in $x^2 + 7x + 10$, *and*, when added together, equal 7, the coefficient of the x-term in $x^2 + 7x + 10$.

Example 1

Factor $x^2 - 5x - 14$ as the product of two binomials.

Solution: $\mathbf{(x + 2)(x - 7)}$

- Rewrite the given expression as follows: $x^2 - 5x - 14 = (x + ?)(x + ?)$.
- Fill in the missing terms of the binomial factors with the two integers whose product is −14, the number term of $x^2 - 5x - 14$, and whose sum is −5, the coefficient of the x-term of $x^2 - 5x - 14$. Since $(+2) \times (-7) = -14$ and $(+2) + (-7) = -5$, the correct factors of −14 are +2 and −7.
- Thus, $x^2 - 5x - 14 = (x + 2)(x - 7)$. Check that the binomial factors are correct by multiplying them together and comparing the product to the original quadratic expression.

Factoring $ax^2 + bx + c$ $(a > 1)$

To factor $3x^2 + 10x + 8$, break down $10x$ into the sum of two terms, ax and bx. Choose a and b so that they add up to 10 and, when multiplied together, give the same product as the product of the coefficients of the first and last terms of $\underline{\underline{3}}x^2 + 10x + \underline{\underline{8}}$. Since $6 + 4 = 10$ and $6 \times 4 = 3 \times 8$, break down $10x$ into $6x$ and $4x$:

$$3x^2 + \overbrace{6x + 4x}^{+10x} + 8$$

- Group pairs of terms with common factors: $(3x^2 + 6x) + (4x + 8)$
- Factor each binomial: $3x\underline{\underline{(x + 2)}} + 4\underline{\underline{(x + 2)}}$
- Factor out $(x + 2)$: $(3x + 4)(x + 2)$

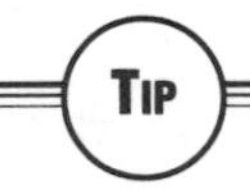

- You can also factor $3x^2 + 10x + 8$ by first factoring $3x^2$ and then using a trial and elimination process to find the correct factors of 8 and their placements:

$$3x^2 + 10x + 8 = (3x + ?)(x + ?) = (3x + 4)(x + 2).$$

- To check that you have factored $3x^2 + 10x + 8$ correctly, graph $Y_1 = 3x^2 + 10x + 8$ and $Y_2 = (3x + 4)\ (x + 2)$ using your graphing calculator (see Section 6.2). If the two graphs exactly coincide, then $Y_1 = Y_2$, so $3x^2 + 10x + 8 = (3x + 4)\ (x + 2)$.

Conjugate Binomials

A pair of binomials that represent the sum and difference of the same two terms, as in $(A + B)$ and $(A - B)$, are called **conjugate binomials**. In general:

$$(A + B)(A - B) = A^2 - B^2.$$

For example,

$$(2y + 5)(2y - 5) = (2y)^2 - (5)^2 = 4y^2 - 25.$$

If you encounter a binomial that has the form $A^2 - B^2$, you can factor it as $(A + B)(A - B)$. For instance,

$$4a^2 - 81b^2 = (2a)^2 - (9b)^2 = (2a + 9b)(2a - 9b).$$

Factoring Completely

A polynomial is **factored completely** when each of its factors cannot be factored further. To factor a polynomial completely, you may need to use more than one factoring method.

Example 2

Factor completely: a. $2x^3 - 50x$ b. $3t^3 + 18t^2 - 48t$

Solutions: a. $\mathbf{2x(x+5)(x-5)}$ b. $\mathbf{3t(t+8)(t-2)}$

$$2x^3 - 50x = 2x(x+5)(x-5)$$
$$2x^3 - 50x = 2x(x^2 - 25)$$
$$= 2x(x+5)(x-5)$$

$$3t^3 + 18t^2 - 48t = 3t(t+8)(t-2)$$
$$3t^3 + 18t^2 - 48t = 3t(t^2 + 6t - 16)$$
$$= 3t(t+8)(t-2)$$

Check Your Understanding of Section 5.2

A. Multiple Choice

1. Which is a factor of $y^2 + y - 30$?
(1) $(y - 6)$ (2) $(y + 6)$ (3) $(y - 3)$ (4) $(y + 3)$

2. If $ax^2 + bx + c = (2x - 3)(x + 5)$, what is the value of b?
(1) -15 (2) 2 (3) 7 (4) 4

3. If $ay - c = d + by$, then $y =$
(1) $\frac{c+d}{a+b}$ (2) $\frac{a-b}{c+d}$ (3) $\frac{c+d}{d-b}$ (4) $\frac{a+c}{d-b}$

4. If the area of a square is $9x^2 - 12xy + 4y^2$, what is the perimeter of the square?
(1) $12x - 8y$ (2) $8x - 12y$ (3) $12x + 8y$ (4) $8y - 12x$

5. Which expression is a factored form of $0.04y^2 - 9$?
(1) $(0.02y + 3)(0.02y - 3)$ (3) $(0.2y + 3)(0.2y - 3)$
(2) $(0.2y + 9)(0.2y - 1)$ (4) $(4y + 3)(0.01y - 3)$

6. If the area of a rectangle is represented by $3x^2 - 7x - 20$ and its width is represented by $x - 4$, which expression represents the length of the rectangle?
(1) $3x + 5$ (2) $3x - 5$ (3) $x + 15$ (4) $x - 15$

7. If $(2y-1)(y+B)=2y^2+Ay+3$, then $A+B=$
(1) -10 (2) -4 (3) 8 (4) 4

B. *In each case, show how you arrived at your answer by clearly indicating all of the necessary steps, formula substitutions, diagrams, graphs, charts, etc.*

8–13. Find each product.

8. $(4b-3)(b+2)$
9. $(x+7)(2x-9)$
10. $(5w-8)(5w+8)$
11. $(0.3y^2+1)(0.3y^2-1)$
12. $(2x-3)^2$
13. $(3x+4y)^2$

14–25. Factor as the product of two binomials.

14. $x^2+8x+15$
15. $x^2-10x+21$
16. y^2+6y+9
17. $100a^2-49b^2$
18. $a^2-4a-45$
19. $b^2+3b-40$
20. $\frac{4}{9}c^2-1$
21. $w^2-13w+42$
22. $0.81-0.25x^2$
23. $4n^2+11n-3$
24. $3x^2+2x-21$
25. $5s^2-14s-3$

26–33. Factor completely.

26. $2y^3-50y$
27. $-5t^2+5$
28. $4m^4-4n^4$
29. $8xy^3-72xy$
30. $2x^3+2x^2-112x$
31. $\frac{1}{2}x^3-18x$
32. $-2y^2+6y+20$
33. $10y^4+50y^3-500y^2$

34. Alisha claims that the difference of the squares of any two consecutive odd integers is exactly divisible by 8.
a. Show that Alisha is correct for 7 and 9.
b. If n is any integer, then which expression *must* represent an odd integer?
(1) $3n$ (2) $2n+1$ (3) $3n-1$ (4) $n+1$
c. Use your answer to part b to prove that Alisha is correct. Explain your reasoning.

5.3 WRITING EQUIVALENT FRACTIONS

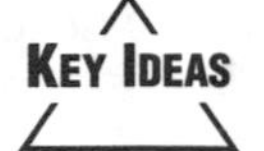

Factoring can be used to find an equivalent form of a fraction. A fraction with a negative sign can be rewritten with the negative sign in different positions:

$$\frac{-b}{5a} = \frac{b}{-5a} = -\frac{b}{5a}.$$

Writing a Fraction in Lowest Terms

To write an algebraic fraction in lowest terms, factor the numerator and the denominator. Then divide out any factor contained in both the numerator and the denominator since the quotient of these factors is 1. For example:

$$\frac{x^2+3x}{x^2-9} = \frac{x\overset{1}{\cancel{(x+3)}}}{(x-3)\cancel{(x+3)}} = \frac{x}{x-3}.$$

When simplifying a fraction, you may need to rewrite the difference of two terms in the opposite order by factoring out -1, as illustrated in Example 1.

Example 1

Write $\frac{ab-b^2}{5ab-5a^2}$ in lowest terms.

Solution: $\mathbf{\frac{-b}{5a}}$

- Factor: $\frac{ab-b^2}{5ab-5a^2} = \frac{b(a-b)}{5a(b-a)}$
- Rewrite $(a-b)$ as $-(b-a)$: $= \frac{-b\cancel{(b-a)}}{5a\cancel{(b-a)}} = \frac{-b}{5a}$

Rewriting a Fraction in Higher Terms

Sometimes it is useful to be able to rewrite a fraction as an equivalent fraction with a specified denominator.

Example 2

Write $\dfrac{x}{x-3}$ as an equivalent fraction whose denominator is $x^2 - 9$.

Solution: $\dfrac{\mathbf{x^2+3x}}{\mathbf{x^2-9}}$

Multiplying a fraction by 1 does not change its value. Since $x^2 - 9 = (x - 3)(x + 3)$, multiplying the numerator and the denominator of $\dfrac{x}{x-3}$ by $\dfrac{x+3}{x+3}$ creates an equivalent fraction whose denominator is $(x - 3)(x + 3) = x^2 - 9$:

$$\frac{x}{x-3} = \frac{x}{x-3}\cdot\overbrace{\frac{x+3}{x+3}}^{1} = \frac{x(x+3)}{(x-3)(x+3)} = \frac{x^2+3x}{x^2-9}.$$

Check Your Understanding of Section 5.3

A. *Multiple Choice*

1. Written in simplest form, $\dfrac{a-a^2b}{b-ab^2}$ is

(1) 1 (2) $\dfrac{a}{b}$ (3) $\dfrac{a-1}{b-1}$ (4) $\dfrac{1-a^2}{1-b^2}$

2. Written in simplest form, $\dfrac{4-(xy)^2}{x^2y^2-4}$ is

(1) 1 (2) 0 (3) $\dfrac{4-x^2y}{x^2y^2-4}$ (4) −1

3. Written in simplest form, $\dfrac{(xy)^3-9xy}{3xy-(xy)^2}$ is

(1) $-(xy+3)$ (2) $\dfrac{xy+3}{xy-3}$ (3) $\dfrac{x+3}{y-3}$ (4) $\dfrac{y(x^2-3)}{1-xy}$

4. Written in simplest form, $\dfrac{3x^3-27xy^2}{12x^2+36xy}$ is

(1) $\dfrac{x+3y}{4}$ (2) $\dfrac{x+4y}{3}$ (3) $\dfrac{x-3y}{4}$ (4) $\dfrac{x-3y}{4(x+3y)}$

B. *In each case, show how you arrived at your answer by clearly indicating all of the necessary steps, formula substitutions, diagrams, graphs, charts, etc.*

5–13. Write each fraction in lowest terms.

5. $\dfrac{6-3x}{x^2-4}$

6. $\dfrac{a-y}{(y-a)^2}$

7. $\dfrac{3x+6}{4-x^2}$

8. $\dfrac{x^2-2x-3}{x^2-5x+6}$

9. $\dfrac{2x^2-18}{x^2-6x+9}$

10. $\dfrac{2x^2-50}{2x^2+5x-25}$

11. $\dfrac{a^3b^2-a}{ab^2-b}$

12. $\dfrac{x^2-4y^2}{x^2+4xy+4y^2}$

13. $\dfrac{(x+y)^2-2xy}{x^4-y^4}$

14. Rewrite $\dfrac{5}{x-1}$ as an equivalent fraction whose denominator is x^2+2x-3.

15. Rewrite $\dfrac{x-2}{x-3}$ as an equivalent fraction whose denominator is x^2-x-6.

5.4 MULTIPLYING AND DIVIDING ALGEBRAIC FRACTIONS

KEY IDEAS

Algebraic fractions are multiplied and divided in much the same way that fractions in arithmetic are multiplied and divided.

Multiplying Algebraic Fractions

Before multiplying algebraic fractions, factor the numerators and also factor the denominators, where possible. Then divide out any factor that is contained in both a numerator and a denominator since the quotient of these factors is 1.

Example 1

Write the product in lowest terms: $\dfrac{x^2-9}{10}\cdot\dfrac{15}{(x+3)^2}$.

Solution: $\mathbf{\dfrac{3(x-3)}{2(x+3)}}$

- Factor: $\dfrac{x^2-9}{10}\cdot\dfrac{15}{(x+3)^2}=\dfrac{(x-3)(x+3)}{5\cdot 2}\cdot\dfrac{5\cdot 3}{(x+3)(x+3)}$
- Cancel common factors: $=\dfrac{(x-3)\overset{1}{\cancel{(x+3)}}}{\cancel{5}\cdot 2}\cdot\dfrac{\overset{1}{\cancel{5}}\cdot 3}{\cancel{(x+3)}(x+3)}$
- Multiply the remaining factors: $=\dfrac{3(x-3)}{2(x+3)}$

Dividing Algebraic Fractions

To divide algebraic fractions, multiply the first fraction by the reciprocal of the divisor.

Example 2

Write the quotient in lowest terms: $\dfrac{6x-3x^2}{9x^3-x}\div\dfrac{x^2-4}{3x^2+5x-2}$

Solution: $\mathbf{-\dfrac{3}{3x+1}}$

- Change to multiplication:

$$\frac{6x-3x^2}{9x^3-x}\div\frac{x^2-4}{3x^2+5x-2}=\frac{6x-3x^2}{9x^3-x}\cdot\frac{3x^2+5x-2}{x^2-4}$$

- Factor completely: $=\dfrac{3x(2-x)}{x(3x+1)(3x-1)}\cdot\dfrac{(3x-1)(x+2)}{(x-2)(x+2)}$
- Cancel like factors: $=\dfrac{3\overset{1}{\cancel{x}}(2-x)}{\cancel{x}(3x+1)\cancel{(3x-1)}}\cdot\dfrac{\overset{1}{\cancel{(3x-1)}}\overset{1}{\cancel{(x+2)}}}{(x-2)\cancel{(x+2)}}$
- Multiply: $=\dfrac{3(2-x)}{(3x+1)(x-2)}$
- Simplify further: $=\dfrac{3\cancel{(2-x)}}{(3x+1)\underset{-1}{\cancel{(x-2)}}}=-\dfrac{3}{3x+1}$

Check Your Understanding of Section 5.4

A. Multiple Choice

1. What is the product $(y+1)\left(\frac{y}{1-y^2}\right)$ expressed in lowest terms?

(1) $\frac{y}{y-1}$ (2) $\frac{y}{1-y}$ (3) $\frac{y+1}{y-1}$ (4) -1

2. If the width of a rectangular garden is represented by $\frac{12}{6a+12}$ and its length is represented by $\frac{a^2-4}{a^2-5a+6}$, which expression represents the area of the garden?

(1) $\frac{-4}{a+6}$ (2) $\frac{-4}{6a(-5a+6)}$ (3) $\frac{2}{a-3}$ (4) $\frac{a+2}{6a(a-3)}$

3. The length of a rectangular garden is represented by $\frac{3x-21}{6x^2-24}$. If the area of the garden can be represented by $\frac{x-7}{3x^2-8x+4}$, which expression represents the width of the garden?

(1) $\frac{2x+2}{3x-2}$ (2) $\frac{2x+2}{3(x-2)}$ (3) $\frac{2(x+2)}{3x-2}$ (4) $\frac{2(x+2)}{3(x-2)}$

B. *In each case, show how you arrived at your answer by clearly indicating all of the necessary steps, formula substitutions, diagrams, graphs, charts, etc.*

4–11. In each case, perform the indicated operation(s) and write the result in simplest form.

4. $\left(\frac{ax^2}{b^3}\right)^3\left(\frac{b^2}{a^2x}\right)^2$

5. $\frac{12y^2}{x^2+7x}\cdot\frac{x^2-49}{2y^5}$

6. $\frac{81-x^2}{6x-54}\div\frac{x^2+9x}{3x}$

7. $\frac{x^2-14x+49}{2x^2-13x-7}\cdot\frac{10x+5}{x^2-49}$

8. $\frac{r^2+2rs+s^2}{rs-r}\div\frac{r^2-s^2}{s^2-s}$

9. $\frac{x^3-9x}{(xy)^2-8y}\cdot\frac{x^3y-8x}{x^2-6x+9}$

10. $\dfrac{2x^2-32}{x^2+x-12}\div\dfrac{20-5x}{2x^2-5x-3}$

11. $\dfrac{t^2-1}{t^2-4}\div\dfrac{9t+9}{4t+12}\cdot\dfrac{2-t}{t^2+2t-3}$

12. A rectangular prism has a length of $\dfrac{2x^2+2x-24}{4x^2+x}$, a width of $\dfrac{x^2+x-6}{x+4}$, and a height of $\dfrac{8x^2+2x}{x^2-9}$. For all values of x for which it is defined, express the volume of the prism in simplest form.

5.5 ADDING AND SUBTRACTING ALGEBRAIC FRACTIONS

KEY IDEAS

To add (or subtract) fractions with like denominators, write the sum (or difference) of the numerators over the common denominator. If the fractions do not have the same denominator, rewrite each fraction with the least common denominator (LCD) as its denominator.

Combining Fractions with Like Denominators

To combine fractions that have the same denominator, write the sum or difference of the numerators over the common denominator and, if possible, simplify:

$$\frac{x}{8y}+\frac{3x}{8y}=\frac{x+3x}{8y}=\frac{\overset{1}{\cancel{4}}x}{\underset{2}{\cancel{8}}y}=\frac{x}{2y}.$$

Combining Fractions with Unlike Denominators

The least common denominator (LCD) of two or more fractions with unlike denominators is the smallest expression into which each of the denominators divides evenly. To determine the LCD, you may need to factor the denominators.

Example 1

Add: $\frac{x}{4y}+\frac{2x}{3y}$.

Solution: $\mathbf{\frac{11x}{12y}}$

The LCD of $4y$ and $3y$ is $12y$ since $12y$ is the smallest expression into which $4y$ and $3y$ divide evenly. Change each fraction into an equivalent fraction having the LCD as its denominator by multiplying the first fraction by 1 in the form of $\frac{3}{3}$ and multiplying the second fraction by 1 in the form of $\frac{4}{4}$:

$$\frac{x}{4y}+\frac{2x}{3y}=\frac{x}{4y}\cdot\frac{3}{3}+\frac{2x}{3y}\cdot\frac{4}{4}=\frac{11x}{12y}.$$

Example 2

When $\frac{4}{x^2-2x-3}$ is subtracted from $\frac{x+3}{x^2-1}$, what is the difference, expressed as a single fraction in lowest terms?

Solution: $\mathbf{\frac{x-5}{(x+3)(x-1)}}$

- Write the difference with the denominators in factored form:

$$\frac{x+3}{x^2-1}-\frac{4}{x^2-2x-3}=\frac{x+3}{(x+1)(x-1)}-\frac{4}{(x-3)(x+1)}$$

- Since the LCD is $(x-3)(x+1)(x-1)$, change each factored fraction into an equivalent fraction with the LCD as its denominator by multiplying the first fraction by 1 in the form of $\frac{x-3}{x-3}$, and multiplying the second fraction by 1 in the form of $\frac{x-1}{x-1}$:

$$\begin{aligned}\frac{x+3}{x^2-1}-\frac{4}{x^2-2x-3}&=\frac{x+3}{(x+1)(x-1)}\cdot\frac{x-3}{x-3}-\frac{4}{(x-3)(x+1)}\cdot\frac{x-1}{x-1}\\&=\frac{x^2-9-4(x-1)}{(x+3)(x+1)(x-1)}\\&=\frac{x^2-4x-5}{(x+3)(x+1)(x-1)}\end{aligned}$$

- Write the difference in lowest terms:

$$\frac{x+3}{x^2-1} - \frac{4}{x^2-2x-3} = \frac{(x-5)\cancel{(x+1)}}{(x+3)\cancel{(x+1)}(x-1)}$$
$$= \frac{x-5}{(x+3)(x-1)}$$

Check Your Understanding of Section 5.5

A. Multiple Choice

1. The sum of $\frac{2}{x+1}$ and $\frac{4}{x^2-1}$ expressed as a single fraction in lowest terms is

(1) $\frac{2}{x-1}$ (2) $\frac{2}{x+1}$ (3) $\frac{2(x+3)}{x^2-1}$ (4) $\frac{2x+3}{(x+1)(x-1)}$

2. When combined into a single fraction in lowest terms, $\frac{3}{x+2} - \frac{2}{x-2}$ is equal to

(1) $\frac{1}{x}$ (2) $\frac{5}{x+2}$ (3) $\frac{x-10}{x^2-4}$ (4) $\frac{x-4}{x^2-4}$

3. Expressed as a single fraction, $\frac{5}{x-3} - \frac{1}{x}$ is equivalent to

(1) $\frac{6x-3}{x^2-3x}$ (2) $\frac{4x+3}{x^2-3x}$ (3) $\frac{4x+3}{2x-3}$ (4) $\frac{4}{x^2-3x}$

4. For all values of x for which the expressions are defined, $\frac{4x}{x^2-4} + \frac{x+6}{4-x^2}$ is equivalent to

(1) $\frac{3}{x+2}$ (2) $\frac{3}{x-2}$ (3) $\frac{x+3}{x-2}$ (4) $\frac{1}{2}$

5. For all values of b for which the expressions are defined, $\frac{2b^2}{b-3} + \frac{18}{3-b}$ is equivalent to

(1) $2b+6$ (2) $\frac{2}{3-b}$ (3) $2b-3$ (4) -2

6. For all values of n for which the expressions are defined, $\left(1+\frac{1}{n}\right) \div \left(\frac{1}{n}+\frac{1}{n^2}\right)$ is equivalent to

(1) $\frac{n+1}{n}$ (2) $\frac{2n}{n+1}$ (3) $\frac{n^2}{n+1}$ (4) n

B. *In each case, show how you arrived at your answer by clearly indicating all of the necessary steps, formula substitutions, diagrams, graphs, charts, etc.*

7–12. In each case, write the indicated sum or difference of the given fractions as a single fraction in lowest terms.

7. $\frac{2x+y}{x+2y}+\frac{x+5y}{x+2y}$

8. $\frac{a^2-5}{a-b}+\frac{b^2-5}{b-a}$

9. $\frac{3}{x^2-4}+\frac{2}{x^2+5x+6}$

10. $\frac{a^2+1}{a^2-1}-\frac{a}{a+1}$

11. $\frac{x+2}{x^2-9}+\frac{2}{x^2+x-6}$

12. $\frac{x}{x^2+3x-4}-\frac{x+1}{2x^2-2}$

5.6 COMPLEX FRACTIONS

KEY IDEAS

A **complex fraction** is a fraction in which the numerator, the denominator, or both contain other fractions. To simplify a complex fraction, multiply the numerator by the reciprocal of the denominator. For example:

$$\frac{\frac{1}{3}}{\frac{4}{5}}=\frac{1}{3}\div\frac{4}{5}=\frac{1}{3}\times\frac{5}{4}=\frac{5}{12}.$$

Simplifying a Complex Fraction Using Division

To simplify the complex fraction $\dfrac{1+\dfrac{2}{x}}{1-\dfrac{4}{x^2}}$:

- Write the numerator as a single fraction:

$$1+\frac{2}{x}=\frac{x}{x}+\frac{2}{x}=\frac{x+2}{x}.$$

- Write the denominator as a single fraction:

$$1-\frac{4}{x^2}=\frac{x^2}{x^2}-\frac{4}{x^2}=\frac{x^2-4}{x^2}.$$

- Multiply the numerator of the complex fraction by the reciprocal of the denominator:

$$\frac{1+\dfrac{2}{x}}{1-\dfrac{4}{x^2}}=\frac{x+2}{x}\cdot\frac{x^2}{x^2-4}$$

$$=\frac{\overset{1}{\cancel{(x+2)}}x^{\cancel{2}}}{\cancel{x}(x-2)\cancel{(x+2)}}$$

$$=\frac{x}{x-2}$$

Simplifying a Complex Fraction by Clearing Fractions

You can also simply a complex fraction by multiplying the numerator and the denominator by the lowest common multiple of the denominators:

$$\frac{x^2}{x^2}\cdot\left(\frac{1+\dfrac{2}{x}}{1-\dfrac{4}{x^2}}\right)=\frac{x^2\left(1+\dfrac{2}{x}\right)}{x^2\left(1-\dfrac{4}{x^2}\right)}=\frac{x^2+2x}{x^2-4}$$

$$=\frac{x\cancel{(x+2)}}{\cancel{(x+2)}(x-2)}$$

$$=\frac{x}{x-2}.$$

Check Your Understanding of Section 5.6

A. Multiple Choice

1. The expression $\dfrac{\dfrac{x}{z}-\dfrac{z}{x}}{\dfrac{1}{z}+\dfrac{1}{x}}$ is equivalent to

(1) $x-z$ (2) $x+z$ (3) xz (4) $\dfrac{x-z}{xz}$

2. When resistors R_1 and R_2 are connected in a parallel electric circuit, the total resistance is $\dfrac{1}{\dfrac{1}{R_1}+\dfrac{1}{R_2}}$. This complex fraction is equivalent to

(1) R_1+R_2 (2) $\dfrac{R_1+R_2}{R_1R_2}$ (3) $\dfrac{R_1}{R_2}+\dfrac{R_2}{R_1}$ (4) $\dfrac{R_1R_2}{R_1+R_2}$

3. The expression $\dfrac{\dfrac{x-y}{y}}{y^{-1}-x^{-1}}$ is equivalent to

(1) x (2) y (3) $\dfrac{1}{y}$ (4) $-\dfrac{x}{y}$

4. The expression $\dfrac{\dfrac{3}{x^2}+\dfrac{1}{x}}{1-\dfrac{9}{x^2}}$ is equivalent to

(1) $\dfrac{1}{3}$ (2) $x-3$ (3) $\dfrac{1}{x-3}$ (4) $\dfrac{x}{3-x}$

B. *In each case, show how you arrived at your answer by clearly indicating all of the necessary steps, formula substitutions, diagrams, graphs, charts, etc.*

5–10. Write each complex fraction in simplest form.

5. $\dfrac{\dfrac{y}{3}-1}{\dfrac{y^2}{3}-3}$

6. $\dfrac{1-\dfrac{1}{w}}{\dfrac{1}{w^2}-\dfrac{1}{w}}$

7. $\dfrac{\dfrac{1}{n}-\dfrac{1}{3n^2}}{1-\dfrac{1}{9n^2}}$

8. $\dfrac{x+y^{-1}}{y+x^{-1}}$ **9.** $\dfrac{m-\dfrac{1}{m}}{m-2+\dfrac{1}{m}}$ **10.** $\dfrac{\dfrac{b}{b-3}+\dfrac{4}{b}}{1-\dfrac{1}{3-b}}$

5.7 OPERATIONS WITH RADICALS

KEY IDEAS

The kth root of x is denoted by $\sqrt[k]{x}$, where the number k is called the **index** of the radical and indicates what root of x, the **radicand**, is to be taken. Radicals with the *same* index can be multiplied (or divided) by writing the radical over the product (or quotient). Radicals with the same index *and* the same radicand can be combined by treating these radicals as like terms.

Simplifying Radicals

To simplify a *square root radical*, factor the radicand into the product of two terms such that one of the terms is the greatest perfect square factor of the radicand. For example:

- $\sqrt{75}=\sqrt{25\cdot 3}=\sqrt{25}\cdot\sqrt{3}=5\sqrt{3}$.
- $\dfrac{1}{2}\sqrt{72}=\dfrac{1}{2}\sqrt{36\cdot 2}=\dfrac{1}{2}\sqrt{36}\cdot\sqrt{2}=\dfrac{1}{2}\cdot 6\cdot\sqrt{2}=3\sqrt{2}$.
- $\sqrt{18x^3}=\sqrt{9x^2\cdot 2x}=\sqrt{9x^2}\cdot\sqrt{2x}=3x\sqrt{2x}$.

To simplify a *cube root radical*, factor the radicand into the product of two terms such that one of the terms is the greatest perfect cube factor of the radicand. For example:

$$\sqrt[3]{54x^6}=\sqrt[3]{27x^6\cdot 2}=\sqrt[3]{27x^6}\cdot\sqrt[3]{2}=3x^2\sqrt[3]{2}.$$

Some radicals can be simplified by using the properties of exponents. For example:

$$\begin{aligned}\sqrt[4]{81x^{12}y^8z^2}&=\left(81x^{12}y^8z^2\right)^{\frac{1}{4}}\\&=\left(81^{\frac{1}{4}}\right)\left(x^{12}\right)^{\frac{1}{4}}\left(y^8\right)^{\frac{1}{4}}\left(z^2\right)^{\frac{1}{4}}\\&=3x^{\frac{12}{4}}y^{\frac{8}{4}}z^{\frac{2}{4}}\\&=3x^3y^2z^{\frac{1}{2}}\quad\text{or}\quad 3x^3y^2\sqrt{z}\end{aligned}$$

Multiplying and Dividing Radicals

To *multiply* two radicals with the same index, write the product of the two radicands underneath the radical sign and, if possible, simplify. For example:

$$\sqrt{8x} \cdot \sqrt{6x^3} = \sqrt{48x^4}$$

Factor the radicand so that one of the factors is the greatest perfect square factor of $48x^4$: $= \sqrt{16x^4} \cdot \sqrt{3} = 4x^2\sqrt{3}$

To *divide* radicals, proceed in a similar way. For example:

$$\frac{\sqrt[4]{80x^{15}}}{\sqrt[4]{5x^3}} = \sqrt[4]{\frac{80}{5} \cdot \frac{x^{15}}{x^3}} = \sqrt[4]{16} \cdot \sqrt[4]{x^{12}}$$

Change $\sqrt[4]{x^{12}}$ to $\left(x^{12}\right)^{\frac{1}{4}}$: $= 2\left(x^{12}\right)^{\frac{1}{4}} = 2x^3$

Combining Radicals

To *combine* radicals with the same index, rewrite the radicals, if possible, so that they have the same radicand. For example:

- $$\begin{aligned}\sqrt{48} + \sqrt{27} &= \left(\sqrt{16} \cdot \sqrt{3}\right) + \left(\sqrt{9} \cdot \sqrt{3}\right)\\ &= 4\sqrt{3} \qquad + 3\sqrt{3}\\ &= 7\sqrt{3}\end{aligned}$$
- $$\begin{aligned}\sqrt{200y} - \sqrt{32y} &= \left(\sqrt{100} \cdot \sqrt{2y}\right) - \left(\sqrt{16} \cdot \sqrt{2y}\right)\\ &= 10\sqrt{2y} \qquad - 4\sqrt{2y}\\ &= 6\sqrt{2y}\end{aligned}$$

Example

Write the difference $4\sqrt[3]{16} - 5\sqrt[3]{54}$ as a single radical.

Solution: $\mathbf{-7\sqrt[3]{2}}$

Factor the radicands of $\sqrt[3]{16}$ and $\sqrt[3]{54}$ so that one of the factors is a perfect cube. Then simplify:

$$\begin{aligned}4\sqrt[3]{16} - 5\sqrt[3]{54} &= 4\sqrt[3]{8} \cdot \sqrt[3]{2} - 5\sqrt[3]{27} \cdot \sqrt[3]{2}\\ &= 4 \cdot 2 \cdot \sqrt[3]{2} - 5 \cdot 3 \cdot \sqrt[3]{2}\\ &= 8\sqrt[3]{2} \qquad -15\sqrt[3]{2}\\ &= -7\sqrt[3]{2}\end{aligned}$$

Multiplying Expressions Containing Radicals

Radical expressions that have the form of a binomial can be multiplied using FOIL. For example:

$$\begin{aligned}(4-2\sqrt{3})^2 &= (4-2\sqrt{3})(4-2\sqrt{3})\\ &= \overbrace{4\cdot 4}^{First}+\overbrace{4(-2\sqrt{3})}^{Outer}+\overbrace{(-2\sqrt{3})4}^{Inner}+\overbrace{(-2\sqrt{3})(-2\sqrt{3})}^{Last}\\ &= 16 \quad +[(-8\sqrt{3}) \quad +(-8\sqrt{3})] \quad +(4)(\sqrt{3})^2\\ &= 16 \qquad\qquad - \ 16\sqrt{3} \quad +12\\ &= 28-16\sqrt{3}\end{aligned}$$

Multiplying Conjugate Radical Expressions

Two radical expressions such as $2 + \sqrt{3}$ and $2 - \sqrt{3}$ are called *conjugate radical expressions*. It is important to be able to recognize conjugate radical expressions since their product is always rational. For example:

$$\begin{aligned}(2+\sqrt{3})(2-\sqrt{3}) &= 4\overbrace{-2\sqrt{3}+2\sqrt{3}}^{0}-(\sqrt{3}\cdot\sqrt{3})\\ &= 4-3\\ &= 1\end{aligned}$$

MATH FACTS

Conjugate radical expressions are two radical expressions that have the form $(A+B\sqrt{C})$ and $(A-B\sqrt{C})$, where A, B, and C are rational. The product of conjugate radical expressions is always rational.

Rationalizing Denominators

To change $\dfrac{3}{7-3\sqrt{5}}$ into an equivalent fraction with a rational denominator, multiply the numerator and the denominator of the fraction by the conjugate of the denominator. For example:

$$\frac{3}{7-3\sqrt{5}} = \frac{3}{7-3\sqrt{5}} \cdot \overbrace{\left(\frac{7+3\sqrt{5}}{7+3\sqrt{5}}\right)}^{1}$$

$$= \frac{21+9\sqrt{5}}{7^2 - 3^2 \cdot 5}$$

$$= \frac{21+9\sqrt{5}}{4}$$

Check Your Understanding of Section 5.7

A. Multiple Choice

1. The fraction $\frac{\sqrt{3}-\sqrt{8}}{\sqrt{2}}$ is equivalent to

(1) $\sqrt{3}-2$ (2) $\frac{\sqrt{6}-4}{2}$ (3) $\frac{\sqrt{3}-2}{2}$ (4) $\sqrt{3}-4$

2. The expression $\frac{2+\sqrt{3}}{2-\sqrt{3}}$ is equivalent to

(1) $11\sqrt{3}$ (2) $7-4\sqrt{3}$ (3) $7+4\sqrt{3}$ (4) $\frac{7+4\sqrt{3}}{7}$

3. Which expression is the multiplicative inverse of $1-\sqrt{3}$?

(1) $1+\sqrt{3}$ (2) $-1+\sqrt{3}$ (3) $-\frac{1}{2}$ (4) $\frac{-1-\sqrt{3}}{2}$

4. The expression $\sqrt{\frac{4}{3}} - \sqrt{\frac{3}{4}}$ is equivalent to

(1) $\frac{2-\sqrt{3}}{2\sqrt{3}}$ (2) $\sqrt{\frac{7}{12}}$ (3) $\frac{\sqrt{3}}{6}$ (4) $2\sqrt{3}$

5. The expression $\frac{\sqrt{7}+\sqrt{2}}{\sqrt{7}-\sqrt{2}}$ is equivalent to

(1) $\frac{9}{5}$ (2) -1 (3) $\frac{9+2\sqrt{14}}{5}$ (4) $\frac{11+\sqrt{2}}{14}$

B. *In each case, show how you arrived at your answer by clearly indicating all of the necessary steps, formula substitutions, diagrams, graphs, charts, etc.*

6–11. Write in simplest form.

6. $\sqrt{40y^{10}}$

7. $\sqrt[3]{64x^{12}}$

8. $\sqrt[3]{-8x^6y^5}$

9. $\left(\dfrac{25x^2}{16y^4}\right)^{-\frac{3}{2}}$

10. $\sqrt[4]{16a^{24}b^{20}c^5}$

11. $\left(\dfrac{8a^9}{27b^6}\right)^{-\frac{2}{3}}$

12–17. Combine, and express the result in simplest form.

12. $2\sqrt{27}+3\sqrt{108}$

13. $3\sqrt{63x}-2\sqrt{28x}$

14. $x\sqrt{48}-2\sqrt{75x^2}$

15. $5\sqrt[3]{2y^3}-\sqrt[3]{16y^3}$

16. $2\sqrt[3]{8y^4}-3y\sqrt[3]{y}$

17. $3\sqrt[4]{32x^8}-x\sqrt[4]{2x^4}$

18–20. Write the product in simplest form.

18. $(8-2\sqrt{5})(8+2\sqrt{5})$

19. $(1+\sqrt{3})(2-\sqrt{12})$

20. $(3\sqrt{2}-2\sqrt{3})^2$

21–23. Combine, and express the result in simplest form.

21. $\dfrac{6}{\sqrt{12}}+\sqrt{27}$

22. $\dfrac{15}{\sqrt{20}}=-\sqrt{45}$

23. $\dfrac{40}{\sqrt{8}}-\sqrt{50}$

24–26. Express each fraction as an equivalent fraction in simplest form with a rational denominator.

24. $\dfrac{\sqrt{5}}{2+\sqrt{5}}$

25. $\dfrac{3+\sqrt{2}}{3-\sqrt{2}}$

26. $\dfrac{\sqrt{5}+3\sqrt{2}}{\sqrt{5}-3\sqrt{2}}$

CHAPTER 6

FUNCTIONS AND GRAPHING CALCULATORS

6.1 UNDERSTANDING FUNCTIONS

KEY IDEAS

Functions arise whenever one quantity depends on another. If your grade on your next Math B test depends on the number of hours you study, then your grade is *a function* of the number of hours studied. In mathematics, however, the term *function* has a narrower meaning. When each possible value of *x*, called the *independent variable*, is paired with only one value of *y*, called the *dependent variable,* we say that *y is a function of x.*

An Example of a Function

Let set X consist of five teenagers and set Y consist of their possible ages:

$$X = \{\text{Alice, Barbara, Chris, Dennis, Enid}\},$$
$$Y = \{13, 14, 15, 16, 17, 18, 19\}.$$

If each teenager in set X is paired with his or her present age in set Y, the result can be written as a set of ordered pairs called a **relation**:

$$\{(\text{Alice}, 17), (\text{Barbara}, 13), (\text{Chris}, 16), (\text{Dennis}, 19), (\text{Enid}, 15)\}.$$

Since each teenager from set X is paired with exactly one age from set Y, the relation is a *function.*

Definition of Function

A **function** is a set of ordered pairs in which no two ordered pairs of the form (x, y) have the same x-value but different y-values. For example, if

$$\text{f} = \{(1, 1), (2, 3), (3, 3), (4, 5)\},$$

then f is a function. However, if

$$\text{g} = \{(1, 2), (2, 5), (3, 8), (1, 4)\},$$

then g is *not* a function since, when $x = 1$, two different y-values are possible.

Representing a Function

Suppose a plumber gets paid \$40 for traveling time to a job plus \$60 for each hour he takes to complete the job. The plumber charges for whole-number hours. The function that shows the relationship between the number of hours, x, that the plumber works and the amount of money, y, that he charges can be represented in different ways.

- **Table Representation**
  The relationship between x and y can be described by the table of values in Figure 6.1 or, equivalently, by the ordered pairs (1, 100), (2, 160), (3, 220), and so forth.
- **Graphical Representation**
  The relationship between x and y can be represented visually by the graph shown in Figure 6.2.
- **Algebraic Representation**
  The relationship between x and y can also be summarized by the equation $y = 40 + 60x$.

x hours	y dollars
1	100
2	160
3	220
...	...

Figure 6.1 A Table as a Function

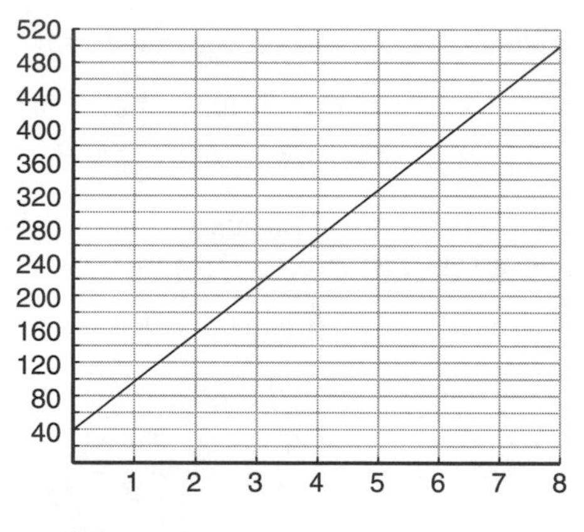

Figure 6.2 A Function as a Graph

Function Notation

Suppose function f is described by the equation $y = 40 + 60x$. The notation f(5) represents the value of y when $x = 5$:

$$y = f(5) = 40 + 60 \times 5 = 340.$$

For any function f, the notation f(x) is read as "f of x" and represents the value of y when x is replaced by the number or expression inside the parentheses.

Example 1

For the equation $2x + 3y = 12$: a. Express y as a function of x. b. Find f(−3).

Solutions: a. $\boldsymbol{y = -\frac{2}{3}x + 4}$

If $2x + 3y = 12$, then

$$\begin{aligned} 3y &= -2x + 12 \\ y &= -\frac{2}{3}x + \frac{12}{3} \\ &= -\frac{2}{3}x + 4 \end{aligned}$$

b. **6**

Since $y = f(x) = -\frac{2}{3}x + 4$

$$f(-3) = -\frac{2}{3}(-3) + 4 = 2 + 4 = 6$$

Example 2

If $f(x) = x^2 + \frac{1}{x}$, find: a. $f(-x)$ b. $f\left(\frac{1}{a}\right)$

Solutions: a. $\boldsymbol{x^2 - \frac{1}{x}}$

- Write the given function: $f(x) = x^2 + \frac{1}{x}$
- Replace x with $-x$: $f(-x) = (-x)^2 + \frac{1}{(-x)}$
- Let $(-x)^2 = (-x)(-x) = +x^2$: $= x^2 - \frac{1}{x}$

b. $\boldsymbol{\frac{1}{a^2} + a}$

- Write the given function: $f(x) = x^2 + \frac{1}{x}$
- Replace x with $\frac{1}{a}$: $= \left(\frac{1}{a}\right)^2 + \frac{1}{1/a}$
- Since the reciprocal of $\frac{1}{a}$ is a: $= \frac{1}{a^2} + a$

Example 3

If $g(x) = 3x$, find in terms of x: a. $2g(x)$ b. $[g(x)]^2$ c. $g(x) + 2$

Solutions: a. $\mathbf{6x}$

$$2g(x) = 2[3x] = 6x$$

b. $\mathbf{9x^2}$

$$[g(x)]^2 = [3x]^2 = (3x)(3x) = 9x^2$$

c. $\mathbf{3x + 2}$

$$g(x) + 2 = 3x + 2$$

Domain and Range

The possible values for x and y for a specific function may be restricted in some way. In the example in which $y = 40 + 60x$ represents the amount of money a plumber charges for working x hours, suppose the plumber has a minimum labor charge of 1 hour and a maximum labor charge of 5 hours. Then the set of possible x-values, called the *domain*, is restricted to $\{1, 2, 3, 4, 5\}$. The corresponding set of y-values, called the *range*, is $\{100, 160, 220, 280, 340\}$.

The domain of a function may need to be restricted to ensure that the function always produces a real number. For example:

- The domain of $(x) = \frac{1}{x}$ excludes 0 since division by 0 is not defined.
- The domain of $f(x) = \sqrt{x}$ excludes negative values of x since $f(x)$ is a real number only when $x \geq 0$.

MATH FACTS

If y "is a function of x," then $y = f(x)$, where f represents the rule for determining y from x. Since the value of y depends on the choice for x, y is called the **dependent variable** and x is the **independent variable**.

- The **domain** of $y = f(x)$ is the set of all possible values of x.
- The **range** of $y = f(x)$ is the set of all values that y takes as x runs through all the values in the domain.
- The domain and range of a function are, unless otherwise indicated, the largest possible sets of real numbers.

Example 4

What are a. the domain and b. the range of $y = \sqrt{x-1}$?

Solutions: a. $\mathbf{x \geq 1}$

- Any value of x less than 1 makes y equal to the square root of a negative number. Hence, $x \geq 1$ so that y is a real number.

b. $\mathbf{y \geq 0}$

- For any value of x in the domain, y evaluates to either 0 or a positive number. Hence, $y \geq 0$.

Vertical-Line Test

You can tell whether an equation represents a function by looking at its graph. A graph describes a function if any vertical line intersects it in *at most* one point. The graph in Figure 6.3(a) represents a function since it passes the vertical-line test, while the graph in Figure 6.3(b) is *not* a function since it fails this test.

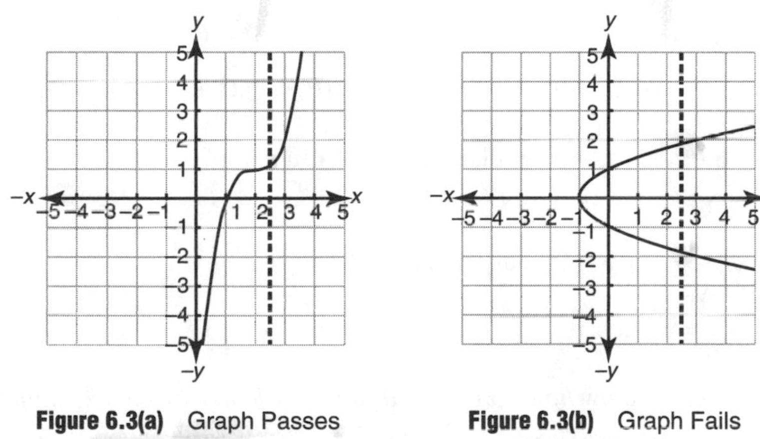

Figure 6.3(a) Graph Passes Vertical-Line Test

Figure 6.3(b) Graph Fails Vertical-Line Test

Check Your Understanding of Section 6.1

A. Multiple Choice

1. If $k(h) = (h + 1)^2$, then $k(x - 2) =$
 (1) $x^2 - x$ (2) $x^2 - 2x$ (3) $x^2 - 2x + 1$ (4) $x^2 + 2x - 1$

2. If $f(x) = x^2 - 3$, then $f(a - b)$ is equivalent to
(1) $a^2 - b^2 - 3$
(2) $a^2 - 2ab - b^2 - 3$
(3) $a^2 - 2ab + b^2 - 3$
(4) $a^2 + b^2 - 3$

3. If $g(t) = t^2 + 1$, then $g(x + 2) =$
(1) $x^2 + 4x + 5$ (2) $x^2 + 5x + 4$ (3) $x^2 + 6x + 9$ (4) $x^2 + 3$

4. If $f(x) = x^3 + x$, which statement is true?
(1) $f(-x) = -f(x)$
(2) $f(-x) = f(x)$
(3) $x - f(x) = f(-x)$
(4) $f(x^2) = [f(x)]^2$

5. If the domain of $g(x) = 1 - 5x$ is $-3 \leq x \leq 2$, what is the smallest value in the range of $g(x)$?
(1) -17 (2) 9 (3) 17 (4) -9

6. If the domain of $f(x) = 9 - x^2$ is $-3 \leq x \leq 4$, what is the largest value in the range of $f(x)$?
(1) 0 (2) 9 (3) 3 (4) 25

7. If $g(x) = \{(-2,4), (-1,1), (0,6), (1,-1), (2,8)\}$, which statement *must* be true?
(1) $g(1) = g(-1)$
(2) $g(2) + g(-2) = 0$
(3) $g(-2) = \frac{1}{2}f(2)$
(4) $g(0) = g(1) + g(-1)$

8. Which of the following relations is *not* a function?
(1) $x = |y|$ (2) $y = \sqrt{x}$ (3) $x = \sqrt{y}$ (4) $2x + 3y = 6$

9. For which of the following functions is $f(-x) = f(x)$?
(1) $f(x) = x^3 - x^2$
(2) $f(x) = x^3 + 1$
(3) $f(x) = x - x^3$
(4) $f(x) = x^2 + 3$

B. *In each case, show how you arrived at your answer by clearly indicating all of the necessary steps, formula substitutions, diagrams, graphs, charts, etc.*

10 and 11. A certain function has the form $f(x) = ax + b$, where $a,b \neq 0$.

10. If $f(2) = 3$ and $f(-2) = 5$, what are the values of a and b?

11. If $f(5) = 1$, $f(-3) = 25$, and $f(c) = 0$, what is the value of c?

12. Given $f(x) = 3x - 1$ and $g(x) = x^2$. If $f(7) - g(2) = g(a)$, find a.

13–15. For each function, express $\frac{f(x+h)-f(x)}{h}$ in simplest form where $h \neq 0$.

13. $f(x) = x^2$ **14.** $f(x) = \frac{1}{x}$ **15.** $f(x) = 5$

6.2 USING A GRAPHING CALCULATOR

KEY IDEAS

If you know the equation of a function, you can use a graphing calculator to display the graph of that function.

The Standard Viewing Window

The viewing window of a graphing calculator shows only a small part of the coordinate plane. In a **standard window** the positive and negative coordinate axes each have 10 tic marks, as shown in Figure 6.4. The current values of the window variables, Xmin, Xmax, Ymin, and Ymax determine the number of tic marks on the coordinate axes. To display the current window variables, press [WINDOW]. Figure 6.5 shows the window variables for a standard window.

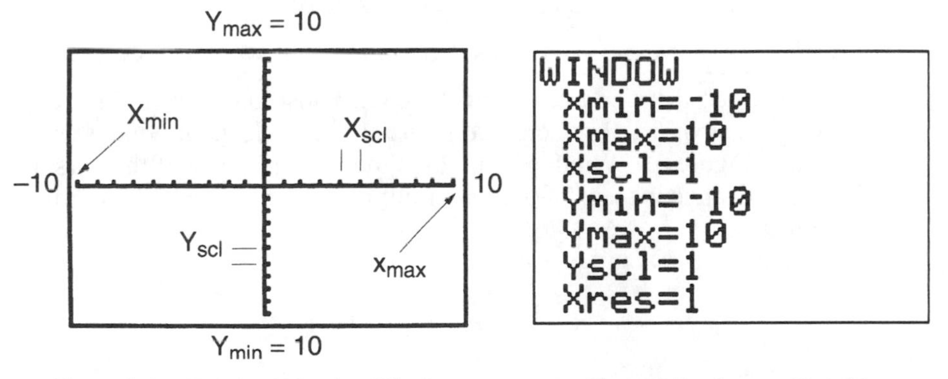

Figure 6.4 Standard Viewing Window

Figure 6.5 Screen Variables

Changing the Window Variables

To change the value of any of the window variables, use one of the four blue cursor arrow keys to move the blinking cursor to the value of the window variable that you want to change. Then overwrite the old value with the new one.

- An arithmetic expression can be entered as the value of a window variable. For example, to double the value of Xmin = –10, you can simply enter ×2 after –10. Pressing ENTER, or moving to another line, sets Xmin = –20.
- The distance between consecutive tic marks on an axis can be set to a number other than 1 by changing the values of Xscl (X scale) or Yscl (Y scale). For example, setting Xscl = 5 scales the *x*-axis so that tick marks are 5 units, instead of 1 unit, apart.

The ZOOM Menu

The ZOOM menu includes options that allow you to quickly resize the viewing window to preset values. When you enter the number of a menu option, the graph is immediately plotted in the resized window.

- Pressing ZOOM 6 creates a *standard window*, in which each coordinate axis has 10 tic marks on either side of 0.
- Pressing ZOOM 4 creates a basic *decimal window* (see Figure 6.6), in which the *x*-coordinate of the cursor position changes in user-friendly steps of 0.1. The screen dimensions for a decimal window are shown in Figure 6.7. Windows in which the *x*-coordinates of the cursor locations change in steps of 0.1, or a multiple of 0.1, are called **friendly windows**.

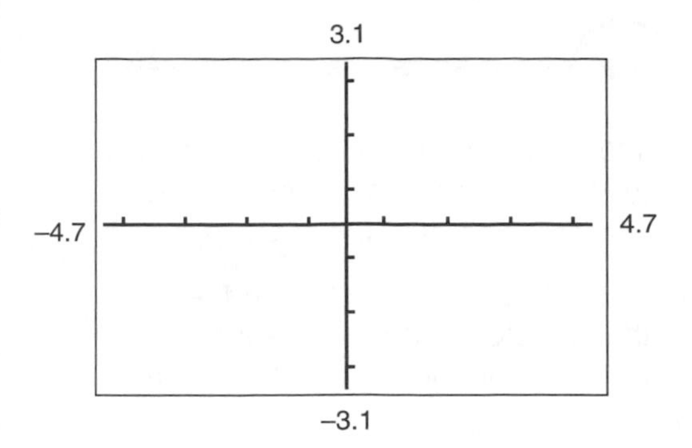

Figure 6.6 Basic Decimal Window

```
WINDOW
 Xmin=-4.7
 Xmax=4.7
 Xscl=1
 Ymin=-3.1
 Ymax=3.1
 Yscl=1
 Xres=1
```

Figure 6.7 Screen Dimensions of a Basic Decimal Window

Decimal Windows

In a **decimal window** the ratio of (Xmax – Xmin) to (Ymax – Ymin) is the same as the ratio of the pixel width to the pixel height of the viewing rectangle. For many graphing calculators, including the TI-83, this ratio is approximately 3 to 2. Graphs viewed in a decimal window maintain their true geometric proportions. In a decimal window, a circle looks like a circle rather than an oval.

- You can create other decimal windows by using positive integer multiples of the settings for [Xmin, Xmax] and [Ymin, Ymax] in Figure 6.7. Adjusting the screen variables in Figure 6.7 so that Xmin = −4.7 × 2 and Xmax = 4.7 × 2 results in the *x*-coordinate of the cursor changing in user-friendly steps of 0.2.
- Setting Xmax and Xmin so that their difference is a multiple of 9.4 will produce a friendly *x*-coordinate readout. For example, if Xmin = 0 and Xmax = 9.4, then the *x*-coordinate of the cursor will change in user-friendly steps of 0.1.

Finding an Appropriate Viewing Window

The window size may need to be adjusted so that the basic shape of a graph and all of its important features can be seen. For example, when graphing a slanted line, you may need to adjust the size of the viewing window so that the *x*-intercept falls within the interval from Xmin to Xmax and the *y*-intercept falls within the interval from Ymin to Ymax.

TIP

To find an appropriate viewing window, some experimenting may be required. Start with a standard viewing window when you suspect that the key part of a graph will fall within the interval $-10 \leq x \leq 10$.

- If the graph doesn't fit within a standard window, change the values of the screen variables in the WINDOW editor as needed.
- If the graph fits easily within a standard window, try a basic decimal window or a multiple of it.

Indicating Window Size

When a graph drawn by a calculator is recreated on paper, it is helpful to provide the [Xmin, Xmax] and [Ymin, Ymax] values of the display window in which the graph is being viewed. For example, the notation $[-4.7, +4.7] \times [-6.2, +6.2]$ under a graph means that the graph is being viewed in a rectangular window that is sized so that $-4.7 \leq x \leq 4.7$ and $-6.2 \leq y \leq 6.2$.

Graphing a Linear Function

A **linear function** is a function whose graph is a line. Any equation that can be put into the form $y = ax + b$, where a and b stand for numbers, except $a \neq 0$, is a linear function. To graph the linear function $y - 2x = 3$, first solve for y which gives $y = 2x + 3$. Then follow these steps:

Step 1. Enter the equation.

Press [Y=] to open the Y = editor. Set Y_1 equal to $2x + 3$ by pressing

[2] [x, T, θ, n] [+] [3].

Step 2. Select a viewing window.

Press [ZOOM] [6] to draw the graph in a standard window, as shown in Figure 6.8.

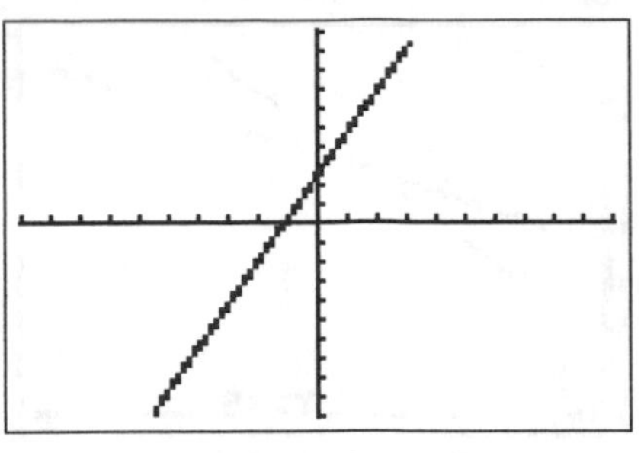

[−10,+10] × [−10,10]

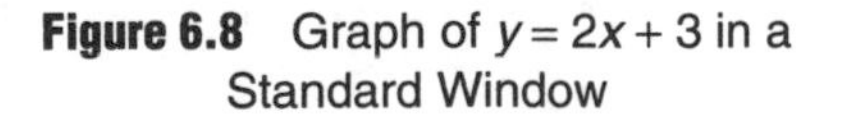

Figure 6.8 Graph of $y = 2x + 3$ in a Standard Window

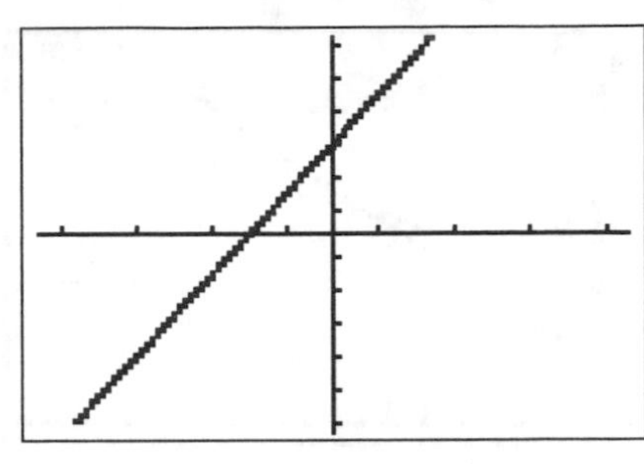

[−4.7,+4.7] × [−6.2,+6.2]

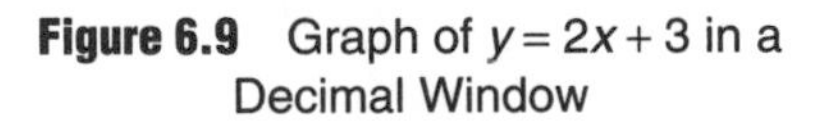

Figure 6.9 Graph of $y = 2x + 3$ in a Decimal Window

Step 3. Show the graph.

Press ZOOM 4 to view the graph in a decimal window. You should verify that the *y*-intercept of the graph now appears to be cut off. Multiplying the decimal window values for Ymin and Ymax by 2 produces the graph in Figure 6.9.

MATH FACTS

To graph an equation:

- Solve for *y*, if necessary, so that *y* is expressed as a function of *x*.
- Enter the equation. If the coefficient of *x* is negative, as in $y = -2x$, press (−) rather than −.
- Adjust the size of the viewing window, if necessary, so that the graph fits.

Graphing a System of Linear Functions

Suppose you are signing up for a cable television service plan. Plan *A* costs \$11 per month plus \$7 for each premium channel. Plan *B* costs \$27 per month plus \$3 for each premium channel. If *x* represents the number of premium channels ordered, the cost of plan *A* can be represented by the function $f(x) = 11 + 7x$ and the cost of plan *B* by the function $g(x) = 27 + 3x$. For what number of premium channels will the two plans have the same cost?

- Recognize that the two plans will have the same cost when their graphs intersect.
- Graph $Y_1 = 11 + 7x$ and $Y_2 = 27 + 3x$ in a friendly window such as the one in Figure 6.10, where Xmin = 0, Xmax = 9.4, Ymin = 0, and Ymax = 60.

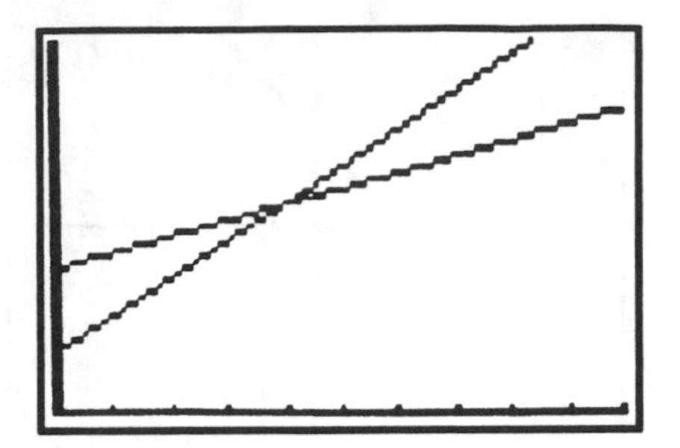

Figure 6.10 Graphs of Plan *A* and Plan *B*

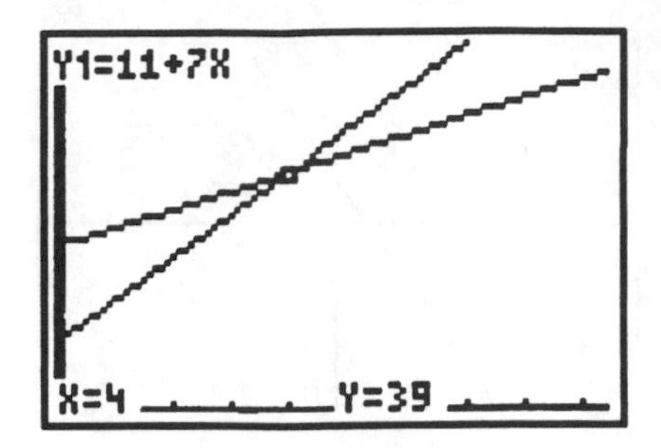

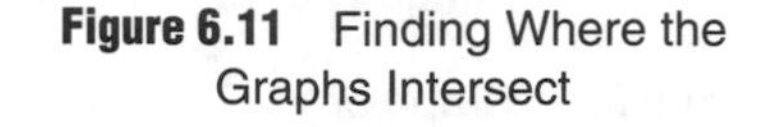

Figure 6.11 Finding Where the Graphs Intersect

- Use TRACE to find the coordinates of the point of intersection of the two graphs.

Step 1. Press [TRACE]. The TRACE cursor appears on the graph of Y_1. As you use the cursor keys to move the TRACE cursor along the line $Y_1 = 11 + 7x$, the x- and y-coordinates of the current position of the TRACE cursor appear at the bottom of the screen.

Step 2. Move the TRACE cursor on the point of intersection of the two lines. The readout is $X = 4$ and $Y = 39$, as shown in Figure 6.11.

step 3. Verify that the coordinates of the point of intersection are correct by using the up and down arrow keys to "toggle" or switch between the two lines at the point of intersection. When you are toggling between the two lines, the equation in the upper left corner of the viewing window alternates between $Y_1 = 11 + 7X$ and $Y_2 = 27 + 3X$, while the trace coordinates stay fixed at (4, 39). This confirms that (4, 39) is a point on both curves.

Thus, the two plans have the same cost, \$39, when four premium channels are ordered.

TIP

- The TRACE feature may not display exact x- or y-coordinates. When this happens, it may help to work in a friendly window.
- To solve the same problem algebraically, set $f(x) = g(x)$. Then $11 + 7x = 27 + 3x$ or, equivalently, $4x = 16$ so $x = 4$.

Creating a Table of Values

The cable television problem can also be solved by using the table-building feature of your graphing calculator to compare the costs of the two plans.

Step 1. Set $Y_1 = 11 + 7x$ and $Y_2 = 27 + 3x$.

Step 2. Press [2nd] followed by [WINDOW] to get into TBLSET mode. If necessary, set TblStart = 1 so that the table starts with $x = 1$. Also, set ΔTb1 = 1 so that x increases in steps of 1.

Step 3. Press [2nd] followed by [GRAPH] to call up the TABLE feature of the calculator, as shown in Figure 6.12. According to the table, when $x = 4$ premium channels, $Y_1 = Y_2 = \$39$.

X	Y_1	Y_2
1	18	30
2	25	33
3	32	36
4	39	39
5	46	42
6	53	45
7	60	48

X=4

Figure 6.12 Table Representation of Plans

Show All Work

If you draw a graph with the help of a graphing calculator, you are expected to include in your handwritten solution all of the following:

- The dimensions of the viewing window and a neat sketch of what appears in it. Label the ends of the axes as shown in Figure 6.6, or use the form [Xmin, Xmax] × [Ymin, Ymax], as shown in Figure 6.9. Scale factors different from 1 should also be noted.
- Clearly labeled x- and y-intercepts and points of intersection, if needed for the solution.

2nd Function Keys and Bracket Notation

Most keys on your graphing calculator have two functions. The second function is printed on top of the key. To activate the second function, you must first press [2nd] and then press the key that has the desired function printed above it. For example, notice that [ON] has the label OFF printed above it. To turn the calculator off, press in succession [2nd] [OFF], where the brackets refer to the key that has OFF *above* it.

Check Your Understanding of Section 6.2

In each case, show how you arrived at your answer by clearly indicating all of the necessary steps, formula substitutions, diagrams, graphs, charts, etc.

1. Solve the equation $1.5x - 5.7 = 4.8 - 2x$ by graphing two lines. Confirm your answer algebraically.

2 and 3. Solve each system of equations graphically. Confirm your answer algebraically.

2. $y = 3x - 4$
$x + 2y = 6$

3. $y + 1 = 4x$
$3x + 2y = 20$

4. The formula to convert from degrees Celsius to degrees Fahrenheit is $F = 1.8C + 32$.
   a. Determine graphically the temperature at which the readings on the Celsius and Fahrenheit scales are the same.
   b. Confirm your answer to part a by creating a table in which C changes in steps of 5. Use the up/down arrow keys to view the hidden parts of the table.

5. Harriet is interested in renting a truck. The Easy Rent-A-Truck company charges \$25 per day plus \$1.75 per mile. The Fast Rent-A-Truck company charges \$49 per day plus \$1 per mile.
   a. For a 1-day rental in which x miles are driven, write:
      (1) a function f that describes the charges for the Easy Rent-A-Truck company,
      (2) a function g that describes the charges for the Fast Rent-A-Truck company.
   b. Determine in two different ways the number of rental miles for which the two companies charge the same amount of money.

6. Mr. Day and Ms. Knight gave the same test consisting of 20 questions, but marked their test papers using different grading systmes. Mr. Day awarded 5 points for each correct answer and then added 10 points to the total. Ms. Knight gave 6 points for each correct answer and subtracted 1.5 points for each incorrect answer.

a. Construct a linear function for each teacher that gives a student's grade, y, when the number of correct answers, x, is known. Represent each function algebraically, graphically, and by a table.
b. Two students in different classes answered the same number of questions correctly and received the same test grade. What test grade did each student receive?

6.3 GRAPHING QUADRATIC FUNCTIONS

KEY IDEAS

The graph of a quadratic equation in two variables is a U-shaped curve called a **parabola**. The parabola in the accompanying figure has a vertical **axis of symmetry** whose equation is

$$x = -\frac{b}{2a}.$$

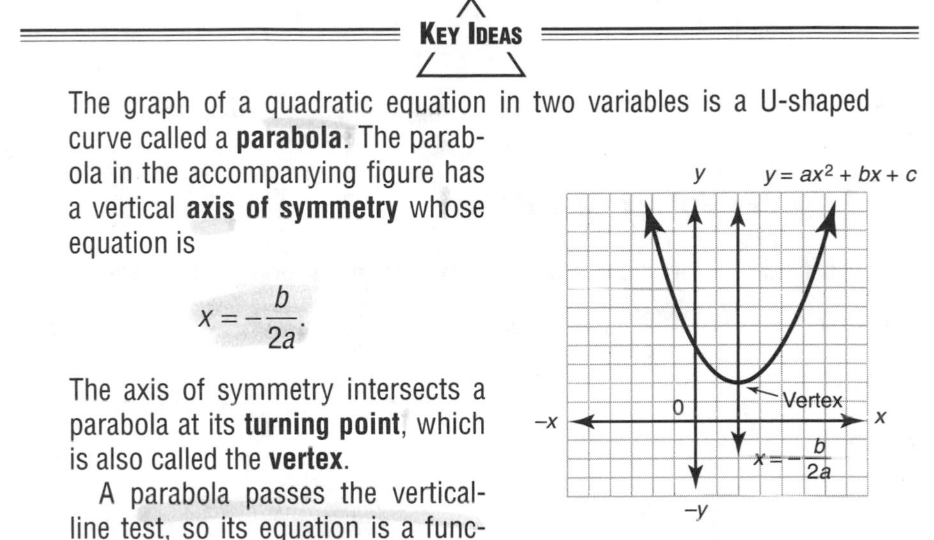

The axis of symmetry intersects a parabola at its **turning point**, which is also called the **vertex**.

A parabola passes the vertical-line test, so its equation is a function. Hence, a **quadratic function** is an equation that has the form $y = ax^2 + bx + c$ $(a \neq 0)$.

Maximum or Minimum Point of a Parabola

The sign of the coefficient of the x^2 – term of $y = ax^2 + bx + c$ determines whether the vertex of a parabola will be a minimum or a maximum point on the curve:

- When $a > 0$, the vertex is the *lowest* point on the graph, as shown in Figure 6.13.
- When $a < 0$, the vertex is the *highest* point on the graph, as shown in Figure 6.14.

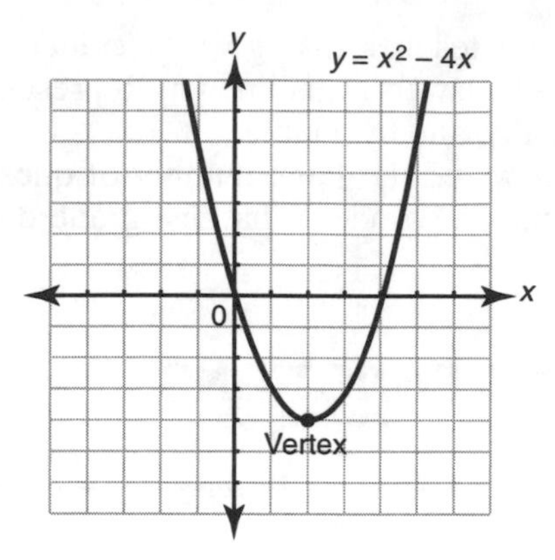

Figure 6.13 $a > 0$: Vertex Is a Minimum

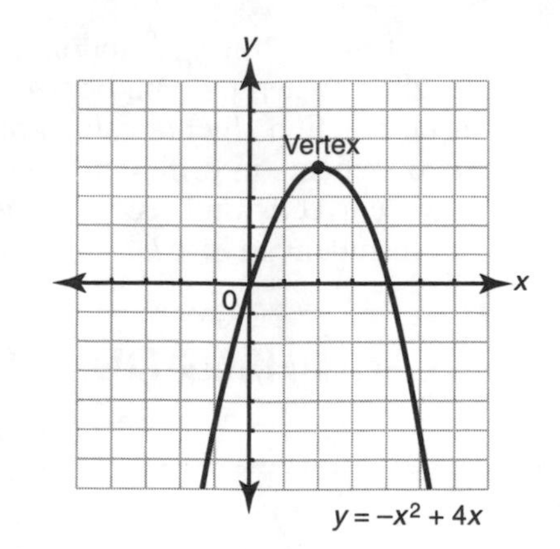

Figure 6.14 $a < 0$: Vertex Is a Maximum

Finding the Vertex of a Parabola

You can find the coordinates of the vertex (turning point) of a parabola by using the formula $x = -\dfrac{b}{2a}$ or by using the minimum or maximum feature in the CALC menu of your calculator.

Example

Find the vertex of $y = 2x^2 - 8x + 7$.

Solution: **(2, −1)**

Method 1. Find the vertex algebraically.

- If $y = 2x^2 - 8x + 7$, then $a = 2$ and $b = -8$, so

$$x = \frac{-b}{2a} = \frac{-(-8)}{2(2)} = \frac{8}{4} = 2.$$

- Since the x-coordinate of the vertex is 2, the y-coordinate of the vertex is

$$y = 2(2)^2 - 8(2) + 7 = -1.$$

Method 2. Find the vertex graphically.

- Set $Y_1 = 2X^2 - 8X + 7$ by pressing:

- Adjust the screen dimensions, if necessary, to view the graph in a friendly window.

- Open the CALC menu by pressing 2nd [CALC]. Since the vertex of $y = 2x^2 - 8x + 7$ is the lowest point on the parabola, select 3:**minimum**. (If the vertex is the highest point on the parabola, select 4:**maximum**.)
- Move the cursor slightly to the left of the vertex and press ENTER. Then, move the cursor slightly to the right of the vertex and press ENTER two times.
- Read the coordinates of the vertex at the bottom of the window, as shown in the accompanying figure. With your calculator, you may not get an exact value for the x-coordinate.

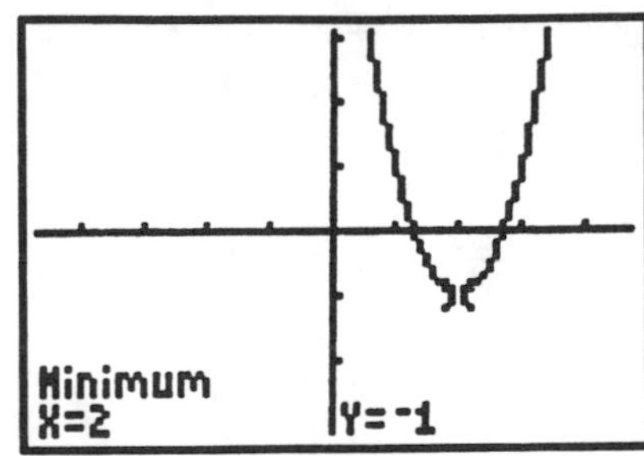

Graphing Using a Table of Values

If you need to draw a parabola on graph paper, plot three points on either side of the vertex. Then connect these points with a smooth curve. To find the coordinates of these points, use the built-in table feature of your graphing calculator. For example, if the equation $y = 2x^2 - 8x + 7$ has already been stored in your calculator, work as follows:

- Press 2nd [TBLSET]. Change the **TblStart** value to −1 since −1 is 3 units less than the x-coordinate of the vertex. If necessary, set ΔTbl = 1 so that x increases in steps of 1 unit.
- Press 2nd [TABLE] to create the table in Figure 6.15. If you need to look at table entries that are not currently in view, use a cursor key to scroll up or down.

X	Y1	
⁻1	17	
0	7	
1	1	
2	⁻1	
3	1	
4	7	
5	17	

X=⁻1

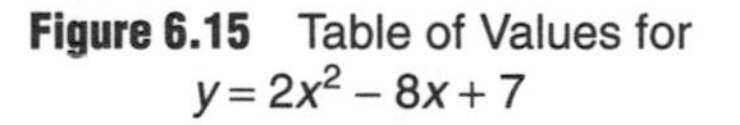

Figure 6.15 Table of Values for $y = 2x^2 - 8x + 7$

It's easy to tell from the table that the coordinates of the vertex are (2, −1) because corresponding points above and below this point have the same y-coordinate.

Check Your Understanding of Section 6.3

A. Multiple Choice

1. Which is an equation of the parabola shown in the accompanying figure?

(1) $y = -x^2 + 2x + 3$ (3) $y = x^2 + 2x + 3$
(2) $y = -x^2 - 2x + 3$ (4) $y = x^2 - 2x + 3$

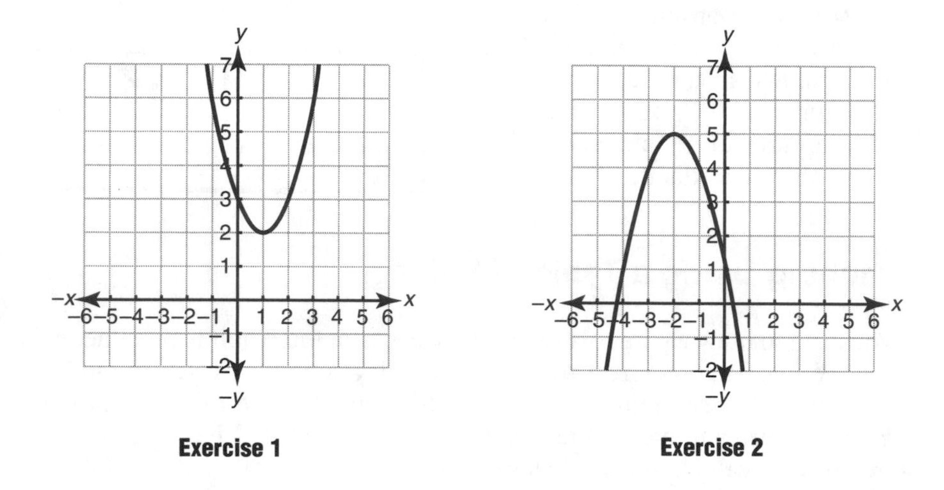

Exercise 1 **Exercise 2**

2. Which is an equation of the parabola shown in the accompanying figure?

(1) $y = -x^2 - 4x + 1$ (3) $y = x^2 - 4x + 1$
(2) $y = -x^2 + 4x + 1$ (4) $y = x^2 + 4x + 1$

B. *In each case, show how you arrived at your answer by clearly indicating all of the necessary steps, formula substitutions, diagrams, graphs, charts, etc.*

3–5. a. Graph each parabola. b. From the graph determine the coordinates of the vertex and an equation of the axis of symmetry. Confirm the equation of the axis of symmetry by obtaining it algebraically using the formula $x = -\dfrac{b}{2a}$.

3. $y = x^2 - 2x - 3$ **4.** $y = -x^2 + 6x - 7$ **5.** $y = x^2 + x + 4$

6. a. Create a table of values for $y = -x^2 - 2x + 8$ that includes three points on either side of the vertex.
b. Use the table of values created in part a to sketch the parabola on graph paper.

6.4 SOLVING QUADRATIC EQUATIONS

KEY IDEAS

If $ax^2 + bx + c$ is factorable, the quadratic equation $ax^2 + bx + c = 0$ (where $a \neq 0$) can be solved algebraically by breaking it down into two linear equations. If a quadratic equation has real roots but is not factorable, then the roots are irrational. These irrational roots can be approximated by examining the x-intercepts of the graph of the quadratic equation.

Solving Quadratic Equations by Factoring

Before solving a quadratic equation by factoring, you may need to rewrite it so that all of the nonzero terms are on the same side of the equation with 0 on the other side. To solve $x^2 + 4x = 5$, for example, first rewrite the equation as $x^2 + 4x - 5 = 0$. After factoring, you have $(x + 5)(x - 1) = 0$. By the zero product rule, if the product of two real numbers is 0, then at least one of these two numbers must be equal to 0. Hence, $x + 5 = 0$ or $x - 1 = 0$, so $x = -5$ or $x = 1$. The check is left for you.

Example 1

Find the solution set: $28 - x^2 = 3x$.

Solution: **{–7,4}**

If $28 - x^2 = 3x$, then $-x^2 - 3x + 28 = 0$. Make the coefficient of the x^2-term positive by multiplying each member of the equation by -1. This is equivalent to changing the sign of each term to its opposite: $x^2 + 3x - 28 = 0$. Then $(x + 7)(x - 4) = 0$, so $x + 7 = 0$ or $x - 4 = 0$. Thus, $x = -7$ or $x = 4$.

Example 2

For what values of x is the fraction $\dfrac{x^2 - 1}{x^2 - x - 12}$ undefined?

Solution: $\mathbf{x = -3, x = 4}$

A fraction is undefined for any value of a variable that makes the denominator evaluate to 0. To find the values of x for which the fraction $\dfrac{x^2 - 1}{x^2 - x - 12}$ is undefined, set the denominator equal to 0 and solve for x. If $x^2 - x - 12 = 0$, then $(x + 3)(x - 4) = 0$, so $x = -3$ or $x = 4$.

Solving Quadratic Equations by Graphing

The real roots of quadratic equations, including those that cannot be factored, can be obtained by looking at the points where their graphs cross the x-axis. In Figure 6.16, the graph of $y = x^2 + 2x - 3$ intersects the x-axis at $(-3, 0)$ and $(1, 0)$. Hence, $x = -3$ and $x = 1$ are the values of x that make $0 = x^2 + 2x - 3$.

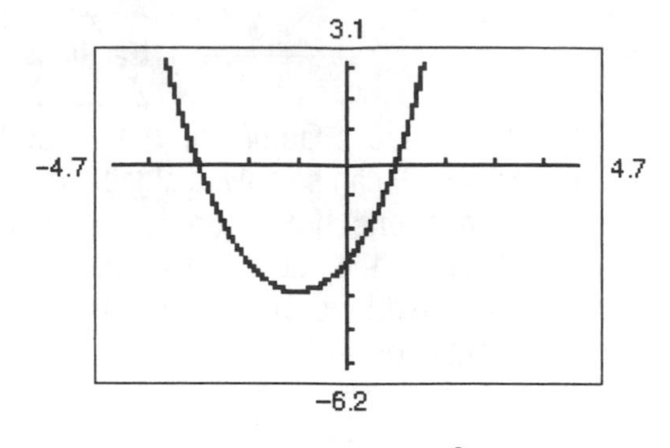

Figure 6.16 Graph of $y = x^2 + 2x - 3$

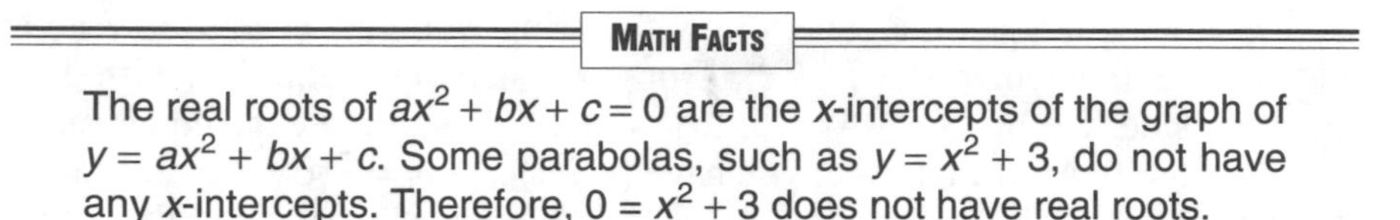

MATH FACTS

The real roots of $ax^2 + bx + c = 0$ are the x-intercepts of the graph of $y = ax^2 + bx + c$. Some parabolas, such as $y = x^2 + 3$, do not have any x-intercepts. Therefore, $0 = x^2 + 3$ does not have real roots.

Example 3

Solve $x^2 - 3x - 2 = 0$ graphically, and approximate the roots to the *nearest hundredth*.

Solution: **–0.56, 3.56**

Graph $Y_1 = x^2 - 3x - 2$ in a friendly window such as $[-4.7, 4.7] \times [-6.2, 6.2]$, as shown in the accompanying diagrams. Since $x^2 - 3x - 2$ is not factorable, the x-intercepts of the parabola represent irrational roots of $x^2 - 3x - 2 = 0$. From the graph at the left we know that one root is between 0 and -1, and the other root is between 3 and 4.

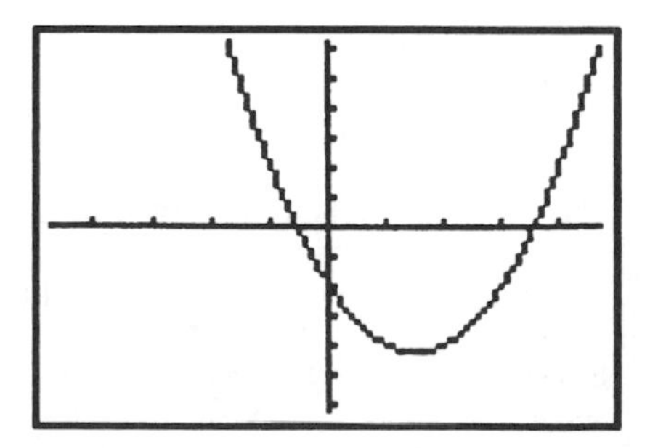

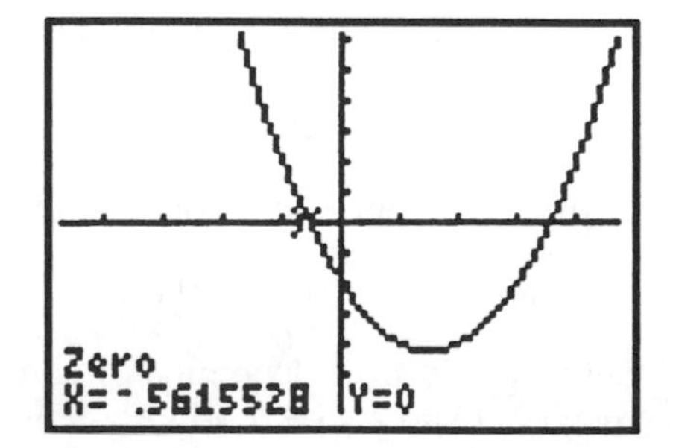

To approximate the negative root, select **2: zero** from the CALC menu.

- Move the cursor to a point on the graph that is slightly to the left of the negative x-intercept. Press ENTER .
- Move the cursor to a point on the graph that is slightly to the right of the negative x-intercept. Press ENTER two times.
- Read from the bottom left corner of the viewing window that the x-intercept is $-.5615528$, as shown in the graph at the right.

By following the same procedure and using the other x-intercept, you can verify that the positive x-intercept is 3.5615528. Hence, the roots of $x^2 - 3x - 2 = 0$, to the *nearest hundredth*, are $-.56$ and 3.56.

Solving a Quadratic Equation with No Middle Term

To solve a quadratic equation such as $2x^2 - 50 = 0$, in which the x-term is missing, solve for the x^2-term. Then take the square root of both sides of the equation:

$$2x^2 - 50 = 0$$

$$2x^2 = 50$$

$$x^2 = \frac{50}{2} = 25$$

Since the solution set of $x^2 = 25$ includes the two square roots of 25,

$$x = +\sqrt{25} = +5 \quad \text{or} \quad x = -\sqrt{25} = -5.$$

Sometimes the notation $x = \pm\sqrt{25}$ is used to abbreviate the statement $x = +\sqrt{25}$ or $x = -\sqrt{25}$.

Solving a Linear-Quadratic System Graphically

A linear-quadratic system of equations can be solved algebraically or graphically by finding the points of intersection of a line and a parabola using your calculator. To solve the system $y = -x^2 + 4x - 3$ and $x + y = 1$ using a calculator, graph the two equations in a friendly window such as $[-4.7, 4.7] \times [-6.2, 6.2]$. Then determine the coordinates of the points of intersection by following these steps:

Step 1. Select **5: intersect** from the CALC menu.
Step 2. Use the cursor keys to move the cursor so that it is close to the first point of intersection of the two graphs.

Step 3. Press ENTER three times. The coordinates of the point of intersection, (1, 0), will appear at the bottom of the screen, as shown in Figure 6.17(a).

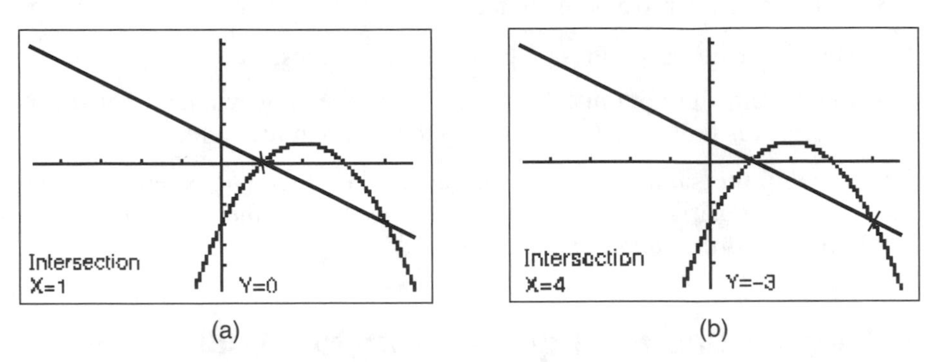

Figure 6.17 Using the Intersect Function in the CALC Menu to Find the Points of Intersection of $y = -x^2 + 4x - 3$ and $x + y = 1$

Step 4. Repeat steps 1–3 to find that the coordinates of the other point of intersection are (4, –3), as shown in Figure 6.17(b).

You can confirm your answer algebraically by solving the linear equation for y and then replacing y with that expression in the quadratic equation. Since $x + y = 1$, then $y = 1 - x$. Therefore, $y = 1 - x = -x^2 + 4x - 3$, which, after simplifying, becomes $x^2 - 5x + 4 = 0$. Thus, $(x - 1)(x - 4) = 0$, so $x = 1$ or $x = 4$.

- If $x = 1$, then $1 + y = 1$, so $y = 0$. Hence, (1, 0) is a solution.
- If $x = 4$, then $4 + y = 1$, so $y = -3$. Hence, (4, –3) is a solution.

TIP

The x-coordinates of the points of intersection of the graphs in Figure 6.17 represent the roots of the equation

$$\underbrace{-x^2 + 4x - 3}_{Y1} = \underbrace{1 - x}_{Y2}$$

To find the real roots of *any* equation that has the form f(x) = g(x), graph $Y_1 = f(x)$ and $Y_2 = g(x)$ in an appropriate viewing window. Then use the **intersect** option from the CALCULATE menu to find the x-coordinate(s) of the point(s) of intersection of the two graphs.

Example 4

Solve $28 - x^2 = 3x$ graphically.

Solution: **−7, 4**

Think of $28 - x^2 = 3x$ as a linear-quadratic system: $Y_1 = 28 - x^2$ and $Y_2 = 3x$. You should verify that the graphs of Y_1 and Y_2 intersect at $x = -7$ and $x = 4$. These roots were obtained algebraically in Example 1.

Check Your Understanding of Section 6.4

A. Multiple Choice

1. If $x^2 + 11x + 30 = 0$, which is the *smaller* of the two roots?
(1) 5 (2) 6 (3) −5 (4) −6

2. If one root of the equation $x^2 - x + q = 0$ is −2, what is the other root?
(1) 4 (2) 2 (3) −6 (4) −4

3. If $f(x) = \frac{1}{2}x$ and $g(x) = x^2$, for what value or values of x does $g(x) = 6f(x)$?
(1) 0, 3 (2) −3 (3) 3 (4) 0, −3

4. For what value(s) of x is the fraction $\frac{x+4}{x^2 - 2x - 3}$ *not* defined?
(1) 1, −3 (2) −1, 3 (3) −3, −1 (4) −4

5. When the graphs of the equations $y = x^2 - 5x + 6$ and $x + y = 6$ are drawn on the same set of axes, at which point do the graphs intersect?
(1) (−1, 13) (2) (5, 1) (3) (4, 2) (4) (2, 4)

***B.** In each case, show how you arrived at your answer by clearly indicating all of the necessary steps, formula substitutions, diagrams, graphs, charts, etc.*

6 and 7. Solve for the variable.

6. $3y^2 - 18 = 9 - y^2$

7. $\frac{9}{4} - 2y^2 = \frac{3}{4} + y^2$

8–13. Find the solution set and check.

8. $y^2 + 3y + 2 = 0$

9. $3x = \frac{x^2}{2}$

10. $3p^2 + 14p = 5$

11. $5r + 3 = 2r^2$ **12.** $9x^2 - 12x + 4 = 0$ **13.** $6t^2 = 7t + 3$

14–16. Solve and check.

14. $\frac{x+5}{x+1} = \frac{x-1}{4}$ **15.** $\frac{x-3}{x-2} = \frac{x+3}{2x}$ **16.** $\frac{x-2}{x} = \frac{x+4}{3x}$

17. a. Express the sum of $\frac{x}{x+1}$ and $\frac{3}{1-x}$ as a single fraction in lowest terms.
b. Find the values of x for which the fraction obtained in part a will be equal to $\frac{9}{4}$.

18. A ball is thrown into the air so that its height, h, in feet after t seconds is given by the equation $h = 144t - 16t^2$.
a. Find the number of seconds that the ball is in the air when it reaches a height of 180 feet.
b. After how many seconds does the ball hit the ground?

19. The manufacturing of a metal box starts with a rectangular sheet of metal that is three times as long as it is wide. Four equal squares are cut from the four corners, and the flap at each corner is turned up to form a box 2 inches high that is open at the top. What should be the width of the rectangular sheet of metal in order for the box to have a volume of 102 cubic inches?

20–22. Using a graphing calculator, approximate the roots of the given quadratic equation correct to the *nearest hundredth*.

20. $x^2 - 6x - 1 = 0$ **21.** $3x^2 - 9x + 4 = 0$ **22.** $2x^2 + 5 = 10x$

23. Michele and her friends are plotting the course for a walkathon to help raise money for a charity. They decide to make the course in the shape of a triangle, ABC, with a right angle at B. Beginning at point A, participants will walk 5 kilometers to B, then walk from B to C, and finally walk from C back to A. Michele wants to position a water station along the path between A and C, at point D, which is the closest point to B. If the distance from C to D exceeds the distance from A to D by 3 kilometers, find to the *nearest tenth* the number of kilometers from A to C.

24–26. Solve each system of equations graphically or algebraically.

24. $y = x^2 - 4x + 9$, $y - x = 5$ **25.** $y = x^2 - 6x + 6$, $y - x = -4$ **26.** $y = x^2 - 6x + 5$, $y + 7 = 2x$

6.5 APPLYING QUADRATIC FUNCTIONS

KEY IDEAS

Often it is possible to find the maximum or minimum value of a real-world quantity, y, by expressing y as a quadratic function of a related variable, x, and then finding the y-coordinate of the vertex of its graph.

Projectile Motion

When an object such as a ball is tossed into the air, its path can be described by a parabola in which the "x-coordinate" represents how much time t, has elapsed after the object was launched and the "y-coordinate" represents the height, h, at time t, as shown in Figure 6.18. The y-coordinate of the vertex of the parabola represents the maximum height of the object.

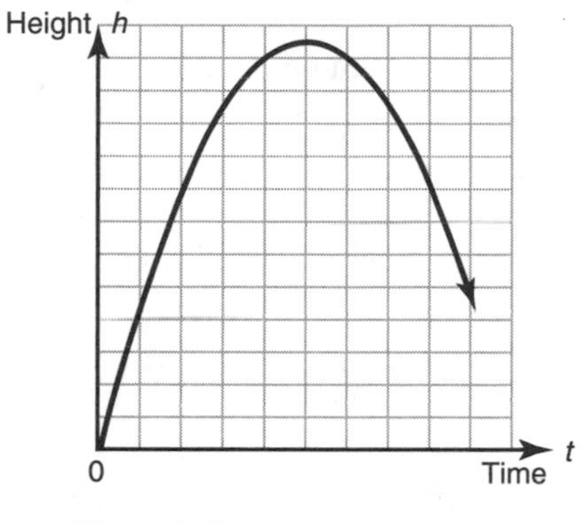

Figure 6.18 A Parabolic Path

Example 1

A ball is tossed straight into the air from ground level and after t seconds reaches a height of h feet, where $h(t) = 88t - 16t^2$. How many seconds after the ball is tossed will it return to the ground?

Solution: **5.5**

When the ball returns to the ground, $h = 0$. Hence, find the value of t that makes $h(t) = 0$ when $t > 0$:

$$h(t) = 88t - 16t^2$$
$$0 = 88t - 16t^2$$
$$0 = 8t(11 - 2t)$$

- If $8t = 0$, then $t = 0$, which represents the time at which the ball is tossed into the air.
- If $11 - 2t = 0$, then $t = \frac{11}{2} = 5.5\,\text{sec}$, which is the number of seconds required for the ball to return to the ground.

Example 2

A model rocket is launched from ground level. At t seconds after it is launched, it is h meters above the ground, where $h(t) = -4.9t^2 + 68.6t$. What is the maximum height, to the *nearest meter*, attained by the model rocket?

Solution: **240**

Since the coefficient of the t^2-term is negative, the vertex of the graph of $h(t) = -4.9t^2 + 68.6t$ is the highest point on the curve. To find the maximum height, find the y-coordinate of the vertex of the parabola.

- The x-coordinate of the vertex is $t = -\frac{b}{2a} = -\frac{68.6}{2(-4.9)} = -\frac{68.6}{-9.8} = 7$.
- To find the y-coordinate of the vertex, evaluate $h(7)$:

$$\begin{aligned} h(7) &= -4.9(7)^2 + 68.6(7) \\ &= -240.1 + 480.2 \\ &= 240.1 \end{aligned}$$

- To the *nearest meter*, the maximum height is 240.

TIP

- To use your calculator to evaluate $h(7)$, enter $Y_1 = -4.9x^2 + 68.6x$. Then press [2nd] [MODE] to quit to the home screen. After you press

[VARS] [▷] [ENTER] [ENTER],

Y_1 appears in the home screen. To find the value of Y_1 at 7, press

- You can solve the problem completely with your calculator by graphing $Y_1 = -4.9X^2 + 68.6X$ in a viewing window in which the graph fits, such as $[-15, 15] \times [-100, 300]$. Then use the maximum feature from the CALC menu of your graphing calculator to see that the y-coordinate of the vertex is 240.1.

Maximizing the Area of a Rectangle

If a fixed amount of fencing is available to enclose a rectangular region, the area of the enclosed rectangle can be expressed as a quadratic function of the length (or width) of the rectangle.

Example 3

Stacy has 30 yards of fencing that she wishes to use to enclose a rectangular garden. If all of the fencing is used, what is the maximum area of the garden that can be enclosed?

Solution: **56.25 yd²**

Let x represent the length of the enclosed rectangular garden, and w represent its width. Since all 30 yd of fencing must be used:

$$2x + 2w = 30$$
$$x + w = 15$$
$$w = 15 - x$$

Let $A(x)$ represent the area of the enclosed rectangle as a function of x.

- Then $A(x) = xw = x(15 - x) = -x^2 + 15x$.
- The maximum value of $A(x)$ occurs at

$$x = -\frac{b}{2a} = -\frac{15}{2(-1)} = 7.5.$$

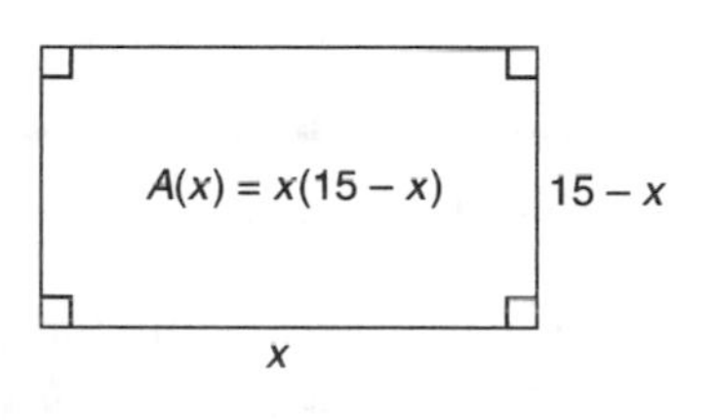

- At $x = 7.5$, $A(7.5) = -7.5^2 + (15)(7.5) = -56.25 + 112.5 = 56.25\text{ yd}^2$.

Solving a Business Problem

Some types of business problems lead to quadratic functions.

Example 4

The marketing department at Sports Stuff found that approximately 600 pairs of running shoes are sold monthly when the average price of each pair is $90. It was also observed that, for each $5 reduction in price, an additional 50 pairs of running shoes are sold monthly. What price per pair will maximize the store's monthly revenue from the sale of running shoes?

Solution: **$75**

If x represents the number of \$5 reductions in the price of a pair of running shoes, then $50x$ represents the number of *additional* pairs of running shoes that are sold during the month. Hence, $600 + 50x$ is the total number of pairs of running shoes that are sold at the reduced price of $90 - 5x$ for each pair of running shoes.

- The store's monthly revenue, R, from the sale of running shoes is the price of each pair times the number sold:

$$R(x) = (90 - 5x)(600 + 50x) = 54{,}000 + 1500x - 250x^2.$$

- The maximum value of $R(x)$ occurs at

$$x = -\frac{b}{2a} = -\frac{1500}{2(-250)} = 3.$$

- When $x = 3$, the reduced price of a pair of running shoes is

$$90 - 5x = 90 - (5 \times 3) = 75.$$

- The price that will maximize the store's revenue from the sale of running shoes is \$75 per pair.

Check Your Understanding of Section 6.5

A. Multiple Choice

1. A ball is thrown into the air in such a way that its height at any time, t, is given by the function $h(t) = -16t^2 + 80t + 10$. What is the maximum height attained by the ball?
(1) 140 (2) 110 (3) 85 (4) 10

2. An archer shoots an arrow into the air in such a way that its height at any time, t, is given by the function $h(t) = -16t^2 + kt + 3$. If the maximum height of the arrow occurs at time $t = 4$, what is the value of k?
(1) 128 (2) 64 (3) 8 (4) 4

3. A toy rocket is launched directly upward so that its height, in meters, above the ground after t seconds is given by the function $h(t) = 147t - 4.9t^2$. How many seconds after the rocket reaches its maximum height does it strike the ground?
(1) 30 (2) 22.5 (3) 15 (4) 7.5

B. *In each case, show how you arrived at your answer by clearly indicating all of the necessary steps, formula substitutions, diagrams, graphs, charts, etc.*

4. A farmer is enclosing a rectangular plot of land that will be bounded on one side by a river and fenced on the other three sides. What is the maximum area that can be enclosed if 880 yards of fencing material are used?

5. The number of feet in the height, h, of a projectile t seconds after it is launched is given by the function $h(t) = -16t^2 + vt + h_0$, where v is the upward initial velocity and h_0 is the height, in feet, from which the projectile is launched. A rocket is launched from a hilltop 2400 feet above a desert with an initial velocity of 400 feet per second.
a. What is the maximum height the rocket will reach?
b. After how many seconds will the rocket land on the desert?

6. A rectangular sheet of aluminum 25 feet long and 12 inches wide is to be made into a rain gutter by folding up the two longer parallel sides the same number of inches at right angles to the sheet. How many inches on each side should be folded up so that the gutter will have its greatest capacity?

7. Student editors are designing the yearbook so that each rectangular page will have a perimeter of 40 inches. The printer advises the student editors that each page must have a margin of $1\frac{1}{2}$ inches on the bottom and 1 inch on the other three sides. What must be the dimensions of the page if the printed area is to be as great as possible?

8. When 14 apple trees are planted per acre in a certain orchard, the average yield per tree is 480 apples per year. For each additional tree planted per acre in the same orchard, the annual yield per tree decreases by 10 apples. How many additional trees should be planted in the orchard so that the maximum number of apples will be produced per year?

9. A rectangular packing box, similar to a shoebox without a lid, is three times as long as it is wide, and half as high as it is long. Each square inch of the bottom of the box costs $0.008 to produce, while each square inch of any side costs $0.003 to produce.
a. Write an equation that expresses the cost, c, of the box as a function of its width, x.
b. Using the function written in part a, determine the dimensions of a box that would cost $0.69 to produce.

10. Two model rockets are launched, one 10 seconds after the first. The height, h, in feet of the first rocket t seconds after it is launched is $h(t) = -16t^2 + 256t$. The height, h, in feet of the second rocket t seconds after it is launched is $h(t) = -16t^2 + 480t - 3200$.

a. What is the height of the second rocket when the first rocket hits the ground?

b. What is the height of the first rocket when the second rocket reaches its maximum height?

c. How many seconds, correct to the *nearest hundredth*, after the second rocket is launched will the two rockets have the same height?

CHAPTER 7
EQUATIONS AND INEQUALITIES

7.1 SOLVING ABSOLUTE-VALUE EQUATIONS AND INEQUALITIES

KEY IDEAS

Since the absolute value of a number is always nonnegative, the graph of $y = |x|$ consists of the rays $y = x$ and $y = -x$ with the origin as their common endpoint. Absolute-value equations and inequalities can be solved either algebraically or graphically.

Definition of Absolute Value

The **absolute value** of a number x is written as $|x|$ and represents the distance of that number from 0 on the number line. Thus, $|x| = x$ when $x \geq 0$ and $|x| = -x$ when $x < 0$. For example, $|3| = 3$ and $|-3| = -(-3) = 3$.

The Absolute-Value Function

To graph $y = |x|$, open the Y = editor and press

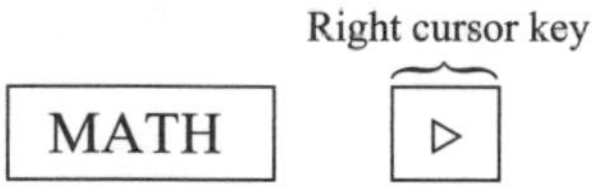

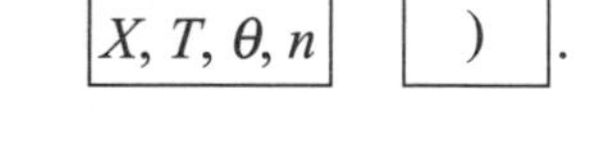

Figure 7.1 shows the graph in a decimal window. From the graph we know that $y = |x|$ is a function since the graph passes the vertical-line test. The domain is the set of real numbers, and the range is the set of all nonnegative real numbers.

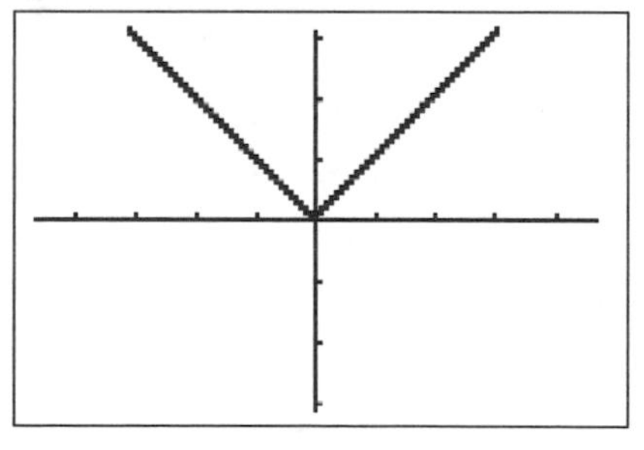

Figure 7.1 Graph of $y = |x|$

Absolute-Value Equations

To solve the equation $|ax - b| = d$, where d is a nonnegative number, write and then solve the equivalent pair of equations $ax - b = -d$ and $ax - b = d$.

Example 1

Solve for x: $|2x + 3| = 1$.

Solution: $\mathbf{x = -2}$ or $\mathbf{-1}$

If $|2x + 3| = 1$, then

$$2x+3=-1 \quad \text{or} \quad 2x+3=1$$

$$2x=-4 \quad \Big| \quad 2x=-2$$

$$x=\frac{-4}{2}=-2 \quad \Big| \quad x=\frac{-2}{2}=-1$$

Example 2 illustrates the importance of checking roots in the original absolute-value equation.

Example 2

Solve for x and check: $|x - 3| = 2x$.

Solution: $\mathbf{x = 1}$

If $|x - 3| = 2x$, then

$$x-3=-2x \quad \text{or} \quad x-3=2x$$

$$x+2x-3=0 \quad \Big| \quad -3=2x-x$$

$$3x=3 \quad \Big| \quad -3=x$$

$$x=\frac{3x}{3}=1 \quad \Big| \quad x=-3$$

Check:

Let $x = 1$:

$|x - 3| = 2x$

$|1 - 3| = 2(1)$

$|-2| = 2$

Let $x = -3$:

$|x - 3| = 2x$

$|-3 - 3| = 2(-3)$

$|-6| \neq -6 \leftarrow$ Reject $x = -3$.

Interpreting Absolute-Value Inequalities

The absolute-value inequality $|x - a| < d$ represents the set of all points x that are less than d units from a. For example:

- The inequality $|t - 68°| < 3°$ states that the temperature, t, is less than 3° from 68°, meaning that t is between 65° and 71°, as shown in Figure 7.2.

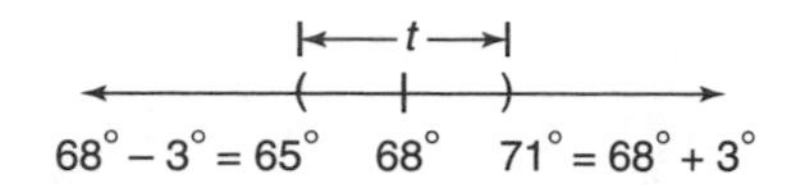

Figure 7.2 Solution of $|t - 68°| < 3°$

- The inequality $|t - 68°| > 3°$ states that the temperature, t, is more than 3° from 68°, meaning that t is lower than 65° or higher than 71°.

Solving Absolute-Value Inequalities

To solve an absolute-value inequality algebraically, remove the absolute-value sign according to the following rules, where d is a positive number:

- if $|ax - b| < d$, then $-d < ax - b < d$.
- if $|ax - b| > d$, then $ax - b < -d$ or $ax - b > d$.

Example 3

Solve and graph the solution set of $|2x - 1| \le 7$.

Solution: $\{x|-3 \le x \le 4\}$

- If $|2x - 1| \le 7$, then $-7 \le 2x - 1 \le 7$.
- Add 1 to each member of the resulting inequality:

$$\begin{array}{l} -7 \le 2x - 1 \le 7 \\ +1 \qquad\quad +1 + 1 \\ \hline -6 \le 2x \qquad \le 8 \end{array}$$

- Divide each member of the inequality by 2:

$$\frac{-6}{2} \le \frac{2x}{2} \le \frac{8}{2}$$
$$-3 \le x \le 4$$

- Graph the solution set:

−3 4 x

Example 4

Solve and graph the solution set of $|3x - 1| > 5$.

Solution: $\left\{x|x < 2-2 \text{ or } x > \frac{4}{3}\right\}$

If $|3x - 1| > 5$, then

$$3x+1<-5 \quad \text{or} \quad 3x+1>5$$
$$3x<-6 \qquad\qquad 3x>4$$
$$x<-2 \qquad\qquad x>\frac{4}{3}$$

The graph is:

−2 0 $\frac{4}{3}$ x

Graphing Calculator Solutions

To solve $|2x - 1| = 7$ using a graphing calculator, graph $Y_1 = \text{abs}(2X - 1)$ and $Y_2 = 7$ in a friendly window in which the graph fits, such as $[-4.7, 4.7] \times [-3.1, 9.3]$, as shown in Figure 7.3. Then use TRACE to find that the x-coordinates of the points of intersection are $x = -3$ and $x = 4$.

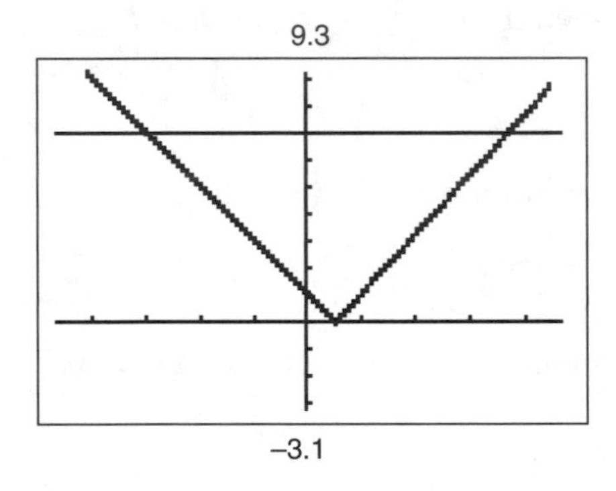

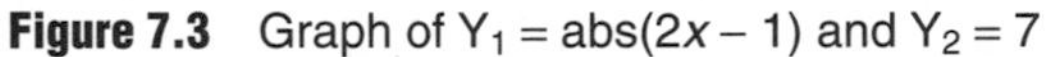
Figure 7.3 Graph of $Y_1 = \text{abs}(2x - 1)$ and $Y_2 = 7$

The solutions of the related inequalities $|2x - 1| \le 7$ and $|2x - 1| > 7$ can also be read from the graph:

- The solution of $|2x - 1| \le 7$ is $-3 \le x \le 4$ since from $x = -3$ to $x = 4$ the graph of Y_1 is *below* (<) the graph of Y_2.
- The solution of $|2x - 1| > 7$ is $x < -3$ or $x > 4$ since this is the interval where the graph of Y_1 is *above* (>) the graph of Y_2.

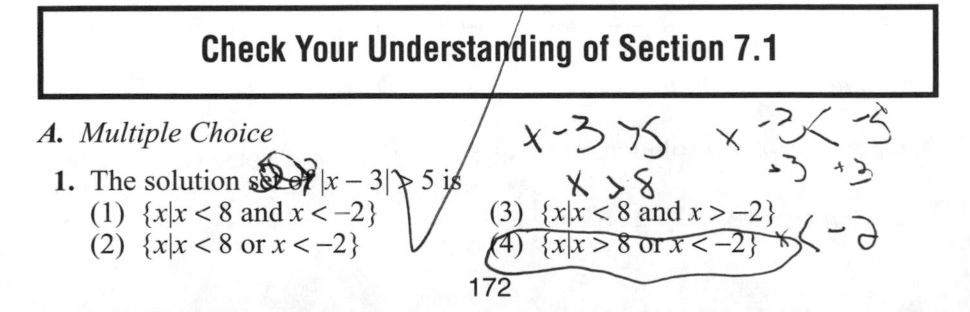

Check Your Understanding of Section 7.1

A. Multiple Choice

1. The solution set of $|x - 3| > 5$ is
   (1) $\{x|x < 8 \text{ and } x < -2\}$
   (2) $\{x|x < 8 \text{ or } x < -2\}$
   (3) $\{x|x < 8 \text{ and } x > -2\}$
   (4) $\{x|x > 8 \text{ or } x < -2\}$

2. Which is the solution set of the inequality $|3x + 6| \le 30$?
(1) $\{x \mid -12 \le x \le 8\}$ (3) $\{x \mid x \le -12 \text{ or } x \ge 8\}$
(2) $\{x \mid -8 \le x \le 12\}$ (4) $\{x \mid x \le -8 \text{ or } x \ge 12\}$

3. Which is the solution set of the equation $|3x - 1| = x + 5$?
(1) $\{-1\}$ (2) $\{-1, 3\}$ (3) $\{3\}$ (4) $\{1, -3\}$

4. The inequality $-3 < x < 7$ is the solution of
(1) $|x - 2| > 5$ (2) $|x - 2| < 5$ (3) $|x + 2| > 5$ (4) $|x + 2| < 5$

5. Which inequality states that the speed of a car, s, is less than 5 miles per hour from the speed limit, L?
(1) $|x - 5| < L$ (2) $|s - L| < 5$ (3) $|L - 5| < s$ (4) $|s| < L - 5$

6. The accompanying graph represents the solution of which inequality?
(1) $|x + 4| > 1$ (3) $|x + 1| \le 4$
(2) $|x + 1| < 4$ (4) $|x + 4| \ge 1$

−5 −4 −3 −2 −1 0 1 2 3

7. Which graph represents the solution of the inequality $|5x - 15| < 10$?
(1) −5 −4 −3 −2 −1 0 1 2 3 4 5
(2) −5 −4 −3 −2 −1 0 1 2 3 4 5
(3) −5 −4 −3 −2 −1 0 1 2 3 4 5
(4) −5 −4 −3 −2 −1 0 1 2 3 4 5

8. Which graph represents the solution of the inequality $|3 - 2x| > 7$?
(1) −3 −2 −1 0 1 2 3 4 5 6
(2) −3 −2 −1 0 1 2 3 4 5 6
(3) −3 −2 −1 0 1 2 3 4 5 6
(4) −3 −2 −1 0 1 2 3 4 5 6

B. *In each case, show how you arrived at your answer by clearly indicating all of the necessary steps, formula substitutions, diagrams, graphs, charts, etc.*

9–11. Solve for x.

9. $11 < |4 - 3x|$ **10.** $\dfrac{|2 - x|}{5} - 1 \le 0$ **11.** $|1 - x| \le x + 2$

12. If $|x - 16| \le 4$ and $|y - 6| \le 2$, what is the greatest possible value of $\dfrac{x}{y}$?

13. An archer shoots an arrow into the air with an initial velocity of 128 feet per second. Because speed is the absolute value of velocity, the arrow's speed, s, in feet per second after t seconds is $|-32t + 128|$. Find the values of t for which s is less than 48 feet per second.

14. In a certain greenhouse for plants, the Fahrenheit temperature, F, is controlled so that it does not vary from 79° by more than 7°.

a. Which absolute-value inequality expresses the possible range in Fahrenheit temperatures of the greenhouse?
(1) $|F-7| \leq 79$ (2) $|F-79| > 7$ (3) $|F-79| \leq 7$ (4) $|F-7| > 79$

b. The formula $F = \frac{9}{5}C + 32$ can be used to convert C degrees Celsius to F degrees Fahrenheit. Using the answer in part a, determine the minimum and maximum Celsius temperatures of the greenhouse to the *nearest tenth of a degree*.

7.2 SOLVING QUADRATIC INEQUALITIES

KEY IDEAS

Quadratic inequalities such as $x^2 - 2x - 3 < 0$ and $x^2 - 2x - 3 > 0$ can be solved graphically or algebraically by solving the related quadratic equations.

Solving Quadratic Inequalities Graphically

The graph of $y = x^2 - 2x - 3$ is shown in Figure 7.4.

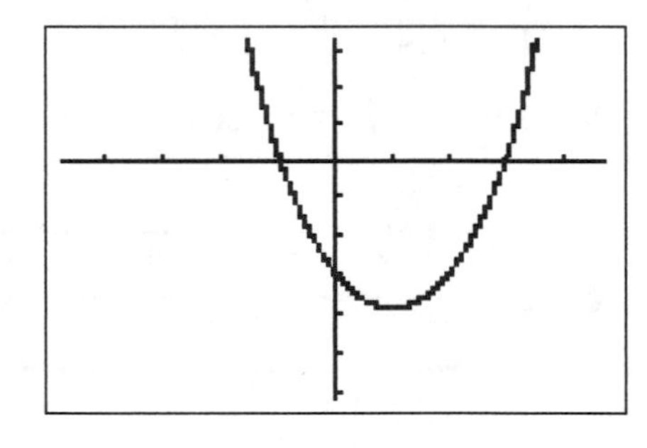

Figure 7.4 Graph of $y = x^2 - 2x - 3$

- To solve $x^2 - 2x - 3 < 0$, find the interval of x for which the graph is below the x-axis. Since $y < 0$ when x is between -1 and 3, the solution is $-1 < x < 3$.
- To solve $x^2 - 2x - 3 > 0$, find the interval of x for which the graph is above the x-axis. The solution is $x < -1$ or $x > 3$.

Solving Quadratic Inequalities Algebraically

From the graphical solutions of $x^2 - 2x - 3 < 0$ and $x^2 - 2x - 3 > 0$, we know that the boundary points of the solution intervals are the roots of the related quadratic equation, $x^2 - 2x - 3 = 0$. We can, therefore, generalize that, if r and R are unequal roots of $ax^2 + bx + c = 0$, the solutions of the related quadratic inequalities will have the forms shown in the accompanying table, provided that a is positive and $r < R$.

GENERAL SOLUTIONS OF QUADRATIC INEQUALITIES WHERE $a > 0$ AND $r < R$

Quadratic Inequality	Solution Interval	Graph of Solution
$ax^2 + bx + c < 0$	$r < x < R$	r R
$ax^2 + bx + c \leq 0$	$r \leq x \leq R$	r R
$ax^2 + bx + c > 0$	$x < x$ or $x > R$	r R
$ax^2 + bx + c \geq 0$	$x \leq r$ or $x \geq R$	r R

For example, to solve $x^2 - 3x - 10 < 0$ algebraically, first solve $x^2 - 3x - 10 = 0$ by factoring the left side of the equation as $(x + 2)(x - 5) = 0$, so $x = -2$ or $x = 5$. Because the inequality is "<," the form of the solution interval is $r < x < R$, where $r = -2$ and $R = 5$. Thus, the solution is $-2 < x < 5$.

Example 1

Find the solution set of $-x^2 + 8 \leq 2x$.

Solution: $\{x|\, x \leq -4$ or $x \geq 2\}$

To solve a quadratic inequality algebraically, rearrange the terms of the inequality so that all of the nonzero terms are on the left side of the inequality and the coefficient of the x^2-term is positive.

- Rewrite $-x^2 + 8 \leq 2x$ as $-x^2 - 2x + 8 \leq 0$. Then make the coefficient of the x^2-term positive by changing the sign of each term to its opposite *and* reversing the direction of the inequality: $x^2 + 2x - 8 \geq 0$.
- Because the inequality is $\geq$, the form of the solution is $x \leq r$ or $x \geq R$. Find r and R by solving the related quadratic equation. If $x^2 + 2x - 8 = 0$, then $(x + 4)(x - 2) = 0$, so $x = -4$ or $x = 2$.
- Since $r = -4$ and $R = 2$, the solution set is $\{x|x \leq -4$ or $x \geq 2\}$.

Example 2

Find the solution set of $3x \geq x^2$.

Solution: $\{x|\, 0 \leq x \leq 3\}$

If $3x \geq x^2$, then $0 \geq x^2 - 3x$ or, equivalently, $x^2 - 3x \leq 0$. Since the inequality is "$\leq$," the form of the solution is $r \leq x \leq R$. Find r and R by solving the related quadratic equation. If $x^2 - 3x = 0$, then $x(x - 3) = 0$, so $x = 0$ or $x = 3$. Because $r = 0$ and $R = 3$, the solution set is $\{x|0 \leq x \leq 3\}$.

Check Your Understanding of Section 7.2

A. Multiple Choice

1. Which is the domain of $f(x) = \sqrt{x^2 - 3x - 28}$?
(1) $-7 \le x \le 4$ (3) $-4 \le x \le 7$
(2) $x \le 7$ or $x \ge -4$ (4) $x \ge 7$ or $x \le -4$

2. Which is the graph of the solution of the inequality $x^2 + 4x - 21 < 0$?
(1) –7 –6 –5 –4 –3 –2 –1 0 1 2 3
(2) –7 –6 –5 –4 –3 –2 –1 0 1 2 3
(3) –7 –6 –5 –4 –3 –2 –1 0 1 2 3
(4) –3 –2 –1 0 1 2 3 4 5 6 7

B. In each case, show how you arrived at your answer by clearly indicating all of the necessary steps, formula substitutions, diagrams, graphs, charts, etc.

3–8. Determine the solution for each inequality.

3. $x^2 + 4x - 5 \ge 0$ **5.** $2x^2 < 5x + 3$ **7.** $2x - 63 \le -x^2$

4. $7x > x^2$ **6.** $6 \le -t + t^2$ **8.** $x^2 + 4 \ge 4x$

9. When a baseball is hit by a batter, the height of the ball, h(t), at time t, $t \ge 0$, is determined by the equation $h(t) = -16t^2 + 64t + 4$. For which interval of time is the height of the ball at least 52 feet?

10. A rocket is launched straight up from ground level. After t seconds its height, h, in feet is given by the function $h(t) = -16t^2 + 376t$. During what interval of time will the height of the rocket exceed 2040 feet?

7.3 SOLVING RADICAL EQUATIONS

KEY IDEAS

An equation in which the variable appears underneath a radical sign, such as $\sqrt{4 - x} - 8 = x$, is called a **radical equation**. To solve a radical equation, isolate the radical and then raise both sides of the equation to the power that eliminates the radical. The resulting equation may, however, include **extraneous roots** that do not satisfy the original equation.

The Square Root Function

To graph $y = \sqrt{x}$ using a calculator, open the Y = editor and press

[X, T, θ, n] [^] [.] [5].

Draw the graph in a decimal window, as shown in Figure 7.5. From the graph you can tell that $y = \sqrt{x}$ is a function since the graph passes the vertical-line test. The domain and range are both restricted to the set of nonnegative real numbers.

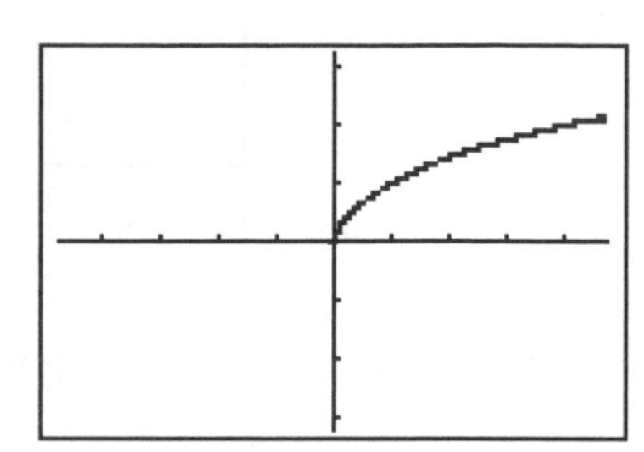

Figure 7.5 Graph of $y = \sqrt{x}$

Solving a Radical Equation Algebraically

To solve $\sqrt{4-x} - 8 = x$:

- Isolate the radical: $\sqrt{4-x} = x + 8$
- Eliminate the square root radical:

$$\left(\sqrt{4-x}\right)^2 = (x+8)^2$$
$$4 - x = x^2 + 16x + 64$$
$$0 = x^2 + 17x + 60$$

- Solve the transformed equation:

$$0 = (x+5)(x+12)$$
$$x = -5 \quad \text{or} \quad x = -12$$

Because extraneous roots are possible, confirm each possible root algebraically by testing the root in the original equation. Since $x = -5$ works in $\sqrt{4-x} - 8 = x$, -5 is a root. If $x = -12$, then $\sqrt{4-(-12)} - 8 = -12$ or, equivalently, $\sqrt{16} - 8 = -12$, which is not a true statement. Reject $x = -12$; then -5 is the only root.

Solving a Radical Equation Graphically

To solve $\sqrt{4-x} - 8 = x$ graphically, graph $Y_1 = \sqrt{(4-x)} - 8$ and $Y_2 = x$ in a window with a friendly x-readout, such as $[-9.4, 9.4] \times [-6.2, 6.2]$. Then use TRACE to determine that the x-coordinate of the point of intersection of the two graphs is -5, as shown in Figure 7.6.

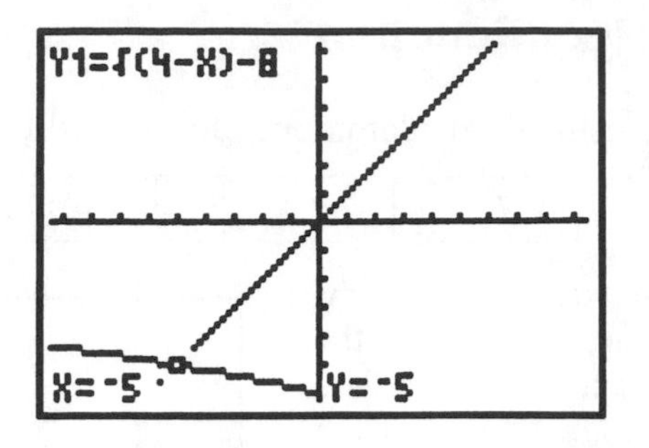

Figure 7.6 Graph of $Y_1 = \sqrt{4-x} - 8$ and $Y_2 = x$

Solving Equations with Rational Exponents

To solve an equation that can be put in the form $x^{\frac{p}{r}} = k$, raise both sides of the equation to the power that makes the exponent of x equal to 1. For example, if $x^{\frac{3}{5}} + 9 = 1$, then $x^{\frac{3}{5}} = -8$. Since the reciprocal of the exponent of x is $\frac{5}{3}$, raise both sides to the $\frac{5}{3}$ power:

$$\left(x^{\frac{3}{5}}\right)^{\frac{5}{3}} = (-8)^{\frac{5}{3}}$$

$$x^1 = \left(\sqrt[3]{-8}\right)^5$$

$$x = (-2)^5 = -32$$

Check Your Understanding of Section 7.3

A. Multiple Choice

1. For the equation $\sqrt{x+21} = x+1$, the solution set for x is
(1) { } (2) {−5} (3) {−5, 4} (4) {4}

2. Which is the solution set of the equation $3 - \sqrt{5-x} = x$?
(1) {1} (2) {4, 1} (3) { } (4) {4}

3. The solution set of the equation $\sqrt{2x+15} = x$ is
(1) {5, −3} (2) {5} (3) {−3} (4) { }

4. Which statement is true of the equation $\sqrt{x^2 = 5x+5} - 1 = 0$?
(1) The only real root is 1. (3) Both 1 and 4 are roots.
(2) The only real root is 4. (4) Neither 1 nor 4 is a root.

B. *In each case, show how you arrived at your answer by clearly indicating all of the necessary steps, formula substitutions, diagrams, graphs, charts, etc.*

5–16. Solve each equation for the variable. Confirm your answer algebraically.

5. $\sqrt{2n} = 4\sqrt{5}$

6. $\sqrt[3]{3x-4} = 2$

7. $\sqrt{5x+1} + 3x = 27$

8. $\dfrac{3+x}{\sqrt{3+x}} = 1$

9. $y - 1 = \sqrt{13 - y^2}$

10. $n - \sqrt{3n+4} = 2$

11. $\sqrt{1-2x} = \sqrt{x^2 - 7}$

12. $2\sqrt{1-3p} = 3 - p$

13. $t - 2\sqrt{4t-7} = 0$

14. $26 - x^{\frac{2}{3}} = 1$

15. $8x^{-\frac{3}{4}} = 27$

16. $\sqrt{13-3y} + y = 1$

17. The number of seconds in the period of a pendulum is the length of time required for the pendulum to make one complete swing back and forth. The formula $T = 2\pi\sqrt{\dfrac{L}{32}}$ gives the period, T, for a pendulum of length L in feet. If Todd wants to build a grandfather clock with a pendulum that swings back and forth every 2 seconds, how long, to the *nearest tenth of a foot*, should he make the pendulum?

18. Solve $\sqrt{x+4} + \sqrt{1-x} = 3$ algebraically or graphically.

7.4 SOLVING FRACTIONAL EQUATIONS

KEY IDEAS

A **fractional equation** is an equation in which a variable appears in the denominator(s) of one or more fractions. To solve a fractional equation, eliminate the denominators by multiplying each member of the equation by the LCD of all the denominators.

Solving a Fractional Equation

When clearing an equation of any fractional terms, make sure you multiply *both* sides of the equation by the LCD of all the denominators.

Example 1

Solve for y: $\dfrac{2}{y}-\dfrac{9}{10}=\dfrac{1}{5y}$.

Solution: $\boldsymbol{y=2}$

Since the LCD of all the denominators is $10y$, clear the equation of its fractions by multiplying each member of the equation by $10y$. Then solve the resulting equation.

$$10y\left(\frac{2}{y}\right)-10y\left(\frac{9}{10}\right)=10y\left(\frac{1}{5y}\right)$$

$$1 \qquad\qquad 1 \qquad\qquad 2$$

$$10\cancel{y}\left(\frac{2}{\cancel{y}}\right)-\cancel{10}y\left(\frac{9}{\cancel{10}}\right)=\cancel{10y}\left(\frac{1}{\cancel{5y}}\right)$$

$$20-9y=2$$

$$-9y=-18$$

$$y=\frac{-18}{-9}=2$$

Checking for Extraneous Roots

Since the equation produced by clearing a fractional equation of its variable denominators may include *extraneous roots*, be sure to check the roots of the transformed equation in the original fractional equation.

Example 2

Find the solution set: $\dfrac{x}{x+2}-\dfrac{1}{x-2}=\dfrac{8}{x^2-4}$.

Solution: **{5}**

- Rewrite the equation with each denominator in factored form:

$$\frac{x}{x+2}-\frac{1}{x-2}=\frac{8}{(x+2)(x-2)}.$$

- Since the LCD of the fractional terms is $(x+2)(x-2)$, multiply each term of the equation by $(x+2)(x-2)$:

$$(x+2)(x-2)\left[\frac{x}{x+2}\right]-(x+2)(x-2)\left[\frac{1}{x-2}\right]=(x+2)(x-2)\left[\frac{8}{(x+2)(x-2)}\right]$$

1 1 1

$$(\cancel{x+2})(x-2)\left[\frac{x}{\cancel{x+2}}\right]-(x+2)(\cancel{x-2})\left[\frac{1}{\cancel{x-2}}\right]=\cancel{(x+2)(x-2)}\left[\frac{8}{\cancel{(x+2)(x-2)}}\right]$$

$$(x-2)x-(x+2)=8$$

$$x^2-2x-x-2=8$$

$$x^2-3x-10=0$$

$$(x+2)(x-5)=0$$

$$x+2=0 \text{ or } x-5=0$$

$$x=-2 \qquad x=5$$

- Check for extraneous roots. If $x = -2$, the first denominator of the original equation is 0, which is not permitted since division by 0 is not defined. Hence, $x = -2$ must be rejected. Since $x = 5$ satisfies the original equation, the solution set of the equation is $\{5\}$.

Solving a Motion Problem

Solving a motion problem depends on the relationship

$$R \times T = D,$$

where R represents the average rate of speed and D represents the distance traveled in time T.

Example 3

On a 75-mile trip Mr. Frank's average rate of driving speed for the first 15 miles was 10 miles per hour less than his average rate of driving speed for the remainder of the trip. If the total driving time for the trip was 2 hours, find the average rate of speed for the first 15 miles of the trip.

Solution: **30 mph**

Mr. Frank drove 15 mi at one rate and 75 – 15 = 60 mi at a faster rate. Hence, if x represents the average rate of speed for the last 60 mi, then $x - 10$ represents the average rate of speed for the first 15 mi.

- Organize the information in a table. Since $R \times T = D$, $T = \frac{D}{R}$. Complete the column headed "Time" by dividing the distance traveled by the average rate of speed over that distance.

	Rate ×	Time =	Distance
First 15 mi of trip	$x - 10$	$T = \frac{D}{R} = \frac{15}{x-10}$	15 mi
Last 60 mi of trip	x	$T = \frac{D}{R} = \frac{60}{x}$	60 mi

- Since the total time for the trip was 2 hrs:

$$\frac{15}{x-10} + \frac{60}{x} = 2.$$

- Clear the fractions by multiplying each member of the equation by $x(x - 10)$:

$$\overset{1}{x\cancel{(x-10)}}\left[\frac{15}{\cancel{x-10}}\right] + \overset{1}{\cancel{x}}(x-10)\left[\frac{60}{\cancel{x}}\right] = 2x(x-10)$$

$$15x + 60x - 600 = 2x^2 - 20x$$

- Simplify and rearrange terms. Then, $2x^2 - 95x + 600 = 0$, so $(2x - 15)(x - 40) = 0$, making $x = 7.5$ or $x = 40$. Since $x - 10$ must be a positive number, x cannot be equal to 7.5. Since $x - 10 = 40 - 10 = 30$, Mr. Frank drove at a rate of 30 mph for the first 15 mi of the trip.

Solving a Work Problem

The solution of a "work" problem depends on the relationship

$$\begin{pmatrix}\text{Rate of}\\ \text{work}\end{pmatrix} \times \begin{pmatrix}\text{Time}\\ \text{worked}\end{pmatrix} = \begin{pmatrix}\text{Part of job}\\ \text{completed}\end{pmatrix},$$

where the rate of work is the reciprocal of the total time required to do the whole job when working alone. For example, if it takes 4 hours to complete a job, then the rate of work is $\frac{1}{4}$ of the job per hour.

Example 4

John takes 2 days longer than Bill to build the scenery for a school play. After John had worked 1 day alone, he was joined by Bill. Working together, they completed the whole job in 4 additional days. How long would Bill working alone have taken to do the whole job?

Solution: **8 days**

- If x represents the number of days Bill working alone would have taken to do the whole job, then $x + 2$ represents the number of days John would require to do the whole job working alone. Since John worked alone for 1 day, he worked a total of 5 days while Bill worked 4 days. Organize the information in a table:

	Rate of work ×	Days worked =	Part of job completed
John	$\frac{1}{x+2}$	5	$\frac{5}{x+2}$
Bill	$\frac{1}{x}$	4	$\frac{4}{x}$

- Since the sum of the fractional parts of the job completed by John and Bill must be equal to 1 whole job:

$$\frac{5}{x+2}+\frac{4}{x}=1$$

- Clear the fractions by multiplying each member of the equation by $x(x+2)$:

$$\overset{1}{\cancel{x(x+2)}}\left[\frac{5}{\cancel{x+2}}\right]+\overset{1}{\cancel{x}}(x+2)\left[\frac{4}{\cancel{x}}\right]=1(x)(x+2).$$

- Simplify and rearrange terms. Then $x^2 - 7x - 8 = 0$, so $(x - 8)(x + 1) = 0$, making $x = 8$ or $x = -1$. Since the number of days cannot be negative, $x = 8$ is the only solution. Hence, Bill working alone would have taken 8 days to do the whole job.

Check Your Understanding of Section 7.4

A. In each case, show how you arrived at your answer by clearly indicating all of the necessary steps, formula substitutions, diagrams, graphs, charts, etc.

1–8. Solve for the variable and check.

1. $\frac{1}{x}+\frac{3}{2x}=\frac{x-4}{2}$

2. $\frac{1}{b-3}-\frac{3}{2b+6}=\frac{b}{b^2-9}$

3. $\frac{1}{y}-\frac{y+1}{8}=\frac{y-1}{4y}$

4. $\frac{t}{t-3}-\frac{t-2}{2}=\frac{5t-3}{4t-12}$

5. $\dfrac{4x}{x+2} = 1 + \dfrac{12}{x}$

6. $\dfrac{1}{x-2} + \dfrac{x+2}{x+5} = \dfrac{3}{x^2+3x-10}$

7. $\dfrac{4}{m^2-16} + \dfrac{m-3}{m+4} = \dfrac{1}{m-4}$

8. $\dfrac{2}{r-6} = \dfrac{10}{r^2-7r+6} - \dfrac{r}{r-1}$

B. *In each case, show how you arrived at your answer by clearly indicating all of the necessary steps, formula substitutions, diagrams, graphs, charts, etc.*

9. Working by herself, Mary requires 16 minutes more than Antoine to solve a mathematics problem. Working together, Mary and Antoine can solve the problem in 6 minutes.

a. If Antoine takes t minutes to solve the problem working alone, write in terms of t an equation that can be used to solve for t.

b. Using the equation written in part a, determine the number of minutes Antoine will take to solve the problem if he works by himself. [*Only an algebraic or graphical solution will be accepted.*]

10. Dave can mow his lawn in 20 minutes less time with the power mower than with his hand mower. One day the power mower broke down 15 minutes after he started mowing, and he took 25 minutes more time to complete the job with the hand mower. How many minutes does Dave take to mow the lawn with the power mower?

11. A teacher drove 280 miles to attend a mathematics conference and arrived 1 hour late. The teacher figured out that, had she increased her average speed by 5 miles per hour, she would have arrived at the time for which the conference was scheduled. What was her average rate of speed?

12. A mechanic's helper requires 4 hours longer than the mechanic to repair a car. The mechanic began the repair job alone and worked on it for 3 hours before he was called away. His helper needed 5 hours to finish the job. How many hours would the mechanic working alone have taken to repair the car?

13. At 9:00 A.M. Beth started from home on a hike to a mountain 12 miles away. After reaching the mountain, she took 1 hour for lunch and then returned over the same route, arriving home at 5:00 P.M. If her average rate returning was 1 mile per hour less than her rate going, find her rate on the return trip.

14. On a business trip a salesman traveled 40 miles per hour for the first third of the distance and 50 miles per hour for the remainder of the distance. If the entire trip took 4 hours and 20 minutes, how many miles did he travel?

15. Sean started to cross a lake by motorboat. After he had traveled 15 miles, the motor failed and he had to row the remaining 6 miles to shore. His average speed by motor was 4 miles per hour faster than his average speed while rowing. If the entire trip took $5\frac{1}{2}$ hours, what was Sean's average speed while rowing?

16. A building contractor knows that a small truck will take 4 days longer than a large one to haul an order of gravel to a construction site. By using eight small and six large trucks working together, the contractor delivered the order in 1 day. Find the number of days that one small truck, working alone, would have taken to deliver the gravel.

17. The members of a running club planned to contribute equally to raise $896 to pay for refreshments and prizes at a race. When it was discovered that $1080 would be needed instead, four members of the club withdrew. To raise the necessary amount, each of the remaining members had to increase his or her contribution by $13. How many members were there in the club originally?

CHAPTER 8
COMPLEX NUMBERS

8.1 THE SET OF COMPLEX NUMBERS

KEY IDEAS

If $x^2 + 1 = 0$, then $x^2 = -1$. What is x equal to? To give meaning to solutions of equations that involve square roots of negative numbers, the **imaginary unit** i is defined so that $i = \sqrt{-1}$. For example, if $x^2 = -1$, then $x = \pm\sqrt{-1} = \pm i$. The set of **complex numbers** includes real numbers, imaginary numbers, and their sums.

The Imaginary Unit

The imaginary unit i is equal to $\sqrt{-1}$ and $i^2 = -1$. Square roots of negative numbers other than -1 can be expressed in terms of i by factoring out $\sqrt{-1}$ and replacing it with i. For example, $\sqrt{-4} = \sqrt{4} \cdot \sqrt{-1} = 2i$. A **pure imaginary number**, such as $2i$ in this example, is the product of any real number and i.

Standard Form of a Complex Number

The sum of a real number and a pure imaginary number, as in $3 + 2i$, is called a **complex number**. The standard form of a complex number is $a + bi$, where a and b are real numbers and $i = \sqrt{-1}$. Since any real number a can be written in the form $a + 0 \cdot i$, the set of complex numbers includes the set of real numbers. For example, 5 is a complex number since $5 = 5 + 0 \cdot i$.

Properties of Complex Numbers

All of the properties of arithmetic that work for real numbers apply also to complex numbers. When performing arithmetic operations with complex numbers, terms involving i are treated as if they are monomials. For example:

- $3i + 5i = 8i$ and $6i - i = 5i$
- $i \cdot i^2 = i^{1+2} = i^3$ and $\dfrac{i^{13}}{i^4} = i^{13-4} = i^9$

To add or subtract complex numbers of the form $a + bi$, combine the real parts and then combine the imaginary parts. For example:

- $(2 + 3i) + (4 - 5i) = (2 + 4) + (3i - 5i) = 6 - 2i$
- $(1 - 2i) - (-4 + 6i) = (1 - 2i) + (4 - 6i)$
  $= (1 + 4) + (-2i - 6i) = 5 - 8i$
- $\sqrt{-50} + 4\sqrt{-18} = \left(\sqrt{25} \cdot \sqrt{2} \cdot \sqrt{-1}\right) + \left(4\sqrt{9} \cdot \sqrt{2} \cdot \sqrt{-1}\right)$
  $= 5\sqrt{2}i \qquad\qquad +12\sqrt{2}i$
  $= 17\sqrt{2}i$

The additive inverse of $-3 + 7i$ is $3 - 7i$ since

$$(-3+7i)+(3-7i) = (-3+3)+(7i-7i) = 0.$$

Graphing Complex Numbers

A complex number in $a + bi$ form can be graphed in the *complex* plane by measuring a units along the horizontal or "real" axis and b units along the vertical or "imaginary" axis.

Example 1

Let $Z_1 = 2 + 5i$ and $Z_2 = -6 - 2i$.

a. Graph Z_1 and Z_2 on the same axes.
b. Use the graph to determine $Z_1 + Z_2$.

Solutions: a. The graph is shown in the accompanying figure. b. $\boldsymbol{Z_1 + Z_2 = -4 + 3i}$

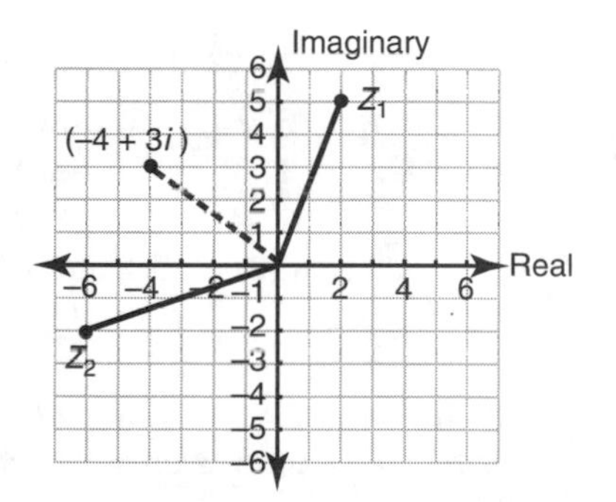

$Z_1 + Z_2$ corresponds to the diagonal of the parallelogram whose adjacent sides are Z_1 and Z_2, as shown in the accompanying figure. You should verify that $Z_1 + Z_2 = (2 + 5i) + (-6 - 2i) = -4 + 3i$.

Simplifying Powers of *i*

The expression i^n, where n is a positive whole number, can be reduced to either ± 1 or $\pm i$, as the following list suggests:

$$i^0 = 1 \qquad i^4 = i^2 \cdot i^2 = (-1)(-1) = 1$$
$$i^1 = 1 \qquad i^5 = i^4 \cdot i = 1 \cdot i = i$$
$$i^2 = -1 \qquad i^6 = i^4 \cdot i^2 = 1 \cdot (-1) = -1$$
$$i^3 = i^2 \cdot i = -i \qquad i^7 = i^4 \cdot i^3 = 1 \cdot (-i) = -i$$

Consecutive positive integer powers of i follow a cyclic pattern that repeats every four integers. Using the list of powers of i, you should be able to predict that $i^8 = 1$, $i^9 = i$, $i^{10} = -1$, $i^{11} = -i, \ldots$, and so forth. Large powers of i can be simplified by dividing the exponent by 4 and using the remainder as the new power of i. For example, to simplify i^{31}:

- Determine that $31 \div 4 = 7$ with a remainder of 3.
- Then write $i^{31} = i^3 = -i$.

Check Your Understanding of Section 8.1

A. Multiple Choice

1. The expression $i^{50} + i^0$ is equivalent to
(1) 1 (2) 2 (3) −1 (4) 0

2. Expressed in simplest form, $2\sqrt{-50} - 3\sqrt{-8}$ is equivalent to
(1) $16i\sqrt{2}$ (2) $3i\sqrt{2}$ (3) $4i\sqrt{2}$ (4) $-i\sqrt{42}$

3. In which quadrant is the sum of $2 - 4i$ and $-3 + 7i$ located?
(1) I (2) II (3) III (4) IV

4. If $Z_1 = 3 - 6i$ and $Z_2 = 2 - 5i$, in which quadrant does $Z_1 - Z_2$ lie?
(1) I (2) II (3) III (4) IV

5. If $-4 + 2i - (a + 4i) = 9 + bi$, what is the value of a?
(1) −2 (2) 2 (3) −13 (4) 13

6. If $2\sqrt{-2}$ is subtracted from $3\sqrt{-18}$, the difference is
(1) $7i\sqrt{2}$ (2) $11i\sqrt{2}$ (3) $-7i\sqrt{2}$ (4) $-11i\sqrt{2}$

***B.** In each case, show how you arrived at your answer by clearly indicating all of the necessary steps, formula substitutions, diagrams, graphs, charts, etc.*

7–15. Write each expression in simplest form in terms of i.

7. $2\sqrt{-9} + 3\sqrt{-25}$

8. $3\sqrt{-1} - 4\sqrt{-4}$

9. $\sqrt{-49} \cdot 2\sqrt{-4}$

10. $\left(\dfrac{i^{80}}{i^{30}}\right)^4$

11. $3\sqrt{-12} - 3\sqrt{-75} + \sqrt{-48}$

12. $(-i^{21})^9$

13. $\sqrt{-121} + 2i^{11}$

14. $i^{37} + i^{90}$

15. $\dfrac{2i^{40}}{i^{25} + i^5}$

8.2 MULTIPLYING AND DIVIDING COMPLEX NUMBERS

KEY IDEAS

Complex numbers in $a + bi$ form are multiplied and divided in much the same way that binomials are multiplied and divided.

Multiplying Imaginary Numbers

To multiply imaginary numbers, first change to the i-form, if necessary. After multiplying, simplify the product using the fact that $i^2 = -1$. For example:

- $2i \cdot 4i = 8i^2 = -8$.
- $(3i)^2 = 3i \cdot 3i = 9i^2 = -9$.
- $2i(3 - 5i) = 6i - 10i^2 = 6i - 10(-1) = 10 + 6i$.
- $\sqrt{-16} \cdot \sqrt{-49} = 4i \cdot 7i = 28i^2 = -28$.

Example 1

If $f(x) = x^3 - x^2 + x$, express $f(2i)$ in standard $a + bi$ form.

Solution: $\mathbf{4 - 6}\boldsymbol{i}$

Write the given function: $f(x) = x^3 - x^2 + x$

Let $x = 2i$:

$$\begin{aligned} f(2i) &= (2i)^3 - (2i)^2 + 2i \\ &= 8i^3 - 4i^2 + 2i \\ &= 8(-i) - 4(-1) + 2i \\ &= -8i + 4 + 2i \\ &= 4 - 6i \end{aligned}$$

Multiplying Complex Numbers

To multiply complex numbers that are in standard $a + bi$ form, use the familiar FOIL method:

$$\begin{aligned} (2+3i)(4-5i) &= \overbrace{(2)(4)}^{\text{First}} + \overbrace{(2)(-5i)}^{\text{Outer}} + \overbrace{(3i)(4)}^{\text{Inner}} + \overbrace{(3i)(-5i)}^{\text{Last}} \\ &= 8 + [-10i + 12i] + (-15)i^2 \\ &= 8 + 2i + (-15)(-1) \\ &= 23 + 2i \end{aligned}$$

TIP

The imaginary unit *i* is located above the [.] key on the bottom row of the TI-83 keypad. To use your calculator to find the product, press [(] [2] [+] [3] [2nd] [*i*] [)] [(] [4] [−] [5] [2nd] [*i*] [)] [ENTER].

Complex Conjugates

The numbers $3 + 4i$ and $3 - 4i$ are **complex conjugates**, because two complex numbers are formed by taking the sum and difference of the same two real and pure imaginary numbers. The product of a pair of complex conjugates is always a positive real number. For example, $(3 + 4i)(3 - 4i) = 3^2 + 4^2 = 25$.

MATH FACTS

The product of two complex conjugates is always a positive real number:

$$(a+bi)(a-bi) = a^2 + b^2.$$

Dividing Complex Numbers

To divide a complex number by a real number, divide each part of the complex number by the real number. For example:

$$\frac{4-6i}{2} = \frac{4}{2} - \frac{6}{2}i = 2 - 3i.$$

To divide $3 + i$ by $1 - 2i$:

- Write the division example as a fraction $\left(\frac{3+i}{1-2i}\right)$. Then multiply the numerator and the denominator of the fraction by the complex conjugate of the denominator $(1 + 2i)$ to get $\frac{1}{5} + \frac{7}{5}i$.
- Alternatively, use your calculator to divide $(3 + i)$ by $(1 - 2i)$ to obtain the equivalent result, $.2 + 1.4i$.

Example 2

Write the multiplicative inverse of $3 + 4i$ in standard $a + bi$ form.

Solution: $\mathbf{\frac{3}{25} - \frac{4}{25}i}$ or **.12 − .16*i***

The multiplicative inverse of a nonzero complex number is its reciprocal.

Method 1: Without a Calculator

The multiplicative inverse of $3 + 4i$ is

$$\frac{1}{3+4i} = \frac{1}{3+4i} \cdot \left(\frac{3-4i}{3-4i}\right) = \frac{3-4i}{3^2+4^2} = \frac{3}{25} - \frac{4}{25}i.$$

Method 2: With a Calculator

Divide 1 by $(3 + 4i)$, obtaining $.12 - .16i$.

Check Your Understanding of Section 8.2

A. Multiple Choice

1. In $a + bi$ form, $(3 + 2i)^2$ is equivalent to
(1) 5 (2) 13 (3) $5 + 12i$ (4) $13 + 12i$

2. In simplest form, $\dfrac{6\sqrt{-72}}{2\sqrt{6}}$ is equivalent to
(1) $6i\sqrt{3}$ (2) $-6i\sqrt{3}$ (3) $3i\sqrt{6}$ (4) $-3i\sqrt{6}$

3. The product of $\left(3\sqrt{-2}+1\right)$ and $\left(3\sqrt{-2}-1\right)$ is
(1) 17 (2) −19 (3) $17 - i\sqrt{2}$ (4) $9\sqrt{2} - 1$

4. The expression $\dfrac{3}{2+3i}$ is equivalent to
(1) $\dfrac{-6+9i}{13}$ (2) $\dfrac{6-9i}{-5}$ (3) $\dfrac{-6-9i}{5}$ (4) $\dfrac{6-9i}{13}$

5. The expression $i^2(2 - i)$ is equivalent to
(1) $-2 - i$ (2) $-2 + i$ (3) $2 - i$ (4) $2 + i$

6. The multiplicative inverse of $5 + 2i$ is equivalent to
(1) $\dfrac{5+2i}{21}$ (2) $\dfrac{5+2i}{29}$ (3) $\dfrac{5-2i}{21}$ (4) $\dfrac{5-2i}{29}$

7. The product of $(-2 + 6i)$ and $(3 + 4i)$ is
(1) $-6 + 24i$ (2) $-6 - 24i$ (3) $18 + 10i$ (4) $-30 + 10i$

8. Expressed in $a + bi$ form, $\frac{5}{3+i}$ is equivalent to

(1) $\frac{15}{8} - \frac{5}{8}i$ (2) $\frac{5}{3} - 5i$ (3) $\frac{3}{2} - \frac{1}{2}i$ (4) $15 - 5i$

9. If $q(x) = (4x + 1)^2$, what is $q\left(\frac{i}{2}\right)$ in $a + bi$ form?

(1) $-3 + 4i$ (2) $5 + 4i$ (3) $2 + 2i$ (4) $\frac{3}{4} + i$

10. If $f(x) = 3x^2 + 3x^{\frac{1}{2}} + 3x$, then $f(-9)$ is equal to

(1) $-270 + 9i$ (2) $216 + 9i$ (3) $216 + 3i\sqrt{3}$ (4) $270 - 3i\sqrt{3}$

B. *In each case, show how you arrived at your answer by clearly indicating all of the necessary steps, formula substitutions, diagrams, graphs, charts, etc.*

11. Express the product of $(4 - 3i)$ and $(2 + i)$ in simplest $a + bi$ form.

12. Express $\frac{4i}{2+i}$ in simplest $a + bi$ form.

13. Express $\left(-3+\sqrt{-36}\right)\left(-3-\sqrt{-36}\right)$ in simplest $a + bi$ form.

14. Express $\frac{3+2i}{1-2i}$ in simplest $a + bi$ form.

15. Express the multiplicative inverse of $\sqrt{2} - 3i$ in $a + bi$ form.

8.3 THE QUADRATIC FORMULA

KEY IDEAS

Any quadratic equation that has the form $ax^2 + bx + c = 0$ can be solved using the **quadratic formula**:

$$x = \frac{-b \pm \sqrt{b^2 - 4ac}}{2a} \quad (a \neq 0).$$

Finding Irrational Roots of Quadratic Equations

If a quadratic equation has irrational roots, its graph can be used to *estimate* its roots. The quadratic formula can be used to find the *exact* roots in radical form.

Example 1

Find the roots of $2x^2 - 3x - 1 = 0$ in radical form.

Solution: $\mathbf{\frac{3+\sqrt{17}}{4}, \frac{3-\sqrt{17}}{4}}$

- Write the quadratic formula: $x = \frac{-b \pm \sqrt{b^2 - 4ac}}{2a}$
- Let $a = 2$, $b = -3$, and $c = -1$: $= \frac{-(-3) \pm \sqrt{(-3)^2 - 4(2)(-1)}}{2(2)}$
- Simplify: $= \frac{3 \pm \sqrt{9+8}}{4} = \frac{3 \pm \sqrt{17}}{4}$

TIP

To *approximate* the roots of $2x^2 - 3x - 1 = 0$, evaluate $\frac{3+\sqrt{17}}{4}$ and $\frac{3-\sqrt{17}}{4}$ in the home screen of your calculator. Then round off the answers to the desired decimal position.

Finding Imaginary Roots of Quadratic Equations

If a quadratic equation has imaginary roots, its graph does not intersect the x-axis. The quadratic formula can be used to find these roots in $a + bi$ form.

Example 2

Find the roots of $x^2 + 5 = 2x$ in $a + bi$ form.

Solution: $\mathbf{1 + 2i, 1 - 2i}$

If $x^2 + 5 = 2x$, then $x^2 - 2x + 5 = 0$, so $a = 1$, $b = -2$, and $c = 5$. Hence:

$$\begin{aligned} x &= \frac{-b \pm \sqrt{b^2 - 4ac}}{2a} \\ &= \frac{-(-2) \pm \sqrt{(-2)^2 - 4(1)(5)}}{2(1)} \\ &= \frac{2 \pm \sqrt{16}}{2} \\ &= \frac{2 \pm 4i}{2} = \frac{2}{2} \pm \frac{4i}{2} = 1 \pm 2i \end{aligned}$$

MATH FACTS

Irrational and imaginary roots of quadratic equations always occur in conjugate pairs. For example, if $3 - \sqrt{2}$ is a root of a quadratic equation, the other root must be $3 + \sqrt{2}$. If $-4 + 3i$ is a root of a quadratic equation, the other root must be $-4 - 3i$.

Using the Discriminant

The expression $b^2 - 4ac$ underneath the radical sign in the quadratic formula is called the **discriminant**. The discriminant can be used to predict the type of roots a quadratic equation may have, as described in the table on the next page.

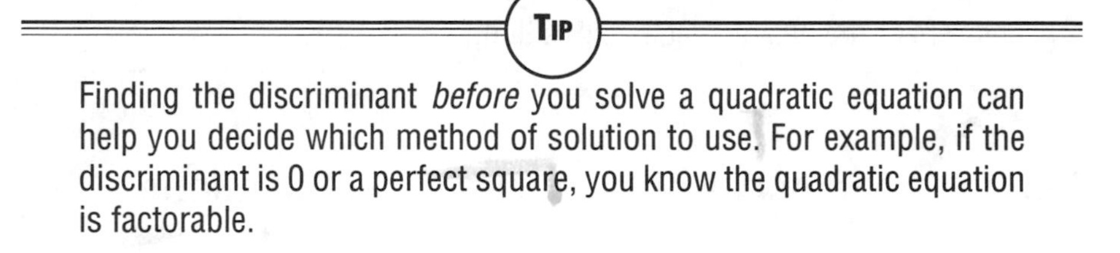

TIP

Finding the discriminant *before* you solve a quadratic equation can help you decide which method of solution to use. For example, if the discriminant is 0 or a perfect square, you know the quadratic equation is factorable.

Example 3

Describe the type of roots of each equation without solving the equation:

a. $3x^2 - 11x = 4$ b. $-2x^2 + 3x - 4 = 0$

Solution: a. **Rational and unequal**

If $3x^2 - 11x = 4$, then $3x^2$
$b = -11$, and $c = -4$:

Discrimin

Since 169 (= 13 × 13) is a
nal and unequal.

b. **Imaginary**
Since $-2x^2 + 3x - 4 = 0$
$b = 3$, and $c = -4$:

Discrim

Since the discriminant is less than 0,

Complex Numbers

Example 4

Find the smallest integer value
imaginary.

Solution: 7

- Since $x^2 - 5x +$
  $b = -5$, and c

- W

PREDICTING THE TYPE OF ROOTS OF $ax^2 + bx + c = 0$ ($a \neq 0$)

Discriminant	Type of Roots	Graph
$b^2 - 4ac > 0$	Real and unequal roots. • If $b^2 - 4ac$ is a perfect square, the roots will be rational. • If $b^2 - 4ac$ is not a perfect square, the roots will be irrational.	Two x-intercepts
$b^2 - 4ac = 0$	Two equal rational roots.	One x-intercept
$b^2 - 4ac < 0$	Two imaginary roots.	No x-intercept

RESENTATIONS

of k for which the roots of $x^2 - 5x + k = 0$ are

$k = 0$, find the value of the discriminant when $a = 1$, $= k$:

$$b^2 - 4ac = (-5)^2 - 4(1)k$$
$$= 25 \quad - 4k$$

hen the roots are imaginary, the discriminant must be less than 0:

$$25 - 4k < 0$$
$$-4k < -25$$

- Divide each side by -4, and reverse the direction of the inequality:

$$k > \frac{-25}{-4} \quad \text{or} \quad k > 6\frac{1}{4}$$

Hence, the smallest integer value of k that makes the inequality a true statement is 7.

Check Your Understanding of Section 8.3

A. Multiple Choice

1. The roots of the equation $ax^2 + 4x = -2$ are real and equal if a has a value of
(1) 1 (2) 2 (3) 3 (4) 4

2. The roots of the equation $5x^2 - 2x = -3$ are
(1) imaginary
(2) real, rational, and equal
(3) real, rational, and unequal
(4) real, irrational, and unequal

3. The roots of $3x^2 - 4x + 2 = 0$ are
(1) $\frac{1 \pm \sqrt{2}}{3}$ (2) $\frac{2 \pm \sqrt{2}i}{3}$ (3) $\frac{2 \pm \sqrt{10}}{3}$ (4) $4 \pm \frac{\sqrt{2}}{3}i$

4. Given the quadratic equation $y = ax^2 + bx + c$, in which a, b, and c are integers. If the graph intersects the x-axis in two distinct points whose abscissas are integers, then $b^2 - 4ac$ may equal
(1) -16 (2) 0 (3) 8 (4) 9

5. Which value of k will make the roots of $2x^2 - 4x + k = 0$ imaginary?
(1) -2 (2) 2 (3) 3 (4) -3

B. *In each case, show how you arrived at your answer by clearly indicating all of the necessary steps, formula substitutions, diagrams, graphs, charts, etc.*

6. For what positive value of k is the graph of $y = x^2 - 2kx + 16$ tangent to the x-axis?

7. What is the smallest integer value that makes the roots of $2x^2 - 5x + k = 0$ imaginary?

8–16. Express irrational roots in simplest radical form and imaginary roots in simplest $a + bi$ form.

8. $3x^2 - 5 = x$

9. $2x^2 + 1 = 10x$

10. $x^2 = 6x - 12$

11. $9x^2 = 2(3x - 1)$

12. $8x^2 + 29 = 28x$

13. $\dfrac{2-3x}{x} = \dfrac{x+6}{2}$

14. $5y + \dfrac{5}{y} = 8$

15. $2 + \dfrac{5}{n^2} = \dfrac{6}{n}$

16. $\dfrac{x+8}{5} + \dfrac{x+5}{x} = 1$

17. a. Solve for x and express in simplest $a + bi$ form: $3x^2 - 6x + 4 = 0$.
b. What is the largest integer value of k for which $3x^2 - 6x + k = 0$ has unequal real roots?

18. a. Using an algebraic method, estimate the roots of $2x^2 - 8x + 1 = 0$ to the *nearest tenth*.
b. Confirm your answer by drawing the graph of the equation $y = 2x^2 - 8x + 1$ and estimating the roots of $2x^2 - 8x + 1 = 0$ from the graph.

19. A homeowner wants to increase the size of a rectangular deck that now measures 15 feet by 20 feet, but the local building code laws state that a deck cannot be larger than 900 square feet. If the length and width are to be increased by the same amount, find, to the *nearest tenth*, the maximum number of feet by which the length of the deck may legally be increased.

20. A manufacturer is designing a cardboard box from a rectangular sheet of cardboard that is twice as long as it is wide. From each of the four corners of the rectangular sheet of cardboard, a square 2 centimeters on a

side is cut off and the flaps turned up to form an uncovered box. Find, to the *nearest hundreth of a centimeter*, the smallest dimensions of the rectangular sheet of cardboard so that the volume of the box formed will be at least 300 square centimeters.

21. A rocket is launched straight up from ground level. After t seconds, its height, h, in feet is given by the function $h(t) = -16t^2 + 376t$.
a. How many seconds after the rocket is launched will it strike the ground?
b. During what interval of time, to the *nearest tenth of a second*, will the height of the rocket exceed 1800 feet?

8.4 ROOTS AND COEFFICIENTS OF QUADRATIC EQUATIONS

KEY IDEAS

Without solving a quadratic equation $ax^2 + bx + c = 0$, you can use the numerical coefficients a, b, and c to figure out the sum and product of the two roots.

Applying the Sum and the Product of Roots Formulas

According to the quadratic formula, the two roots of $ax^2 + bx + c = 0$ are

$$\frac{-b - \sqrt{b^2 - 4ac}}{2a} \quad \text{and} \quad \frac{-b + \sqrt{b^2 - 4ac}}{2a}.$$

The sum of these two roots is $-\frac{b}{a}$, and their product is $\frac{c}{a}$.

Example 1

By what amount does the product of the roots of $2x^2 - 7x + 11 = 0$ exceed the sum of the roots?

Solution: **2**

- The sum of the roots is $-\frac{b}{a} = -\frac{-7}{2} = \frac{7}{2}$.

- The product of the roots is $\frac{c}{a} = \frac{11}{2}$.

- Since $\frac{11}{2} - \frac{7}{2} = \frac{4}{2} = 2$, the product of the roots exceeds the sum of the roots by 2.

Example 2

If one root of $2x^2 - 6x + k = 0$ is -1, find: a. the other root, b. the value of k

Solution: a. $\mathbf{x = 4}$

- The sum of the two roots of $2x^2 - 6x + k = 0$ is $-\frac{b}{a} = -\frac{-6}{2} = 3$.
- Since one root is -1, the sum of the roots is $-1 + x = 3$. Hence, $x = 4$.

b. $\mathbf{k = 8}$

The solution $x = -1$ must satisfy $2x^2 - 6x + k = 0$:

$$\begin{aligned} 2(-1)^2 - 6(-1) + k &= 0 \\ 2(1) \quad +6 \qquad +k &= 0 \\ k &= -8 \end{aligned}$$

Forming a Quadratic Equation from Its Roots

If you know the two roots of a quadratic equation $x^2 + Bx + C = 0$, you can form the quadratic equation by letting $B = -$(sum of roots) and $C =$ product of roots. For example, if the roots of a quadratic equation are 5 and -2, then the sum of the roots is +3 and the product of the roots is -10, so the quadratic equation is $x^2 - 3x - 10 = 0$.

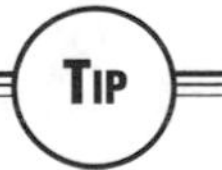

You can also form the quadratic equation by writing $(x - 5)(x - (-2)) = 0$ or, equivalently, $(x - 5)(x + 2) = 0$ and multiplying together the factors on the left side of the equation.

Example 3

Form a quadratic equation that has $5 + 2i$ as one of its roots.

Solution: $\mathbf{x^2 - 10x + 29 = 0}$

Since imaginary roots of quadratic equations occur in conjugate pairs, the other root is $5 - 2i$. The sum of the roots is $(5 + 2i) + (5 - 2i) = 10$, and the product of the roots is $(5 + 2i)(5 - 2i) = 25 - 4i^2 = 29$. Hence, the quadratic equation is $x^2 - 10x + 29 = 0$.

MATH FACTS

- The sum of the roots of $ax^2 + bx + c = 0$ is $-\frac{b}{a}$.
- The product of the roots of $ax^2 + bx + c = 0$ is $\frac{c}{a}$.
- $x^2 + [-(\text{sum of roots})]x + (\text{product of roots}) = 0$.

Check Your Understanding of Section 8.4

A. Multiple Choice

1. If -1 and 7 are the roots of the equation $x^2 + kx - 7 = 0$, then k must be equal to
(1) -7 (2) -6 (3) 6 (4) 8

2. For which equation does the sum of the roots equal 3 and the product of the roots equal 4.5?
(1) $x^2 + 3x - 9 = 0$ (3) $2x^2 + 6x + 9 = 0$
(2) $x^2 - 3x - 9 = 0$ (4) $2x^2 - 6x + 9 = 0$

3. One root of the equation $x^2 - 3x + q = 0$ is n. The other root, in terms of n, is
(1) $n - 3$ (2) $3 - n$ (3) $\frac{n}{3}$ (4) $\frac{3}{n}$

4. If the roots of $x^2 + px + q = 0$ are $1 + \sqrt{2}$ and $1 - \sqrt{2}$, then
(1) $p = 2$ and $q = -1$ (3) $p = -2$ and $q = -1$
(2) $p = -2$ and $q = 1$ (4) $p = 2$ and $q = 1$

5. Which quadratic equation has roots $3 + i$ and $3 - i$?
(1) $x^2 - 6x + 10 = 0$ (3) $x^2 - 6x + 8 = 0$
(2) $x^2 + 6x - 10 = 0$ (4) $x^2 + 6x - 8 = 0$

B. *In each case, show how you arrived at your answer by clearly indicating all of the necessary steps, formula substitutions, diagrams, graphs, charts, etc.*

6–9. For each solution set, write a quadratic equation with integer coefficients.

6. $\{1, -2\}$ **7.** $\left\{\frac{2}{3}, -1\right\}$ **8.** $\{2 \pm \sqrt{3}\}$ **9.** $\{-1 \pm 2i\}$

Unit Three TRANSFORMATIONS, FUNCTIONS, AND GROWTH

CHAPTER 9

TRANSFORMATIONS AND FUNCTIONS

9.1 REFLECTIONS AND TRANSLATIONS

KEY IDEAS

Flipping a figure over a line is an example of a *transformation* called a **reflection**. Sliding a figure to the left or to the right, up or down, or both horizontally and vertically is a **translation**. Under a transformation, each point of the original figure has exactly one *image point* and each image point corresponds to exactly one point of the original figure.

Reflecting Points in the Coordinate Axes

The notation $r_{x\text{-axis}}(2, 4) = (2, -4)$ means that the reflected image of point (2, 4) in the x-axis is (2, −4), as shown in Figure 9.1. You can determine the reflected image of point $P(x, y)$ in the coordinate axes and the origin by using these rules:

- $r_{x\text{-axis}}\ A(x, y) = A'(x, -y)$.
- $r_{y\text{-axis}}\ A(x, y) = A''(-x, y)$.
- $r_{\text{origin}}\ A(x, y) = A'''(-x, -y)$.

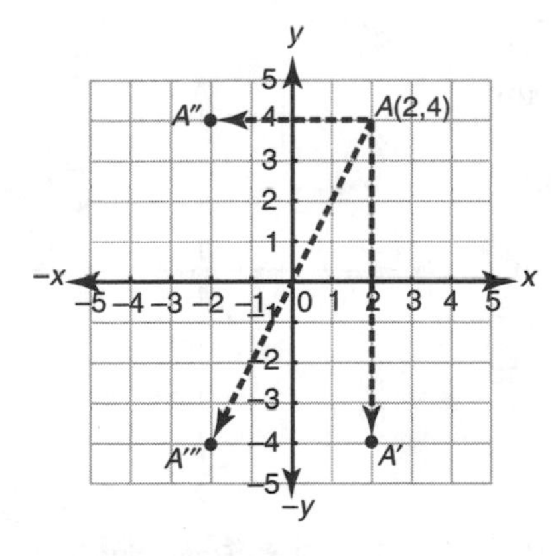

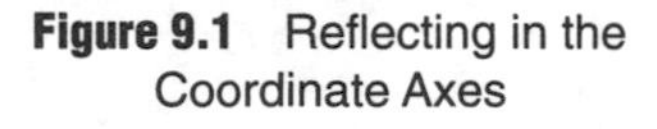
Figure 9.1 Reflecting in the Coordinate Axes

Figure 9.2 Reflecting in $y = x$ and $y = -x$

Reflecting Points in the Lines $y = \pm x$

Figure 9.2 shows the reflection of point (2, 4) in the lines $y = x$ and $y = -x$. In general:

$$r_{y=x}\, A(x, y) = A'(y, x) \qquad \text{and} \qquad r_{y=-x}\, A(x, y) = A''(-y, -x).$$

Reflecting Graphs

Graphs can be reflected in a coordinate axis or in the origin by using rules similar to those for reflecting a point.

- To reflect $y = f(x)$ in the x-axis, replace y with $-y$. For example, the reflection of $Y_1 = \sqrt{x}$ in the x-axis is $Y_2 = -\sqrt{x}$, as shown in Figure 9.3.

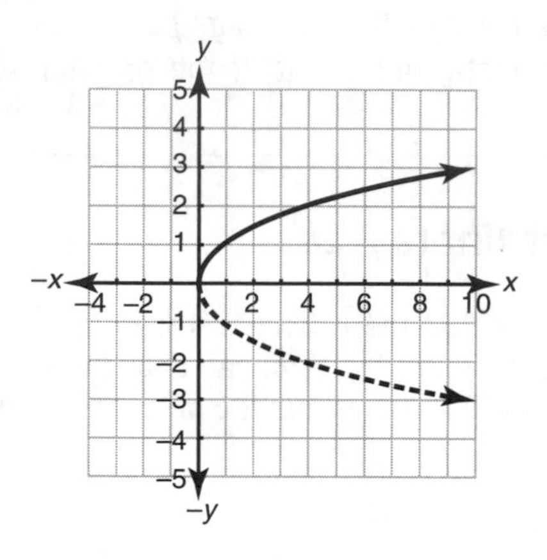

Figure 9.3 Reflecting in the *x*-Axis

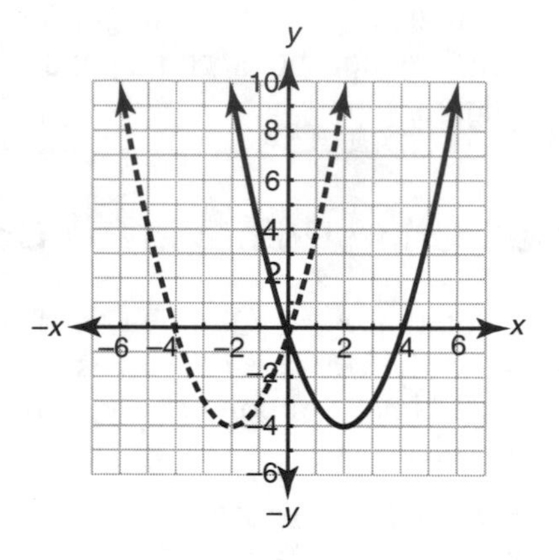

Figure 9.4 Reflecting in the *y*-Axis

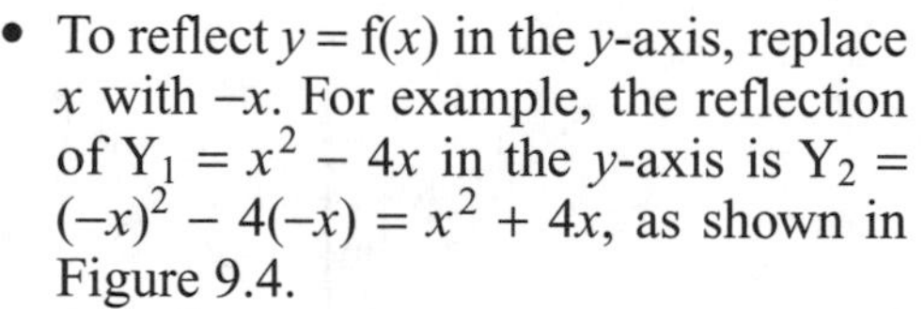

- To reflect $y = f(x)$ in the y-axis, replace x with $-x$. For example, the reflection of $Y_1 = x^2 - 4x$ in the y-axis is $Y_2 = (-x)^2 - 4(-x) = x^2 + 4x$, as shown in Figure 9.4.
- To reflect $y = f(x)$ in the origin, replace x with $-x$ and y with $-y$. For example, the reflection of $Y_1 = \sqrt{4-(x-2)^2}$ in the origin is $Y_2 = -\sqrt{4-(-x-2)^2}$, as shown in Figure 9.5.

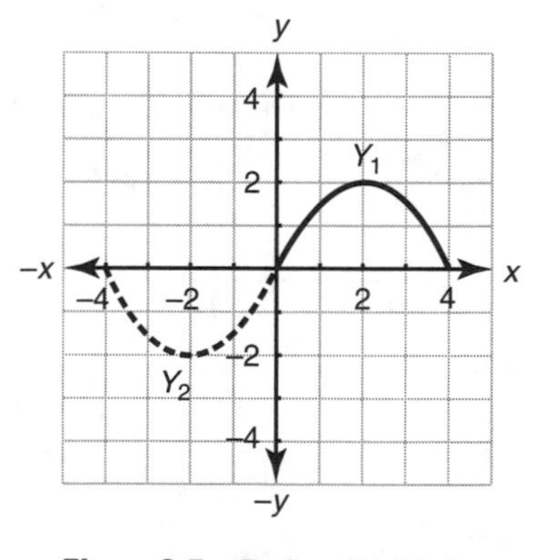

Figure 9.5 Reflecting Y_1 in the Origin

The rules for obtaining an equation of a reflected graph are summarized in the accompanying table.

REFLECTION RULES FOR THE GRAPH OF $y = f(x)$

Type of Reflection	What to Do	Function Rule
In the x-axis	Replace y with $-y$.	$f(x) \to -f(x)$
In the y-axis	Replace x with $-x$.	$f(x) \to f(-x)$
In the origin	Replace x with $-x$ and y with $-y$.	$f(x) \to -f(-x)$

Example 1

Multiple Choice:
The graph of $y = f(x)$ is shown in the accompanying diagram. Which can be the graph of $y = -f(x)$?

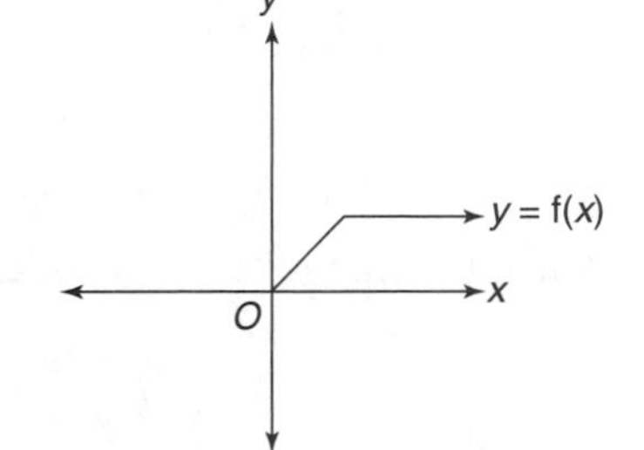

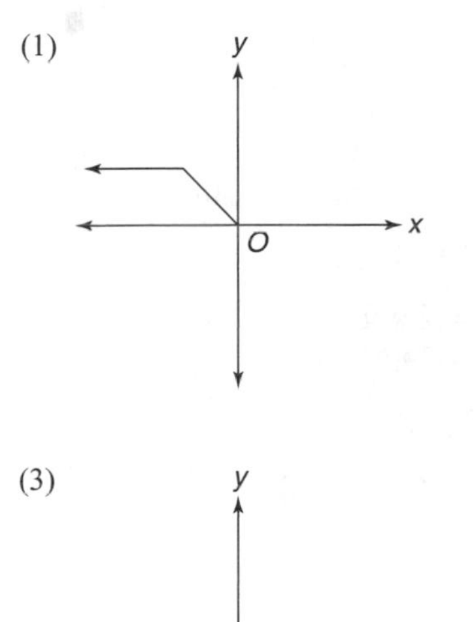

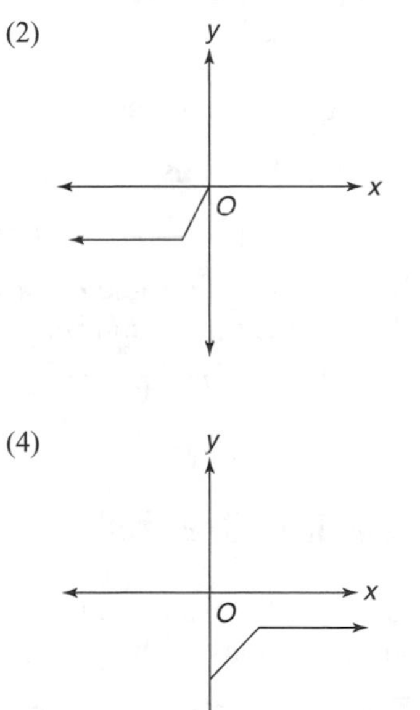

(3)

Solution: **(3)**

The graphs of $y = f(x)$ and $y = -f(x)$ are reflections of each other in the x-axis. Since choice (3) is the image of the original graph $y = f(x)$ flipped over the x-axis, its equation is $y = -f(x)$.

Translating Points

A **translation** is a transformation that "slides" each point of a figure the same distance in a given direction. The shorthand notation $T_{h,k}$ is used to represent a translation of h units horizontally and k units vertically. For example:

$$T_{2,-3}(3,4) = (3+2, 4-3) = (5,1).$$

Thus, the image of (3,4) under translation $T_{2,-3}$ is (5,1). In general,

$$T_{h,k}(x,y) = (x+h, y+k).$$

Example 2

Under a certain translation, the image of (–1,4) is (1,0). What is the image of (3,–2) under the same translation?

Solution: **(5,–6)**

Let $T_{h,k}(x, y)$ represent the given translation.

- If under this translation the image of (–1, 4) is (1, 0), then

$$T_{h,k}(-1,4) = (-1+h, 4+k) = (1,0).$$

- Hence, $-1 + h = 1$ and $4 + k = 0$, so $h = 2$ and $k = -4$.
- If the same translation is applied to (3,–2), then

$$T_{2,-4}(3,-2) = (3+2, -2-4) = (5,-6).$$

Translating Graphs

Graphs, as well as individual points, can be translated. Figure 9.6 shows vertical translations of the graph of $y = x^2$:

- $T_{0,3}$: If $Y_1 = f(x) = x^2$ is translated up 3 units, the image is $Y_2 = f(x) + 3 = x^2 + 3$.
- $T_{0,-3}$: If $Y_1 = f(x) = x^2$ is translated down 3 units, the image is $Y_3 = f(x) - 3 = x^2 - 3$.

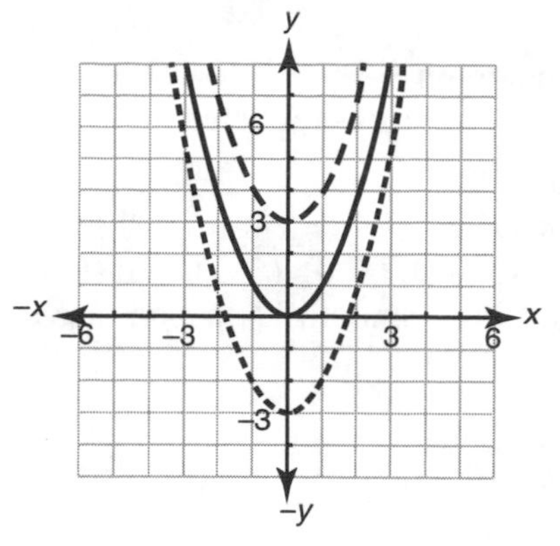

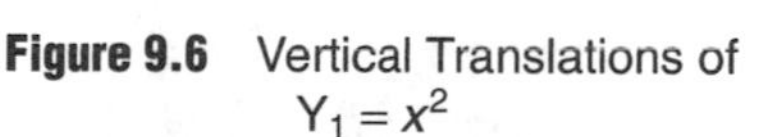

Figure 9.6 Vertical Translations of $Y_1 = x^2$

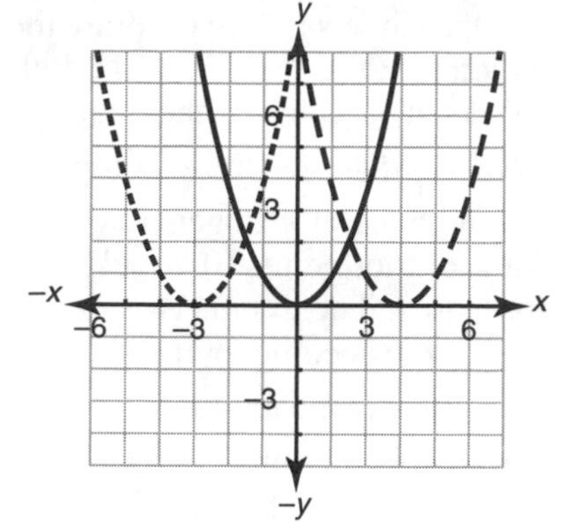

Figure 9.7 Horizontal Translations of $Y_1 = x^2$

Figure 9.7 shows horizontal translations of the graph of $y = x^2$:

- $T_{0,3}$: If $Y_1 = f(x) = x^2$ is translated 3 units to the right, the image is $Y_4 = f(x - (3)) = (x - 3)^2$.
- $T_{0,3}$: If $Y_1 = f(x) = x^2$ is translated 3 units to the left, the image is $Y_5 = f(x - (-3)) = (x + 3)^2$.

MATH FACTS

Under the translation $T_{h,k}$, the image of $y = f(x)$ is $y = f(x - (h)) + k$.

- If $h > 0$, the original graph is shifted h units to the right; if $h < 0$, the shift is $|h|$ units to the left.
- If $k > 0$, the original graph is shifted k units up; if $k < 0$, the shift is $|k|$ units down.

Vertex Form of the Parabola Equation

Under the translation $T_{h,k}$, the image of $y = x^2$ is $y = (x - (h))^2 + k$. The vertex of the parabola shifts from (0, 0) to (h, k). When an equation of a parabola is in the form $y = (x - (h))^2 + k$, the vertex of the parabola is (h, k). For example, the vertex of the parabola $y = (x + 2)^2 + 5$ is (−2,5) since $y = (x + 2)^2 + 5 = (x - (-2))^2 + 5$, so $h = -2$ and $k = 5$. Thus, $y = (x + 2)^2 + 5$ is the image of $y = x^2$ under the translation $T_{-2,5}$.

Example 3

a. On graph paper, draw the graph of the equation $y = x^2 - 6x + 8$ for all values of x in the interval $0 \le x \le 6$. Label this graph ***a***.

b. On the same set of axes, draw the image of $y = x^2 - 6x + 8$ after the translation $(x, y) \to (x - 3, y + 1)$. Label this graph ***b***.

c. Write an equation of the image of graph ***b*** after it is reflected in the x-axis.

Solution: a. See the accompanying figure.

Use your calculator to generate a table of values for $y = x^2 - 6x + 8$ from $x = 0$ to $x = 6$. Plot points (0, 8), (1, 3), (2, 0), (3, –1), (4, 0), (5, 3), and (6, 8) on graph paper. Then connect these points with a smooth U-shaped curve, as shown in the accompanying graph.

b. See the accompanying figure.

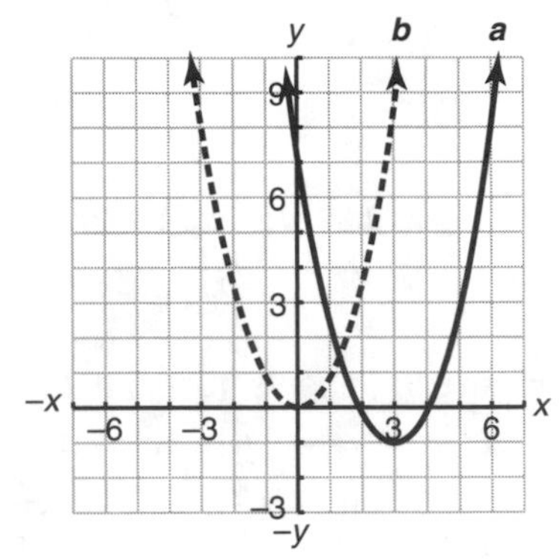

Under the translation, $(x,y) \to (x - 3, y + 1)$. Then:

$$(0, 8) \to (0-3, 8+1) = (-3, 9)$$
$$(1, 3) \to (1-3, 3+1) = (-2, 4)$$
$$(2, 0) \to (2-3, 0+1) = (-1, 1)$$
$$(3, -1) \to (3-3, -1+1) = (0, 0)$$
$$(4, 0) \to (4-3, 0+1) = (1, 1)$$
$$(5, 3) \to (5-3, 3+1) = (2, 4)$$
$$(6, 8) \to (6-3, 8+1) = (3, 9)$$

Plot the image points and connect them with a broken curve. Label this graph ***b***, as shown in the accompanying figure.

c. $\boldsymbol{y = -x^2}$

Since an equation of graph ***b*** is $y = x^2$, an equation of its reflection in the x-axis is $y = -x^2$.

TIP

Since the original parabola has its vertex at (3, –1), its equation can be rewritten as $y = (x - 3)^2 - 1$. Under the translation $(x, y) \to (x - 3, y + 1)$, $h = -3$ and $k = 1$, so the equation of the image is $y = ((x - 3) - \boxed{-3}\,)^2 - 1 + \boxed{+1} = x^2$.

Check Your Understanding of Section 9.1

A. Multiple Choice

1. The graph of which equation is the reflection of the graph of $y = -2x^2 + 3$ in the x-axis?

(1) $y = 2x^2 - 3$ (2) $y = -2x^2 - 3$ (3) $y = 2x^2 + 3$ (4) $y = -2x^2 + 3$

2. Which equation shifts the graph of $y = x^2$ horizontally to the left 4 units and vertically down 3 units?
(1) $y = (x - 4)^2 + 3$ (3) $y = (x + 4)^2 - 3$
(2) $y = (x - 3)^2 + 4$ (4) $y = (x + 3)^2 - 4$

3. The graph of $y = x^2 - x$ is reflected in the x-axis to produce the graph of Y_1. The graph of Y_1 is reflected in the y-axis to produce the graph of Y_2. What is an equation of Y_2?
(1) $y = -x^2 + x$ (2) $y = x^2 + x$ (3) $y = -x^2 + x$ (4) $y = -x^2 - x$

4. A translation maps $A(-2, 1)$ onto $A'(2, 2)$. What are the coordinates of B', the image of $B(-4, -5)$, under the same translation?
(1) (4, 0) (2) (0, –4) (3) (4, –10) (4) (–8, –6)

5. What are the coordinates of the vertex of the parabola $y = \frac{1}{2}(x - 3)^2 - 2$?
(1) (3, –2) (2) (–3, 2) (3) (–3, –2) (4) $\left(\frac{3}{2}, -2\right)$

B. *In each case, show how you arrived at your answer by clearly indicating all of the necessary steps, formula substitutions, diagrams, graphs, charts, etc.*

6. Determine an equation of the line that contains the origin and the image of $T_{2,3}(1, -4)$.

7. Determine an equation of the line that contains the reflected images of $(-1, 4)$ and $(3, -4)$ in the line $y = x$.

8. Determine an equation of the line that contains the reflected images of $P(1, -2)$ in the line $y = x$ and in the line $y = -x$.

9–12. Given the graph of $y = h(x)$ in the accompanying diagram, sketch the graph of:

y
y = h (x)
–x
x
O
–y

9. $y = 2h(x)$ 11. $y = -h(x)$

10. $y = h(-x)$ 12. $y = -h(-x)$

13. a. Using graph paper, draw the graph of $y = (x - 1)^2 - 3$ on the domain $-2 \le x \le 4$.
a. Label this graph ***a***.
b. On the same set of axes, draw the reflection of graph ***a*** in the x-axis. Label this graph with its equation.
c. On the same set of axes, draw the image of graph ***a*** under $T_{-3,2}$. Label this graph with its equation.

14. Two parabolic arches are to be built. An equation of the first arch is $y = -x^2 + 9$, with a range of $0 \leq y \leq 9$. The second arch is created by the transformation $T_{7,0}$.

a. Using graph paper, graph the equations of the two arches on the same set of axes.

b. Graph the line that is symmetric to the parabola and its transformation. Label the line of symmetry with its equation.

9.2 ROTATIONS AND DILATIONS

KEY IDEAS

A figure may be transformed by turning it. A figure can also be transformed by enlarging it or shrinking it while keeping its shape.

Rotating Figures

A **rotation** is a transformation that turns a figure through an angle about a fixed point called the **center**. In Figure 9.8, rectangle $AB'C'D'$ is the image of rectangle $ABCD$ under a 90° counterclockwise rotation about the origin. Compare the coordinates of corresponding vertices of the two rectangles. For example,

$$R_{90°}(6, 3) = (-3, 6),$$

which states that the image of point (6, 3) under a 90° counterclockwise rotation about the origin is (−3, 6).

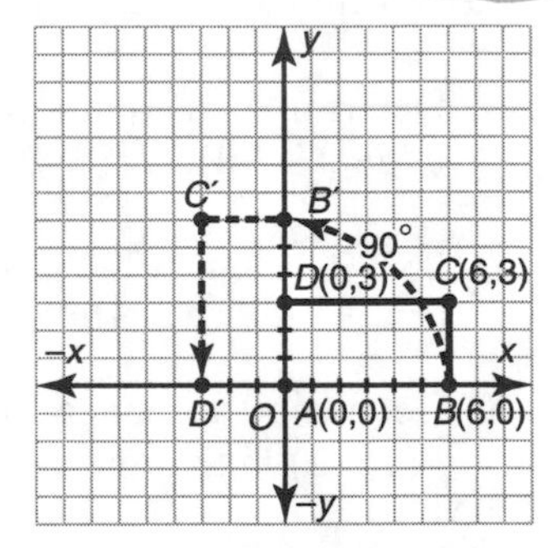

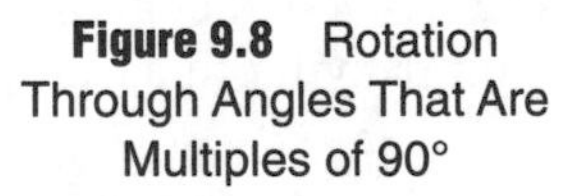
Figure 9.8 Rotation Through Angles That Are Multiples of 90°

The coordinates of the images of points rotated about the origin through angles that are multiples of 90° can be determined by using these rules:

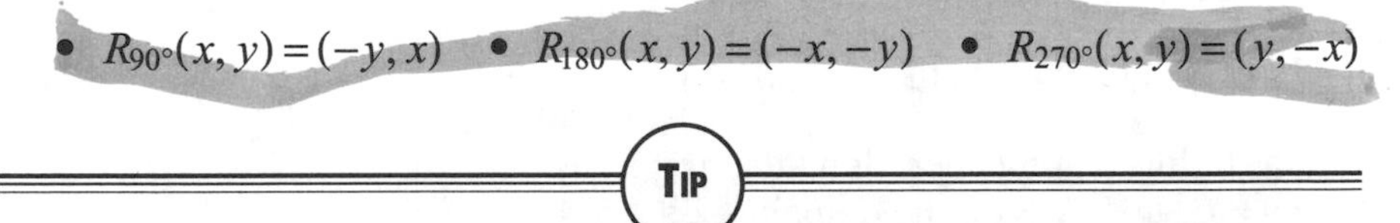

- $R_{90°}(x, y) = (-y, x)$
- $R_{180°}(x, y) = (-x, -y)$
- $R_{270°}(x, y) = (y, -x)$

TIP

It will always be assumed that the center of the rotation is the origin and that a positive angle of rotation turns a figure in the counterclockwise direction. A negative angle of rotation turns a figure in the clockwise direction.

Dilating Figures

A **dilation** is a transformation that "shrinks" or "stretches" a figure according to a given scale factor while maintaining the figure's shape. Reflections, translations, and rotations are "rigid" transformations since they produce *congruent* images, while dilations are "nonrigid" transformations because lengths are not preserved. A figure and its dilated image are similar. The notation D_k means a dilation with scale factor k. Under a dilation with a nonzero scale factor k:

- The image of each point (x, y) is (kx, ky). For example, the coordinates of the image of $(-3, 1)$ under D_2 are $(-3 \times 2, 1 \times 2) = (-6, 2)$.
- An equation of the dilated graph is obtained from the equation of the original graph by replacing x with $\frac{x}{k}$ and y with $\frac{y}{k}$. For example, the dilation of $Y_1 = x^2$ with a scale factor of 3 is $\frac{Y_2}{3} = \left(\frac{x}{3}\right)^2 = \frac{x^2}{9}$ or $Y_2 = 3\left(\frac{x^2}{9}\right) = \frac{1}{3}x^2$, as shown in Figure 9.9.

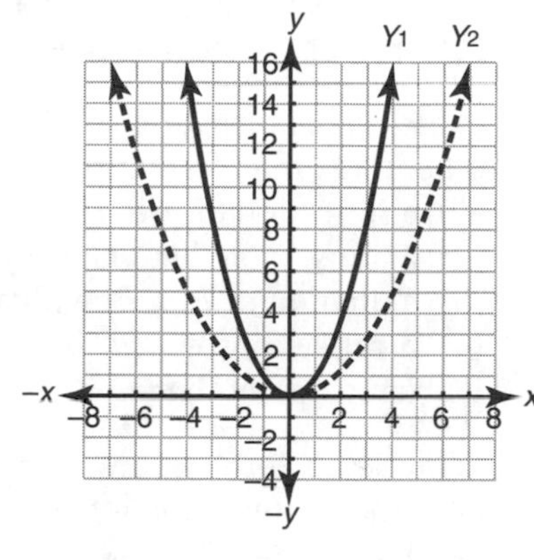

Figure 9.9 Dilation of $Y_1 = x^2$

Example

What is an equation of the image of the graph of the circle $x^2 + y^2 = 9$ under a dilation with a scale factor of 2?

Solution: $\mathbf{x^2 + y^2 = 36}$

To find an equation of the image of $x^2 + y^2 = 9$ under D_2, replace x with $\frac{x}{2}$ and y with $\frac{y}{2}$. Then $\left(\frac{x}{2}\right)^2 + \left(\frac{y}{2}\right)^2 = 9$ or, equivalently, $\frac{x^2}{4} + \frac{y^2}{4} = 9$. Multiplying each member of the equation by 4 gives $x^2 + y^2 = 36$.

Check Your Understanding of Section 9.2

A. Multiple Choice

1. What is image of point (−3, −1) under a rotation of 90° about the origin?
(1) (1, −3) (2) (3, −1) (3) (3, 1) (4) (1, 3)

2. If the dilation of (9, 3) is (6, 2), what are the coordinates of the image of (12, −6) under the same dilation?
(1) (8, −2) (2) (18, −9) (3) (8, −4) (4) (9, −18)

3. Which transformation of $P(a, b)$, where $a, b \neq 0$, is *not* equivalent to the other three transformations?
(1) a reflection of $P(a, b)$ in the origin
(2) a translation that maps $P(a, b)$ onto $P'(a-1, b-1)$
(3) a dilation of $P(a, b)$ with a scale factor of −1
(4) a reflection of $P(a, b)$ in the x-axis followed by a reflection of the image point in the y-axis

4. Which transformation when applied to point $P(a, b)$, where $a, b \neq 0$, does *not* produce the same image point as the other three transformations?
(1) $r_{(0,0)}$ (2) $R_{180°}$ (3) $T_{-2a,-2b}$ (4) $r_{y=-x}$

***B.** In each case, show how you arrived at your answer by clearly indicating all of the necessary steps, formula substitutions, diagrams, graphs, charts, etc.*

5. Given: $J = -2 + 5i$ and $K = 3 + 2i$, where i is the imaginary unit.
a. On graph paper, plot the sum of J and K and label it L.
b. On the same set of axes, plot the image of L after a counterclockwise rotation of 270° and label it L'.
c. Express L' as a complex number in $a + bi$ form.

6. Given $f(x) = x^2 - 2x$.
a. Determine an equation for $g(x)$, the function that represents the rotation of $f(x)$ 180° about the origin.
b. If $f(x)$ is rotated 90° about the origin, determine the y-coordinate of each image point on the rotated graph for which $x = -3$.

9.3 COMPOSITE FUNCTIONS

KEY IDEAS

Functions f and g may be combined to form a new function by using the output of either function as the input of the other function. The new function is called a **composite function**.

Forming Composite Functions

The process of using the output of one function as the input to a second function is called the **composition of functions**. Two functions can be composed in either order. For example, suppose $f(x) = x - 1$ and $g(x) = 5x$.

The composite function formed by using the output of function g as the input of function f is denoted as $f \circ g$. The value of this composite function is represented by $f(g(x))$, read as "f of g of *x*." If $x = 2$, then $g(2) = 5 \cdot 2 = 10$, so

$$f(g(2)) = f(10) = 10 - 1 = 9.$$

The composite function formed by using the output of function f as the input of function g is denoted as $g \circ f$. The value of this composite function is represented by $g(fx))$, read as "g of f of *x*." If $x = 2$, then $f(2) = 2 - 1 = 1$, so

$$g(f(2)) = g(1) = 5 \cdot 1 = 5 = 5.$$

Notice that $f(g(2)) \neq g(f(2))$. In general, $f \circ g$ and $g \circ f$ do not give the same composite function.

When working with composite functions, keep these things in mind:

- To evaluate $f(gx))$ and $g(f(x))$, you should first evaluate the "inside" function.
- The order in which two functions are composed matters.
- The expressions $f(g(x))$, $(f \circ g)(x)$, and $f \circ g(x)$ are equivalent.

Example 1

If $f(x) = 2x + 1$ and $g(x) = |x|$, find:
a. $f(g(-3))$ b. $(g \circ f)(-3)$

Solution: a. **7**

Since $g(-3) = |-3| = 3$, then $f(g(-3)) = f(3) = 2(3) + 1 = 7$.

b. **5**

Since $f(-3) = 2(-3) + 1 = -5$, then $(g \circ f)(-3) = g(f(-3)) = g(-5) = |-5| = 5$.

Finding an Equation for a Composite Function

If $f(x) = \dfrac{1}{x^2}$ and $g(x) = 3x$, an equation for $f(g(x))$ can be obtained as follows:

- Write the "outside" function: $f(x) = \frac{1}{x^2}$
- Replace x with the "inside" function: $f(g(x)) = [g(x)]^2$
- Let $g(x) = 3x$: $= \frac{1}{(3x)^2}$
- Simplify: $= \frac{1}{9x^2}$

You should verify that $g(f(x)) = \frac{3}{x^2}$. Since the composition of two functions is not a commutative operation, $f(g(x))$ and $g(f(x))$ can have different equations.

Example 2

The retail price of a new computer is $\$p$. The store is offering an instant rebate, authorized by the manufacturer, of \$250 as well as a store discount of 10% of the sale price.

a. Write a function f in terms of p that gives the cost of the computer after the instant rebate.
b. Write a function g in terms of p that gives the cost of the computer after receiving the store discount.
c. Form the composite functions $f(g(p))$ and $g(f(p))$. If the price of the computer is \$1200, how much money is saved if the cost is computed using $f(g(p))$ instead of $g(f(p))$?

Solutions: a. $\mathbf{f(p) = p - 250}$

b. $\mathbf{g(p) + 0.9p}$

$$g(p) = p - 0.10p = 0.9p$$

c. **\$25**

$$f(g(p)) = 0.9p - 250 \quad \text{and} \quad g(f(p)) = 0.9(p - 250)$$

$$f(g(1200)) = 0.9(1200) - 250 \quad \text{and} \quad g(f(1200)) = 0.9(1200 - 250)$$

$$= 830 \qquad\qquad = 855$$

The saving is \$855 – \$830 = \$25.

Domain of a Composite Function

The domain of the composite function $f \circ g$ consists of the numbers in the domain of g for which $g(x)$ is in the domain of f.

Example 3

Let $f(x) = \dfrac{1}{x-1}$ and $g(x) = \sqrt{x}$.

a. Find g(f(x)), and state its domain.
b. Find f(g(x)), and state its domain.

Solution: a. $\sqrt{\dfrac{\mathbf{1}}{\boldsymbol{x}-\mathbf{1}}}$, **where $\boldsymbol{x} > 1$**

$$g(f(x)) = g\left(\frac{1}{x-1}\right) = \sqrt{\frac{1}{x-1}}$$

For g(f(x)) to be a real-valued function, x must be restricted so that $x > 1$.

b. $\dfrac{\mathbf{1}}{\sqrt{\boldsymbol{x}}-\mathbf{1}}$, **where $\boldsymbol{x} \geq 0$ except $\boldsymbol{x} \neq 1$**

$$f(g(x)) = f(\sqrt{x}) = \frac{1}{\sqrt{x}-1}$$

The domain of g(x) = $\sqrt{x}$ is $x \geq 0$. Since $f(g(x)) = \dfrac{1}{g(x)-1}$, x must be restricted so that $x \geq 0$ and, in order to avoid division by 0, g(x) = $\sqrt{x} \neq 1$. If $\sqrt{x} \neq 1$, then $x \neq 1$. Hence, the domain of g(f(x)) is all $x \geq 0$ except $x \neq 1$.

Check Your Understanding of Section 9.3

A. Multiple Choice

1. If f(x) = $3x$ and g(x) = $7x - 1$, what is $(f \circ g)(4)$?
(1) 81 (2) 83 (3) 63 (4) 324

2. If f(x) = x^2 and g(x) = $-x$, what is f(g($1 - i$))?
(1) $2i$ (2) $-2i$ (3) $2 + 2i$ (4) $2 - 2i$

3. If f(x) = $4x - 1$ and $g(x) = \dfrac{x+1}{4}$, what is g(f(x))?
(1) x (2) $x + 4$ (3) $x - 4$ (4) $x - \dfrac{1}{4}$

4. If f(x) = $x^2 - 1$, g(x) = $\sqrt{2x}$, and h(x) = $2 - x$, what is $h \circ f \circ g(x)$?
(1) $3 - 4x$ (2) $4x - 2$ (3) $3 - 2x$ (4) $2x + 3$

5–7. Let f = {(0, 2), (1, −1), (4, 6), (2, 5)} and g = {(−1, 6), (1, 0), (2, 4), (4, 1)}.

5. What is the value of f(g(4))?
(1) 1 (2) 2 (3) −1 (4) 5

6. What is the value of $g \circ f(1)$?
(1) 0 (2) −1 (3) 6 (4) 4

7. For what value of x is $f(g(x))$ *not* defined?
(1) 1 (2) 2 (3) −1 (4) 4

8. If $f(x) = 2x + k$ and $g(x) = \frac{x-5}{2}$, for what value of k is $f(g(x)) = g(f(x))$?
(1) 15 (2) −5 (3) −15 (4) 5

9. Which statement *must* be true?
(1) The composite of two linear functions is a linear function.
(2) The composite of two quadratic functions is a quadratic function.
(3) $f(g(x)) = f(x) \cdot g(x)$
(4) $f(f(x) = [f(x)]^2$

B. *In each case, show how you arrived at your answer by clearly indicating all of the necessary steps, formula substitutions, diagrams, graphs, charts, etc.*

10. If $f(x) = 2x$ and $g(x) = x^2 + 8$, find all values of x such that $f(g(x)) = g(f(x))$.

11. The function $f(r) = \pi r^2$ expresses the area of a circular garden as a function of the radius r of the garden.
a. Express the radius r as a function g of the circumference C of the garden.
b. Find an equation of the composite function $f(g(C))$.
c. If the circumference of the garden is decreased by 10%, by what percent does the area of the garden decrease?

12. In a certain section of the country the average level of air pollution is $0.78\sqrt[3]{P} + 15{,}000$ parts per million, where P is the population. The population is estimated to be $7500 + 45t^2$, where t is the number of years after 1990.
a. Express the pollution level t years after 1990 as a composite function expressed in terms of t.
b. To the nearest tenth of a part per million, what level of air pollution can be expected in 2006?

9.4 COMPOSITE TRANSFORMATIONS

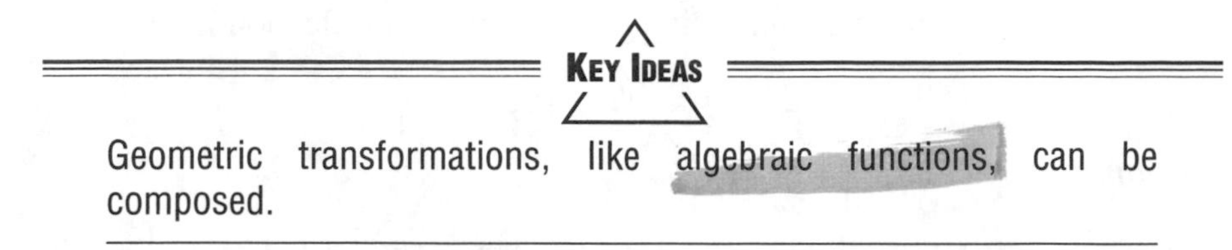

Geometric transformations, like algebraic functions, can be composed.

Composing Transformations

The order in which geometric transformations are composed matters.

Example 1

Find the coordinates of the image of $P(-2, 4)$ under each composite transformation:

a. $r_{y=x} \circ r_{y\text{-axis}}$ b. $r_{y\text{-axis}} \circ r_{y=x}$

Solutions: a. **(4, 2)**

In general, $r_{y\text{-axis}}(x, y) = (-x, y)$ and $r_{y=x}(x, y) = (y, x)$. Evaluate $r_{y=x} \circ r_{y\text{-axis}}(-2, 4)$ by performing $r_{y\text{-axis}}(-2, 4)$ first and then applying $r_{y=x}$ to the image point:

$$r_{y=x} \circ r_{y\text{-axis}}(-2, 4) = (-2, 4) \xrightarrow{r_{y\text{-axis}}} (2, 4) \xrightarrow{r_{y=x}} (4, 2).$$

b. **(–4, –2)**

Evaluate $r_{y\text{-axis}} \circ r_{y=x}(-2, 4)$ by performing $r_{y=x}(-2, 4)$ first and then applying $r_{y\text{-axis}}$ to the image point:

$$r_{y\text{-axis}} \circ r_{y=x}(-2, 4) = (-2, 4) \xrightarrow{r_{y=x}} (4, -2) \xrightarrow{r_{y\text{-axis}}} (-4, -2).$$

Example 2

In the accompanying figure, p and q are lines of symmetry for regular hexagon $ABCDEF$ intersecting at point O, the center of the hexagon. Determine the image of each composite transformation:

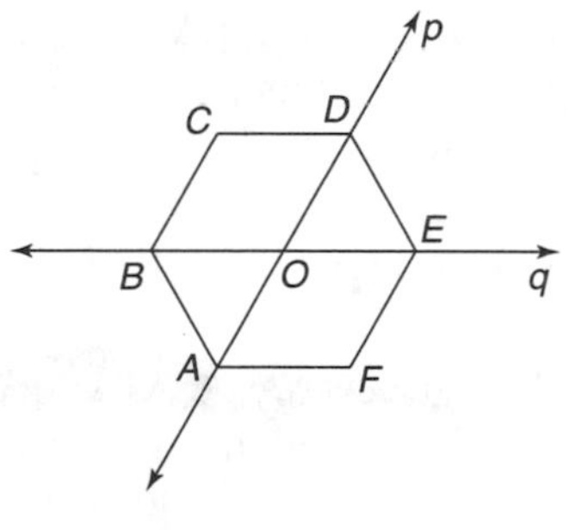

a. $r_p \circ r_q(\overline{AB})$ b. $r_q \circ r_p \circ r_q(D)$

Solution: a. $\overline{\boldsymbol{EF}}$

Reflect $\overline{AB}$ in line q, followed by the reflection of its image in line p:

$$\overline{AB} \xrightarrow{r_q} \overline{CB} \xrightarrow{r_p} \overline{EF}.$$

b. ***B***

When the composite is evaluated from right to left, point D is reflected in line q, followed by a reflection in line p, followed by a reflection in line q:

$$D \xrightarrow{r_q} F \xrightarrow{r_p} B \xrightarrow{r_q} B.$$

Example 3

In the accompanying diagram, regular hexagon $ABCDEF$ is inscribed in circle O. Line p is a line of symmetry. What is the image of $r_p \circ R_{-240°}(F)$?

Solution: ***B***

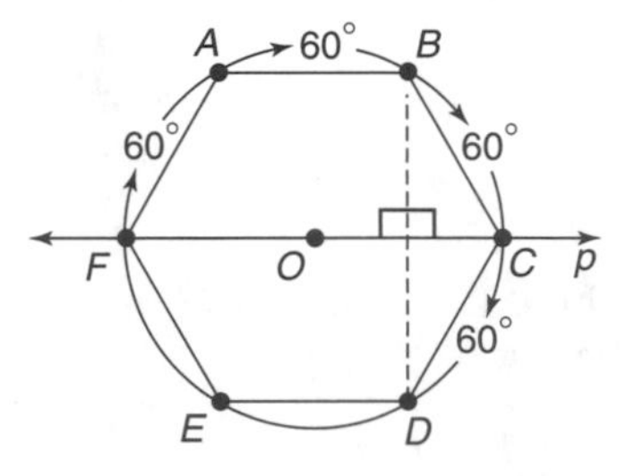

The given notation $r_p \circ R_{-240°}(F)$ means that point F must be rotated 240° in the *clockwise* direction, followed by a reflection of the image point in line p. In a regular hexagon, each of the six sides has the same length, and therefore circle O is divided into six equal arcs.

Since a circle measures 360°, the measure of each of the six arcs is $\frac{360°}{6}$, or 60°. Thus "moving" from one vertex to an adjacent vertex corresponds to a rotation of 60° about center O. Hence:

$$r_p \circ R_{-240°}(F) = F \xrightarrow{r_{-240°}} D \xrightarrow{r_p} B.$$

Glide Reflection

A **glide reflection** is a composite transformation in which a figure is reflected in a line and then the image is translated parallel to the reflecting line. In Figure 9.10, $\Delta A''B''C''$ is the image of ΔABC under a glide reflection that reflects ΔABC in line ℓ and translates its image, $\Delta A'B'C'$, a fixed distance parallel to line ℓ.

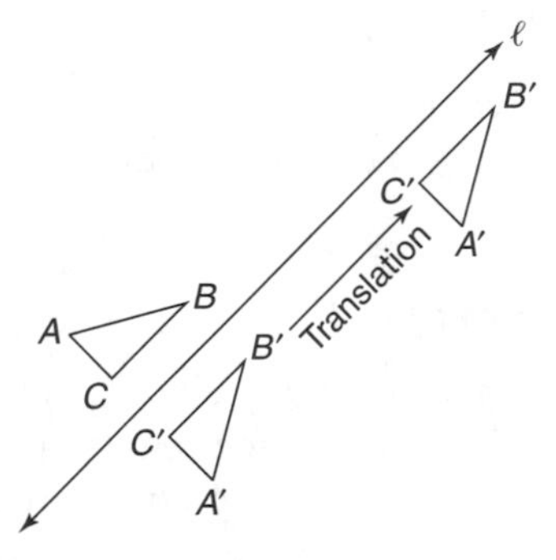

Figure 9.10 Glide Reflection

Isometry

An **isometry** is a transformation that preserves distance. Line reflections, rotations, and translations are isometries. A dilation, however, is *not* an isometry since it does not produce images that are the same size as the original figures.

An isometry that also preserves orientation is a **direct isometry**, and an isometry that reverses orientation is an **opposite isometry**. When a triangle is rotated or translated, for example, the clockwise order of the vertices does not change. Therefore, translations and rotations preserve orientation and, as a result, are *direct* isometries. A line reflection reverses orientation and is, therefore, an *opposite* isometry.

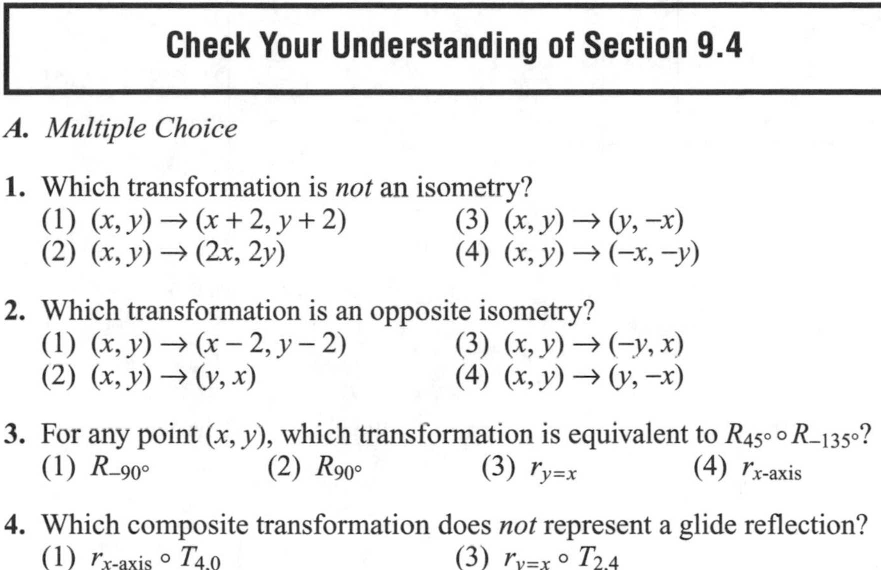

Check Your Understanding of Section 9.4

A. Multiple Choice

1. Which transformation is *not* an isometry?
(1) $(x, y) \to (x + 2, y + 2)$ (3) $(x, y) \to (y, -x)$
(2) $(x, y) \to (2x, 2y)$ (4) $(x, y) \to (-x, -y)$

2. Which transformation is an opposite isometry?
(1) $(x, y) \to (x - 2, y - 2)$ (3) $(x, y) \to (-y, x)$
(2) $(x, y) \to (y, x)$ (4) $(x, y) \to (y, -x)$

3. For any point (x, y), which transformation is equivalent to $R_{45^\circ} \circ R_{-135^\circ}$?
(1) R_{-90° (2) R_{90° (3) $r_{y=x}$ (4) $r_{x\text{-axis}}$

4. Which composite transformation does *not* represent a glide reflection?
(1) $r_{x\text{-axis}} \circ T_{4,0}$ (3) $r_{y=x} \circ T_{2,4}$
(2) $r_{y\text{-axis}} \circ T_{0,4}$ (4) $r_{x=1} \circ r_{y=3} \circ r_{x=5}$

5–7. In the accompanying diagram, ℓ and m are lines of symmetry.

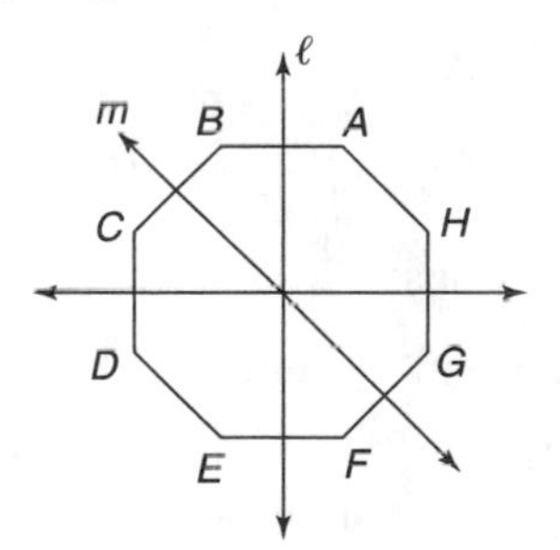

Exercises 5–7

5. What is $r_m \circ r_\ell(\overline{FG})$?
(1) $\overline{CD}$ (2) $\overline{AH}$ (3) $\overline{HG}$ (4) $\overline{BC}$

6. What is $r_\ell \circ r_m(\overline{AB})$?
(1) $\overline{CD}$ (2) $\overline{AH}$ (3) $\overline{HG}$ (4) $\overline{BC}$

7. What is $r_m \circ r_\ell \circ r_m(H)$?
(1) A (2) E (3) F (4) G

8. In the accompanying diagram, p and q are lines of symmetry for figure $ABCDEF$. What is $r_p \circ r_q \circ r_p(A)$?
(1) B (2) D (3) E (4) F

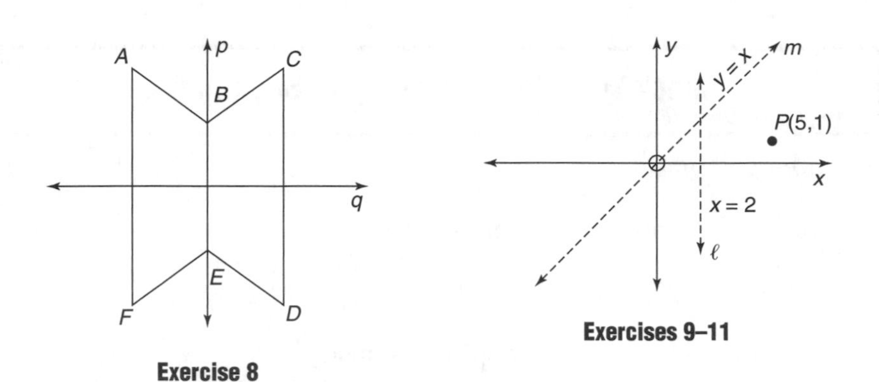

Exercise 8

Exercises 9–11

9–11. In the accompanying diagram, ℓ is the line $x = 2$ and m is the line $y = x$.

9. What are the coordinates of the image of $r_\ell \circ r_m P$?
(1) (3, 5) (2) (−1, 1) (3) (3, −5) (4) (1, −1)

10. What are the coordinates of the image of $r_m \circ r_\ell p$?
(1) (−3, 5) (2) (−1, 1) (3) (3, −5) (4) (1, −1)

11. What are the coordinates of the image of $r_{\text{origin}} \circ r_\ell \circ r_{x\text{-axis}} P$?
(1) (−1, −1) (2) (1, 1) (3) (3, 5) (4) (−5, −3)

12. Given these transformations:

$$R(x, y) \to (-x, y) \quad \text{and} \quad S(x, y) \to (y, x).$$

What is $(R \circ S)(5, -1)$?
(1) (1, 5) (2) (1, −5) (3) (−1, 5) (4) (−1, −5)

13. Which transformation does *not* have the same image as $r_{x\text{-axis}} \circ r_{y\text{-axis}}$ $P(a, b)$, where $a, b \neq 0$?
(1) $r_{x\text{-axis}} \circ r_{y\text{-axis}} P(b, a)$
(2) $r_{\text{origin}} P(a, b)$
(3) $r_{y\text{-axis}} \circ r_{x\text{-axis}} P(a, b)$
(4) $r_{y=x} P(-b, -a)$

14. The composite transformation that reflects point P through the origin, the x-axis, and the line $y = x$, in the order given, is equivalent to which rotation?
(1) $R_{90°}$ (2) $R_{180°}$ (3) $R_{270°}$ (4) $R_{360°}$

B. *In each case, show how you arrived at your answer by clearly indicating all of the necessary steps, formula substitutions, diagrams, graphs, charts, etc.*

15. If $(-4, 8)$ is the image under the composite transformation $T_{h,3} \circ T_{-2,k}(-3, 0)$, what are the coordinates of the image of $(2, -1)$ under the same composite transformation?

16. a. On graph paper, graph and label the triangle whose vertices are $A(0, 0)$, $B(8, 1)$, and $C(8, 4)$. Then graph and state the coordinates of $\Delta A'B'C'$ under the composite transformation $r_{x=0} \circ r_{y=x}(\Delta ABC)$.

b. Which single type of transformation maps ΔABC onto $\Delta A'B'C'$?
(1) rotation (2) dilation (3) glide reflection (4) translation

17. a. On graph paper, graph and label the triangle whose vertices are $A(0, 0)$, $B(8, 1)$, and $C(8, 4)$. Then graph and state the coordinates of $\Delta A'B'C'$ under the composite transformation $r_{y=-4} \circ r_{y=0}(\Delta ABC)$.

b. Which single type of transformation maps ΔABC onto $\Delta A'B'C'$?
(1) rotation (2) dilation (3) glide reflection (4) translation

18. In the accompanying diagram, P is a line of symmetry of regular hexagon $ABCDEF$. Name the point that is the image under each transformation:

a. $R_{120°} \circ r_p(C)$
b. $r_p \circ R_{-240°}(B)$
c. $R_{120°} \circ r_p \circ R_{-60°}(A)$
d. $r_p \circ R_{-240°} \circ r_p(F)$

9.5 INVERSES OF FUNCTIONS

KEY IDEAS

Interchanging the domain and range of a function forms an **inverse relation** that may or may not be a function.

One-to-One Functions

A function is a **one-to-one function** if the same value of y is *never* associated with two different values of x. A "marriage function," which pairs "husbands" to their "wives," consists of ordered pairs of the form (husband, wife). This is a one-to-one function since a wife cannot legally have two husbands. A geometric transformation, in which each point of the original figure is paired with exactly one point of the image, is also a one-to-one function since each point of the image is paired with exactly one point of the original figure. Not all functions, however, are one to one. For example, since the function $y = x^2$ includes (2, 4) and (–2, 4), it is *not* a one-to-one function.

Horizontal-Line Test

The graphs in Figures 9.11 and 9.12 represent functions. If a horizontal line intersects a graph in at most one point, as in Figure 9.11, the graph represents a one-to-one function. The graph in Figure 9.12 fails the horizontal-line test, so it is *not* a one-to-one function.

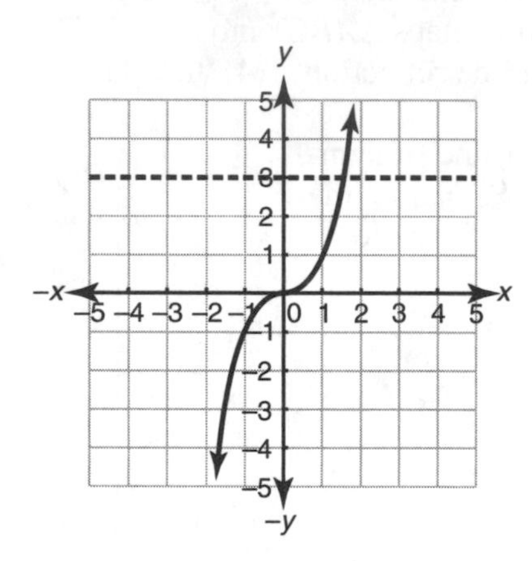

Figure 9.11 Graph Passes the Horizontal-Line Test

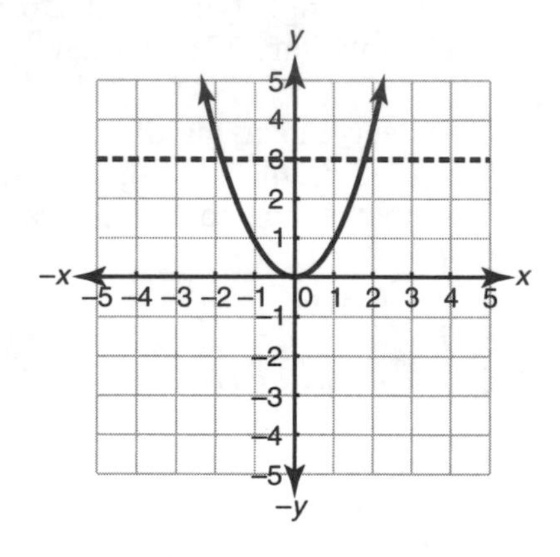

Figure 9.12 Graph Fails the Horizontal-Line Test

Inverse Functions

If a function is described by a set of ordered pairs, its inverse is formed by interchanging x and y in each ordered pair. For example, the inverse of f = {(0, –1), (1, 0), (2, 1)} is {(–1, 0), (0, 1), (1, 2)}. In this case the inverse of function f is also a function.

An inverse relation is not necessarily a function. For example, if g = {(–2, 4), (1, 1), (3, 4)}, then g does not have an inverse function because interchanging the x and y in each ordered pair produces the set of ordered pairs {(4, –2), (1, 1), (4, 3)}, which is *not* a function. Since function g is not a one-to-one function, it does not have an inverse function. Only one-to-one functions have inverse functions.

MATH FACTS

If function f is a one-to-one function, then it has an inverse function. The inverse function is denoted by f^{-1}, read as "the inverse of function f." For example, if f = {(0, –1), (1, 0), (2, 1)}, then f^{-1} = {(–1, 0), (0, 1), (1, 2)}. Do not confuse the inverse function notation f^{-1} with exponent notation, where $f^{-1} = \frac{1}{f}$.

Finding Equations of Inverse Functions

When a function is expressed as an equation, its inverse can be obtained by interchanging x and y. For example, to form the inverse of $y = 2x - 3$:

- Write the original function: $y = 2x - 3$
- Interchange x and y: $x = 2y - 3$
- Solve for y: $y = \frac{1}{2}x + \frac{3}{2}$

Compare the tables of values in Figures 9.13 and 9.14, where $Y_1 = 2x - 3$ and $Y_2 = \frac{1}{2}x + \frac{3}{2}$, respectively. For Figure 9.14, ΔTbl = 2. Corresponding ordered pairs of values have their x and y members interchanged.

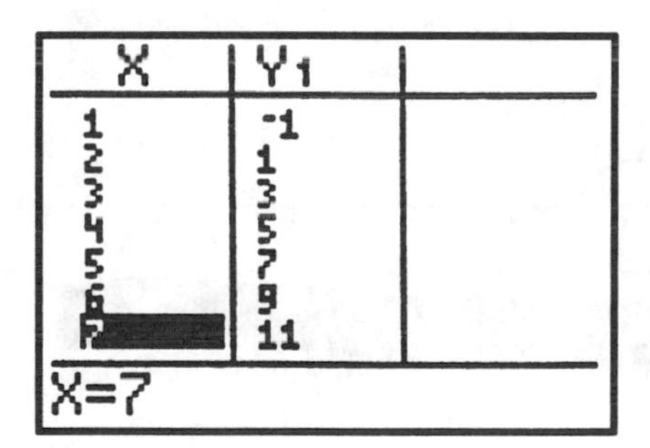

X	Y1
1	-1
2	1
3	3
4	5
5	7
6	9
7	11

X=7

Figure 9.13 Table for $Y_1 = 2x - 3$

X	Y2
-1	1
1	2
3	3
5	4
7	5
9	6
11	7

X=11

Figure 9.14 Table for $Y_2 = \frac{1}{2}x + \frac{3}{2}$

Example 1

Find the inverse of $f(x) = x^3 - 1$.

Solution: $\mathbf{f^{-1}(x) = (x + 1)^{\frac{1}{3}}}$

- Write the original function: $y = f(x) = x^3 - 1$
- Interchange x and y: $x = y^3 - 1$
- Solve for y: $y = (x+1)^{\frac{1}{3}}$

TIP

You can verify, using your graphing calculator, that $Y_1 = x^3 - 1$ passes the horizontal-line test and, as a result, has an inverse function.

Example 2

If $f(x) = x^2$: a. Determine the inverse of function f.
b. Determine whether the inverse is a function.

Solution: a. $\boldsymbol{y = \pm\sqrt{x}}$

Since $y = x^2$, the inverse relation is $x = y^2$. Solving for y gives $y = \pm\sqrt{x}$.

b. **No** Use your graphing calculator to obtain the graph in the accompanying figure. Since the graph of $x = y^2$ is comprised of the graphs of $Y_1 = \sqrt{x}$ and $Y_2 = -\sqrt{x}$, it fails the vertical-line test. Hence, $x = y^2$ or, equivalently, $y = \pm\sqrt{x}$ is *not* a function.

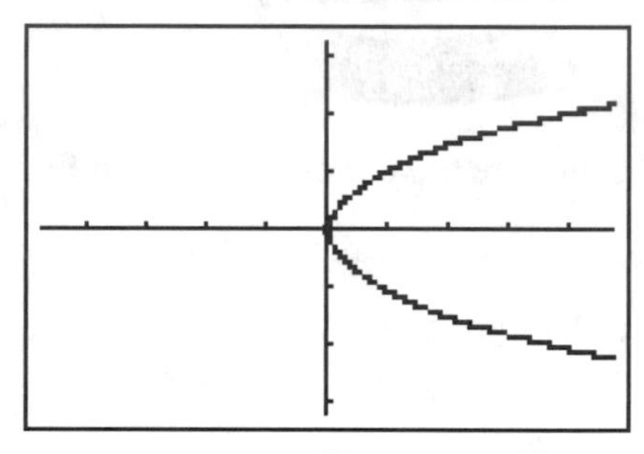

Graphs of $Y_1 = \sqrt{x}$ and $Y_2 = -\sqrt{x}$

TIP

Since the parabola $y = x^2$ fails the horizontal-line test, its inverse is *not* a function. If the domain of $f(x) = x^2$ is restricted so that $x \geq 0$, then $f^{-1}(x) = \sqrt{x}$.

Symmetry and Inverse Functions

Figure 9.15 shows the graphs of $Y_1 = 2x - 3$ and its inverse, $Y_2 = \frac{1}{2}x + \frac{3}{2}$. Figure 9.16 shows the graphs of $Y_1 = f(x) = x^3 - 1$ and its inverse, $Y_2 = f^{-1}(x+1)^{\frac{1}{3}}$. In each figure, the broken line is the line of symmetry, $y = x$.

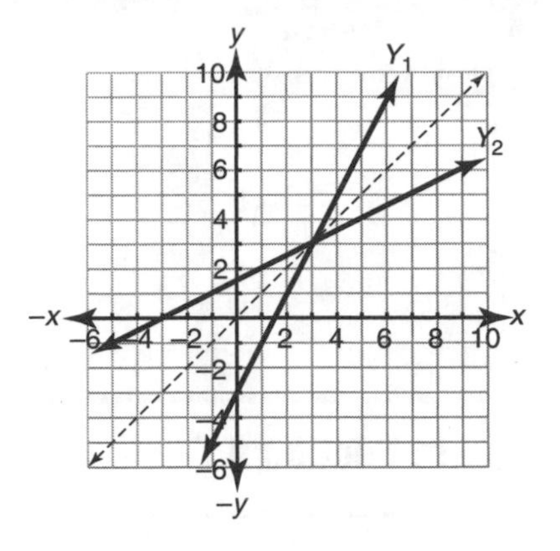

Figure 9.15 Symmetry of Linear Functions with Respect to $y = x$

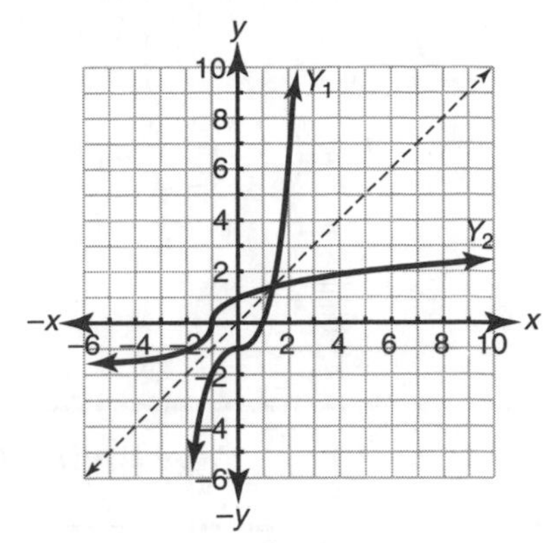

Figure 9.16 Symmetry of Nonlinear Functions with Respect to $y = x$

On the basis of Figures 9.15 and 9.16 we can say that:

- A function and its inverse are symmetric to the line $y = x$.
- Either graph may be obtained by reflecting the other graph in the line $y = x$.
- If point (a, b) is on the graph of Y_1, then point (b, a) is on the graph of Y_2.

Composing Inverse Functions

Composing inverse functions in either order always produces the starting value, x, since one function undoes the effect of the other.

MATH FACTS

To prove that functions f and g are *inverse* functions, show that

$$f(g(x)) = x \quad \text{and} \quad g(f(x)) = x$$

for all values of x for which these functions are defined.

To prove algebraically that $f(x) = x^3 - 1$ and $g(x) = (x+1)^{\frac{1}{3}}$ are inverse functions, show that $f(g(x)) = x$ and $g(f(x)) = x$:

$$\begin{aligned} f(g(x)) &= f\left((x+1)^{\frac{1}{3}}\right) \\ &= \left((x+1)^{\frac{1}{3}}\right)^3 - 1 \\ &= (x+1)-1 \\ &= x \end{aligned}$$

and

$$\begin{aligned} g(f(x)) &= g\left(x^3-1\right) \\ &= \left(\left(x^3-1\right)+1\right)^{\frac{1}{3}} \\ &= \left(x^3\right)^{\frac{1}{3}} \\ &= x \end{aligned}$$

Check Your Understanding of Section 9.5

A. Multiple Choice

1. Which equation is the inverse of the linear function $y = -3x + 4$?

(1) $y = \frac{1}{3}x - \frac{4}{3}$
(2) $y = -\frac{1}{4}x + \frac{3}{4}$
(3) $y = -\frac{1}{3}x + \frac{4}{3}$
(4) $y = -\frac{1}{4}x - \frac{3}{4}$

2. Under which type of transformation is the image of a one-to-one function the inverse of that function?

(1) rotation of 180°
(2) reflection in the y-axis
(3) reflection in the origin
(4) reflection in $y = x$

3. If $f(x) = 2x^3 + \frac{1}{2}$, what is the value of $f^{-1}\left(f\left(\frac{1}{2}\right)\right)$?

(1) $\frac{3}{4}$ (2) 2 (3) $\frac{1}{2}$ (4) $\frac{4}{3}$

4. If a function is defined by the equation $y - 3 = 2x$, which equation defines the inverse of this function?

(1) $y = \frac{1}{2}x - \frac{3}{2}$
(2) $x = \frac{1}{2}y - \frac{3}{2}$
(3) $y = -2x - 3$
(4) $y = \frac{1}{2}x + \frac{1}{3}$

5. If $y = 7x + 2$ and $y = \frac{x}{7} + k$ are inverse functions, what is the value of k?

(1) -2 (2) $-\frac{2}{7}$ (3) $\frac{2}{7}$ (4) $\frac{7}{2}$

6. If $g(x)=\frac{1}{x-2}$, where $x \neq 2$, what is $g^{-1}\left(\frac{1}{2}\right)$?

(1) $-\frac{2}{3}$ (2) 2 (3) $\frac{3}{2}$ (4) 4

7. Which function does *not* have an inverse function?

(1) $y=|x|$ (2) $y=\frac{2}{x}$ (3) $y-x^3=0$ (4) $y=\frac{2x-1}{4}$

B. *In each case, show how you arrived at your answer by clearly indicating all of the necessary steps, formula substitutions, diagrams, graphs, charts, etc.*

8. Given the function $f(x)=\frac{\sqrt{x+3}}{2}$.

a. State the domain and range of function f.

b. Find f^{-1}, the inverse of function f, and state its domain and range.

9. a. On the same set of axes, draw on graph paper $f(x) = 2x^2$ and $f^{-1}(x)$ in the interval $0 \leq x \leq 2$.

b. State the coordinates of the points of intersection.

9.6 GRAPHS OF QUADRATIC RELATIONS

KEY IDEAS

Circles, ellipses, and hyperbolas are graphs of quadratic equations.

Circle: $(x - h)^2 + (y - k)^2 = r^2$

The graph of the equation $x^2 + y^2 = r^2$ is a *circle* whose center is at the origin with radius r. Under the translation $T_{h,k}$, the image of $x^2 + y^2 = r^2$ is $(x - h)^2 + (y - k)^2 = r^2$, which is a circle with center at (h, k) and radius r, as shown in Figure 9.17. Since a circle fails the vertical-line test, its equation does *not* represent a function.

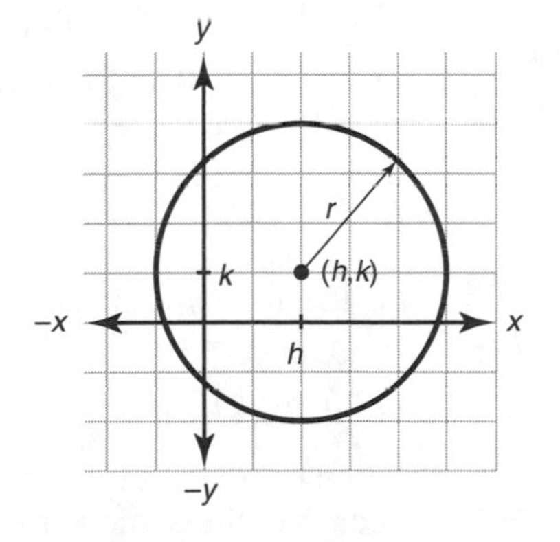

Figure 9.17 Graph of $(x - h)^2 + (y - k)^2 = r^2$

Square Viewing Windows

A **square window** is a window whose dimensions have the same ratio (approximately 3 to 2) as the actual pixel dimensions of the screen. In a square window, graphs have their true shapes. To create a square window, do one of the following:

- Select a decimal window (option 4) in the ZOOM menu.
- Set the screen variables to a fixed multiple of a basic decimal window.
- Select a square window (option 5) in the ZOOM menu.

Figure 9.18 compares the graphs of $x^2 + y^2 = 4$ in square and nonsquare windows.

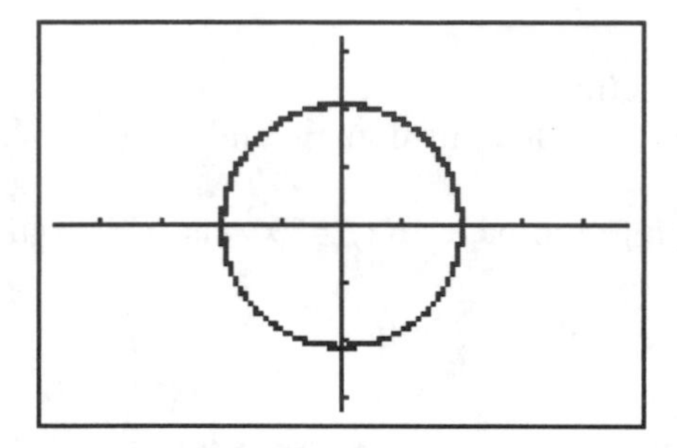

Figure 9.18(a) $x^2 + y^2 = 4$ in a Decimal Window

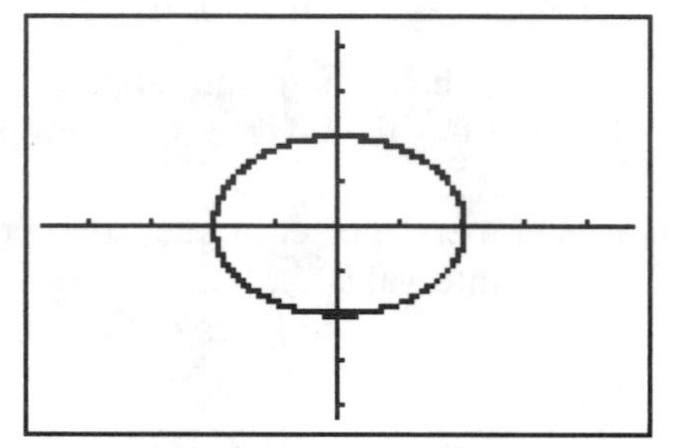

Figure 9.18(b) $x^2 + y^2 = 4$ in a $[-4, 4] \times [4, 4]$ Window

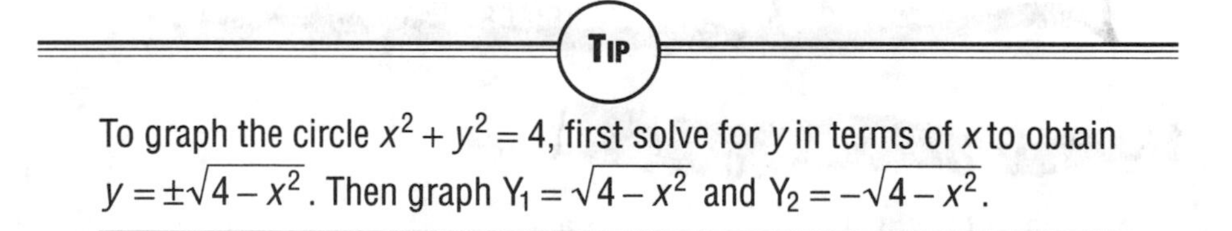

TIP

To graph the circle $x^2 + y^2 = 4$, first solve for y in terms of x to obtain $y = \pm\sqrt{4 - x^2}$. Then graph $Y_1 = \sqrt{4 - x^2}$ and $Y_2 = -\sqrt{4 - x^2}$.

Example

What are the radius and the coordinates of the circle whose equation is $(x - 2)^2 + (y + 3)^2 = 49$?

Solution: **7**; **(2,–3)**

The equation $(x - 2)^2 + (y + 3)^2 = 49$ can be rewritten as $(x - 2)^2 + (y - [-3])^2 = 7^2$, which has the same form as $(x - h)^2 + (y - k)^2 = r^2$, where $h = 2$, $k = -3$, and $r = 7$. Hence, the center of the circle is at $(2, -3)$, and the radius is 7.

Ellipse: $ax^2 + by^2 = c$ ($a \neq b$)

In the standard form of an equation of a circle, the numerical coefficients of the x^2- and y^2-terms are the same. When these numerical coefficients are unequal, as in $4x^2 + 9y^2 = 36$, the graph is an ellipse, as shown in Figure 9.19. In general:

- The graph of $ax^2 + by^2 = c$, where $a \neq b$ and a, b, and c have the same sign, is an ellipse whose center is at the origin.
- An ellipse fails the vertical-line test, so its equation is *not* a function.
- An ellipse has horizontal and vertical lines of symmetry.
- If an equation of an ellipse is written in the form $\frac{x^2}{a^2} + \frac{y^2}{b^2} = 1$, the ellipse has its center at the origin with x-intercepts at $(\pm a, 0)$ and y-intercepts at $(0, \pm b)$.

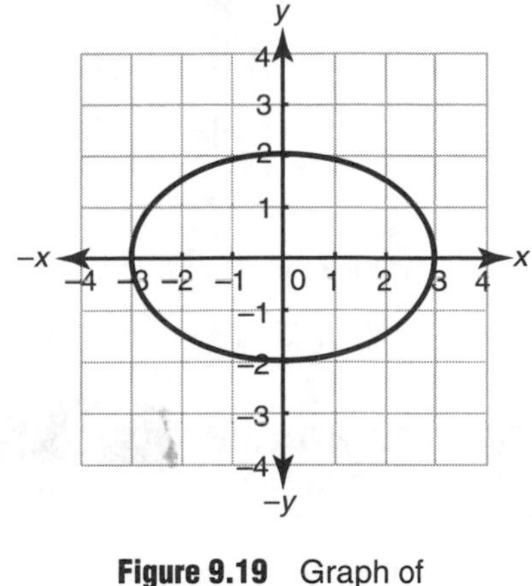

Figure 9.19 Graph of $4x^2 + 9y^2 = 36$

Rectangular or Equilateral Hyperbola: $xy = k$ ($k \neq 0$)

In the equation $xy = 8$, the product of x and y is equal to a nonzero constant. The graph of this type of equation is a *rectangular* or *equilateral hyperbola*. The rectangular hyperbola in Figure 9.20 has two disconnected branches located in opposite quadrants. As x increases along the branch in the first quadrant, the corresponding value of y decreases so that the product of x and y is always 8 for each point on the curve.

A rectangular hyperbola has these properties:

- The graph passes the vertical-line test, so $xy = k$ or, equivalently, $y = \frac{k}{x}$ is a function.
- The coordinate axes are *asymptotes* of the graph—lines that frame the graph and to which the graph gets closer and closer, but never touches, as x increases or decreases without bound.
- The graph is located in Quadrants I and III when $k > 0$, and in Quadrants II and IV when $k < 0$. Each branch of the hyperbola is the reflection of the other in the origin.

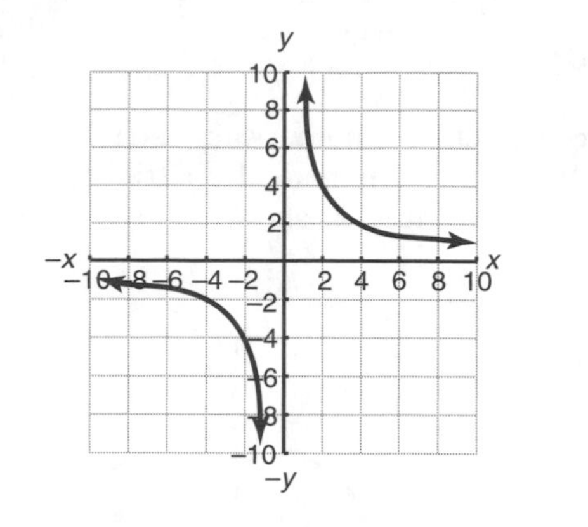

Figure 9.20 Graph of $xy = 8$

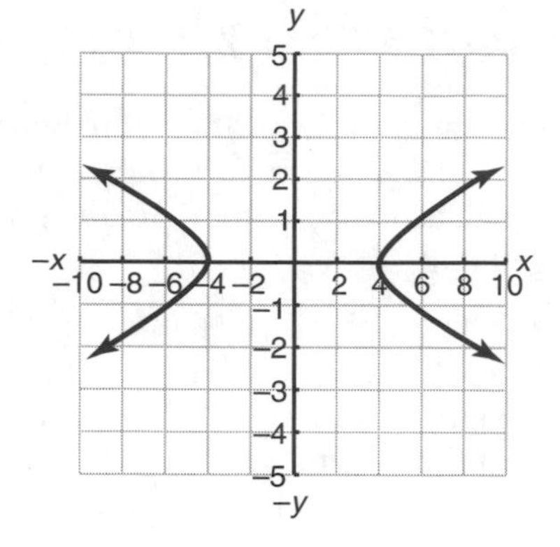

Figure 9.21 Graph of $x^2 - 4y^2 = 16$

Nonrectangular Hyperbola: $ax^2 + by^2 = c$ ($ab < 0$ and $c \neq 0$)

The signs of the x^2- and y^2-terms of the equation $x^2 - 4y^2 = 16$ are different. The graph of this type of equation is a *nonrectangular hyperbola*, as shown in Figure 9.21.

Systems of Quadratic Equations

The solution set of a system of two quadratic equations contains, at most, four ordered pairs of real numbers. The graphs of

$$x^2 + y^2 = 13$$
$$xy = 6$$

appear in a standard coordinate grid in Figure 9.22. To solve this system of equations, using your graphing calculator, first solve each equation for y in terms of x:

$$Y_1 = \sqrt{13 - x^2}, \qquad Y_2 = -\sqrt{13 - x^2}, \qquad Y_3 = \frac{6}{x}.$$

Then use the **intersect** feature of your graphing calculator to find that the coordinates of the points of intersections of the graphs of Y_1, Y_2, and Y_3 are $(-3, -2)$, $(3, 2)$, $(-2, -3)$, and $(2, 3)$. Figure 9.22 shows the graphs in a $[-9.4 \times 9.4]$ by $[-6.2 \times 6.2]$ square window.

To confirm your answers algebraically, substitute each ordered pair solution into both of the original equations and verify that the values for x and y work in each equation. This check is left for you to complete.

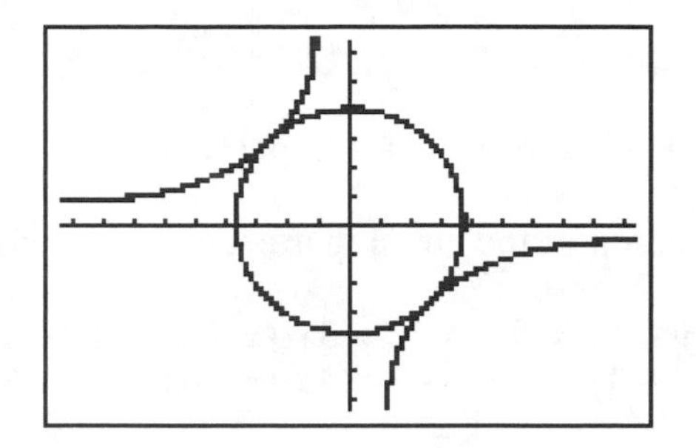

Figure 9.22 Solving $x^2 + y^2 = 13$ and $xy = 6$ Graphically

TIP

The graphs of two quadratic equations may intersect in fewer than four points. A system of quadratic equations may have imaginary roots that occur in conjugate pairs (see Figure 9.23) or can have duplicate real roots if the graphs are tangent to each other (see Figure 9.24).

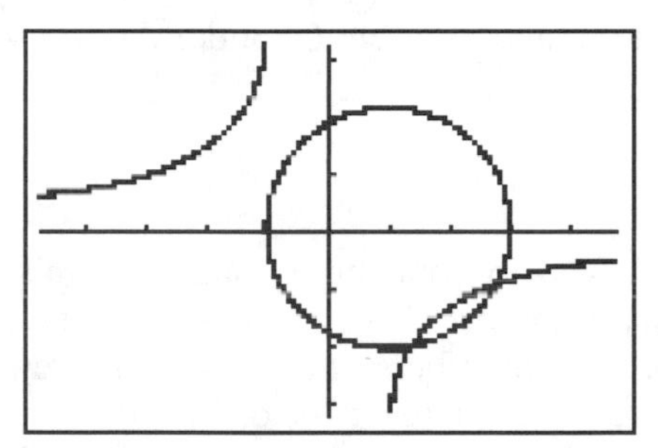

Figure 9.23 Graph of $(x - 1)^2 + y^2 = 4$ and $xy = -3$

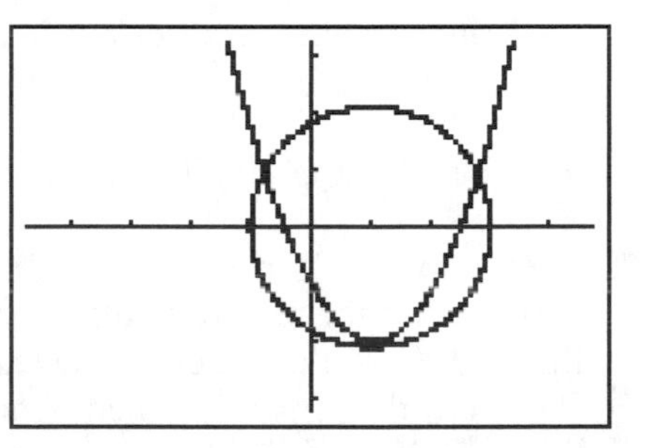

Figure 9.24 Graph of $(x - 1)^2 + y^2 = 4$ and $y = (x - 1)^2 - 2$

Check Your Understanding of Section 9.6

A. Multiple Choice

1. The graph of the equation $4x^2 + 4y^2 = 28$ is
(1) an ellipse (2) a circle (3) a parabola (4) a hyperbola

2. The graph of which equation is an ellipse?
(1) $2x^2 - 3y^2 = 6$
(2) $3x^2 + 3y^2 = 12$
(3) $y^2 = 8 - 2x^2$
(4) $3x^2 - 9 = y$

3. The graph of $\frac{x}{2} = \frac{3}{y}$ is
(1) a circle (2) an ellipse (3) a hyperbola (4) a line

4. Which is an equation of the circle that is tangent to the x-axis and whose center is at $(-4,3)$?
(1) $(x+4)^2 + (y-3)^2 = 9$ (3) $(x-4)^2 + (y+3)^2 = 9$
(2) $(x+4)^2 + (y-3)^2 = 16$ (4) $(x-4)^2 + (y+3)^2 = 16$

5. Which is an equation of the image of the graph of $xy = 4$ under the transformation D_2?
(1) $xy = 8$ (2) $xy = 2$ (3) $xy = 16$ (4) $xy = 1$

B. *In each case, show how you arrived at your answer by clearly indicating all of the necessary steps, formula substitutions, diagrams, graphs, charts, etc.*

6. a. Write an equation of circle O with center at $(-1, 2)$ and a radius length of 6.
b. Write an equation of circle O', the image of circle O under the transformation $T_{-1,3}$.
c. Write an equation of circle O'', the image of circle O' under the transformation D_2.

7 and 8. Solve using a system of two equations.

7. The members of a running club planned to contribute equally to raise \$900 to pay for refreshments at a marathon run. When it was discovered that \$1200 would be needed instead, three members of the club withdrew. If each of the remaining members needed to increase his or her contribution by \$40 to raise the necessary amount, how many members were in the club originally? [*Only an algebraic or graphical solution will be accepted.*]

8. A pet shop bought a litter of puppies for \$320. All except four were sold, and the total sum received from the sale was also \$320. If each puppy was sold for \$40 more than was paid for it, how many puppies were in the litter? [*Only an algebraic or graphical solution will be accepted.*]

9. a. On graph paper, draw the graph of $xy = -10$ using at least eight values of x in the interval $-10 \le x \le 10$ $(x \ne 0)$.
b. On the same set of axes, draw and label the graph of the image of $xy = -10$ after a rotation of 90°.
c. What is an equation of the graph drawn in part b?

10 and 11. Solve for x and y, and check the solutions algebraically.

10. $(x-3)^2+(y+2)^2=10$
$x+y=3$

11. $2x+y=6$
$2x^2-3y^2=2$

12 and 13. Estimate to the *nearest tenth* the coordinates of all points that satisfy each system of equations.

12. $x^2+y^2=49$
$xy=18$

13. $x^2+y^2=16$
$y=x^2+2$

9.7 TYPES OF VARIATION

KEY IDEAS

A variable may be related to a power, a product, or a combination of powers and products of other variables.

Basic Types of Variation

There are three basic types of variation:

- **Direct variation**. If y *varies directly as* x, then $y = kx$. The graph of a direct variation is a line through the origin whose slope, k, is the constant of variation.
- **Inverse variation**. If y *varies inversely as* x, then $y=\frac{k}{x}$ or, equivalently, $xy = k$. When two variables are inversely related, their product is constant. Therefore, as the value of one variable increases, the value of the other must decrease so that their product does not change. The graph of an inverse variation is a rectangular hyperbola.
- **Joint variation**. If z *varies jointly as* x and y, then $z = kxy$.

Example

If y varies inversely as x, and $y = 33$ when $x = 6$, what is y when $x = 9$?

Solution: $\boldsymbol{y = 22}$

Since y varies inversely as x, the product of the variables must be the same for any two ordered pairs that belong to the relation:

$$x_1 \cdot y_1 = x_2 \cdot y_2$$

Let $x_1 = 6$, $y_1 = 33$, and $x_2 = 9$: $(6)(33) = 9(y)$

$$198 = 9y$$

$$\frac{198}{9} = \frac{9y}{9}$$

$$22 = y$$

Power and Combined Variation

A variable may vary directly, inversely, or jointly as the *power* of another variable. For example, if z varies jointly as x and the square of y, then $z = kxy^2$. Three or more variables may be related by a *combination* of types of variation. For example, the volume, V, of a contained gas varies directly as the temperature, T, and inversely as the pressure, P, so $V = \frac{kT}{P}$.

Check Your Understanding of Section 9.7

A. Multiple Choice

1. The illumination of a light source varies inversely as the square of the distance from that light source. When the distance from the light source is doubled, the amount of illumination provided by the light source is multiplied by

(1) 2 (2) $\frac{1}{2}$ (3) 4 (4) $\frac{1}{4}$

2. The centripetal force on an object moving in a circular path varies directly as the square of its speed and inversely as the radius of the circle. If both the speed and the radius are halved, the centripetal force on the object is

(1) halved (2) doubled (3) divided by 8 (4) unchanged

3. Which table of values does *not* show that x varies inversely as y?

(1)

x	y
2	50
4	25
10	10

(2)

x	y
4	12
3	16
0.5	48

(3)

x	y
–3	10
5	–6
–15	2

(4)

x	y
$\sqrt{8}$	$\sqrt{6}$
4	$\sqrt{3}$
$\sqrt{24}$	$\sqrt{2}$

4. Given that y varies inversely as x and x varies directly as z. If z is doubled, then y is
   (1) halved (2) doubled (3) divided by 4 (4) unchanged

B. *In each case, show how you arrived at your answer by clearly indicating all of the necessary steps, formula substitutions, diagrams, graphs, charts, etc.*

5. The pressure, P, exerted by the wind on the wall of a house varies directly as the square of the velocity, v, of the wind. If $P = 250$ when $v = 10$, find P when $v = 40$.

6. The price per person to rent a limousine for a prom varies inversely as the number of passengers. If five people rent the limousine, the cost is $70 each. How many people are renting the limousine when the cost *per couple* is $87.50?

7. The acceleration due to Earth's gravitational attraction varies inversely as the square of the distance from the center of Earth. The acceleration is 32 feet per second per second at 4000 miles from the center. What is the acceleration, in feet per second per second, of a satellite 8000 miles from the center of Earth?

8. A law of physics states that t varies directly as v and inversely as d^3. If $t = \frac{1}{2}$ when $v = 25$ and $d = 10$, find t when $v = 40$ and $d = 15$.

9. The electrical resistance of a cable varies directly as its length and inversely as the square of its diameter. If the resistance of 5000 feet of cable, 0.5 inch in diameter, is 0.25 ohm, find the number of ohms in the resistance of a cable of the same material that is 2700 feet long and 0.3 inch in diameter.

10. The equation $I = \frac{64}{d^2}$ models the amount of light, I, provided by a 60-watt bulb at a distance of d meters from the bulb, where I is measured in lux.
    a. Using graph paper, sketch a graph showing how the amount of light changes as the distance from the bulb increases for $d \geq 1$.
    b. Using the graph, find to the *nearest lux*, the amount of light that falls on a desk that is 1.5 meters from the bulb.
    c. By what percent does the amount of light decrease when the distance from the bulb increases from 2 to 4 meters?

CHAPTER 10

EXPONENTIAL AND LOGARITHMIC FUNCTIONS

10.1 THE EXPONENTIAL FUNCTION

KEY IDEAS

The equation $y = b^x$ is an **exponential function** provided that b is a positive number other than 1. Exponential functions have variables as exponents.

Graphs of $y = 2^x$ and $y = \left(\frac{1}{2}\right)^x$

As x increases, the graph of $y = 2^x$ rises while the graph of $y = \left(\frac{1}{2}\right)^x$ falls, as shown in Figure 10.1. The equations are functions since their graphs pass the vertical-line test.

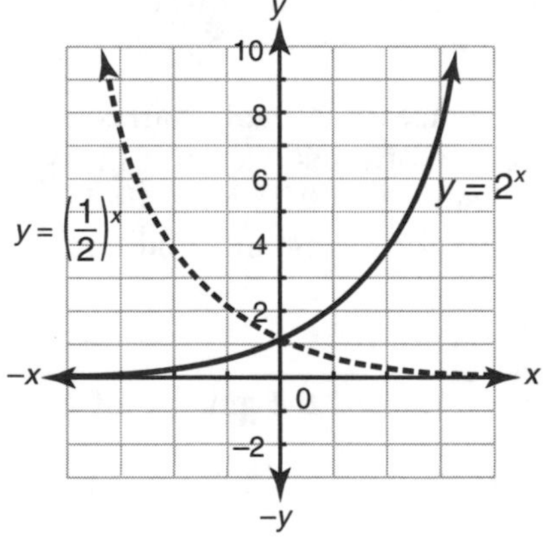

Figure 10.1 Graphs of $y = 2^x$ and $y = \left(\frac{1}{2}\right)^x$

Generalizations About Exponential Functions

The graphs of $y = 2^x$ and $y = \left(\frac{1}{2}\right)^x$ can be used as a basis for generalizations about $y = b^x$, where b is a positive number other than 1:

- The domain is the set of real numbers, and the range is the set of positive real numbers.
- If $b > 1$, the graph of $y = b^x$ has the same shape as the graph of $y = 2^x$. The graph rises from left to right and intersects the y-axis at (0, 1). As x decreases, the negative x-axis is a horizontal asymptote of the graph.
- If $0 < b < 1$, the graph of $y = b^x$ has the same shape as the graph of $y = \left(\frac{1}{2}\right)^x$. The graph falls from left to right and intersects the y-axis at (0, 1). As x increases, the positive x-axis is a horizontal asymptote of the graph.
- If b is a positive number other than 1, the graphs of $y = b^x$ and $y = \left(\frac{1}{b}\right)^x$ are reflections of each other in the y-axis. For example, $y = \left(\frac{1}{2}\right)^x$ is the reflection of the graph of $y = 2^x$ in the y-axis. You can verify this statement by replacing x with $-x$ in $y = 2^x$:

$$y = 2^{-x} = \left(2^{-1}\right)^x = \left(\frac{1}{2}\right)^x.$$

Example

The transformation $T_{-1,-2}$ is applied to $y = 3^x$.

a. What is an equation of the translated graph?

b. What are (1) the domain and (2) the range of the translated graph?

Solution: a. $\mathbf{y = 3^{x+1} - 2}$

The notation $T_{h,k}$ represents a translation or shift of $|h|$ units horizontally and $|k|$ units vertically. In general, when the translation $T_{h,k}$ is applied to the graph of $y = f(x)$, an equation of the translated graph is $y = f(x - h) + k$. Hence, under the translation $T_{-1,-2}$, $h = -1$ and $k = -2$, so the image of $y = 3^x$ is $y = 3^{x-(-1)} + (-2) = 3^{x+1} - 2$.

The graph is shown in the accompanying figure in a $[-4.7, 4.7] \times [-4, 4]$ window. An equation of this graph is $y = 3^{x+1} - 2$.

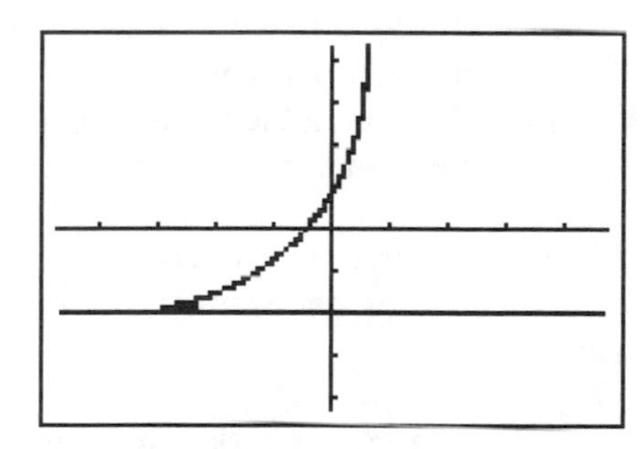

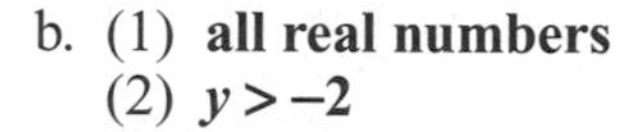

b. (1) **all real numbers**
   (2) $\mathbf{y > -2}$

The domain is the set of real numbers. Since the line $y = -2$ is a horizontal asymptote, the range is $y > -2$.

Check Your Understanding of Section 10.1

A. Multiple Choice

1. At what point do the graphs of $y = 5^x$ and $y = 5^{-x}$ intersect?
(1) (0, 1) (2) (1, 0) (3) (−5, 0) (4) (5, 0)

2. Which is an equation of the reflection of the graph of $y = 3^x$ in the origin?
(1) $y = -3^x$ (2) $y = \left(\frac{1}{3}\right)^x$ (3) $y = -\left(\frac{1}{3}\right)^x$ (4) $x = 3^y$

3. The graph in the accompanying figure can be represented by which equation?
(1) $y = 2^x$
(2) $y = 2^x - 1$
(3) $y = 2^{x+1}$
(4) $y = 2^x + 1$

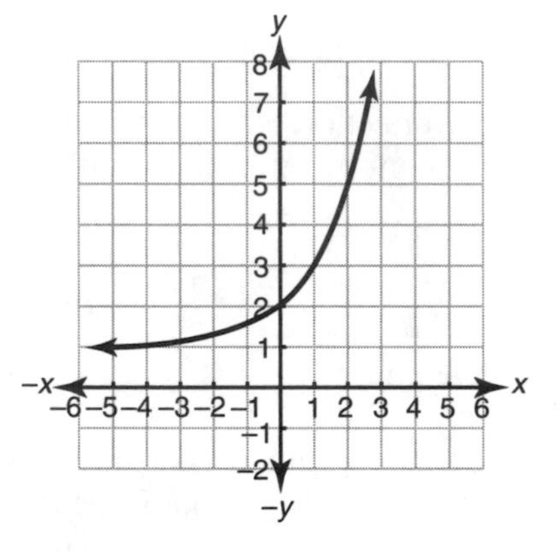

4. If, in the equation $y = 3^x$, the variable x is increased by 2, then y is
(1) increased by 2
(2) multiplied by 6
(3) increased by 9
(4) multiplied by 9

5. If $y = (0.25)^x$ and $y = 4^x$ are graphed on the same set of axes, which transformation will map one of these curves onto the other?
(1) reflection in the y – axis
(2) reflection in the x – axis
(3) reflection in the line $y = x$
(4) reflection in the origin

B. *In each case, show how you arrived at your answer by clearly indicating all of the necessary steps, formula substitutions, diagrams, graphs, charts, etc.*

6. a. On graph paper, sketch the graph of $y = 2^x$ using integer values of x in the domain $-1 \le x \le 3$.
b. On the same set of axes, sketch the reflection of $y = 2^x$ in the origin. Label this graph **b**.
c. On the same set of axes, sketch the image of $y = 2^x$ under the translation $T_{6,2}$. Label this graph **c**.

7. a. On graph paper, sketch the graph of $f(x)=\left(\frac{1}{2}\right)^x$ using integer values of x in the domain $-3 \le x \le 1$.

b. On the same set of axes, sketch the image of function f after the transformation $T_{2,-1} \circ r_{x\text{-axis}}$. Label this graph **b**.

c. On the same set of axes, sketch the image of the graph drawn in part b after D_{-2}. Label this graph **c**.

10.2 THE LOGARITHMIC FUNCTION

KEY IDEAS

The inverse of the exponential function $y = b^x$ is $x = b^y$. There is no algebraic method for solving $x = b^y$ for y in terms of x. A new function, called the *logarithmic function*, allows y to be expressed in terms of x.

Definition of the Logarithmic Function

Let b represent a positive number other than 1. The inverse of the exponential function $y = b^x$ is called the **logarithmic function with base *b*** and is written as $y = \log_b x$, where $x > 0$. The equation $y = \log_b x$ is read as "y is the logarithm of x with base b." In the equation $y = \log_b x$, y is the exponent to which base b must be raised to produce x. For example, $\log_8 64 = 2$ means that $64 = 8^2$.

Definition of the Common Logarithm

The expression $\log_b x$ is a number called the **logarithm of *x* with base *b***. For example, $\log_2 8$ is 3 since $2^3 = 8$. Thus, a logarithm is an *exponent*.

A **common logarithm** is a logarithm with base 10. Common logarithms are indicated when the base is omitted. Thus, $\log x$ is understood to mean $\log_{10} x$.

- The common logarithm of a power of 10 is simply the exponent of 10. For example, $\log 100 = 2$ since $100 = 10^2$. Also, $\log\left(\frac{1}{10}\right) = -1$ since $\frac{1}{10} = 10^{-1}$.
- To find the value of a common logarithm such as $\log 45$, quit to the home screen of your graphing calculator and press LOG. Enter 45, and close the parentheses. Pressing ENTER displays 1.653212514. Thus, $\log(45) = 1.653212514$. Therefore, $10^{1.653212514} \approx 45$.

Graph of $y = \log_b x$

Since the exponential function $y = b^x$ and the logarithmic function $y = \log_b x$ are inverse functions, their graphs are symmetric to the line $y = x$, as shown in Figures 10.2 and 10.3. The graph of every exponential function of the form $y = b^x$ contains (0, 1) and has no x-intercept. The graph of every logarithmic function of the form $y = \log_b x$ contains (1, 0) and has no y-intercept.

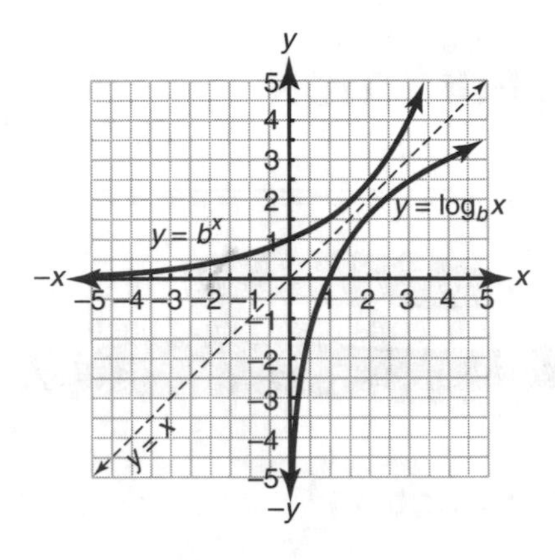

Figure 10.2 Graphs of $y = b^x$ and $y = \log_b x$ with $b > 1$

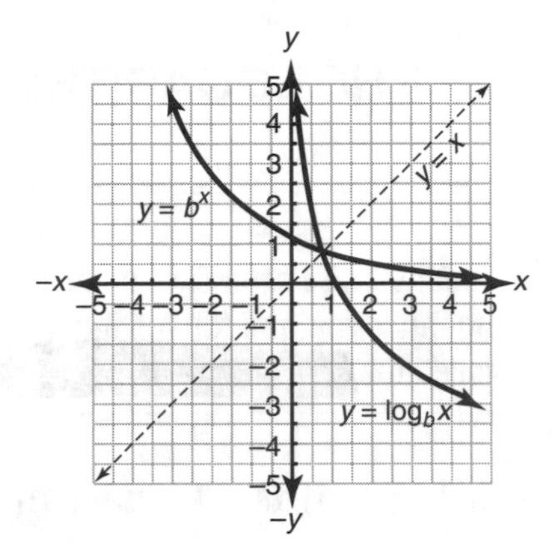

Figure 10.3 Graphs of $y = b^x$ and $y = \log_b x$ with $0 < b < 1$

Properties of Logarithmic Functions

Table 10.1 compares the properties of the graphs of $y = b^x$ and $y = \log_b x$.

TABLE 10.1 PROPERTIES OF EXPONENTIAL AND LOGARITHMIC FUNCTIONS

Property	Graph of $y = b^x$	Graph of $y = \log_b x$
Quadrants	I and II	I and IV
Domain	All real numbers	Positive real numbers
Range	Positive real numbers	All real numbers
x-intercept	None	(1, 0)
y-intercept	(0, 1)	None
Asymptote	x-axis	y-axis
Rises as x increases	$b > 1$	$b > 1$
Falls as x increases	$0 < b < 1$	$0 < b < 1$

Example 1

a. On graph paper, sketch the graph of $y = \log_2 x$ using integer values of x in the domain $-1 \le x \le 3$. Label this graph **a**.
b. On the same set of axes, sketch the reflection of $y = \log_2 x$ in the line $y = x$. Label this graph **b**.

Solutions: a. See the unbroken graph in the accompanying figure. Since $y = \log_2 x$ and $y = 2^x$ are inverse functions, either graph is the reflection of the other in the line $y = x$. Using integer values of x in the domain $-1 \le x \le 3$, prepare a table of values like the accompanying one for points on $y = 2^x$ and $y = \log_2 x$. Obtain the ordered pairs for $y = \log_2 x$ by interchanging x and y in each ordered pair that satisfies $y = 2^x$.

x	$y = 2^x$	**Point on** $y = 2^x$	**Point on** $y = \log_2 x$
−1	$y = 2^{-1} = \frac{1}{2}$	$\left(-1, \frac{1}{2}\right)$	$\left(\frac{1}{2}, -1\right)$
0	$y = 2^0 = 1$	(0, 1)	(1, 0)
1	$y = 2^1 = 2$	(1, 2)	(2, 1)
2	$y = 2^2 = 4$	(2, 4)	(4, 2)
3	$y = 2^3 = 8$	(3, 8)	(8, 3)

Plot the set of points for $y = \log_2 x$, and then connect them with a smooth curve. Label this graph **a**, as shown in the accompanying diagram.

b. See the broken graph in the accompanying figure. Plot the set of points for $y = 2^x$, and then connect them with a smooth curve. Label this graph **b**, as shown in the accompanying diagram.

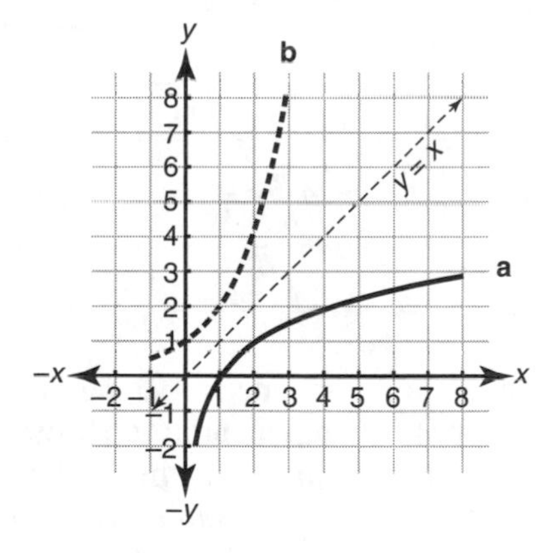

Example 2

a. Write an equation of the graph that is the image of $y = \log_3 x$ under the translation $T_{2,-1}$.
b. Determine (1) the domain and (2) the range of the translated graph.

Solution: a. $\mathbf{y = \log_3 (x - 2) - 1}$

The image of $y = f(x)$ under $T_{2,-1}$ is $y = f(x - h) + k$. Hence, under the translation $T_{2,-1}$, the image of $y = \log_3 x$ is $y = \log_3 (x - 2) - 1$.

b. (1) $\boldsymbol{x > 2}$ (2) ***y* is any real number.**

The logarithm function is defined only for positive numbers. Hence, for $y = \log_3 (x - 2) - 1$, $x - 2 > 0$, so $x > 2$. Since the range of $y = \log_3 x$ is the set of real numbers, the range of any translation of that graph is also the set of real numbers.

Change-of-Base Formula

The formula

$$\log_b x = \frac{\log x}{\log b}$$

changes $\log_b x$ into an equivalent base-10 logarithm. For example, $\log_3 x = \frac{\log x}{\log 3}$. To graph $y = \log_3 x$ using your calculator, graph $Y_1 = \frac{\log (x)}{\log (3)}$.

Writing Equivalent Equations

To change between exponential and logarithmic forms, use the fact that $y = \log_b x$ and $x = b^y$ are equivalent equations.

Exponential Form	Logarithmic Form
• $3^2 = 9$	$\log_3 9 = 2$
• $10^3 = 1000$	$\log_{10} 1000 = 3$
• $4^{-2} = \frac{1}{16}$	$\log_4 \left(\frac{1}{16}\right) = -2$

Example 3

In a laboratory one amoeba divides into two amoebas every hour.

a. Write an exponential function that models the reproduction of amoebas in which N represents the total number of amoebas present after t hours.

b. To the *nearest tenth of an hour*, how long will one amoeba take to multiply to a total of 10,000 amoebas?

Solution: a. $\boldsymbol{N = 2^t}$

t	0	1	2	3	. . .	t
N	1	2	4	8	. . .	2^t

b. **13.3**

- Use $N = 2^t$ to find t when $N = 10{,}000$: $\quad 10{,}000 = 2^t$
- Change to logarithmic form: $\quad t = \log_2 10{,}000$
- Use the change-of-base formula: $\quad = \dfrac{\log 10{,}000}{\log 2}$
- Let $\log 10{,}000 = \log 10^4 = 4$: $\quad \approx \dfrac{4}{0.301}$

$$\approx 13.3$$

One amoeba will take 13.3 hr to multiply to 10,000 amoebas.

Example 4

For each equation, find the value of x:

a. $\log_2 x = -5$ b. $\log_x 8 = \dfrac{3}{4}$

Solution: a. $\boldsymbol{x = \dfrac{1}{32}}$

If $\log_2 x = 5$, then in exponential form $2^{-5} = x$. Hence, $x = \dfrac{1}{2^5} = \dfrac{1}{32}$.

b. $\boldsymbol{x = 16}$

If $\log_x 8 = \dfrac{3}{4}$, then in exponential form $x^{\frac{3}{4}} = 8$. Thus:

$$\left(x^{\frac{3}{4}}\right)^{\frac{4}{3}} = (8)^{\frac{4}{3}}$$

$$x = \left(\sqrt[3]{8}\right)^4 = 2^4 = 16$$

Example 5

If $\log\left(\dfrac{x}{4}\right) = 1.6$, find x to the *nearest hundredth*.

Solution: $\boldsymbol{x = 159.24}$

- If $\log\left(\dfrac{x}{4}\right) = 1.6$, then $\dfrac{x}{4} = 10^{1.6}$.
- Use your graphing calculator to find that $10^{1.6} = 39.81071706$.
- Thus, $\dfrac{x}{4} = 39.81071706$, so $x = 39.81071706 \times 4 \approx 159.24$.

Check Your Understanding of Section 10.2

A. Multiple Choice

1. If $f(x) = \dfrac{1}{\log x}$, what is f(0.01)?

(1) 100 (2) 2 (3) $-\dfrac{1}{2}$ (4) −2

2. If the graph of $y = \log_{10} x$ contains point $(10, k)$, what is the value of k?
(1) 1 (2) 2 (3) 10 (4) 100

3. What is the domain of $y = \log_b (x + 2)$?
(1) $x > -2$ (2) $x \geq -2$ (3) $0 < x < 2$ (4) $-2 < x < 0$

4. If $\log_3 (x - 2) = 2$, what is the value of x?
(1) 7 (2) 8 (3) 11 (4) 9

5. If $\log_9 x = \dfrac{3}{2}$, what is the value of x?

(1) $\dfrac{3}{2}$ (2) 8 (3) $\dfrac{27}{2}$ (4) 27

6. The expression $\log_2 (x - 4)$ is undefined if
(1) $x > 2$ (2) $x > 0$ (3) $x \leq 4$ (4) $x > 4$

7. For what value of x does $\log_{10} (36 - x^3) = 2$?
(1) −8 (2) 8 (3) 4 (4) −4

8. Which function is the inverse of $y = \log (x - 1)$?

(1) $y = x^{10} + 1$ (2) $y = 10^x - 1$ (3) $y = 10^x + 1$ (4) $y = \left(\dfrac{1}{10}\right)^x - 1$

9. If the graph of $x = 2^y$ is reflected in the line $y = x$, which is an equation of the image?

(1) $y = \log_2 x$ (2) $y = 2^x$ (3) $y = \left(\dfrac{1}{2}\right)^x$ (4) $xy = 2$

10. What is the value of $3 \log_2 16 + \log_2 \dfrac{1}{16}$?

(1) 12 (2) 2 (3) 8 (4) 4

11. Which is an equation of the graph in the accompanying diagram?
(1) $y = \log_2 (x - 3) + 1$
(2) $y = \log_2 (x - 3) - 1$
(3) $y = \log_2 (x + 3) - 1$
(4) $y = \log_2 (x + 3) + 1$

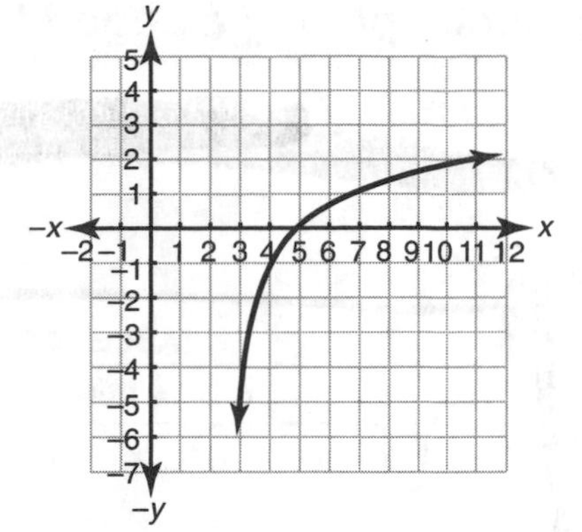

B. *In each case, show how you arrived at your answer by clearly indicating all of the necessary steps, formula substitutions, diagrams, graphs, charts, etc.*

12. a. Using graph paper, draw the graphs of $y = 3^x$ and $y = \log_3 x$ on the same set of coordinate axes.
b. Describe the transformation that maps either of these graphs onto the other.

13. a. Using graph paper, sketch and label the graphs of the equations $xy = 8$ and $y = \log_2 x$ on the same set of axes in the interval $-8 \leq x \leq 8$.
b. Using the graphs sketched in part a, find the smallest integer value of x for which $\log_2 x > \frac{8}{x}$.

14. a. Using graph paper, sketch and label the graph of $f(x) = \log_2 x$.
b. On the same set of axes, rotate the graph drawn in part a 90° counterclockwise about the origin. Sketch this rotation, and label it **b**.
c. Write an equation of the function graphed in part b.

10.3 LOGARITHM LAWS

KEY IDEAS

The laws for finding the logarithms of a product, quotient, and power follow the laws of exponents for these operations. For example, to multiply powers of the same base, we *add* the exponents. Since logarithms are exponents, the logarithm of a product is the *sum* of the logarithms of the factors of that product.

Decomposing a Logarithm Expression

Logarithms of products, quotients, powers, and roots can be broken down into their component parts by using one or more of the three logarithm laws in Table 10.2.

TABLE 10.2 LOGARITHM LAWS

Law	General Rule	Examples
Product	$\log_b(xy) = \log_b x + \log_b y$	• $\log 21 = \log(7 \times 3) = \log 7 + \log 3$ • $\log 100a = \log 100 + \log a = 2 + \log a$
Quotient	$\log_b\left(\frac{x}{y}\right) = \log_b x - \log_b y$	• $\log \frac{pq}{2} = \log p + \log q - \log 2$ • $\log\left(\frac{k}{10}\right) = \log k - \log 10 = \log k - 1$
Power	$\log_b(x^n) = n \log_b x$	• $\log 25 = \log 5^2 = 2 \log 5$ • $\log \sqrt{x} = \log x^{\frac{1}{2}} = \frac{1}{2} \log x$

Example 1

Express $\log\left(\sqrt[3]{\frac{m}{n}}\right)$ in terms of $\log m$ and $\log n$.

Solution: $\mathbf{\frac{1}{3}(\log m - \log n)}$

$$\log\left(\sqrt[3]{\frac{m}{n}}\right) = \log\left(\frac{m}{n}\right)^{\frac{1}{3}} = \frac{1}{3}\log\left(\frac{m}{n}\right) = \frac{1}{3}(\log m - \log n)$$

Example 2

If $\log 2 = x$ and $\log 3 = y$, express each of the following in terms of x and y:

a. $\log \sqrt{6}$ b. $\log 24$

Solution:

a. $\mathbf{\frac{1}{2}(x - y)}$ b. $\mathbf{3x - y}$

$$\log \sqrt{6} = \log (2 \times 3)^{\frac{1}{2}}$$
$$= \frac{1}{2} \log (2 \times 3)$$
$$= \frac{1}{2}(\log 2 + \log 3)$$
$$= \frac{1}{2}(x + y)$$

$$\log 24 = \log (8 \times 3)$$
$$= \log \left(2^3 \times 3\right)$$
$$= \log 2^3 + \log 3$$
$$= 3 \log 2 + \log 3$$
$$= 3x + y$$

Solving Equations Containing Logarithms

When each term of an equation is a logarithm with the same base, solve the equation by writing it in the form $\log_b A = \log_b B$. Then write $A = B$ and, if necessary, solve for the variable.

Example 3

If $\log N = 2 \log x + \log y$, solve for N in terms of x and y.

Solution: $\boldsymbol{N = x^2 y}$

Write the right side of $\log N = 2 \log x + \log y$ as a single logarithm:

- Undo the power law: $\log N = \log x^2 + \log y$
- Undo the product law: $\log N = \log x^2 y$
- Set the numbers equal: $N = x^2 y$

Example 4

Solve for x: $\log x - \frac{1}{3} \log 8 = \log 7$.

Solution: $\boldsymbol{x = 14}$

- Undo the power law: $\log x - \log 8^{\frac{1}{3}} = \log 7$
  $\log x - \log 2 = \log 7$
- Undo the quotient law: $\log \frac{x}{2} = \log 7$
- Set the numbers equal: $\frac{x}{2} = 7$
  $x = 14$

The check is left for you.

Solving Log Equations by Exponentiating

When only some of the terms of an equation are logarithms, consolidate the logarithms to put the equation into the form $\log_b N$ = number. Then write the equation in exponential form and solve the resulting equation.

Example 5

Solve for x if $\log_4 (x-3) + \log_4 (x+3) = 2$.

Solution: $\mathbf{x = 5}$

- Undo the product law: $\log_4 [(x-3)(x+3)] = 2$
- Write in exponential form: $(x-3)(x+3) = 4^2$
- Multiply: $x^2 - 9 = 16$
- Solve for x: $x = \pm\sqrt{25} = \pm 5$

Check each root in the original equation. If $x = -5$, then $\log_4 (-8) + \log_4 (-2) = 2$. Since logarithms of negative numbers are not defined, $x = -5$ is an extraneous root and must be rejected. You should verify that $x = 5$ works and is, therefore, the only root of the equation.

Check Your Understanding of Section 10.3

A. Multiple Choice

1. If $x = \dfrac{\sqrt{r}}{s}$, which expression is equivalent to $\log x$?

(1) $2 \log r$ (2) $2 \log r - \log s$ (3) $\dfrac{1}{2} \log r - \log s$ (4) $\dfrac{\log r - \log s}{2}$

2. The expression $3 \log x - \dfrac{1}{2} \log y$ is equivalent to

(1) $\log\left(\dfrac{x^3}{y^2}\right)$ (2) $\log\left(\dfrac{x^3}{\sqrt{y}}\right)$ (3) $\log\sqrt{\dfrac{3x}{y}}$ (4) $\dfrac{\log 3x}{\frac{1}{2}\log y}$

3. Which statement is true?

(1) $\log (ab) = (\log a)(\log b)$
(2) $\log\left(\dfrac{1}{a}\right) = -\log a$
(3) $\log (a+b) = \log a + \log b$
(4) $\log\left(\dfrac{a}{b}\right) = \dfrac{\log a}{\log b}$

4. The expression $\log 12$ is equivalent to

(1) $\log 6 + \log 6$
(2) $\log 3 + 2 \log 2$
(3) $\log 3 - 2 \log 2$
(4) $(\log 3)(\log 4)$

5. If $\log 3 = x$ and $\log 5 = y$, then $\log 45$ can be expressed as
(1) $2xy$ (2) x^2y (3) $(xy)^2$ (4) $2x + y$

6. If $\log N = \frac{1}{2}(\log r - 2\log t) + \log s$, then $N =$

(1) $\frac{\sqrt{rs}}{t}$ (2) $\frac{s\sqrt{r}}{t}$ (3) $\sqrt{\frac{r+s}{t^2}}$ (4) $\sqrt{\frac{r}{t^2}+s}$

7. If $f(x) = \log(x^2)$ and $g(x) = \frac{x}{10}$, then $f(g(x))$ can be expressed as

(1) $2(\log x - 1)$ (2) $\frac{\log x}{50}$ (3) $\frac{1}{2}(\log x + 1)$ (4) $2\log x + 1$

8. The magnitude, R, of an earthquake is related to its intensity, I, by the equation $R = \log\left(\frac{I}{T}\right)$, where T is the threshold below which the earth quake is noticed. If the intensity is doubled, the earthquake's magnitude can be expressed as
(1) $2(\log I - \log T)$
(2) $\log I - \log T$
(3) $2\log I - \log T$
(4) $\log 2 + \log I - \log T$

9. If $\log 30 = b$, then $(b - 1)^2$ is equivalent to
(1) $(\log 29)^2$ (2) $2(\log 30 - 1)$ (3) $(\log 3)^2$ (4) $\log 899$

10. If $\log 5 = a$, then $\log 250$ can be expressed as
(1) $50a$ (2) $2a + 1$ (3) $10 + 2a$ (4) $25a$

B. *In each case, show how you arrived at your answer by clearly indicating all of the necessary steps, formula substitutions, diagrams, graphs, charts, etc.*

11. a. If $\log N = x\log y + 2\log y$, solve for N in terms of x and y.
b. Solve for x: $\log_3(7x + 4) - \log_3 2 = 2\log_3 x$.

12. a. If $\log 3 = x$ and $\log 4 = y$, express $\log\left(\frac{2}{3}\right)$ in terms of x and y.

b. Solve for x: $\log_9 x + \log_9(x - 6) = \frac{3}{2}$.

13. a. Given: $\log_b 3 = p$ and $\log_b 5 = q$.

(1) Express $\log_b\left(\frac{9}{\sqrt{5}}\right)$ in terms of p and q.

(2) Express $\log_b \sqrt[3]{0.6}$ in terms of p and q.

b. Solve for x: $\frac{1}{2}\log(x+2)=2$.

14. Solve for x: $\log_4(x^2+3x)=1+\log_4(x+5)$.

15. Solve for x to the *nearest tenth*: $\log_3 x=2-\log_3(x+2)$.

10.4 SOLVING EXPONENTIAL EQUATIONS

KEY IDEAS

An **exponential equation** is an equation, such as $3^{x-4}=9$, in which the variable is in an exponent. To solve an exponential equation, write each side as a power of the same base. For example, if $3^{x-4}=9$, then $3^{x-4}=3^2$, so $x-4=2$ and $x=6$. If this approach is not possible, as in $2^{3x}=6$, use logarithms to obtain an equivalent equation without exponents.

Exponential Equations with Like Bases

When each side of an exponential equation can be written as a power of the same base, set the exponents equal to each other and solve the resulting equation.

Example 1

Solve for x: a. $9^{x+1}=27^x$ b. $64^{1-x}=\frac{1}{16^{2x}}$

Solution: Write each side as a power of the same base. Then equate the exponents and solve for x.

a. $\boldsymbol{x=2}$

$$9^{x+1}=27^x$$
$$(3^2)^{x+1}=(3^3)^x$$
$$3^{2(x+1)}=3^{3x}$$
$$2(x+1)=3x$$
$$2x+2=3x$$
$$2=x$$

b. $\boldsymbol{x=-3}$

$$64^{1-x}=\frac{1}{16^{2x}}$$
$$(4^3)^{1-x}=(4^2)^{-2x}$$
$$4^{3(1-x)}=4^{-4x}$$
$$3(1-x)=-4x$$
$$3-3x=-4x$$
$$3=-x \quad \text{so} \quad x=-3$$

Exponential Equations with Different Bases

If you cannot write each side of an exponential equation with the same base, eliminate the exponent by taking the logarithm of each side of the equation. For example, if $2^{3x} = 6$, then $\log(2^{3x}) = \log 6$ or $3x \log 2 = \log 6$, so

$$x = \frac{\log 6}{3\log 2} \approx \frac{0.7782}{3(0.3010)} \approx 0.8618.$$

Example 2

Solve $\log_3 21 = x$ for x to the *nearest hundredth*.

Solution: $\mathbf{x = 2.77}$

- Write the equation in exponential form: $3^x = 21$
- Take the log of each side of the equation: $x \log 3 = \log 21$
- Solve for x: $x = \dfrac{\log 21}{\log 3}$
- Evaluate the logarithms: $\approx \dfrac{1.322}{0.4771} \approx 2.77$

TIP

The equation $3^x = 21$ can also be solved by graphing $Y_1 = 3^x$ and $Y_2 = 21$ in an appropriate viewing rectangle, such as $[0, 5] \times [0, 30]$, and then using the intersect feature of your calculator to estimate the x-coordinate of the point of intersection of the two graphs.

Check Your Understanding of Section 10.4

A. Multiple Choice

1. Solve for x: $64^{x-2} = 256^{2x}$.

(1) $-\frac{6}{11}$ (2) $-\frac{6}{5}$ (3) $-\frac{1}{5}$ (4) 0

2. Solve for x: $4^{3x-4} = \left(\frac{1}{8}\right)^{x-1}$.

(1) $\frac{5}{3}$ (2) $\frac{7}{9}$ (3) $\frac{11}{9}$ (4) $-\frac{1}{3}$

3. If $27^x = 9^{y-1}$, what is y in terms of x?
(1) $\frac{3}{2}x+1$ (2) $\frac{3}{2}x+2$ (3) $\frac{3}{2}x+\frac{1}{2}$ (4) $\frac{1}{2}x+\frac{2}{3}$

4. Determine the values of x and y if $2^y = 8^x$ and $3^y = 3^{x+4}$.
(1) $x=6, y=2$
(2) $x=-2, y=-6$
(3) $x=2, y=6$
(4) $x=1, y=3$

B. *In each case, show how you arrived at your answer by clearly indicating all of the necessary steps, formula substitutions, diagrams, graphs, charts, etc.*

5. Solve the equation $9^{(x^2+x)} = 3^4$ for all values of x. [*Only an algebraic solution will be accepted.*]

6. a. Find all real values of x that satisfy the equation $x^9 = x^{(2x-1)^2}$.
b. Explain how you obtained each solution.

7. Determine the values of x and y if $8^{\frac{x}{3}} = 16^{y+8}$ and $5^{x+y} = 125^{x-5}$.

8. a. Solve for p: $7^{p^2-p} = 1$.
b. Solve for x to the *nearest hundredth*: $2^x = \frac{3}{16}$.

9. a. Solve for x: $2^{1-3x} - \left(\frac{1}{4}\right)^2 = 0$.
b. Solve for x to the *nearest hundredth*: $\log_7 75 = x$.

10. A rubber ball is dropped from 7 feet above the gymnasium floor and is allowed to bounce up and down. At each bounce it rises 75% of the height from which it fell.
a. Write a function that expresses the relationship between the height, h, of the ball and the number, n, of bounces.
b. After how many bounces will the ball be at a height of, *at most*, 1 inch? [*Only an algebraic solution will be accepted.*]

10.5 EXPONENTIAL GROWTH AND DECAY

KEY IDEAS

Many real-life growth processes can be represented by an exponential function that has the form

$$y = \text{initial amount} \times (\text{constant growth factor})^{\text{time variable}}.$$

Linear Versus Exponential Rates of Change

Linear and exponential functions do *not* behave in the same way.

- For the *linear* function $y = 2x + 3$, the change in y for each unit change in x is always 2 since the slope of the line $y = 2x + 3$ is 2.
- The accompanying table shows that, for the *exponential* function $y = 3 \cdot 2^x$, y does *not* change at a constant rate. Instead, y increases by the same percent of its previous value for each unit change in x:

x	$3 \cdot 2^x =$	y
0	$3 \cdot 2^0 =$	3
1	$3 \cdot 2^1 =$	6
2	$3 \cdot 2^2 =$	12
3	$3 \cdot 2^3 =$	24

$y = 6 = 3 + 100\%$ of 3,
$y = 12 = 6 + 100\%$ of 6,
$y = 24 = 12 + 100\%$ of 12.

- Over equal intervals of x, a linear function changes by the same amount but an exponential function changes at a constant *percent* of its previous value.

Modeling Growth and Decay

If a substance increases (or decreases) at the constant rate of $r\%$ of its previous value per year, then after t years an initial amount, A, of that substance has increased (or decreased) to an amount, y, where

$$y = A(1 + r)^t$$

and r is expressed as a decimal. For instance, suppose \$320 is deposited in a bank account that earns 7% interest compounded annually.

- After 1 year, the balance in the account is:

$$\underbrace{320}_{\text{Old amount}} + \underbrace{320 \times 0.07}_{\text{Interest}} = 320(1.07)^1.$$

- After the second year, the balance in the account is

$$\underbrace{320(1.07)}_{\text{Old amount}} + \underbrace{[320(1.07)] \times (0.07)}_{\text{Interest}} = 320(1.07)^2.$$

- After t years, the balance, B, has increased by a factor of 1.07 for a total of t times, so

$$B(t) = 320\underbrace{(1.07)(1.07)\ldots(1.07)}_{t \text{ factors}} = 320(1.07)^t.$$

In the exponential model $B(t) = 320(1.07)^t$, the initial amount is 320 and the growth factor is 1.07. Since $1.07 = 1 + 0.07$, the rate of increase per year is 7%.

Example 1

Using the model $B(t) = 320(1.07)^t$, where $t = 0$ represents the year 2000, determine:

a. the balance, correct to the *nearest hundredth*, in the account at the end of the year 2005
b. the first year in which the amount saved will be at least \$700

Solution: a. **\$448.82**

Since $t = 0$ represents the year 2000, $t = 5$ is 2005. Use your calculator to find that $320(1.07)^5 = 448.81655$, which, correct to the *nearest hundredth*, is 448.82.

b. **2011**

To find the first year in which the amount saved will be at least \$700, solve the exponential equation $700 = 320(1.07)^t$:

$$\log 700 = \log\left[320(1.07)^t\right]$$

$$\log 700 = \log 320 + t\log 1.07$$

$$t = \frac{\log 700 - \log 320}{\log 1.07} \approx 11.57$$

Since $t = 0$ represents the year 2000, any value of t between 11 and 12 is the year 2011.

TIP

The equation $700 = 320(1.07)^t$ can also be solved by graphing $Y_1 = 700$ and $Y_2 = 320(1.07)^x$ on the same set of axes and then using the intersect feature of your calculator to estimate the x-coordinate of the point of intersection of the two graphs.

Example 2

Raymond buys a new car for \$21,500. The car depreciates by about 11% per year.

a. Write an exponential decay model.
b. What is the value of the car after 5 years?

Solution: a. $\mathbf{y = 21{,}500(0.89)^x}$

The general form of an exponential decay model is $y = A(1 + r)^t$, where $r < 0$. Let $r = -11\% = -0.11$ and $A = 21{,}500$. Then

$$\begin{aligned} y &= A(1+r)^t \\ &= 21{,}500(1-0.11)^t \\ &= 21{,}500(0.89)^t \end{aligned}$$

b. **$12,005.73**

Find y when $t = 5$:

$$y = 21{,}500(0.89)^5 = 12{,}005.73$$

The value of the car after 5 years is $12,005.73.

MATH FACTS

The exponential function $y = Ab^t$ can be used to model a growth process in which A is the initial amount of a substance, y is the amount present after t units of time, and b is the growth or decay factor.

- If $b > 1$, b is the *growth* factor since y increases as t increases.
- If $0 < b < 1$, b is the *decay* factor since y decreases as t increases.

Rates of Growth and Decay

In the exponential model $y = Ab^t$, the rate of growth or decay, denoted as r, can be determined by rewriting b:

- If $b > 1$, rewrite b as $1 + r$. For example, if $y = 35(1.085)^t$, where t is measured in years, then $b = 1.085$, so $1.085 = 1 + r$ and $r = 0.085 = 8.5\%$. Hence, the rate of growth per year is 8.5%.
- If $0 < b < 1$, rewrite b as $1 - r$. For example, if $y = 47.3(0.87)^t$, where t is measured in years, then $b = 0.87$, so $1 - r = 0.87$ and $r = 0.13 = 13\%$. Hence, the rate of decay per year is 13%.

Radioactive Decay

The *half-life* of a radioactive element is the amount of time required for one-half of an initial amount of the element to decay. This process can be modeled by the function

$$y(t) = y_0(0.5)^{\frac{\text{time } t}{\text{half-life}}},$$

where y_0 is the initial amount of the radioactive substance and $y(t)$ is the amount that remains after t units of time have elapsed. The time, t, and the half-life must be expressed in the same units. For example, since the half-life of radioactive strontium-90 is 29 years, the decay of 250 grams of strontium-90 after t years can be modeled by the equation $y(t) = 250(0.5)^{\frac{t}{29}}$.

Example 3

In approximately how many years will 30% of a 250-gram mass of strontium-90 remain?

Solution: **50.4**

When 30% of the original amount of strontium-90 remains, $y = 0.30 \times 250$. If x represents the number of years after which 30% of a 250-gram mass of strontium-90 will remain, then:

$$y = 250(0.5)^{\frac{x}{29}}$$

$$0.30(250) = 250(0.5)^{\frac{x}{29}}$$

$$0.30 = (0.5)^{\frac{x}{29}} \quad \left\{\begin{array}{l}\text{The solution does not depend on} \\ \text{the initial amount of the substance}\end{array}\right.$$

If $0.30 = (0.5)^{\frac{x}{29}}$, then $\log 0.30 = \dfrac{x}{29} \log 0.5$

$$x = \frac{29 \log 0.30}{\log 0.5} \approx 50.4\text{yr}$$

Check Your Understanding of Section 10.5

A. Multiple Choice

1. The population of Clarkstown was 3,381,000 in 2000 and is growing at an annual rate of 1.8%. If it is assumed this growth rate will continue, what is the projected population of Clarkstown in the year 2006?
(1) 3,696,000 (2) 3,763,000 (3) 3,798,000 (4) 3,831,000

2. A culture of 5000 bacteria triples every 20 minutes. If $P(t)$ represents the size of the bacteria population after t minutes have elapsed, which equation can be used to model the growth of bacteria?

(1) $P(t) = 5000(20)^{3t}$ (3) $P(t) = 5000\left(3^{\frac{t}{20}}\right)$

(2) $P(t) = 5000(3)^{20t}$ (4) $P(t) = 5000\left(20^{\frac{t}{3}}\right)$

B. *In each case, show how you arrived at your answer by clearly indicating all of the necessary steps, formula substitutions, diagrams, graphs, charts, etc.*

3. The amount, A, in milligrams, of a 10-milligram dose of a drug remaining in the body after t hours is given by the exponential function $A(t) = 10(0.8)^t$. Find, to the *nearest tenth*, the number of hours after which half of the original drug dose will be left in the body.

4. The number of endangered species of animals, S, in the United States has grown since 1980 according to the exponential model $S(t) = 103(2.71)^{0.084t}$, where t is the number of years since 1980.

a. Use the model to estimate the number of endangered species at the end of the year 2000.

b. Use a graph to estimate the year in which the number of endangered species will reach 1000.

5. The Department of Health in a developing country has issued a warning that, according to the statistics it has gathered, the number, N, of reported cases of a certain disease is rising at a rate of 14% per year. The first year it gathered statistics, there were 5000 reported cases.

a. If the number of reported cases of this disease can be modeled by the exponential function $N = ab^t$, where $t = 0$ represents the first year in which statistics were gathered, find the values of a and b.

b. Estimate the number of cases reported after 7 years.

c. Estimate the number of years, to the *nearest tenth*, required for the number of reported cases to rise to 12,000.

6. Radioactive iodine has a half-life of 60 days. The decay of a sample of radioactive iodine can be modeled by the function $y(t) = 80(0.5)^{\frac{t}{60}}$, where y grams of the radioactive element remain after t days.

a. What is the number of grams of radioactive iodine present in the initial sample?

b. To the *nearest tenth of a percent*, what percent of the original mass of the element will be present after 42 days?

c. In approximately how many days, correct to the *nearest tenth*, will 15% of the original mass be present?

7. If P dollars are invested at r% annual interest compounded n times during the year, the amount, A, of the investment after t years is given by the equation

$$A = P\left(1 + \frac{r}{n}\right)^{nt}.$$

Mary invests \$7500 at an annual rate of 8% compounded quarterly. What is the *least* number of years and months needed for Mary's initial investment to double?

8. The growth in the number of snowboarders can be modeled by the function $y = 0.5(1.21^x)$, where y represents the number of snowboarders in millions and x represents the number of years since 1989.

a. By what percent is the number of snowboarders increasing each year?

b. In which year will the number of snowboarders reach 10 million for the first time? [*Only an algebraic or graphical solution will be accepted.*]

9. A pharmaceutical company has tested a new time-release cold pill. It finds that the amount in milligrams of the active ingredients of the pill left in the bloodstream t hours after it is taken can be estimated using the exponential function $A(t) = 35(0.68)^t$.

a. What is the number of milligrams of the drug left in the bloodstream 12 hours after the drug is taken?

b. What percent of the drug leaves the body each hour?

c. How many hours will elapse before less than 0.75 milligram of the drug is left in the body?

10. A radioactive substance decays at the rate of 9% per hour. Find, to the *nearest minute*, how long the initial amount of the element will take to decay to one-half of that amount.

11. Depreciation on a new car can be determined by the formula $V = C(1 - r)^t$, where V is the value of the car after t years, C is the original cost, and r is the rate of depreciation. A car purchased 5 years ago, when it was new, now has a value of $6000. If the car's original cost was $24,000, what is the rate of depreciation, correct to the *nearest tenth* of a percent?

Unit Four

REGRESSION AND PROBABILITY

CHAPTER 11

FITTING LINES AND CURVES TO DATA

11.1 LINEAR MODELING

KEY IDEAS

After height (x) and weight (y) measurements are collected for a group of people, the data can be organized as a set of ordered pairs of the form (height, weight) for each person. If a *scatterplot* or graph of these data points closely approximates a line, the height and weight measurements are said to be *linearly related* and, as a result, can be represented by a linear function of the form $y = ax + b$. This linear function *models* the relationship between height and weight. Mathematical and statistical methods can be used to figure out specific functions that model the relationships between sets of real-world data.

Mathematical Models

A **model** is a mathematical representation of the relationship between two real-world quantities. Typically, a mathematical model reflects some simplifying assumptions that allow a real-world situation to be reduced to an equation that is easy to work with yet provides useful information. Linear functions are used to model situations that are characterized by *constant* rates of change. When a linear function is written in the form $y = ax + b$, the constant a gives the rate at which y changes. This constant rate of change is the slope of the line $y = ax + b$.

Example

After 2 hours of driving, Tyrone finds that 13 gallons of gas are left in his car's fuel tank, and after 3 hours of driving 10.5 gallons are left.

a. Construct a linear model in which the number of gallons, g, left in the tank is a function of hours, h, of driving.
b. At what rate is the car consuming gas?
c. How many gallons of gasoline were in the tank before the car was driven?

Solutions: a. $\mathbf{g = -2.5h + 18}$

Find a linear function of the form $g(h) = ah + b$ that contains (2, 13) and (3, 10.5):

- Since the slope of the line is $\frac{10.5-13}{3-2} = -2.5$, the model has the form $g = -2.5h + \text{b}$.
- To find b, substitute (2, 13) into $g = -2.5h + b$; then $13 = -2.5(2) + b$, so $b = 18$.
- The linear model is $g = -2.5h + 18$.

b. **2.5 gal/hr**

Since the slope of $g = -2.5h + 18$ is -2.5, the car is consuming gas at the constant rate of 2.5 gal/hr of driving.

c. **18**

If $h = 0$, then $g = -2.5(0) + 18 = 18$ gal.

Drawing a Scatterplot

The accompanying table shows the number of applications for admission that a college received for certain years from 1991 to 2000. Graphing a set of data points formed by pairing a set of x-data values with a set of y-data values creates a visual model called a **scatterplot**. You can draw a scatterplot of the data points in the accompanying table either by hand, using graph paper, or by using your graphing calculator. To use your calculator, follow these steps:

Step 1. Identify the independent (x) variable and the dependent (y) variable. Since the number of applicants depends on the year, let x represent the specific year and y represent the number of applicants for that year. To make the data easier to work with, let $x = 1$ represent 1991, $x = 3$ represent 1993, and so forth, with the year 2000 represented as $x = 10$.

COLLEGE APPLICATIONS FOR CERTAIN YEARS

Year	Number of Applications
1991	297
1993	331
1995	409
1996	482
1999	647
2000	615

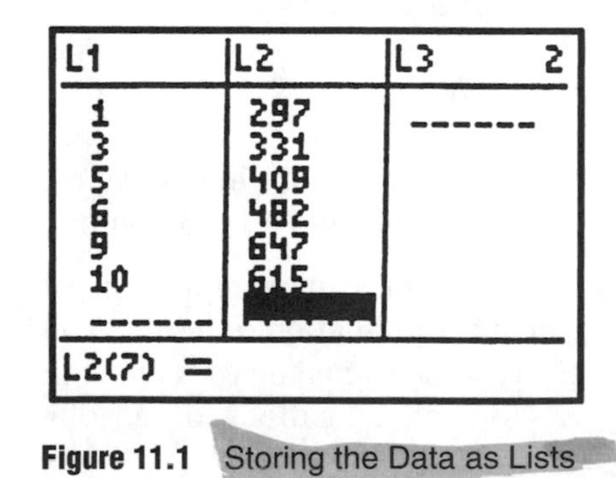

Figure 11.1 Storing the Data as Lists

Step 2. Store the data.

Press [STAT] [ENTER]. Enter the *x*-data values in list L1 and the corresponding *y*-data values in list L2, as shown in Figure 11.1.

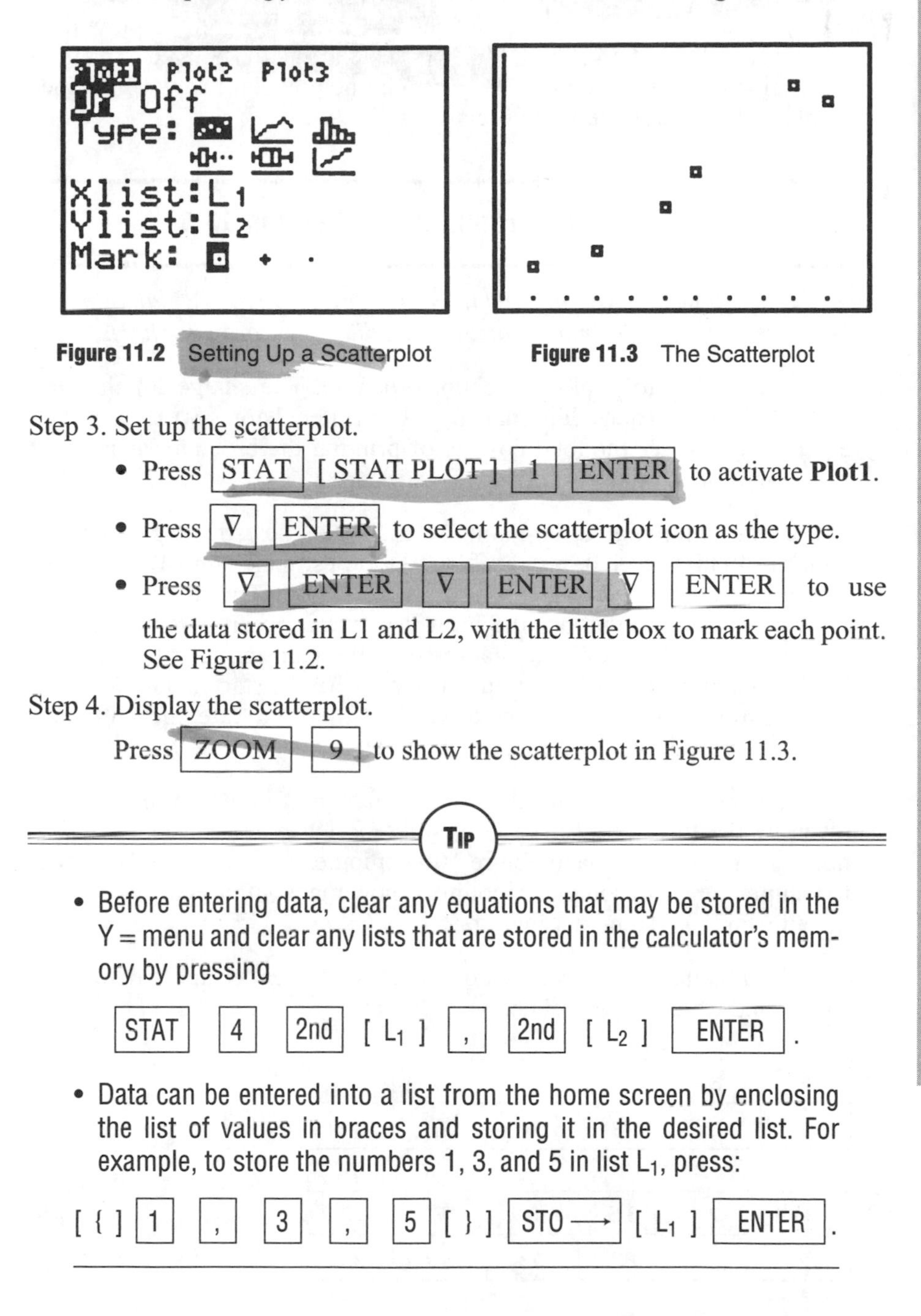

Figure 11.2 Setting Up a Scatterplot

Figure 11.3 The Scatterplot

Step 3. Set up the scatterplot.

- Press [STAT] [STAT PLOT] [1] [ENTER] to activate **Plot1**.
- Press [∇] [ENTER] to select the scatterplot icon as the type.
- Press [∇] [ENTER] [∇] [ENTER] [∇] [ENTER] to use the data stored in L1 and L2, with the little box to mark each point. See Figure 11.2.

Step 4. Display the scatterplot.

Press [ZOOM] [9] to show the scatterplot in Figure 11.3.

TIP

- Before entering data, clear any equations that may be stored in the Y = menu and clear any lists that are stored in the calculator's memory by pressing

  [STAT] [4] [2nd] [L_1] [,] [2nd] [L_2] [ENTER].

- Data can be entered into a list from the home screen by enclosing the list of values in braces and storing it in the desired list. For example, to store the numbers 1, 3, and 5 in list L_1, press:

  [{] [1] [,] [3] [,] [5] [}] [STO ⟶] [L_1] [ENTER].

Significance of a Scatterplot

A scatterplot may suggest how x and y are related. If a line can be drawn through all of the data points, there is a perfect linear relationship between x and y. If, however, all of the data points do not fall exactly on the same line, but are clustered about a line, x and y may still be linearly related. Errors in the collection or measurement of the data can help explain why a line does not contain all of the scatterplot points.

Check Your Understanding of Section 11.1

In each case, show how you arrived at your answer by clearly indicating all of the necessary steps, formula substitutions, diagrams, graphs, charts, etc.

1. A printer agrees to publish a school brochure. The charge for the first copy is $300. For each additional copy the printer charges $0.15. Write an equation that gives the total cost, y, of printing the brochure as a linear function of the number of copies printed, x.

2. Ari begins painting at 12:00 noon. At 12:30 P.M. he estimates that 15.75 gallons of paint are left, and at 2:00 P.M. he estimates that 10.5 gallons remain. Assume that the paint is being used at a constant rate.
a. Write a function that has the form $y = ax + b$, where y represents the number of gallons of paint that remain after painting for x hours.
b. How many gallons of paint did Ari have when he started the job?
c. Assuming that Ari continues to work at the same rate and without a break, at what time will no paint be left?

3. The number of hours, H, needed to manufacture X computer monitors is given by the function $H = kX + q$, where k and q are constants. If 270 hours are required to manufacture 100 computer monitors and 410 hours to manufacture 160 computer monitors, how many *minutes* are needed to manufacture one computer monitor?

4–7. Make a scatterplot for each set of data. Use the scatterplot to determine whether x and y are linearly related.

4.

x	3	5	7	9	10
y	3	13	51	207	399

5.

x	2	3.5	5	7	8
y	6	32	93	244	358

6.

x	3	7	15	23	31
y	15	18	26	32	37

7.

x	1	2	3	5	7	8	10
y	10	7.8	9.5	3.9	7.1	9.2	4.9

8. A cellular telephone company has two plans. Plan A charges \$11 a month and \$0.21 per minute. Plan B charges \$20 a month and \$0.10 per minute.

a. Write a linear function for each plan that describes the monthly cost, y, when x minutes are charged.

b. After how much time, to the *nearest minute*, will the monthly cost of plan A be equal to the monthly cost of plan B?

(1) 1 hr 22 min (3) 81 hr 8 min

(2) 1 hr 36 min (4) 81 hr 48 min

9. The cost of a long-distance telephone call is determined by a flat fee, C, for the first 5 minutes and a fixed amount, r, for each additional minute. Let y represent the total cost of a call that lasts x minutes, where x is a whole number greater than 5.

a. Using the variables x and y and the constants r and C, write an equation that expresses y as a linear function of x.

b. If a 15-minute telephone call costs \$3.25 and a 23-minute call costs \$5.17, find the cost of a 30-minute call.

11.2 FITTING LINES TO DATA

KEY IDEAS

A linear model or function that best approximates a set of data points can be constructed using the statistics capabilities of your graphing calculator. The line that best describes the data is called the **regression line**. Typically, not all of the data points in a plotted set lie on the same line. The **coefficient of linear correlation** is a number from −1 to +1 that measures how well the line "fits" the plotted data points.

Understanding Least-Squares Regression

Different lines can be drawn by hand that appear to the eye to fit a scatterplot of data points equally well. According to the **least-squares criterion**, the best-fitting line is the line for which the sum of the squares of the vertical distances from the individual data points to the fitted line is as small as possible. In Figure 11.4, the line that best fits the five data points is the line that makes the sum $(d_1)^2 + (d_2)^2 + (d_3)^2 + (d_4)^2 + (d_5)^2$ as small as possible. The fitted line is called the **regression line**. A graphing calculator can be used to find an equation of the regression line.

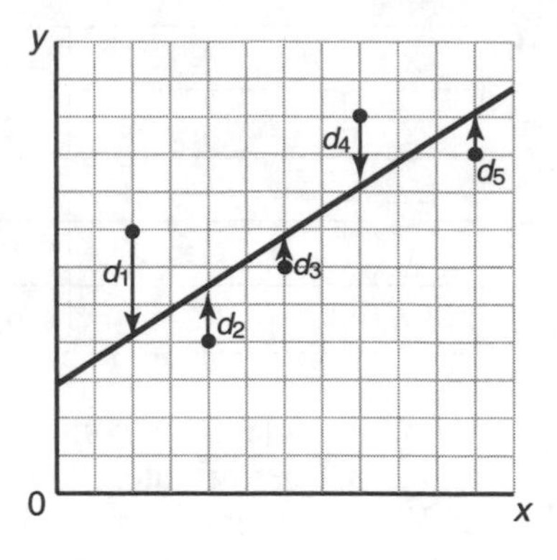

Figure 11.4 Least-Squares Regression Line

Calculating a Regression Line

If the data in the table on page 260 has already been stored in lists L1 and L2, the regression line can be calculated by following these steps:

Step 1. Press STAT ▷ 4 to choose the **LinReg**(ax + b) option.

Step 2. Press VARS ▷ 1 1 to store the regression equation as Y_1.

Step 3. Press ENTER to get the display in Figure 11.5, where a is the slope of the regression line and b is the y-intercept.

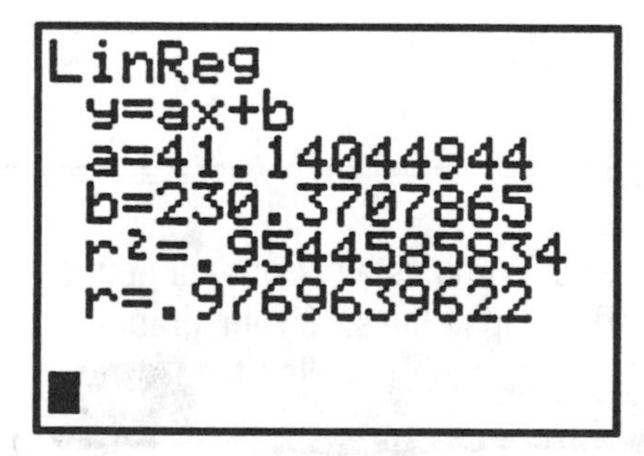

Figure 11.5 Regression Line Coefficients

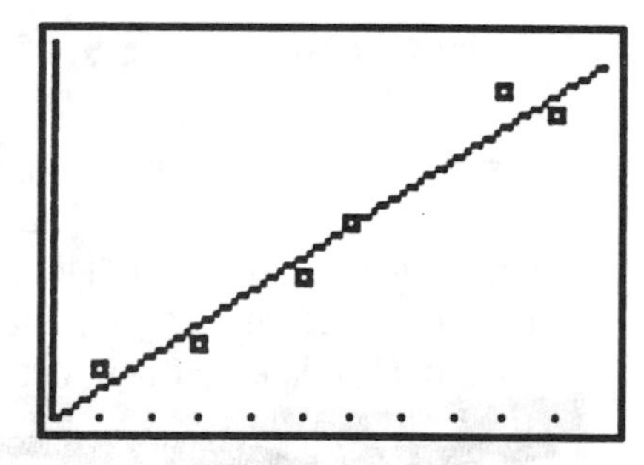

Figure 11.6 Regression Line with Scatterplot

The equation of the regression line is approximately $y = 41.140x + 230.371$. Pressing ZOOM 9 displays the regression line with the scatterplot, as shown in Figure 11.6.

Each regression line has an accompanying r-value that can range from -1 to $+1$. The closer the absolute value of r is to 1, the more closely the regression line fits the data. Figure 11.5 shows an r-value of approximately 0.977, indicating that the calculated regression line fits the data very closely. If the r-value does not appear with the regression equation, you can turn on this feature by pressing:

- 2nd [CATALOG].
- x^{-1} to scroll down to the catalog entries that begin with **D**.
- The down arrow cursor key until **DiagnosticsOn** is highlighted.
- ENTER two times.

When the regression line is calculated, the r-value will also be displayed.

TIP

- The regression line always contains point $(\bar{x}, \bar{y})$, where $\bar{x}$ is the mean of the set of x-values and $\bar{y}$ is the mean of the set of y-values.
- Regression analysis can be used to find an equation of the line through two given points. You can easily verify that the regression line obtained for points (2,13) and (3,10.5) in the example on page 259 is $y = -2.5x + 18$.

Comparing Coefficients of Correlation

Figure 11.7 illustrates that, if:

- $r > 0$, then, as x increases, y increases. The closer r is to 1, the better a rising line fits the data and the stronger the linear relationship. For example, we would expect to find a strong *positive* correlation between ages of students and their school grade levels.
- $r \approx 0$, then there is no significant *linear* relationship between x and y. The closer r is to 0, the weaker the linear relationship. For example, we would expect the coefficient of correlation between men's shoe sizes and the zip codes of their home addresses to be very close to 0.
- $r < 0$, then, as x increases, y decreases. The closer r is to -1, the better a falling line fits the data and the stronger the linear relationship. For example, we would expect to find a strong *negative* correlation between the set of year numbers from 1990 to 1999 and the average prices of a home computer system for those years.

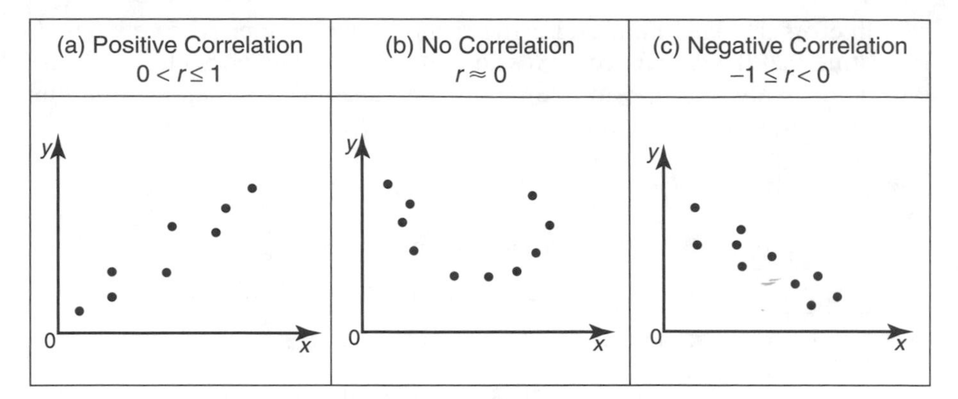

Figure 11.7 Comparing Coefficients of Correlation

TIP

- A correlation coefficient of approximately 0 does not imply that the data are not related. It simply means that no *linear* relationship exists between the paired items of data.
- A strong linear correlation between x and y does not imply that x *causes* y. For example, we would expect to find a high positive correlation between the shoe sizes and reading levels of children. This does not mean, however, that larger feet cause higher reading levels. Rather, reading level and foot size both increase as the child grows older.

Making Predictions

When there is strong linear correlation ($r \approx \pm 1$), the regression equation becomes a useful model for predicting y, using a new value for x.

- Estimating *within* the range of observed measurements for x is called **interpolation**. To estimate or *interpolate* the number of college applications for the year 1997, let $x = 7$. Then:

$$y = 41.140(7) + 230.371 \approx 518.$$

- **Extrapolation** is estimation *outside* the range of observed measurements for x. To predict or *extrapolate* the number of college applications for the year 2009, let $x = 19$. Then:

$$y = 41.140(19) + 230.371 \approx 1012.$$

Calculating a Function Value

If the regression equation for the data in the table on page 260 is stored in memory as Y_1, the predicted function value of y when $x = 7$ can be calculated by evaluating $Y_1(7)$ or by creating a table.

- To evaluate $Y_1(7)$, press

- To create a table, simply press 2nd [GRAPH] and, if necessary, scroll down to the line on which $x = 7$, as shown in Figure 11.8.

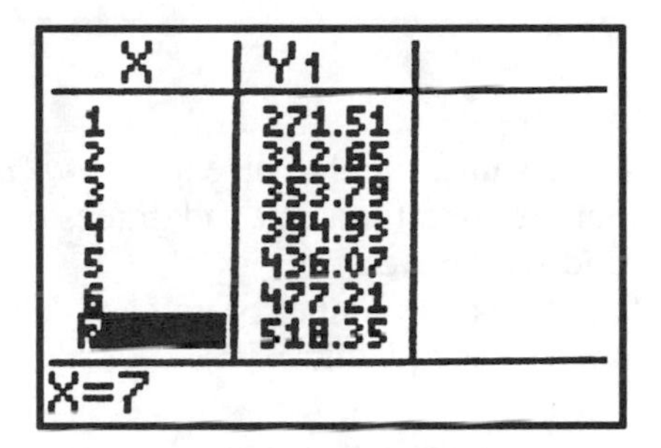

Figure 11.8 Using a Table to Calculate a Function Value

TIP

- Extrapolation should be done with caution since the linear pattern of the data may change outside the interval of *x*-values included in the sample data.
- Rather than creating a table that includes unnecessary values for *x* and *y*, it is sometimes convenient to select **Indpnt: Ask** in the TABLE SETUP screen. Although this option produces an empty table, it allows you to manually enter your own value for *x*. The corresponding value of *y* will be calculated automatically and displayed in the table.

MATH FACTS

To fit a **least-squares regression line** to collected data:

- Store the x-values as list L1 and the y-values as list L2.
- Make a scatterplot. If the data points appear to trace a straight path, fit a regression line to the data.
- Press STAT ▷ 4 ENTER to obtain the regression equation.
- Use the r-value to determine how well the regression line fits the data.

Check Your Understanding of Section 11.2

A. Multiple Choice

1. For which pair of data values would you expect a *negative* correlation?
(1) number of hours studied for a test and grades on that test
(2) ages of husbands and wives
(3) sale price of an item and number of units of that item sold
(4) income and shoe sizes of adults

2. For which pair of measurements would you expect *no* significant correlation?
(1) hand size and shoe size
(2) income and education
(3) car weight and number of miles the car can travel on 1 gal of gasoline
(4) bowling scores and number of traffic tickets.

3. The points in the accompanying scatterplot represent the ages of automobiles and their values. On the basis of this scatterplot, it would be reasonable to conclude that
(1) age and value have a coefficient of correlation that is less than 0
(2) age and value have a coefficient of correlation that is equal to 0
(3) age and value have a coefficient of correlation that is between 0 and 0.5
(4) age and value have a coefficient of correlation that is greater than 0.5

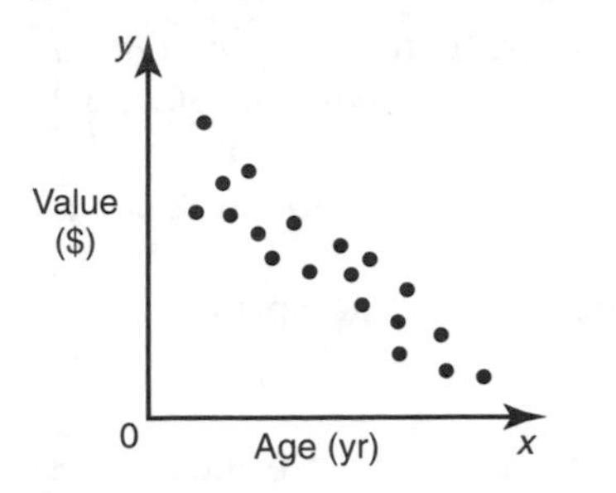

4. The relationship of a man's hat size and shoe size is given in the accompanying table:

Hat size	7.5	8	8.5	9
Shoe size	8	8.5	9	9.5

The linear correlation coefficient for this relationship is
(1) 1 (2) −1 (3) 0.5 (4) 0

5. Which statement is *always* true?
   (1) The regression line contains at least one of the data points.
   (2) If the slope of the regression line is negative, then the coefficient of correlation is less than 0.5.
   (3) The regression line $y = ax + b$ contains point $(\overline{x}, \overline{y})$.
   (4) The slope of the regression line is equal to the coefficient of correlation.

B. *In each case, show how you arrived at your answer by clearly indicating all of the necessary steps, formula substitutions, diagrams, graphs, charts, etc.*

6. When determining an equation that describes the relationship between the weight, w, in pounds added to a spring and the length, L, in inches of the spring, as shown in the accompanying figure, Cynthia recorded the measurements shown in the accompanying table.

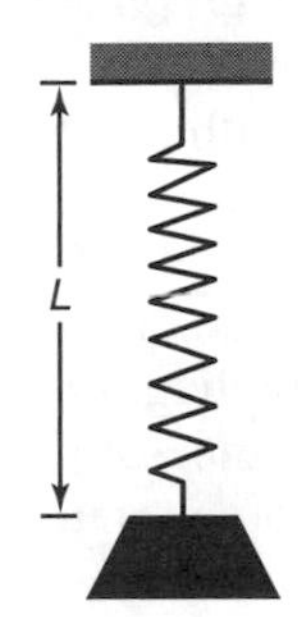

Length, L	Weight, w
10.75	10
11.8	20
13.0	30
13.9	40
15.7	50

Unit Four REGRESSION AND PROBABILITY

a. Estimate a and b correct to the *nearest hundredth*. Then write an equation of the line $w = aL + b$ that best fits the data Cynthia recorded.
b. Graph the regression line with a scatterplot of the data.
c. Determine the load to the *nearest tenth of a pound* when $L = 12.5$ inches.
d. Determine the length of the spring to the *nearest tenth of an inch* when $w = 68$ pounds.

7. The data in the accompanying table show the SAT scores for 14 students.
a. Use least-square regression to find the line $y = ax + b$ that best fits the data. Then estimate a and b correct to the *nearest hundredth*.
b. Find the mean math SAT score, $\overline{x}$, and the mean verbal SAT score, $\overline{y}$ correct to the nearest point.
c. Using the regression equation obtained in part a and the value of $\overline{x}$ calculated in part b, estimate the value of $\overline{y}$ correct to the nearest point.

Math SAT, x	480	540	590	600	600	620	640	640	650	660	680	700	700	740
Verbal SAT, y	510	500	550	600	560	580	600	620	600	620	680	640	720	700

8. The accompanying table shows the boiling points of water at different altitudes.

Location	Altitude, h (km)	Boiling Point, t (°C)
Wellington, New Zealand	0	100
Banff, Alberta, Canada	1.38	95
Quito, Ecuador	2.85	90
Mt. Logan, Canada	5.95	80

a. Make a scatterplot of the data.
b. Estimate a and b correct to the *nearest hundredth*. Then write an equation of the line $y = ax + b$ that best fits the data.
c. Predict to the *nearest tenth of a degree* the boiling point of water at Lhasa, Tibet, where the altitude is 3680 meters.

9. A projector x meters from a screen throws an image on the screen that is y meters in width, as shown in the accompanying diagram. To determine how the width of the image is related to the distance of the screen from the projector, measurements were taken and recorded in the accompanying table.

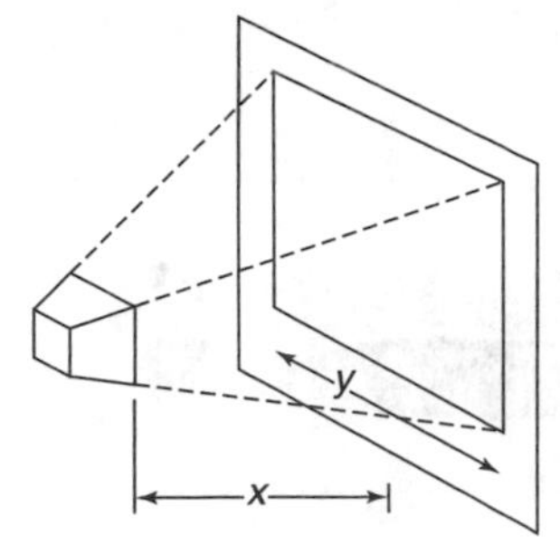

Distance, x, from screen to projector (m)	1.0	1.4	2.7	3.9	5.0
Width, y, of image (m)	0.6	0.9	1.8	2.7	3.5

a. Make a scatterplot of the data.
b. Estimate a and b correct to the *nearest hundredth*. Then write the linear equation $y = ax + b$ that best fits the data.
c. Predict to the *nearest tenth of a meter* the width of the image when the projector is 3.0 meters from the screen.
d. Find to the *nearest tenth of a meter* the distance from the projector to the screen when the width of the image is 3.0 mcters.

10. The accompanying table shows the number of milligrams of active ingredients of a cold medication that remain in the bloodstream x hours after the medication has been ingested.

a. Determine a linear relationship for x (hr) versus y (mg) that best fits the data given. Estimate constants to the *nearest hundredth.*
b. To the *nearest tenth of a milligram*, how many milligrams of the active ingredients remain in the bloodstream after 6 hours?
c. After approximately how many hours and minutes are the active ingredients no longer present in the bloodstream?

x (hr)	y (mg)
1	123
2	71
5	60
7	34
8	15
10	11

11. The availability of leaded gasoline in New York State is decreasing, as shown in the accompanying table.

Year	1984	1988	1992	1996	2000
Gallons available (thousands)	150	124	104	76	50

a. Determine the linear relationship for x (yr) versus y (gal available) that best fits the data given. Estimate constants to the *nearest tenth*.
b. If this relationship continues, determine the number of gallons of leaded gasoline that will be available in New York State in the year 2005.
c. If this relationship continues, determine the first year that leaded gasoline will become unavailable in New York State.

11.3 QUADRATIC MODELING

KEY IDEAS

If you know at least three data points (x,y) that satisfy $y = ax^2 + bx + c$, you can find the values of a, b, and c using algebraic or statistical methods.

Constructing a Quadratic Function

Algebraic methods can be used to construct a quadratic function that contains three ordered pairs of numbers.

Example

Frank's test grade on a Math B test is a function of the number of hours he studies. From past experience Frank knows that 0 hour of studying will result in a grade of 55; 2 hours, in a grade of 83; and 5 hours, in a grade of 95.

a. If Frank's grade is a quadratic function f of x hours studied, determine an equation for $f(x)$.
b. Use the function to predict Frank's grade when he studies 3 hours.

Solution: a. $\mathbf{f(x) = -2x^2 + 18x + 55}$

Let $f(x) = ax^2 + bx + c$. Then $f(0) = 55$, $f(2) = 83$, and $f(5) = 95$.

- Evaluate the function using the given values of x:

$$f(0) = c = 55,$$

$$f(2) = a(2)^2 + b(2) + c = 4a + 2b + c = 83,$$

$$f(5) = a(5)^2 + b(5) + c = 25a + 5b + c = 95.$$

- Since $c = 55$, then:

$$4a + 2b + 55 = 83, \quad \text{so} \quad 2a + b = 14 \qquad \text{(equation 1)}$$
$$25a + 5b + 55 = 95, \quad \text{so} \quad 5a + b = 8 \qquad \text{(equation 2)}$$

Solving equations (1) and (2) simultaneously gives $a = -2$ and $b = 18$.

- The quadratic function is $f(x) = -2x^2 + 18x + 55$.

b. **91**

Since $f(x) = -2x^2 + 18x + 55$, then:

$$f(3) = -2(3)^2 + 18(3) + 55 = 91.$$

Quadratic Regression

A calculator can be used to fit a quadratic function to three or more data points. The data are stored in lists L1 and L2, and then **QuadraticRegression** option 5 is selected from the STAT-CALC menu by a procedure similar to the one followed for linear regression in Section 11.2. Using the data provided in the preceding example, you should verify that the quadratic regression equation is $y = -2x^2 + 18x + 55$.

Check Your Understanding of Section 11.3

In each case, show how you arrived at your answer by clearly indicating all of the necessary steps, formula substitutions, diagrams, graphs, charts, etc.

1. A certain quadratic function has the form $f(x) + ax^2 + bx + c$, where a, b, and $c \neq 0$. If $f(-2) = -18$, $f(0) = -8$, and $f(b) = c$, determine algebraically the values of a, b, and c.

2. The numbers of feet, d, needed to stop a car for different speeds, x, where x is in miles per hour, are shown in the accompanying table. Assume that the equation $d = ax^2 + bx + c$ can be used to model the relationship represented by the statistics in the table.

x (mph)	10	40	60
d (ft)	15	114	225

a. Find a, b, and c.
b. Estimate to the *nearest foot* the distance needed to stop a car traveling at 55 miles per hour.

3. Kai realizes that a test grade is a function of the number of hours studied and knows from past experience that 1 hour of studying will result in a grade of 66; 2 hours, in a grade of 75; and 5 hours in a grade of 90.

a. If Kai's grade is a quadratic function, f, of the number of hours studied, x, determine an equation for $f(x)$.
b. What is the *minimum* amount of time Kai must study to get a grade of *at least* 80?
c. According to the function determined in part a, what is the maximum grade that Kai can get on the exam?

11.4 FITTING CURVES TO DATA

KEY IDEAS

If a set of data points do not appear to form a linear relationship, then different nonlinear functions can be constructed from the data using the regression feature of your graphing calculator. An **exponential regression** model has the form $y = ab^x$, a **power regression** model, the form $y = ax^b$; and a **logarithmic regression** model, the form $y = a\ln x + b$. A graphing calculator determines the constants a and b for each type of regression model.

Fitting an Exponential Curve

The graph of an exponential function $y = ab^x$ may be a good fit to data that describes growth.

Example 1

The data in the accompanying table show the growth of cellular phone subscriptions in the United States from 1993 to 1999.

a. Fit an exponential curve $y = ab^x$ that best fits the data, where $x = 0$ represents the year 1990 and y is the number of cellular phone subscribers. Approximate a and b to the *nearest thousandth.*

b. Use the model to estimate the number of cellular phone subscribers in 1998.

Year	Subscriptions (millions)
1993	16.0
1995	33.79
1996	44.04
1997	55.31
1999	86.05

Solution: a. $\mathbf{y = 7.746(1.319)^x}$

a. Enter the data by storing the year number in list L1 and the corresponding numbers of cellular phone subscriptions in list L2. Press

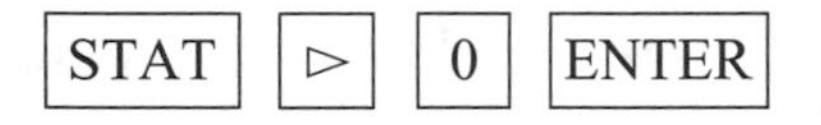

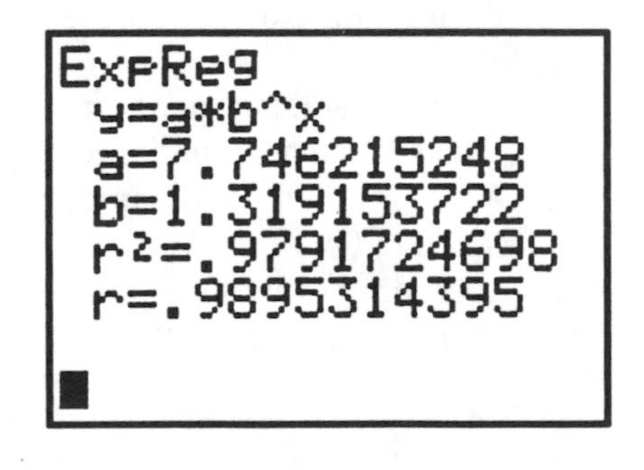

to find that the exponential regression equation, as seen from the accompanying figure, is approximately $y = 7.746(1.319)^x$.

b. **71.032 million**

In 1998 there were approximately $7.746(1.319)^8 \approx 71.032$ million cellular phone users.

Fitting a Power Curve

A power function $y = ax^b$ may fit certain types of scientific measurements better than a linear or an exponential function.

Example 2

The accompanying table shows that the number of days, y, needed for a planet to revolve around the Sun depends on the average distance, x, in millions of kilometers, of the planet from the sun. Fit a power function $y = ax^b$ that best fits the data, and approximate a and b to the *nearest ten-thousandth*.

Planet	Average Distance from Sun, x ($\times 10^6$ km)	Time of Orbit, y (days)
Mercury	57.9	88
Venus	108.2	225
Earth	149.6	365
Mars	227.9	687
Jupiter	778.3	4,333
Saturn	1,427	10,759
Uranus	2,871	30,685
Neptune	4,497	60,189
Pluto	5,913	90,800

Solution: $\mathbf{y = 0.1999(1.4998)^x}$

Enter the data by storing the distances in list L1 and the corresponding orbit times in list L2. Press STAT ▷ ALPHA [MATH] ENTER to find that the power regression equation, as seen from the accompanying figure, is approximately $y = 0.1999(1.4998)^x$.

```
PwrReg
 y=a*x^b
 a=.199858111
 b=1.499827121
 r²=.9999999397
 r=.9999999698
```

Natural Logarithmic Function

The **natural logarithmic function** is the logarithmic function whose base is the special irrational number e, called *Euler's constant.* A graphing calculator has a LN key with its inverse, e^x, above it. You should use your calculator to verify that $e \approx 2.7182818 \ldots$ The natural logarithmic function has all the properties of the common logarithmic function except that its base is e rather than 10. A logarithmic function of the form $y = a + b \ln x$ may better fit nonlinear data that show a leveling off in growth over time.

Example 3

Age, x (yr)	Amount of Sleep Required, y (hr)
2	13
6	11
12	10
16	9
25	8
50	6

a. Fit a logarithmic regression equation to the data in the accompanying table. Estimate the regression coefficients to the *nearest tenth*.
b. Find to the *nearest tenth of an hour* the number of hours of sleep required by a 35-year-old.

Solution: a. $\mathbf{y = 14.8 - 2.1 \ln x}$

Enter the ages in list L1 and the corresponding hours of sleep in list L2. Choose option **9:LnReg** from the regression menu of your graphing calculator.

```
LnReg
 y=a+blnx
 a=14.7908375
 b=-2.134363133
 r²=.9804020481
 r=-.9901525378
```

b. **7.3**

If $x = 35$, then:

$$\begin{aligned} y &= 14.8 - 2.1(\ln x) \\ &= 14.8 - 2.1(\ln 35) \\ &= 14.8 - (2.1)(3.56) \\ &\approx 7.3 \end{aligned}$$

A 35-year-old requires 7.3 hr of sleep.

Choosing a Regression Model

When you are analyzing a set of data points and do not know what type of regression model to use, draw a scatterplot to see whether the points suggest a particular type of graph. Comparing the values of r for different types of regression models can help you decide which type of regression equation best fits the data. For example, if you fit a line to the data in Example 2, you will find that $r \approx 0.9887$, which is less than the value of r obtained by fitting a power function to the data. Here are some additional guidelines that may prove useful:

- Fit an exponential function when modeling population growth or an analogous situation.
- Fit a power function to area or volume measurements. Power functions are often needed when working with scientific measurements.
- Fit a power function when the scatterplot indicates that both coordinate axes are asymptotes.
- Fit a power function rather than an exponential function if (0, 0) is a possible data point.

Check Your Understanding of Section 11.4

In each case, show how you arrived at your answer by clearly indicating all of the necessary steps, formula substitutions, diagrams, graphs, charts, etc.

1. The populations of the United States in the years 1900, 1950, 1980, and 1999 are shown in the accompanying table, where $t = 0$ represents the year 1900.

a. Determine an exponential function $y = ab^t$ that best fits the data in the table. Estimate a and b to the *nearest thousandth*.

b. If this growth relationship continues, predict the U.S. population in 2020, to the *nearest tenth* of a million.

c. If this growth relationship continues, in what year will the U.S. population exceed 400 million for the first time?

Year	1900	1950	1980	1999
Population (millions)	76.2	151.3	226.5	272.7

2. Under laboratory conditions, the number of insects increases exponentially so that after 3 days the number of insects is 112, after 5 days is 167, and after 8 days is 302.

a. Fit these data to an exponential function given by $y = ab^x$, where y represents the number of insects present after x days. Find a and b to the *nearest tenth*.

b. Estimate the number of insects in the initial population.

c. Find to the *nearest hour* the time when the number of insects is 210.

3. The period of a pendulum is the length of time required for the pendulum to make one complete swing back and forth. Rene took measurements that showed how different lengths, L, in centimeters, of pendulums affected the period, T, in seconds, as shown in the accompanying table.

L (cm)	9	15	28	35	42	50	65	80
T (s)	0.54	0.83	1.06	1.19	1.32	1.45	1.61	1.82

a. Draw a scatterplot in which L is the independent variable on the horizontal axis and T is the dependent variable on the vertical axis.
b. Fit an exponential function, power function, and logarithmic function to the data. Find the regression constants to the *nearest hundredth*.
c. Determine which function best fits the data, and give a reason for your answer.

4. The accompanying table shows the number of feet, d, needed to stop a car for different speeds, s, where s is in miles per hour.

s (mph)	10	20	40	60
d (ft)	15	40	114	225

a. Fit an exponential function, power function, and logarithmic function to the data. Find the regression constants to the *nearest hundredth*.
b. Determine which function best fits the data, and give a reason for your answer.
c. Using the function that gives the best fit, estimate the distance, correct to the *nearest foot*, needed to stop a car traveling at 55 miles per hour.

5. The volumes, V, of a particular gas were determined at various pressures, P, as shown in the accompanying table.

P (atms)	0.1	0.3	0.5	0.7	0.9	1.1	1.5	1.7	1.9	2.1	2.3
V (L)	225	75.0	45	32.1	25	20.5	15	13.2	11.8	10.7	9.9

a. Draw a scatterplot in which P is the independent variable on the horizontal axis, and V is the dependent variable on the vertical axis.
b. Determine an equation of the curve of best fit. Find the regression constants to the *nearest tenth*.

6. a. Determine an exponential function $y = ab^x$ that best fits the data given in Exercise 11 on page 271, where x represents the number of years after 1980. Approximate the constants a and b correct to the *nearest hundredth*.
b. Assuming this relationship continues, determine the first year in which fewer than 25,000 gallons of leaded gasoline will be available.

7. The accompanying table shows the growth in the number of transistors that can be placed on an Intel computer processor chip.

Name of Intel Chip	Year of Introduction	Number of Transistors (millions)
386	1985	0.275
486 DX	1989	1.18
Pentium	1993	3.1
Pentium II	1997	7.5
Pentium III	1999	24
Pentium IV	2000	42

a. Determine an exponential function $y = ab^x$ that best fits the data, where $x = 1$ represents the year 1985 and y represents the corresponding number of transistors in millions. Approximate the constants a and b correct to the *nearest hundredth*.
b. Assuming that the exponential relationship continues, predict the first year in which more than 200,000,000 transistors can be placed on an Intel computer processor chip.
c. Explain why an exponential function is a better model of the data than a power function.

8. The accompanying table shows the number of milligrams, y, of a dose of a medicinal drug that remains in the body x hours after it has been ingested.

x hr	2	3	5	7	9
y mg	11	6.7	2.9	1.5	0.9

a. Determine a linear function $y = ax + b$ that best fits the given data. Estimate constants a and b to the *nearest tenth*.
b. Using the regression model calculated in part a, find to the *nearest tenth* of an hour when the amount of the drug that remains in the body is one half of the original amount taken.
c. Determine an exponential function $y = ax^b$ that best fits the given data. Estimate constants a and b to the *nearest tenth*.
d. Using the regression model calculated in part c, find to the *nearest minute* when the amount of the drug that remains in the body first falls below 0.5 milligrams.
e. Which of the regression models best fits the given data? Justify your answer.

CHAPTER 12

PROBABILITY AND NORMAL CURVES

12.1 BINOMIAL PROBABILITIES

KEY IDEAS

When a probability experiment has exactly two possible outcomes, a formula can be used to find the probability that a particular outcome will occur r times when the same experiment is repeated n times, where $r \leq n$.

Probabilities of Two-Outcome Experiments

Each time a coin is tossed, there are two possible outcomes: a head ("success") and a tail ("failure"). Finding the probability of getting 3 heads in 5 tosses of a coin is an example of a *binomial probability experiment* with 5 repetitions called **trials**. A **binomial probability experiment** has these features:

- There is a fixed number of repeated trials.
- Each trial has two possible outcomes: success and failure. If p is the probability of a success, then q is the probability of a failure, where $q = 1 - p$.
- The probability of a success is the same for each trial. Therefore, the outcome of each trial has no effect on the outcome of any other trial.

Combinations Formula

Consider tossing a coin and getting a head (H) success and getting a tail (T) failure. In 4 trials, there are 4 combinations of possible outcomes that include exactly 3 successes:

H, H, H, T   H, H, T, H   H, T, H, H   T, H, H, H

When working with larger numbers of trials, making lists can be time-consuming and can lead to error. Therefore, to figure out the number of combinations of r successes in n trials, denoted by ${}_nC_r$, use the formula

$$ {}_nC_r = \frac{n!}{r!(n-r)!}. $$

For example, to find the number of combinations of 3 wins in 5 games, evaluate ${}_nC_r$ for $n = 5$ and $r = 3$:

$$
{}_5C_3 = \frac{5!}{3!(5-3)!} = \frac{5 \times \overset{2}{\cancel{4}} \times \cancel{3 \times 2 \times 1}}{\underset{1}{\cancel{3!}}\cancel{(2 \times 1)}} = \frac{5 \times 2}{1} = 10.
$$

MATH FACTS

The number of combinations of r successes in n trials ($r \leq n$) is given by the formula

$$
{}_nC_r = \frac{n!}{r!(n-r)!}.
$$

The notation ${}_nC_r$ is read as "the combination of n objects taken r at a time." Time and effort can be saved by remembering these relationships:

- ${}_nC_0 = 1.$ For example, ${}_7C_0 = 1.$
- ${}_nC_1 = n.$ For example, ${}_7C_1 = 7.$
- ${}_nC_n = 1.$ For example, ${}_7C_7 = 1.$

Evaluating ${}_nC_r$ Using a Calculator

To evaluate ${}_nC_r$ using your calculator:

- Press [2nd] [QUIT] to return to the home screen. Then enter the value of n.
- Press [MATH] [▷] [▷] [▷] [3] to select ${}_nC_r$ from the MATH PRB menu.
- Enter the value of r. Then press [ENTER].

Binomial Probabilities When r Successes Occur First

Suppose a coin is weighted so that the probability of tossing a head is $\frac{2}{3}$ and the probability of tossing a tail is $\frac{1}{3}$. If the coin is tossed 5 times, the probability of obtaining 3 heads (successes) followed by 2 tails (failures) is

$$
P(\text{H, H, H, T, T}) = \frac{2}{3} \times \frac{2}{3} \times \frac{2}{3} \times \frac{1}{3} \times \frac{1}{3} = \left(\frac{2}{3}\right)^2 \left(\frac{1}{3}\right)^2.
$$

Thus, $P(\text{H, H, H, T, T}) = p^r q^{n-r}$, where $p = \frac{2}{3}$, $q = 1 - \frac{2}{3} = \frac{1}{3}$, $r = 3$ heads or successes, and $n = 5$ trials.

In general, if p represents the probability of a success in each of n independent trials, then the probability of r consecutive successes followed by $n - r$ consecutive failures is given by the expression $p^r q^{n-r}$, where $q = 1 - p$.

Binomial Probabilities of *r* Successes in *n* Trials

Now suppose that the 3 heads do not necessarily occur first, as in T, H, T, H, H. Since the number of combinations of the 3 heads in 5 tosses is ${}_5C_3$, the probability of obtaining exactly 3 heads in 5 tosses is

$$\begin{aligned} P(3\text{ H in 5 tosses}) &= {}_5C_3\left(\frac{2}{3}\right)^3\left(\frac{1}{3}\right)^2 \\ &= 10 \times \frac{8}{27} \times \frac{1}{9} \\ &= \frac{80}{243} \end{aligned}$$

Math Facts

If p represents the probability of a success in each of n independent trials, then

$$P(r \text{ successes in } n \text{ trials}) = {}_nC_r p^r q^{n-r},$$

where $p + q = 1$.

Binomial Probability Distribution

The expansion of $(p + q)^n$ gives the complete probability distribution for a binomial experiment with n trials. For example, if there are 2 trials, $n = 2$. Then

$$(p+q)^2 = p^2 + 2pq + q^2$$

or, equivalently,

$$(p+q)^2 = p^2 + 2p^1q + p^0q^2.$$

The exponent of p in each term of the binomial expansion represents the number of successes. Hence:

- The probability of 2 successes in 2 trials is p^2.
- The probability of 1 success in 2 trials is $2p^1q$.
- The probability of 0 success in 2 trials is p^0q^2.
- Since the sum of the probabilities of all the possible outcomes of a probability experiment is always 1, $p^2 + 2p^1q + p^0q^2 = 1$.

Binomial Probabilities with Conditions

Binomial probabilities may need to be added:

- The probability of obtaining "*at least*" r successes in n trials is the sum of the probabilities of obtaining r successes, $r + 1$ successes, . . . , continuing up to n successes.
- The probability of obtaining "*at most*" r successes in n trials is the sum of the probabilities of 0 success, 1 success, . . . , continuing up to r successes.

Example 1

In a baseball game, the probability that Peter will get on base safely when he comes to bat is $\frac{3}{7}$. What is the probability that he will get on base safely *at least* 3 of the 4 times he comes to bat?

Solution: $\mathbf{\dfrac{513}{2401}}$

Consider the probability of getting on base a binomial experiment with exactly two possible outcomes: Peter gets on base safely, and Peter does *not* get on base safely. The probability of getting on base safely r times out of n times at bat is given by the expression ${}_nC_rp^rq^{n-r}$, where $p = \dfrac{3}{7}$ and $q = 1 - p$ $= 1 - \dfrac{3}{7} = \dfrac{4}{7}$.

The probability that Peter will get on base safely *at least* 3 times is the sum of the probabilities that he will get on base safely exactly 3 times or exactly 4 times. Thus:

- If $r = 3$ and $n = 4$, then:

$$P(\text{gets on base 3 of 4 at bat}) = {}_4C_3\left(\frac{3}{7}\right)^3\left(\frac{4}{7}\right)^{4-3}$$
$$= 4 \cdot \frac{3^3}{7^3} \cdot \frac{4}{7}$$
$$= \frac{432}{2401}$$

- If $r = 4$ and $n = 4$, then:

$$P(\text{gets on base 4 of 4 at bat}) = {}_4C_4\left(\frac{3}{7}\right)^3\left(\frac{4}{7}\right)^{4-4}$$
$$= 1 \cdot \frac{3^4}{7^3} \cdot 1$$
$$= \frac{81}{2401}$$

Hence, the probability that Peter will get on base safely *at least* 3 times is

$$\frac{432}{2401} + \frac{81}{2401} \quad \text{or} \quad \frac{513}{2401}.$$

Example 2

In the month of February at a ski resort, the probability of snow on any day is $\frac{1}{4}$. What is the probability that snow will fall on *at most* 2 days of a 5-day trip to that ski resort in February?

Solution: $\mathbf{\frac{918}{1024}}$

Finding the probability that it will snow any day in the 5-day trip is a binomial probability experiment in which the two possible outcomes are it snows and it does not snow. The probability that it will snow exactly r days of n days is given by the expression ${}_nC_r p^r q^{n-r}$, where $p = \frac{1}{4}$ and $q = 1 - p = 1 - \frac{1}{4} = \frac{3}{4}$.

The probability that snow will fall on *at most* 2 days is the sum of the probabilities that it will snow on exactly 0, 1 or 2 days. Thus:

- If $r = 0$ and $n = 5$, then:

$$P(\text{snows 0 day}) = {}_5C_0\left(\frac{1}{4}\right)^0\left(\frac{3}{4}\right)^5 = 1 \cdot 1 \cdot \frac{243}{1024} = \frac{243}{1024}.$$

- If $r = 1$ and $n = 5$, then:

$$P(\text{snows 1 day}) = {}_5C_1\left(\frac{1}{4}\right)^1\left(\frac{3}{4}\right)^4 = 5 \cdot \frac{1}{4} \cdot \frac{81}{256} = \frac{405}{1024}.$$

- If $r = 2$ and $n = 5$, then:

$$P(\text{snows 2 days}) = {}_5C_2\left(\frac{1}{4}\right)^2\left(\frac{3}{4}\right)^3 = 10 \cdot \frac{1}{16} \cdot \frac{27}{64} = \frac{270}{1024}.$$

Hence, the probability that it will snow on *at most* 2 days is

$$\frac{243}{1024} + \frac{405}{1024} + \frac{270}{1024} = \frac{918}{1024}.$$

Check Your Understanding of Section 12.1

A. Multiple Choice

1. The probability of Rick getting an A on any test is $\frac{2}{5}$. Which expression represents the probability that he will earn an A on exactly 3 of 4 tests?

(1) ${}_5C_4\left(\frac{2}{5}\right)^4\left(\frac{3}{5}\right)$ (3) ${}_5C_4\left(\frac{3}{5}\right)^4\left(\frac{2}{5}\right)$

(2) ${}_4C_3\left(\frac{2}{5}\right)^3\left(\frac{3}{5}\right)$ (4) ${}_4C_3\left(\frac{3}{5}\right)^3\left(\frac{2}{5}\right)$

2. The probability of Gordon's team winning any given game in a 5-game series is 30%. Which expression represents the probability that Gordon's team will win exactly 2 games in the series?

(1) $(0.3)^2(0.7)^3$ (3) $10(0.3)^2(0.7)^3$

(2) $5(0.3)^3(0.7)^2$ (4) $5(0.3)^2(0.7)$

3. A box contains 8 good and 4 bad light bulbs. Daniel randomly picks a bulb from the box, tests it, and then replaces it. If Daniel picks 4 light bulbs with replacement, what is the probability he will pick exactly 2 bad light bulbs?

(1) $\frac{4}{27}$ (2) $\frac{1}{9}$ (3) $\frac{8}{27}$ (4) $\frac{4}{9}$

4. If the probability that Jamal will successfully make a foul shot is 80%, what is the probability that he will successfully make exactly 3 of his next 4 foul shots?

(1) $\frac{64}{125}$ (2) $\frac{64}{625}$ (3) $\frac{192}{625}$ (4) $\frac{256}{625}$

5. Each day the probability of rain on a tropical island is $\frac{7}{8}$. Which expression represents the probability that it will rain on the island exactly x days in the next 3 days?

(1) ${}_3C_x\left(\frac{7}{8}\right)^x\left(\frac{1}{8}\right)^{3-x}$ (3) ${}_xC_3\left(\frac{7}{8}\right)^3\left(\frac{1}{8}\right)^x$

(2) ${}_3C_3\left(\frac{7}{8}\right)^3\left(\frac{1}{8}\right)^x$ (4) ${}_xC_3\left(\frac{7}{8}\right)^{3-x}\left(\frac{1}{8}\right)^x$

6. In a family of 6 children, what is the probability that exactly 1 child is female, assuming that $P(\text{male}) = P(\text{female})$?

(1) $\frac{32}{64}$ (2) $\frac{7}{64}$ (3) $\frac{6}{64}$ (4) $\frac{58}{64}$

B. *In each case, show how you arrived at your answer by clearly indicating all of the necessary steps, formula substitutions, diagrams, graphs, charts, etc.*

7. As shown in the accompanying diagram, a circular target with a radius of 9 inches has a bull's-eye with a radius of 3 inches. If 5 arrows randomly hit the target, what is the probability that at least 4 hit the bull's-eye?

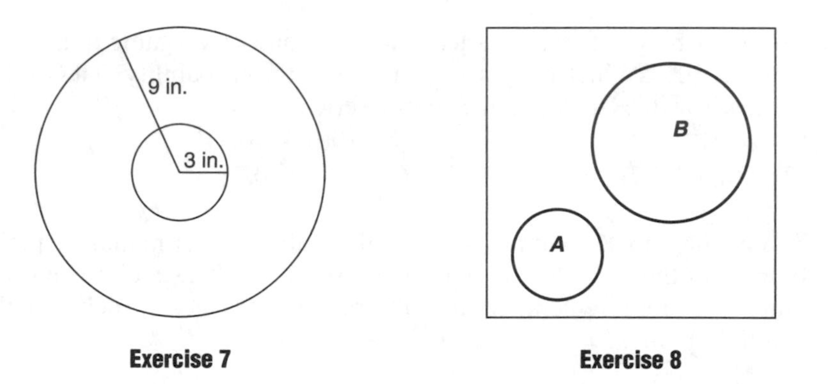

Exercise 7 **Exercise 8**

8. The perimeter of a square dartboard is 40 inches. Circle A, with an area of 9 square inches, and circle B, with an area of 16 square inches, lie inside the square and do not overlap, as shown in the accompanying figure. If a dart hits the board 3 times, find the probability that it lands outside both circles *at most* once.

9. A mathematics quiz has 5 multiple-choice questions that are all worth the same number of points. There are 4 possible responses for each question. Jennifer selects her responses at random on every question. What is the probability she will select correct responses that will give her a grade of *at least* 80% on the quiz?

10. In the month of January at a ski resort, the probability of snow on any day is $\frac{3}{7}$.

a. What is the probability that snow will *not* fall on any day during a 5-day trip to that resort in January?

b. What is the probability that snow will fall on *at least* 3 days of that 5-day trip in January?

11. At a certain intersection, the light for eastbound traffic is red for 30 seconds, yellow for 5 seconds, and green for 45 seconds. Find the probability that, of the next 5 eastbound cars that arrive randomly at the intersection, more than half will be stopped by a red light.

12. The probability that a new calculator battery will fail within the first 100 hours of use is 30%.

a. A graphing calculator will power on only if each of its 4 batteries does not fail. If 4 new batteries are put into a graphing calculator, what is the probability that the calculator will *always* power on in the next 100 hours of calculator use?

b. What is the probability that *no more than 1* of the original 4 batteries will fail in the first 100 hours of calculator use?

12.2 THE BINOMIAL THEOREM

KEY IDEAS

The **binomial theorem** tells how to expand a binomial raised to a power, such as $(a + b)^5$, without multiplying the factors together.

Expanding $(a + b)^n$

By repeated multiplication of $(a + b)$, you can determine that

$$(a+b)^5 = \underbrace{1a^5}_{\text{Term 1}} + \underbrace{5a^4b}_{\text{Term 2}} + \underbrace{10a^3b^2}_{\text{Term 3}} + \underbrace{10a^2b^3}_{\text{Term 4}} + \underbrace{5a^1b^4}_{\text{Term 5}} + \underbrace{1b^5}_{\text{Term 6}}$$

Observe that:

- The first term is a^5, and the last term is b^5. In each term, the sum of the exponents of a and b is 5.
- The number of terms in the expansion is 6, which is 1 more than the exponent of $(a + b)^5$.
- The numerical coefficients of the terms are the values of ${}_5C_r$ as r increases from 0 (the first term) to 5 (the last term). For example, the numerical coefficient of the fourth term is ${}_5C_3 = 10$, where $n = 5$ and $r = 3$ (1 less than 4). Similarly, the numerical coefficient of the fifth term is ${}_5C_4 = 5$, and the numerical coefficient of the sixth term is ${}_5C_5 = 1$.

The binomial theorem is an algebraic formula that generalizes these observations for the expansion of any binomial that has the form $(a + b)^n$, where n is a positive whole number.

MATH FACTS

BINOMIAL THEOREM: *If n is a positive whole number, then*

$$(a+b)^n = {}_nC_0a^n + {}_nC_1a^{n-1}b^1 + {}_nC_2a^{n-2}b^2 + {}_nC_3a^{n-3}b^3 + \cdots + {}_nC_{n-1}a^1b^{n-1} + {}_nC_nb^n.$$

Example 1

Write the expansion of $(x - 3y)^4$.

Solution: $\mathbf{x^4 - 12x^3y + 54x^2y^2 - 108xy^3 + 81y^4}$

Use the binomial theorem to expand $(a + b)^n$, where $a = x$, $b = -3y$, and $n = 4$:

$$\begin{aligned}(x-3y)^4 &= {}_4C_0x^4 + {}_4C_1x^3(-3y)^1 + {}_4C_2x^2(-3y)^2 + {}_4C_3x^1(-3y)^3 + {}_4C_4x^0(-3y)^4 \\ &= 1x^4 + 4x^3(-3y) + 6x^2(-3y)^2 + 4x(-3y)^3 + 1(-3y)^4 \\ &= x^4 - 12x^3y + 54x^2y^2 - 108xy^3 + 81y^4\end{aligned}$$

Writing the *k*th Term of $(a + b)^n$

Sometimes the only thing you want to know about the power of a binomial is what a particular term looks like.

MATH FACTS

The *kth term of* $(a+b)^n$ is ${}_nC_{k-1}a^{n-(k-1)}b^{k-1}$.

From Example 1, we know that the third term of the expansion of $(x - 3y)^4$ is $54x^2y^2$. This particular term can also be obtained by using the expression ${}_nC_{k-1}a^{n-(k-1)}b^{k-1}$, where $n = 4$, $k = 3$, $a = x$, and $b = -3y$:

$$\begin{aligned}{}_nC_{k-1}a^{n-(k-1)}b^{k-1} &= {}_4C_{3-1}x^{4-(3-1)}(-3y)^{3-1} \\ &= {}_4C_2x^{4-2}(-3y)^2 \\ &= 6(x^2)(9y^2) \\ &= 54x^2y^2\end{aligned}$$

Example 2

What is the middle term of the expansion of $\left(\frac{x}{3}+y^2\right)^4$?

Solution: $\mathbf{\frac{2x^2y^4}{3}}$

Since the expansion of $\left(\frac{x}{3}+y^2\right)^4$ consists of 4 + 1 or 5 terms, the middle term of the expansion is the third term. To find the third term of the expansion of $\left(\frac{x}{3}+y^2\right)^4$, evaluate ${}_nC_{k-1}a^{n-(k-1)}b^{k-1}$, where $n = 4$, $k = 3$, $a = \frac{x}{3}$, and $b = y^2$:

$$
\begin{aligned}
{}_nC_{k-1}a^{n-(k-1)}b^{k-1} &= {}_4C_{3-1}\left(\frac{x}{3}\right)^{4-(3-1)}\left(y^2\right)^{3-1} \\
&= {}_4C_2\left(\frac{x}{3}\right)^2\left(y^2\right)^2 \\
&= 6\frac{x^2}{9}\left(y^4\right) \\
&= \frac{6x^2y^4}{9} \\
&= \frac{2x^2y^4}{3}
\end{aligned}
$$

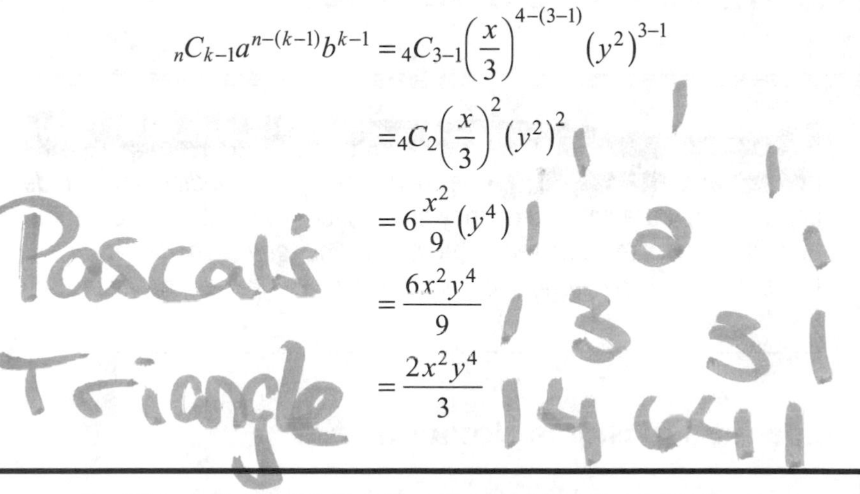

Check Your Understanding of Section 12.2

Multiple Choice

1. What is the third term in the expansion of $(x + 2y)^5$?
(1) $10x^3y^2$ (2) $40x^3y^2$ (3) $80x^2y^3$ (4) $20x^2y^3$

2. What is the fourth term in the expansion of $(2x - y)^7$?
(1) $16x^4y^3$ (2) $35x^3y^4$ (3) $-560x^4y^3$ (4) $-560x^3y^4$

3. The fifth term in the expansion of $(3a - b)^6$ is
(1) $135a^2b^4$ (2) $540a^3b^3$ (3) $-18ab^5$ (4) $-135a^2b^4$

4. What is the middle term in the expansion of $(3x - 2y)^4$?
(1) $-6x^2y^2$ (2) $36x^2y^2$ (3) $-216x^2y^2$ (4) $216x^2y^2$

5. What is the numerical coefficient of the term containing x^3y^2 in the expansion of $(x + 2y)^5$?
(1) 10 (2) 20 (3) 40 (4) 80

6. The expression $(1 - 2i)^5$, where $i = \sqrt{-1}$, is equivalent to
(1) $41 + 38i$ (2) $121 + 102i$ (3) $-39 + 38i$ (4) $-119 + 102i$

7. What is the middle term in the expansion of $\left(x^2 + \frac{1}{x}\right)^6$?

(1) $10x^3$ (2) $30x^2$ (3) $20x^3$ (4) $\frac{10}{x^2}$

12.3 DESCRIPTIVE STATISTICS

KEY IDEAS

Individual numbers that are used to describe and compare sets of data are called **statistics**. The *mean* (average), *median*, and *mode* are measures of central tendency since each gives information about how the data are centered. The *range* and *standard deviation* are measures of dispersion since each suggests how widely scattered the data are.

Subscript and Sigma Notation

If x represents an ordered list of data values consisting of six test scores, the *subscripted variables*

$$x_1, x_2, x_3, x_4, x_5, \text{ and } x_6$$

represent the individual scores in the list; in each case, the number written to the right and one-half line below x, called a **subscript**, indicates the position number of the score in the list. For example, if $x = \{15, 7, 4, 13, 11, 10\}$, then

$$x_1 = 15, x_2 = 7, x_3 = 4, x_4 = 13, x_5 = 11, \text{ and } x_6 = 10.$$

The Greek letter Σ, read as "sigma," can be used to indicate the sum of an ordered list of subscripted variables. For example, $\sum_{i=1}^{6} x_i$ represents the sum of the values of x_i as i successively takes on integer values from 1 to 6. Thus:

$$\sum_{i=1}^{6} x_i = x_1 + x_2 + x_3 + x_4 + x_5 + x_6$$
$$= 15 + 7 + 4 + 13 + 11 + 10$$
$$= 60$$

In the summation $\sum_{i=1}^{6} x_i$, i is called the **index variable**, 1 is the first value of the index variable, and 6 is the last value. Here are some additional facts about summation notation you should know:

- The index variable may start at a whole number different from 1. For example:

$$\sum_{i=3}^{6} x_i = x_3 + x_4 + x_5 + x_6.$$

- If a constant factor appears inside the summation sign, it can be passed through it. For example:

$$\sum_{i=1}^{6} 2x_i = 2(x_1 + x_2 + x_3 + x_4 + x_5 + x_6).$$

- The index variable can be used to represent the numbers to be added. For example:

$$\sum_{1}^{5} i^2 = 1^2 + 2^2 + 3^2 + 4^2 + 5^2$$
$$= 1 + 4 + 9 + 16 + 25$$
$$= 55$$

Example 1

Evaluate: a. $\sum_{k=1}^{3}(2k-1)^2$ b. $\sum_{n=3}^{5}\frac{1}{7}(2^n)$

Solution: a. **35**

The notation $\sum_{k=1}^{2}(2k-1)^2$ means to take the sum of the terms of the form $(2k-1)^2$ for $k = 1$, 2, and 3:

$$\sum_{k=1}^{3}(2k-1)^2 = [2(1)-1]^2 + [2(2)-1]^2 + [2(3)-1]^2$$
$$= 1^2 \quad + 3^2 \quad + 5^2$$
$$= 35$$

b. **8**

To evaluate $\sum_{n=3}^{5}\frac{1}{7}(2^n)$, pass $\frac{1}{7}$ through the summation sign and then multiply it by the sum of the terms of the form 2^n for $n = 3$, 4, and 5, inclusive:

$$\sum_{n=3}^{5}\frac{1}{7}(2^n) = \frac{1}{7}\sum_{n=3}^{5}(2^n)$$
$$= \frac{1}{7}[2^3 + 2^4 + 2^5]$$
$$= \frac{1}{7}[2^3 + 2^4 + 2^5]$$
$$= \frac{1}{7}[56]$$
$$= 8$$

Mean, Median, and Mode

The mean (average), median, and mode are measures of central tendency since these statistics indicate where the data are centered.

- The **mean** (average) of a set of scores is the sum of the scores divided by the number of scores. If $\overline{x}$ represents the mean of a set of n scores, then

$$\text{Mean} = \overline{x} = \frac{1}{n}\sum_{i=1}^{n}x_i.$$

For example, if $x = \{15,7,4,13,11,10\}$, then

$$\overline{x} = \frac{1}{6}(15+7+4+13+11+10) = \frac{1}{6}(60) = 10.$$

- The **median** of a set of scores is the middle score when the numbers are arranged in size order. If there is an even number of data values, as in $x = \{15,7,4,13,11,10\}$, the median is the average of the two middle scores:

$$4, 7 \qquad \underbrace{10, 11}_{\text{Median}=\frac{10+11}{2}=10.5.} \qquad 13, 15$$

- The **mode** is the most frequently occurring score in a set. The mode of {2, 3, 1, 2, 4, 3, 5, 6, 2, 5} is 2, since 2 appears more times than any other value in the list.

Comparing Statistics

Consider two sets of scores:

$$A = \{49, 53, 51, 55\} \qquad \text{and} \qquad B = \{1, 2, 5, 200\}.$$

The mean of each set is 52. For set A, the mean is representative of how the scores are distributed in value since the scores are clustered about the mean. The same is not true for set B. Each of the scores in set B varies widely from the mean. In this case, the mean does not indicate where the data are centered. Other statistics are needed to describe the extent to which the data are scattered or dispersed about the mean.

Range and Standard Deviation

The range and standard deviation are measures of dispersion since these statistics describe how spread out the data are.

- The **range** of a set of scores is the difference between the largest and smallest scores. In the scores considered above, the range of set A is $55 - 49 = 6$ while the range of set B is $200 - 1 = 199$. When the range is large compared to the mean, the mean may not be a good measure of central tendency.
- The **standard deviation** of a set of scores is a statistic, usually denoted by the Greek letter σ (lower case sigma), that measures how far apart the individual scores are from the mean. The standard deviation, σ, of a set of n scores whose mean is $\overline{x}$ can be calculated using the formula

$$\sigma = \sqrt{\frac{1}{n}\sum_{i=1}^{n}(\overline{x} - x_i)^2}.$$

Calculating Standard Deviation

To calculate the standard deviation of set A = {49, 53, 51, 55}:

Step 1. Press STAT ENTER. Enter the individual scores in list L1.

Step 2. Press [STAT] [▷] [ENTER] [ENTER] to get the display in Figure 12.1. The readout for the standard deviation is $\sigma_x = 2.236067977$. Since the standard deviation is small compared to the mean, the individual scores tend to be centered about the mean.

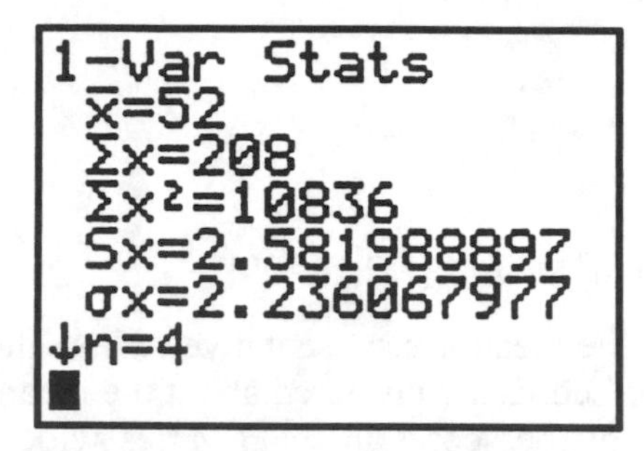

Figure 12.1 Statistics for Set *A*

```
1-Var Stats
 x̄=52
 Σx=208
 Σx²=40030
 Sx=98.68130522
 σx=85.4605172
↓n=4
```

Figure 12.2 Statistics for Set *B*

Repeat Step 1, overwriting the previous scores stored in L1 with the scores in set $B = \{1, 2, 5, 200\}$. Then repeat step 2 to get the display in Figure 12.2. Since $\sigma_x = 85.4605172$, the standard deviation is large compared to the mean which suggests that the individual scores vary widely from the mean.

Grouping Data

Sometimes equal data values are grouped together and summarized in a two-column table that shows how many times each particular data value, x_i, occurs. The number of times x_i appears is called its **frequency**, denoted as f_i. To find the standard deviation of grouped data, enter the individual data values in list L1 of your graphing calculator and their corresponding frequencies in list L2.

Example 2

The accompanying table summarizes the midterm exam results of 30 students in a Math B class.

Grade, x_i	Frequency, f_i
100	2
95	3
94	1
88	3
85	4
79	5
74	3
65	9

a. Find (1) the mean score and (2) the standard deviation, correct to the *nearest tenth*.

b. Find the percent of the class that had scores more than one standard deviation above the mean.

c. What is the probability that a score picked at random will fall within one standard deviation of the mean?

Solutions: a. (1) **79.5** (2) **11.7**

- Enter the set of scores in list L1 and the corresponding frequencies in list L2. Press [STAT] [▷] [ENTER]. On the line that reads "1-Var Stats," press [2nd] [1] [,] [2nd] [2].

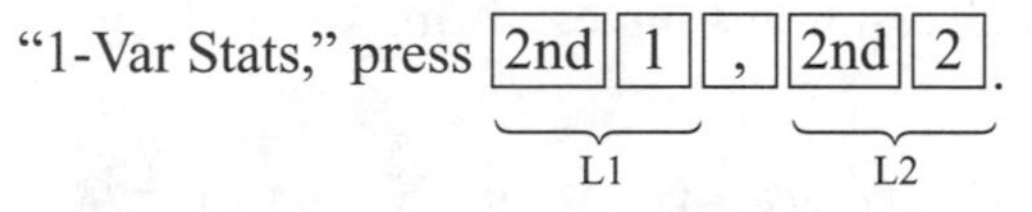

The screen should now look like the one in the figure at the right.

```
1-Var Stats L1,L
2
```

- Press [ENTER] to get the display shown in the figure at the right. Before you continue, verify that the number of scores, n, is 30.
- According to the display, $\bar{x} = 79.5$ and $\sigma_x \approx 11.7$.

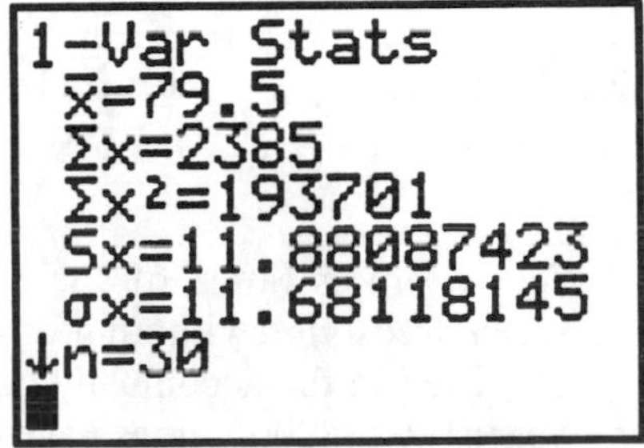

b. **20%**

The score $\bar{x} + 1\sigma_x = 79.5 + 11.7 = 91.2$ is one standard deviation above the mean. Hence,

- Any score that is greater than 91.2 is more than one standard deviation above the mean.
- According to the table, 6 of the 30 student test scores are greater than 91.2 and, as a result, are more than one standard deviation above the mean.
- Thus, $\frac{6}{30} \times 100\% = 20\%$ of the students in the class scored more than one standard deviation above the mean.

c. $\mathbf{\frac{15}{30}}$ or $\mathbf{\frac{1}{2}}$

The score $\bar{x} - 1\sigma = 79.5 - 11.7 = 67.8$ is one standard deviation below the mean, and the score $\bar{x} + 1\sigma = 79.5 + 11.7 = 91.2$ is one standard deviation above the mean.

- Any score that falls in the interval from 67.8 to 91.2 is within one standard deviation of the mean.
- According to the table, 15 of the 30 scores fall between 67.8 and 91.2.
- Hence, the probability that a score picked at random will fall within one standard deviation of the mean is $\frac{15}{30}$ or $\frac{1}{2}$.

Check Your Understanding of Section 12.3

In each case, show how you arrived at your answer by clearly indicating all of the necessary steps, formula substitutions, diagrams, graphs, charts, etc.

1–9. Evaluate, and express each result in simplest form.

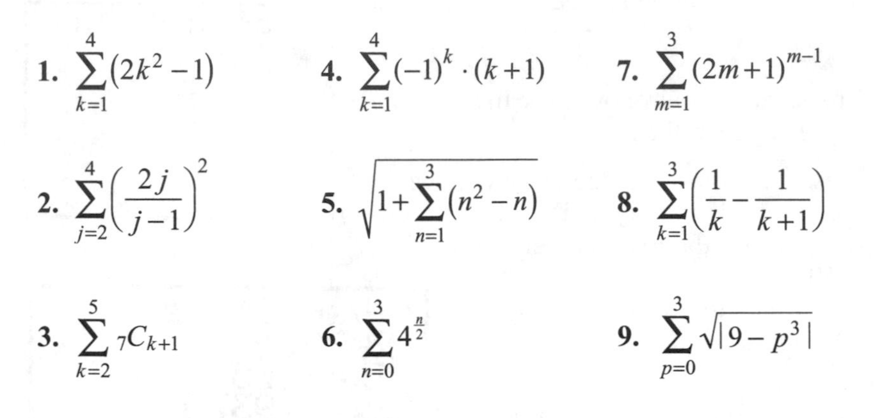

1. $\sum_{k=1}^{4}(2k^2-1)$

2. $\sum_{j=2}^{4}\left(\frac{2j}{j-1}\right)^2$

3. $\sum_{k=2}^{5}{}_7C_{k+1}$

4. $\sum_{k=1}^{4}(-1)^k\cdot(k+1)$

5. $\sqrt{1+\sum_{n=1}^{3}(n^2-n)}$

6. $\sum_{n=0}^{3}4^{\frac{n}{2}}$

7. $\sum_{m=1}^{3}(2m+1)^{m-1}$

8. $\sum_{k=1}^{3}\left(\frac{1}{k}-\frac{1}{k+1}\right)$

9. $\sum_{p=0}^{3}\sqrt{|9-p^3|}$

10. The winning times of the women's 400-meter freestyle swimming at the Olympics are listed in the accompanying table for the years 1964–2000. Times have been rounded to the nearest hundredth of a minute.

a. Find (1) the mean and (2) the standard deviation of the winning times correct to the *nearest hundredth*.

b. Determine an equation of the regression model that best fits the data where $y = 0$ corresponds to the year 1960. Estimate regression coefficients to the *nearest thousandth*.

c. Using the regression model, estimate to the *nearest hundredth of a minute* the winning time at the Olympics in 2004.

Year, y	Time, x (min)
1964	4.72
1968	4.53
1972	4.32
1976	4.17
1980	4.15
1984	4.12
1988	4.06
1992	4.12
1996	4.12
2000	4.10

11. The accompanying table shows the scores on a writing test in an English class.
   a. Find (1) the mean and (2) the standard deviation correct to the *nearest tenth*.
   b. Find, to the *nearest tenth*, the percent of scores that are within one standard deviation of the mean.
   c. What is the probability that a test score picked at random will be more than one standard deviation below the mean?

Score, x_i	Frequency, f_i
95	4
85	13
75	11
70	6
65	2

12. The accompanying table represents the PSAT scores of a group of 20 students.
   a. Find (1) the mean and (2) the standard deviation correct to the *nearest tenth*.
   b. Find the percent of scores that are more than one standard deviation above the mean.
   c. Find the probability that a test score picked at random will be within one standard deviation of the mean.

Score	Frequency
48	2
50	6
53	2
54	4
57	2
62	3
68	1

13. The accompanying table shows the frequency of the average daily temperatures during the month of June.
   a. Find (1) the mean and (2) the standard deviation. Estimate the standard deviation to the *nearest tenth*.
   b. If the temperatures for two different days are chosen at random, what is the probability that *at least* one of the temperatures is equal to the mean?

Temperature	Frequency
63	5
70	3
78	4
79	3
80	6
84	4
96	5

Exercise 13

Height (in.)	Frequency
72	3
71	2
70	1
69	2
68	4
67	2
66	4
65	2

Exercise 14

14. The accompanying table (see page 297) shows the heights in inches of a group of 20 students.

a. Find (1) the mean and (2) the standard deviation correct to the *nearest tenth*.

b. If three students' heights are chosen at random, what is the probability that *at most* one of the heights falls within one standard deviation of the mean?

12.4 THE NORMAL CURVE

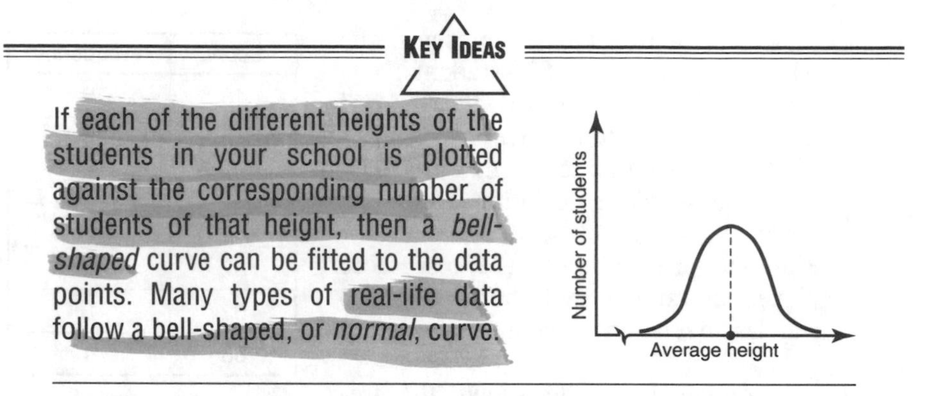

If each of the different heights of the students in your school is plotted against the corresponding number of students of that height, then a *bell-shaped* curve can be fitted to the data points. Many types of real-life data follow a bell-shaped, or *normal*, curve.

Normally Distributed Data

If a set of data conforms to a bell-shaped curve, the data are said to be **normally distributed**. Figure 12.3 shows a normal distribution of ages that has a mean of 35 and a standard deviation of 7. The numbers to the left and right of 35 show intervals that are one, two, and three standard deviations above and below the mean.

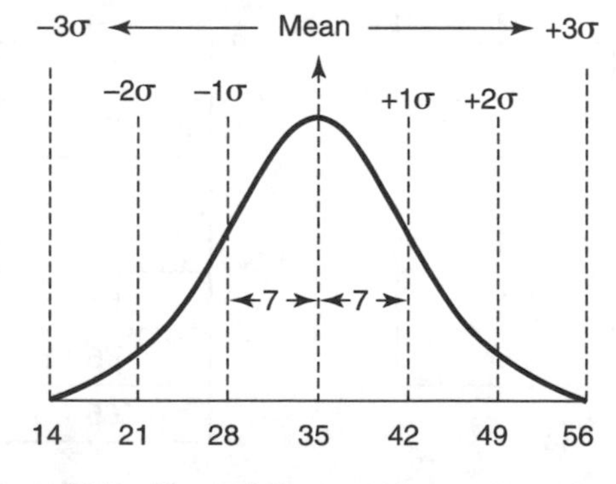

Figure 12.3 Normal Curve with $\overline{x} = 35$ and $\sigma = 7$

Every normal curve has these properties:

- The curve is centered about the mean data value and is asymptotic to the horizontal axis.
- The mean and median are equal.
- The total area of the region between the normal curve and the horizontal axis is 1. Hence, the area of any interval under the normal curve represents the probability that a randomly selected value will fall in that interval.
- The distribution of values from the mean for every normal curve is the same. For example, about 34% of the values from a normal distribution will always lie within one standard deviation above the mean. For example, if the normal curve in Figure 12.3 represents a population of 5000 people, then approximately $0.34 \times 5000 = 1700$ people in that population will range in age from 35 to 42 years old. If a person is picked at random from this population, the probability that he or she will be from 35 to 42 years old is 34%. Similarly, the probability that the person will be from 28 to 35 years old is also 34% since the interval from 28 to 35 is *one* standard deviation below the mean.

Standard Normal Curve

The normal curve shown in Figure 12.4 is a *standard* normal curve. A **standard normal curve** is a normal curve that is centered on the y-axis, so its mean is 0 and its standard deviation is 1. Since all normal curves have the same percent distribution of data values, the percentages shown are true for all normal curves. For example, approximately 19.1% of the values fall between the mean and the value that is 0.5 standard deviation above the mean. Also, approximately 19.1% + 15.0% = 34.1% of the values fall between the mean and the value that is one standard deviation above the mean.

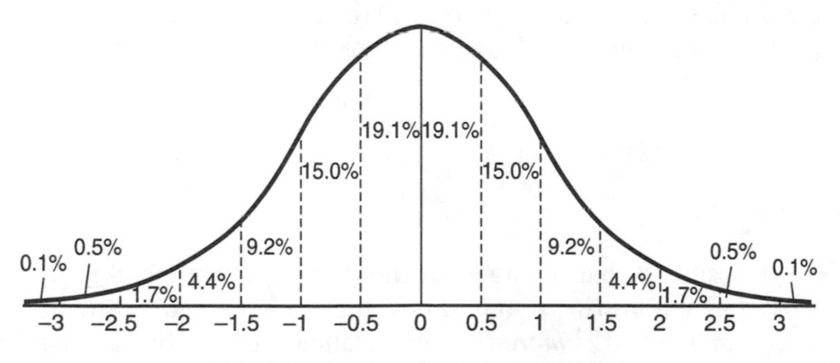

Figure 12.4 Standard Normal Curve

From the curve in Figure 12.4 you can tell that:

- Approximately 34.1% + 34.1% = 68.2% of the values lie within an interval that is one standard deviation below and above the mean.
- Approximately 47.7% + 47.7% = 95.4% of the values lie within an interval that extends two standard deviations below and above the mean.
- Approximately 49.9% + 49.9% = 99.8% of the values lie within an interval that extends three standard deviations below and above the mean.

Example 1

The mean score on the mathematics section of a standardized examination was 483, and the standard deviation was 97. If 10,000 students took the exam, approximately how many students had scores from 386 to 580?

Solution: **6820**

It is given that $\overline{x} = 483$ and $\sigma_x = 97$. Then:

- Since $\overline{x} - 1\sigma_x = 483 - 97 = 386$, 386 is one standard deviation below the mean. Because $\overline{x} + 1\sigma_x = 483 + 97 = 580$, 580 is one standard deviation above the mean.
- The scores 386 and 580 are the endpoints of an interval that extends from one standard deviation below the mean to one standard deviation above the mean. According to the curve in Figure 12.3, this interval includes approximately 68.2% of the scores.
- Since there were 10,000 test takers, approximately $0.682 \times 10{,}000 =$ 6820 students had scores from 386 to 580.

Example 2

Multiple Choice:
The ages of people who watched a popular television program are normally distributed. The mean age is 25 years, and the standard deviation is 2.2. If an age is picked at random, which age can be expected to be selected less than 2.3% of the time?

(1) 20 (2) 29 (3) 23 (4) 25

Solution: **(1)**

- If it is assumed that the ages are normally distributed, approximately 95.4% fall within two standard deviations above and below the mean. Hence, an age that is *more than* two standard deviations away from the mean can be expected to be picked less than 2.3% of the time.

- It is given that $\sigma_x = 2.2$, so $2\sigma_x = 2 \times 2.2 = 4.4$. Also, since $\bar{x} = 25$,

  $\bar{x} - 2\sigma_x = 25 - 4.4 = 20.6$ and $\bar{x} + 2\sigma_x = 25 + 4.4 = 29.6$.
- The next step is to look in the choices for an age that is either less than 20.6 or greater than 29.6. The correct choice is (1).

The Cumulative Normal Distribution Function

You can use the **normalcdf** function of your graphing calculator to find the percent of values that lie between two given values for a normal distribution in which the mean and standard deviation are known.

Example 3

Mr. Yee has 184 students in his mathematics class. The scores on the final examination are normally distributed and have a mean of 72.3 and a standard deviation of 8.9. How many students in the class can be expected to receive a score between 82 and 90?

Solution: **21**

Use your graphing calculator to find the area of the normal curve that lies between the lower bound of 82 and the upper bound of 90:

- Press [2nd] [VARS] [2] to select the normal cumulative density function.
- Enter the lower bound, the upper bound, the mean, the standard deviation, and the right parenthesis. Separate numerical values with commas. The display should show: normalcdf (82, 90, 72.3, 8.9).
- Press [ENTER] to get .1145178018. This value represents the percent of scores between 82 and 90. Since there are 184 students in the class, $184 \times .1145 \approx 21$ students can be expected to receive scores between 82 and 90.

Percentiles

The pth percentile of a set of data values is the value at or below which $p\%$ of the total number of values fall. If a set of values are normally distributed, then:

- The mean is the 50th percentile since 50% of the values lie at or below the mean.
- $\bar{x} + 1\sigma_x \approx$ 84th percentile since 50% + 34.1%, or 84.1%, of the total number of values lie at or below the value that is one standard deviation above the mean, as shown in Figure 12.5.

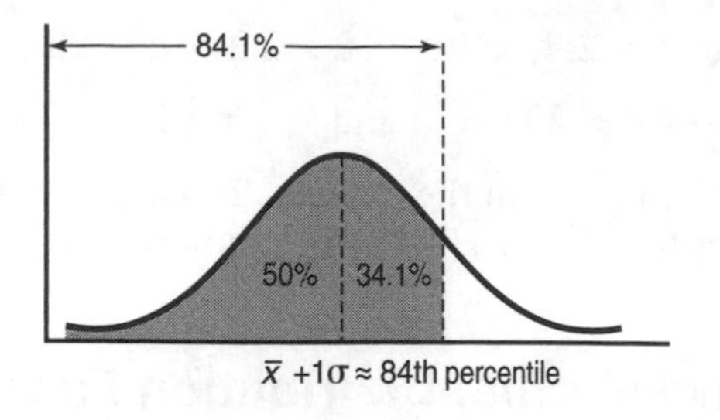

Figure 12.5 Percentiles

- $\bar{x} + 2\sigma_x \approx$ 97th percentile since 50% + 47.7%, or 97.7%, of the total number of values lies at or below the value that is two standard deviations above the mean.
- Similarly, the 99th percentile is located at approximately three standard deviations above the mean.

Example 4

Multiple Choice:
A standardized test with a normal distribution of scores has a mean score of 43 and a standard deviation of 6.3. Which range contains the score of a student in the 90th percentile?

(1) 30.4–36.7 (2) 36.7–43.0 (3) 43.0–49.3 (4) 49.3–55.6

Solution: **(4)**

- 84th percentile $\approx \bar{x} + 1\sigma_x = 43 + 6.3 = 49.3$.
- 97th percentile $\approx \bar{x} + 2\sigma_x = 43 + 2(6.3) = 55.6$.
- Hence, the 90th percentile must be a score in the range of 49.3–55.6. The correct choice is (4).

Check Your Understanding of Section 12.4

A. Multiple Choice

1. A survey of the soda-drinking habits of the student population in a high school revealed that the mean number of cans of soda a student drinks per week is 20, with a standard deviation of 3.5. If a normal distribution is assumed, which interval represents the total number of cans of soda that approximately 95% of the population of this school will drink in a week?
(1) 13–27 (2) 20–27 (3) 20–34 (4) 6–34

2. The heights of the members of a high school class are normally distributed. If the mean height is 65 inches and a height of 72 inches represents

the 84th percentile, which value best approximates the standard deviation for this distribution?
(1) 3.5 (2) 7 (3) 12 (4) 137

3. If Tamara scored in the 95th percentile on a standardized mathematics test, which statement *must* be true?
(1) Tamara's score was between two and three standard deviations above the mean.
(2) Tamara's score was more than three standard deviations above the mean.
(3) Tamara's score was higher than the scores of 95% of the students who took the test.
(4) Tamara's score was lower than the scores of 5% of the students who took the test.

4. A set of 418 test scores follow a normal distribution with a mean score of 72 and a standard deviation of 8. Approximately how many students taking the test received a score greater than 84?
(1) 390 (2) 28 (3) 66 (4) 9

5. On a quiz, the mean score is 72 and the standard deviation is 3.4. Which score can be expected to occur less than 5% of the time?
(1) 65 (2) 67 (3) 72 (4) 78

6. The scores on an exam are normally distributed with a mean of 77. If the interval from 71 to 83 contains approximately 95% of the scores, which value can be the standard deviation?
(1) 1.5 (2) 6 (3) 3 (4) 12

7. A student scores 84 on a standardized test. The mean for the test is 75, and the standard deviation is 4. If a normal distribution of scores is assumed, in what percentile does this student rank?
(1) below the 75th percentile
(2) between the 75th and 85th percentiles
(3) above the 95th percentile
(4) between the 85th and 95th percentiles

8. The heights of a group of 600 women are normally distributed. The mean height of the group is 172 centimeters with a standard deviation of 10 centimeters. What is the best approximation of the number of women who are between 157 centimeters and 172 centimeters tall?
(1) 205 (2) 260 (3) 408 (4) 115

9. The scores on a 100-point exam are normally distributed with a mean of 80 and a standard deviation of 6. A student's score places him between the 69th and 70th percentiles. Which value best represents his score?
(1) 75 (2) 81 (3) 84 (4) 86

10. Which inequality states that score x is less than two standard deviations, σ, from the mean, $\overline{x}$?
(1) $|x-\overline{x}|<2\sigma$ (2) $|\overline{x}-2\sigma|<x$ (3) $x<|2\sigma-\overline{x}|$ (4) $|x-2\sigma|<\overline{x}$

B. *In each case, show how you arrived at your answer by clearly indicating all of the necessary steps, formula substitutions, diagrams, graphs, charts, etc.*

11. Ms. Atkins has 146 students in her mathematics class. The scores on the final examination are normally distributed and have a mean of 80.6 and a standard deviation of 7.9. How many students in the class can be expected to receive a score between 89 and 95?

12. A test was given to 180 students. A statistical analysis of the test results showed that the scores were normally distributed and that approximately 95% of the scores ranged from 52 to 96, as shown in the accompanying diagram. How many students can be expected to receive scores between 77 and 84?

13. Twenty high school students took an examination and received the following scores:

70, 60, 75, 68, 85, 86, 78, 72, 82, 88, 88, 73, 74, 79,
86, 82, 90, 92, 93, 73

Determine what percent of the students scored within one standard deviation of the mean. Do the results of the examination approximate a normal distribution? Justify your answer.

14. A set of normally distributed test scores has a mean of 77 and a standard deviation of 6.
a. What is the probability that a randomly selected score will be between 74 and 86?
b. What percent of the test scores fall below a grade of 65?

Unit Five TRIGONOMETRIC FUNCTIONS AND LAWS

CHAPTER 13

TRIGONOMETRIC FUNCTIONS AND THEIR GRAPHS

13.1 TRIGONOMETRIC FUNCTIONS AND ANGLES

KEY IDEAS

In a right triangle, the ratios of the lengths of selected pairs of sides can be related to the degree measure of either of the acute angles using trigonometric functions.

The Six Trigonometric Functions

The three basic trigonometric functions, **sin**e, **cos**ine, and **tan**gent, are defined in Figure 13.1, together with the corresponding reciprocal functions, **cos**e**c**ant (csc), **sec**ant (sec), and **cot**angent (cot). The Greek letter θ (theta) is used to represent the unknown measure of $\angle A$.

Basic Three Trigonometric Functions	Reciprocal Trigonometric Functions
$\sin\theta = \frac{a}{c}$	$\csc\theta = \frac{c}{a}$
$\cos\theta = \frac{b}{c}$	$\sec\theta = \frac{c}{b}$
$\tan\theta = \frac{a}{b}$	$\cot\theta = \frac{b}{a}$

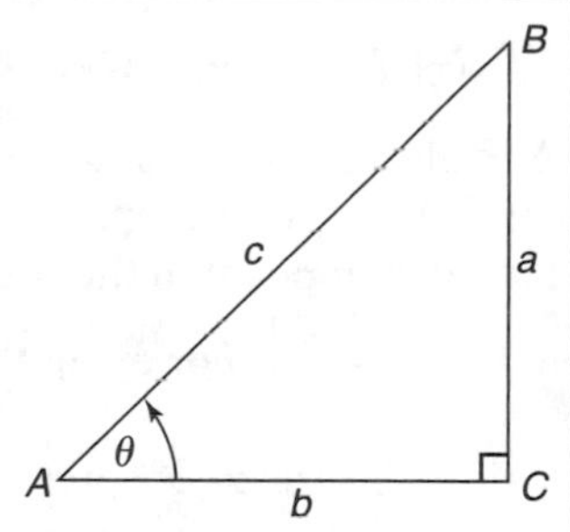

Figure 13.1 The Six Trigonometric Functions

For all values of θ except those that make the denominator 0, the reciprocal relationships of the trigonometric functions are as follows:

$$\csc\theta = \frac{1}{\sin\theta}, \qquad \sec\theta = \frac{1}{\cos\theta}, \qquad \cot\theta = \frac{1}{\tan\theta}.$$

Converting Minutes to Degrees

Each of the 60 equal parts of a degree is called a **minute**. For example, an angle measurement of 28°30′ is read as "28 degrees, 30 minutes." Since 60 minutes is equivalent to 1 degree, dividing 30 minutes by 60 changes 30 minutes to a fractional part of a degree. Thus:

$$28°\,30' = 28° + \left(\frac{30}{60}\right)^{\circ} = 28.5°.$$

In general:

$$x°y' = \left(x + \frac{y}{60}\right)^{\circ}.$$

Finding Trigonometric Function Values

To find the value of $\sin\theta$, $\cos\theta$, or $\tan\theta$, where θ is measured in degrees, use your graphing calculator as follows:

- Set the angular mode to Degree, if necessary, by pressing MODE, highlighting **Degree**, pressing ENTER, and then returning to the home screen by pressing 2nd [MODE].
- Press the key labeled with the name of the appropriate trigonometric function, enter the value of θ, then press) ENTER. You should verify that $\sin 54° \approx 0.8090169944$.

To evaluate $\sec x$, $\csc x$, or $\cot x$, find the value of the reciprocal function ($\cos x$, $\sin x$, or $\tan x$). Then find the reciprocal of the calculated function value that appears in the display window by pressing the calculator's reciprocal key, x^{-1}. For example:

- To find sec 60°, press

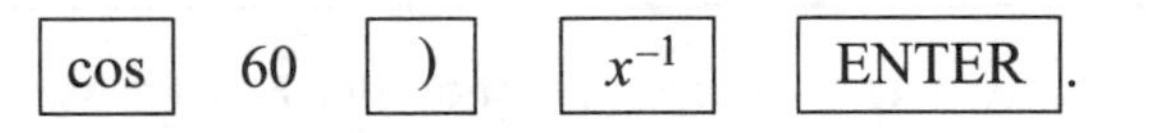

Thus, sec 60° = 2.

- To find csc 35°20′ correct to four decimal places, press

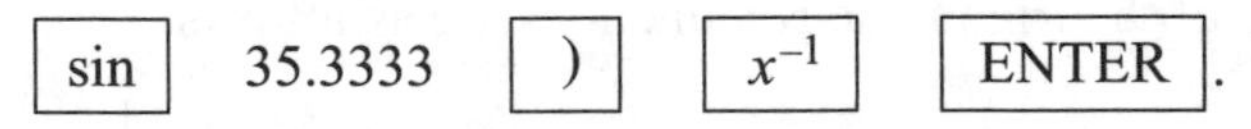

Thus, $\csc 35°20' \approx 1.7291$.

Finding Angles Using Inverse Trigonometric Functions

Your graphing calculator has $\sin^{-1}$ printed above the SIN key, $\cos^{-1}$ printed above the COS key, and $\tan^{-1}$ printed above the TAN key. These second or inverse calculator functions are used to find the measure of an angle when the value of a trigonometric function of that angle is known. For example, if $\tan x = 2.197$, then $x = \tan^{-1} 2.197$, which is read as "x is an angle whose tangent is 2.197."

To find the degree measure of $\angle x$, make sure the angular mode of your calculator is set to degrees. Then press:

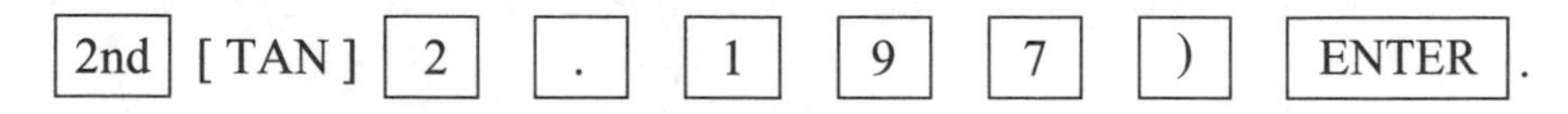

You should verify that $x \approx 65.52657916°$.

Cofunction Relationships

The prefix *co-* in *co*sine, *co*secant, and *co*tangent represents *co*mplementary. Pairs of cofunctions have equal values when their angles are complementary:

- $\sin \theta° = \cos (90 - \theta)°$ *Example*: $\sin 50° = \cos 40°$
- $\sec \theta° = \csc (90 - \theta)°$ *Example*: $\sec 24° = \csc 66°$
- $\tan \theta° = \cot (90 - \theta)°$ *Example*: $\tan 20°50' = \cot 69°10'$

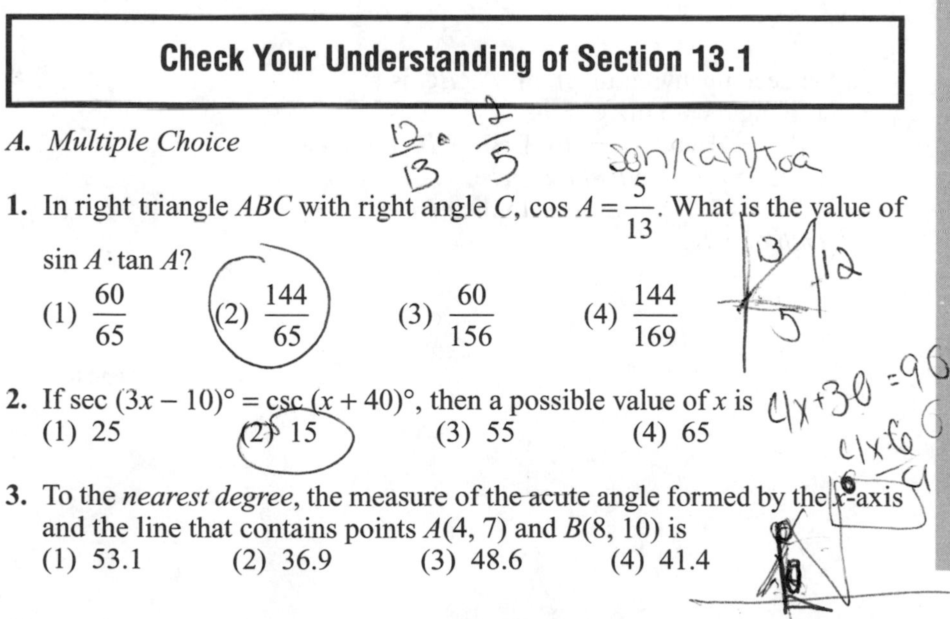

Check Your Understanding of Section 13.1

A. *Multiple Choice*

1. In right triangle ABC with right angle C, $\cos A = \frac{5}{13}$. What is the value of $\sin A \cdot \tan A$?
(1) $\frac{60}{65}$ (2) $\frac{144}{65}$ (3) $\frac{60}{156}$ (4) $\frac{144}{169}$

2. If $\sec (3x - 10)° = \csc (x + 40)°$, then a possible value of x is
(1) 25 (2) 15 (3) 55 (4) 65

3. To the *nearest degree*, the measure of the acute angle formed by the x-axis and the line that contains points $A(4, 7)$ and $B(8, 10)$ is
(1) 53.1 (2) 36.9 (3) 48.6 (4) 41.4

4. If $\dfrac{\sin (x-3)^\circ}{\cos (2x+6)^\circ} = 1$, then the value of x is
(1) -9 (2) 26 (3) 29 (4) 64

5. At Slippery Ski Resort, the beginner's slope is inclined at an angle of 12.3°, while the advanced slope is inclined at an angle of 26.4°. If Rudy skis 1000 meters down the advanced slope while Valerie skis the same distance down the beginner's slope, how much greater is the horizontal distance that Valerie covers?
(1) 81.3 m (2) 231.6 m (3) 895.7 m (4) 977.0 m

B. *In each case, show how you arrived at your answer by clearly indicating all of the necessary steps, formula substitutions, diagrams, graphs, charts, etc.*

6. The perimeter of regular pentagon *ABCDE*, shown in the accompanying diagram, is 100 inches.

a. Find, to the *nearest tenth of an inch*, the length of the perpendicular from center *O* of the pentagon to side $\overline{AB}$.

b. Find the area of the pentagon to the *nearest square inch*.

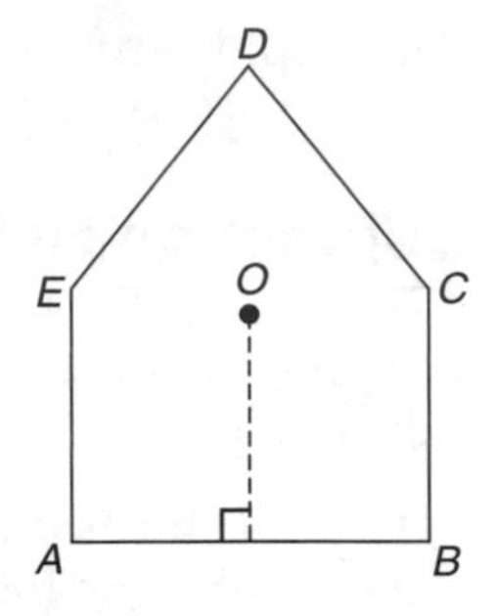

7. In the accompanying diagram, $\triangle ABC$ is an isosceles right triangle with $m\angle C = 90$. If $AB = 10$ and $\overline{AD}$ bisects $\angle BAC$, find BD to the *nearest tenth*.

8. In the accompanying diagram, $\triangle ABC$ is a right triangle with $m\angle C = 90$.

a. If $AD = 41$ and $AC = 40$, find $m\angle CAD$ to the *nearest tenth of a degree*.

b. If $BD = 87$, find $m\angle BAD$ to the *nearest tenth of a degree*.

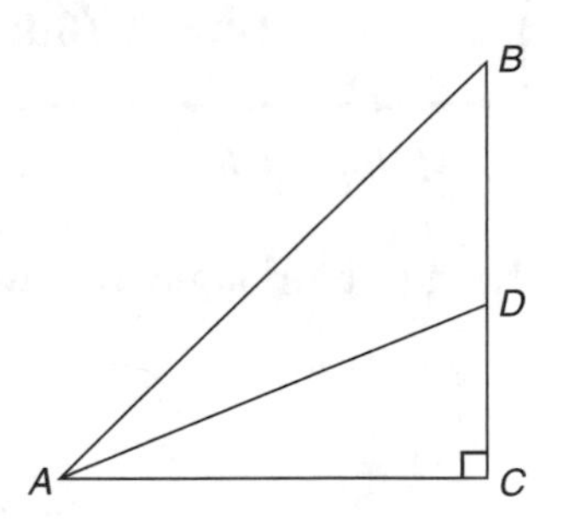

Exercises 7 and 8

13.2 RADIAN MEASURE AND ARC LENGTH

KEY IDEAS

A *radian* is a unit of angle measure that is defined for a central angle of a circle by comparing the length of the intercepted arc of the circle with the radius.

Definition of a Radian

One **radian** is the measure of a central angle of a circle that intercepts an arc whose length equals a radius of the circle.

In Figure 13.2, central angle θ intercepts an arc whose length, s, is 6π inches in a circle that has a radius length of 12 inches. The measure of $\angle\theta$ is $\frac{6\pi}{12} = \frac{\pi}{2}$ radians. In general:

$$\theta = \frac{s}{r} \qquad \text{or} \qquad s = r \cdot \theta.$$

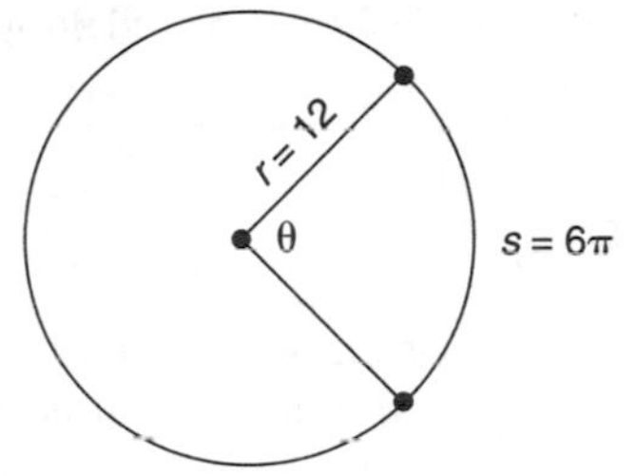

Figure 13.2 Radians and Arc Length

Converting Between Radians and Degrees

Since the circumference of a circle is $2\pi r$, the number of radians that represents a complete rotation of 360° is $\frac{2\pi r}{r} = 2\pi$. Hence, 2π radians $= 360°$, so π radians $= \frac{360°}{2} = 180°$. This relationship provides a way to convert between radian and degree measures:

- To change from *degrees* to *radians*, multiply the number of degrees by $\frac{\pi}{180°}$. For example:

$$60° = \overset{1}{\cancel{60°}} \times \frac{\pi}{\underset{3}{\cancel{180°}}} = \frac{\pi}{3} \text{ radians.}$$

- To change from *radians* to *degrees*, multiply the number of radians by $\frac{180°}{\pi}$. For example:

$$\frac{7}{12}\pi \text{ radians} = \frac{7\cancel{\pi}}{\cancel{12}} \times \frac{\overset{15°}{\cancel{180°}}}{\cancel{\pi}} = 105°.$$

TIP

To evaluate a trigonometric function of an angle expressed in radians, set the angular mode of your calculator to radians. Then return to the home screen and enter the trigonometric function followed by the radian measure of the angle. Using your calculator, verify that

$$\tan 1.42 \approx 6.581119456 \qquad \text{and} \qquad \cos\frac{\pi}{4} \approx 0.7071067812$$

Example 1

The radius of circle O, shown in the accompanying diagram, is 12 centimeters. What is the number of centimeters in the length of the minor arc intercepted by a central angle that measures 135°?

Solution: **9π**

Use the formula $s = r \cdot \theta$, where s is the length of the arc intercepted by a central angle of θ radians in a circle with radius r.

- Change 135° to radian measure:

$$135° = \overset{3}{\cancel{135°}} \times \frac{\pi}{\cancel{180°}} = \frac{3\pi}{4} \text{ rads.}$$

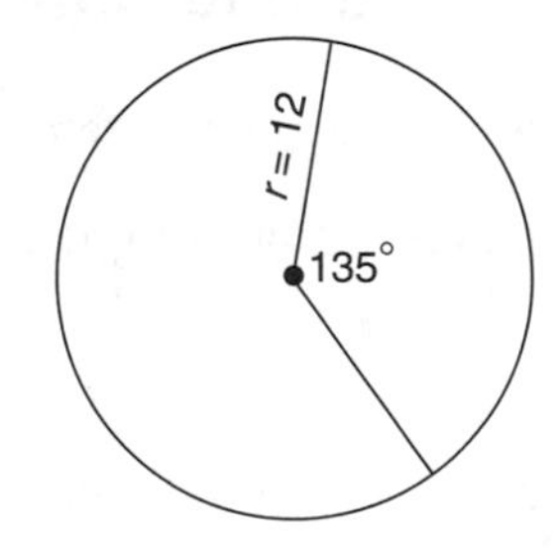

- Since $r = 12$ cm and $\theta = \frac{3}{4}\pi$,

$$s = \overset{3}{\cancel{12}} \text{ cm} \times \frac{3\pi}{\cancel{4}} = 9\pi \text{ cm}.$$

Example 2

A wedge-shaped piece is cut from a circular pizza. The radius of the pizza is 6 inches. The rounded edge of the crust of the piece measures 4.2 inches. What is the number of degrees in the angle of the pointed end of the piece of pizza?

Solution: **40.1°**

Use the formula $s = r \cdot \theta$, where r is the radius of the circular pizza, θ is the number of radians in the angle of the pointed end, and s is the length of the crust of the piece. Hence, let $s = 4.2$, $r = 6$, and solve for θ. Since $4.2 = 6\theta$,

$$\theta = \frac{4.2}{6} = 0.7 \text{ rad}$$

The number of *degrees* in the angle of the pointed end is

$$0.7 \times \frac{180°}{\pi} \approx 0.7 \times 57.3° \approx 40.1°$$

Check Your Understanding of Section 13.2

A. Multiple Choice

1. In radian measure, 225° is

(1) $\frac{3}{4}\pi$ (2) $\frac{4}{5}\pi$ (3) $\frac{4}{3}\pi$ (4) $\frac{5}{4}\pi$

2. In degree measure, 1.2π radians is

(1) 144 (2) 216 (3) 220 (4) 240

3. Which angle represents the *greatest* amount of rotation?

(1) 105° (2) $\frac{3}{5}\pi$ rad (3) 1.8 rad

(4) a central angle that intercepts a $\frac{3\pi}{4}$-in. arc in a circle whose radius is 1.5 in.

4. Which angle represents the *least* amount of rotation?

(1) 125° (2) $\frac{3}{4}\pi$ rad (3) 2.2 rad

(4) the angle through which the minute hand of a clock turns in 20 min

5. The bottom of a pendulum traces an arc 3 feet in length when the pendulum swings through an angle of $\frac{1}{2}$ radian. Find the number of feet in the length of the pendulum.

(1) 1.5 (2) 6 (3) $\frac{1.5}{\pi}$ (4) 6π

6. Through how many radians does the minute hand of a clock turn in 24 minutes?
(1) 0.2π (2) 0.4π (3) 0.6π (4) 0.8π

7. A wedge-shaped piece is cut from a circular pizza. The radius of the pizza is 14 inches. If the angle of the pointed end of the pizza measures 0.35 radian, what is the approximate number of inches in the length of the rounded edge of the crust?
(1) 4.9 (2) 4.0 (3) 7.5 (4) 5.7

8. A ball is rolling in a circular path that has a radius of 10 inches. What distance, to the *nearest hundredth of an inch*, has the ball rolled when it has traveled along an arc that measures 54°?

B. *In each case, show how you arrived at your answer by clearly indicating all of the necessary steps, formula substitutions, diagrams, graphs, charts, etc.*

9. The equatorial diameter of Earth is approximately 8000 miles. A communications satellite makes a circular orbit around Earth at a distance of 1600 miles from the planet. If the satellite completes one orbit every 5 hours, how many miles does the satellite travel in 1 hour?

10. A rod pivoted at one end is 6 inches long. If the free end rotates in a machine at the rate of 165 revolutions per minute, express in radians the angle through which the free end rotates in 1 second.

13.3 ANGLES IN STANDARD POSITION

KEY IDEAS

Trigonometric functions of angles of rotation greater than 90° or less than 0° can be given meaning by using coordinates to define the six trigonometric functions.

Standard Position

An angle is in **standard position** when its vertex is at the origin and one of its sides, called the **initial side**, remains fixed on the x-axis. The side of the angle that rotates is called the **terminal side**.

- If the terminal side rotates in a counterclockwise direction, as shown in Figure 13.3(a), the angle of rotation is *positive.*
- If the terminal side rotates in a clockwise direction, as shown in Figure 13.3(b), the angle of rotation is *negative*.

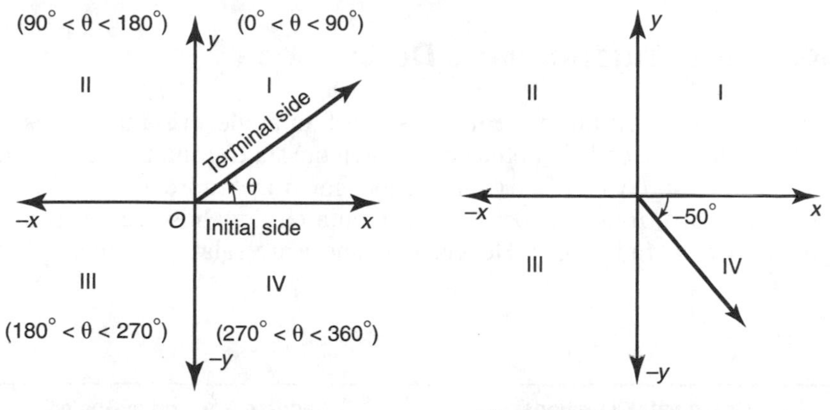

Figure 13.3(a) Positive Angle

Figure 13.3(b) Negative Angle

Coterminal Angles

Angles of rotation of $-150°$ and $210°$, as shown in Figure 13.4, are **coterminal angles**, that is, angles whose terminal sides coincide when placed in standard position. A negative and a positive angle are coterminal if the sum of the absolute values of their degree measures is 360 or a multiple of 360.

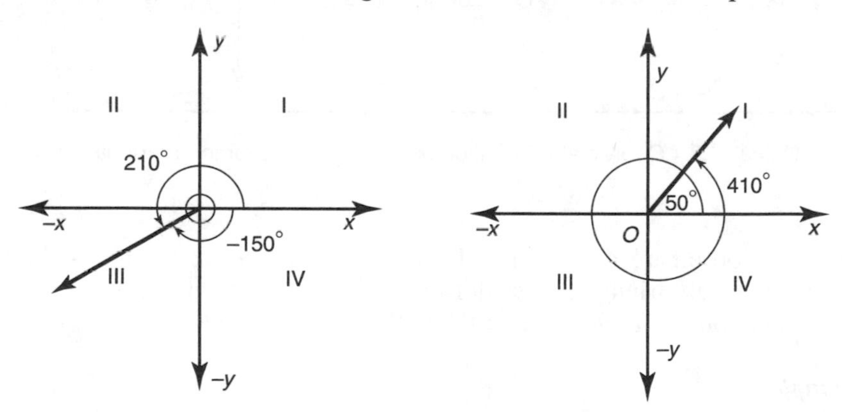

Figure 13.4 Coterminal Angles

Figure 13.5 Angles Greater Than 360°

Angles Greater Than 360°

An angle of rotation of $410°$ will complete one full rotation and its terminal side will lie in Quadrant I since $410° - 360° = 50°$, as shown in Figure 13.5.

To reduce an angle greater than 360° to a coterminal angle between 0° and 360°, successively subtract 360° from the given angle until the difference is between 0° and 360°. For example, since 870° – 360° = 510° and 510° – 360° = 150°, angles of rotation of 870° and 150° are coterminal.

Coordinate Trigonometric Definitions

If $P(x, y)$ is any point on the terminal side of an angle in standard position and r is the distance of P from the origin, then the six trigonometric functions can be defined in terms of x, y, and r, as shown in Figure 13.6. When the terminal side of θ rotates 360°, $P(x, y)$ is on a circle whose center is at the origin and whose radius is r. Hence, x, y, and r are related by the equation $x^2 + y^2 = r^2$.

Coordinate Definitions	Standard Position of Angle π
• $\sin\theta = \frac{y}{r}$ and $\csc\theta = \frac{r}{y}$ • $\cos\theta = \frac{x}{r}$ and $\sec\theta = \frac{r}{x}$ • $\tan\theta = \frac{y}{x}$ and $\cot\theta = \frac{x}{y}$	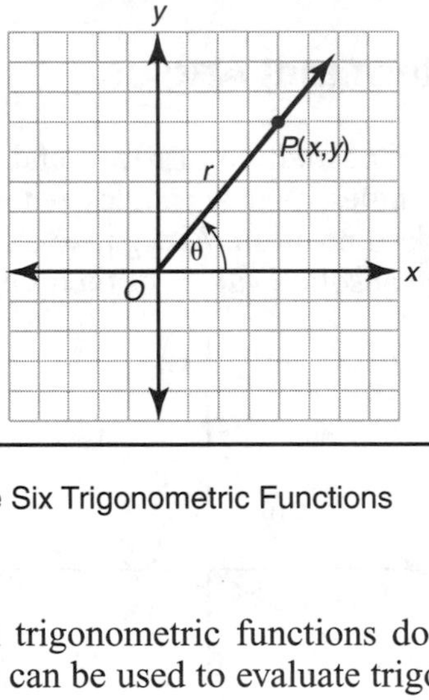 y P(x,y) r θ O x

Figure 13.6 Coordinate Definitions of the Six Trigonometric Functions

Since the coordinate definitions of the six trigonometric functions do not depend on a right triangle, these definitions can be used to evaluate trigonometric functions of angles greater than 90° and less than 0°.

Example 1

If $P(\sqrt{7}, -3)$ is a point on the terminal side of an $\angle\theta$, find the values of $\sin\theta$ and $\cos\theta$.

Solution: $\sin\theta = -\frac{3}{4}$, $\cos\theta = \frac{\sqrt{7}}{4}$

- Determine the quadrant in which the terminal side of $\angle\theta$ lies. Since $P(\sqrt{7}, -3) = P(x,y)$, $x = +\sqrt{7}$ and $y = -3$. Hence, $\angle\theta$ terminates in Quadrant IV, where $x > 0$ and $y < 0$.
- Find the value of r. Since $x^2 + y^2 = r^2$:

$$\begin{aligned}(\sqrt{7})^2 + (-3)^2 &= r^2\\ 7 \quad + \quad 9 \quad &= r^2\\ 16 \quad &= r^2\end{aligned}$$

Because r is always positive, $r = +\sqrt{16} = 4$.

- Use the coordinate definitions of $\sin\theta$ and $\cos\theta$:

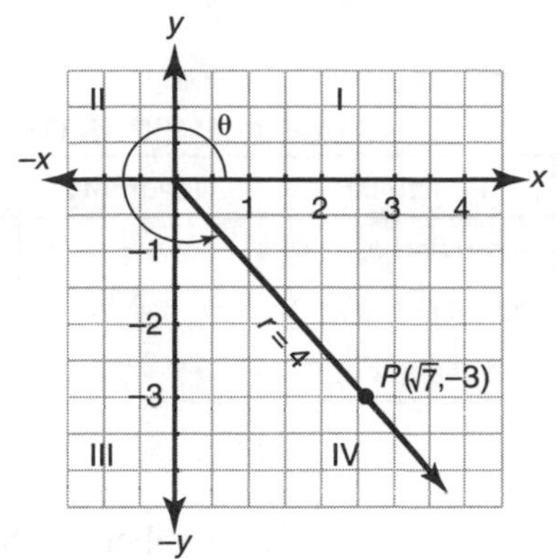

$$\sin\theta = \frac{y}{r} = \frac{-3}{4} \quad \text{and} \quad \cos\theta = \frac{x}{r} = \frac{+\sqrt{7}}{4}.$$

Unit Circle

A **unit circle** is a circle with center at the origin and a radius length of 1. If the terminal side of an $\angle\theta$ intersects a unit circle at $P(x, y)$, as shown in Figure 13.7, then the coordinates of P can be represented in terms of cosine and sine:

$$\cos\theta = \frac{x}{r} = \frac{x}{1} = x,$$

$$\sin\theta = \frac{y}{r} = \frac{y}{1} = y.$$

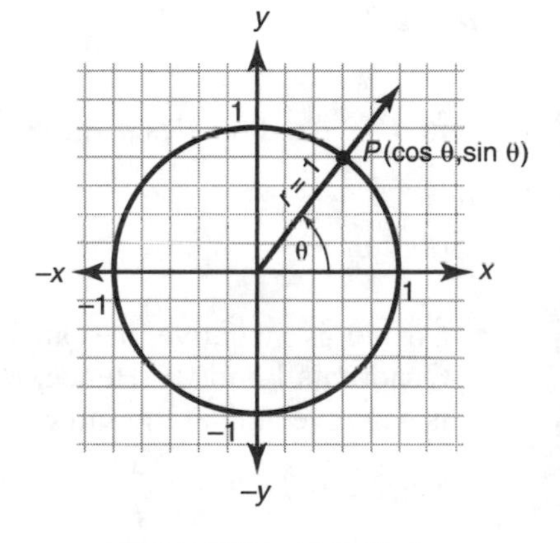

Figure 13.7 Unit Circle

Signs of Trigonometric Functions in Different Quadrants

The signs of the six trigonometric functions of $\angle\theta$ depend on the signs of x and y in the quadrant in which the terminal side of θ lies, as shown in Table 13.1. Since x and y are both positive in Quadrant I, the first quadrant is the only quadrant in which all of the six trigonometric functions are positive at the same time. In each of the other quadrants, only one of the basic trigonometric functions and its reciprocal function are positive.

TABLE 13.1 SIGNS OF THE TRIGONOMETRIC FUNCTIONS

Functions	Quadrant I	Quadrant II	Quadrant III	Quadrant IV
sin θ and csc θ	+	+	−	−
cos θ and sec θ	+	−	−	+
tan θ and cot θ	+	−	+	−

Another way of remembering the quadrants in which the different trigonometric functions are positive is to memorize the sentence "***A***ll ***S***tudents ***T***ake ***C***alculus," where the first letter of each word has special meaning:

A = ***A***ll are positive in Quadrant I,
S = ***S***ine (and cosecant) are positive in Quadrant II,
T = ***T***angent (and cotangent) are positive in Quadrant III,
C = ***C***osine (and secant) are positive in Quadrant III.

Example 2

If $\cos\theta = -\frac{4}{5}$ and $\tan\theta$ is positive, what is the exact value of $\sin\theta$?

Solution: $\mathbf{-\frac{3}{5}}$

- Cosine is negative in Quadrants II and III. Tangent is positive in Quadrants I and III. Hence, θ terminates in Quadrant III, where cosine is negative and, at the same time, tangent is positive.
- Since $\cos\theta = -\frac{4}{5} = \frac{x}{r}$,
  $x = -4$ and $r = 5$.
- Find y:

$$x^2 + y^2 = r^2$$
$$(-4)^2 + y^2 = 5^2$$
$$y = \pm\sqrt{9} = \pm 3$$

  In Quadrant III, y is negative, so $y = -3$.
- Hence, $\sin\theta = \frac{y}{r} = -\frac{3}{5}$.

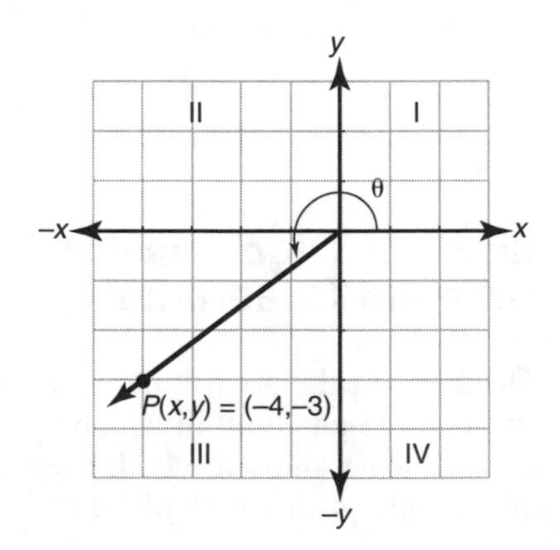

Example 3

If the sine of an angle is $\frac{5}{13}$ and the angle is *not* in Quadrant I, what is the exact value of the cosine of the angle?

Solution: $-\frac{12}{13}$

- Sine is positive in Quadrants I and II. Since it is given that the angle is *not* in Quadrant I, the angle must terminate in Quadrant II.
- If A is the unknown angle, then $\sin A = \frac{5}{13} = \frac{y}{r}$, so $y = 5$ and $r = 13$.
- Since $x^2 + y^2 = r^2$, then $x^2 + 5^2 = 13^2$, so $x = \pm\sqrt{144} = \pm 12$.
- Let $x = -12$ since the angle terminates in Quadrant II, where x is negative. Hence, $\cos A = \frac{x}{r} = -\frac{12}{13}$.

Check Your Understanding of Section 13.3

A. Multiple Choice

1. If $\cos A > 0$ and $\csc A < 0$, in which quadrant does the terminal side of $\angle A$ lie?
(1) I (2) II (3) III (4) IV

2. If $\cos\theta = -\frac{3}{4}$ and $\tan\theta$ is negative, the value of $\sin\theta$ is
(1) $-\frac{4}{5}$ (2) $-\frac{\sqrt{7}}{4}$ (3) $\frac{4\sqrt{7}}{7}$ (4) $\frac{\sqrt{7}}{4}$

3. If $\cos A = \frac{4}{5}$ and $\angle A$ is *not* in Quadrant I, what is the value of $\sin A$?
(1) -0.2 (2) 0.75 (3) -0.6 (4) 0.6

4. If $\sin A < 0$ and $\sec A < 0$, in which quadrant does the terminal side of $\angle A$ lie?
(1) I (2) II (3) III (4) IV

5. An angle that measures $\frac{7}{6}\pi$ radians is drawn in standard position. In which quadrant does the terminal side of the angle lie?
(1) I (2) II (3) III (4) IV

6. The value of $\sin \frac{3\pi}{2} + \cos \frac{2\pi}{3}$ is

(1) $\frac{1}{2}$ (2) $\frac{3}{2}$ (3) $-\frac{3}{2}$ (4) $-\frac{1}{2}$

7. If $f(x) = 2 \sin 7x + \frac{1}{2} \cos 2x$, what is $f\left(\frac{\pi}{6}\right)$?

(1) $\frac{5}{4}$ (2) $\frac{3}{4}$ (3) $-\frac{1}{4}$ (4) $-\frac{3}{4}$

8. In the accompanying diagram, point *P*(–0.6, –0.8) is on unit circle *O*. What is the measure of $\angle\theta$ to the *nearest degree*?
(1) 143 (2) 217 (3) 225 (4) 233

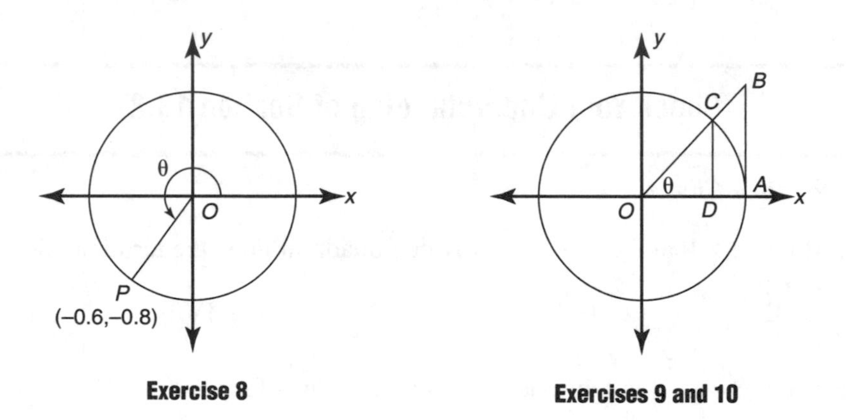

Exercise 8 **Exercises 9 and 10**

9 and 10. In the accompanying diagram of a unit circle, $\overline{BA}$ is tangent to circle *O* at *A*, $\overline{CD}$ is perpendicular to the *x*-axis, and $\overline{OC}$ is a radius.

9. Which distance represents $\sin \theta$?
(1) *OD* (2) *CD* (3) *BA* (4) *OB*

10. Which distance represents $\tan \theta$?
(1) *OD* (2) *CD* (3) *BA* (4) *OB*

11. If *x* is a positive acute angle and $\cos x = a$, an expression for $\tan x$ in terms of *a* is

(1) $\frac{1-a}{a}$ (2) $\sqrt{1-a^2}$ (3) $\frac{\sqrt{1-a^2}}{a}$ (4) $\frac{1}{1-a}$

B. In each case, show how you arrived at your answer by clearly indicating all of the necessary steps, formula substitutions, diagrams, graphs, charts, etc.

12. If $\tan\theta = -\dfrac{5}{12}$ and $\sin\theta$ is negative, what is the exact value of $\cos\theta$?

13. If $\sin\theta = -\dfrac{1}{\sqrt{5}}$ and $\cos\theta$ is negative, what is the exact value of $\tan\theta$?

14. If $\cot\theta = -\dfrac{15}{8}$ and $\csc\theta$ is positive, what is the exact value of $\sec\theta$?

15. If $\sin x = \dfrac{8}{17}$, where $0° < x < 90°$, determine the exact value of $\cos(x + 180°)$.

13.4 WORKING WITH TRIGONOMETRIC FUNCTIONS

KEY IDEAS

A trigonometric function of an angle whose terminal side lies in Quadrant II, III, or IV can be expressed as a trigonometric function of a positive *acute* angle. Familiar operations can be performed with trigonometric functions by treating the trigonometric function and its angle as a single variable. For example, $2\tan x + 3\tan x = 5\tan x$.

Finding Reference Angles

For each $\angle\theta$ in standard position, there is a corresponding acute angle called the *reference angle*. The **reference angle**, θ_R, is the acute angle whose vertex is the origin and whose sides are the terminal side of $\angle\theta$ and the x-axis, as shown in Figure 13.8. The right triangle that contains the reference angle is called the **reference triangle**. If $\angle\theta$ is a Quadrant I angle, then θ and θ_R are the same angle.

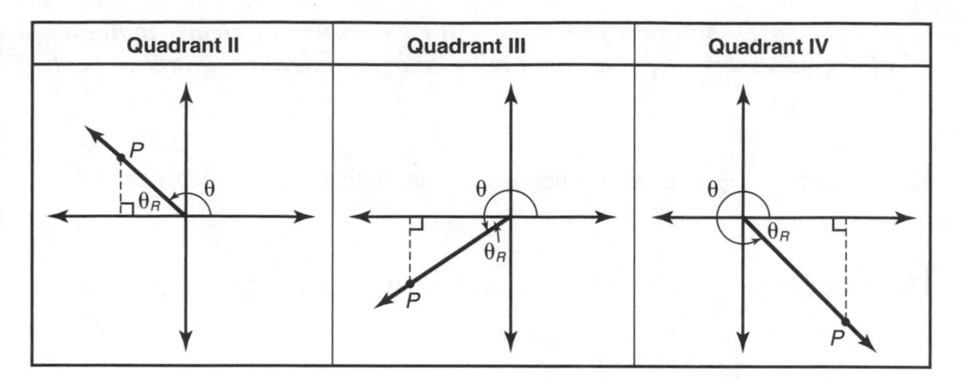

Figure 13.8 Reference Angle

Reducing Angles of Trigonometric Functions

The trigonometric function of *any* $\angle\theta$ can be expressed as either plus or minus the same trigonometric function of its reference angle, θ_R.

For example, to reduce cos 135°:

- Find the reference angle. As shown in Figure 13.9, $\theta_R = 45°$.
- Determine the sign of the cosine function in the quadrant in which θ_R is located. In Quadrant II, cosine is negative.

Hence, $\cos 135° = -\cos 45°$.

Because in Quadrant II sine is positive while tangent is negative:

$$\sin 135° = \sin 45°$$
and
$$\tan 135° = -\tan 45°.$$

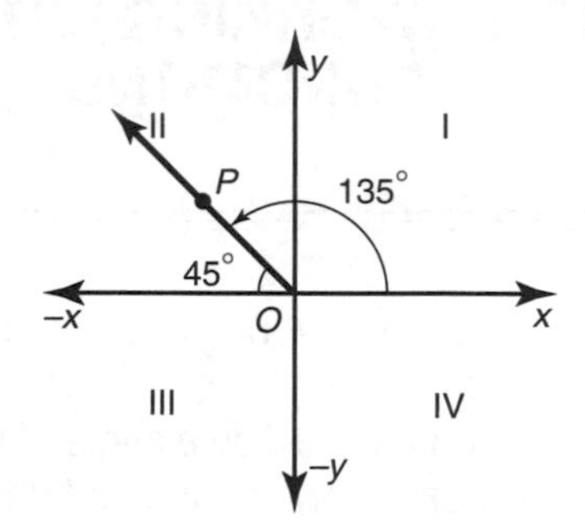

Figure 13.9 Reducing a Trigonometric Function of an Angle

Sometimes it is necessary to reduce functions of angles greater than 360° or less than 0°. You should draw a diagram and verify that:

- $\cos 570° = \cos (570 - 360)° = \cos 210° = -\cos 30°$.
- $\sin (-140°) = \sin 220° = -\sin 40°$.
- $\tan 650° = \tan 290° = -\tan 70°$.

Operations with Trigonometric Functions

When a trigonometric function is raised to a power, the exponent is written next to the function and a half line above it, as in $\sin^2 x$. Thus, $\sin^2 x$ means

$(\sin x)\cdot(\sin x)$, while $\sin x^2$ represent the sine of $\angle x^2$. The following examples illustrate how to work with trigonometric functions:

1. Multiplying and Dividing

- $(2\cos^2 x)\cdot(4\cos x) = 8\cos^3 x$
- $\dfrac{15\sin^3 x}{3\sin x} = 5\sin^2 x$

2. Factoring

- $\tan^2 - \tan x - 6 = (\tan x - 3)(\tan x + 2)$
- $4\sin x - \sin^3 x = \sin x\,(4 - \sin^2 x) = \sin x\,(2 - \sin x)(2 + \sin x)$

3. Combining Fractions

$$\begin{aligned}\frac{2\tan^2 x - 5}{\tan x + 1} + \frac{\tan^2 x + 2}{\tan x + 1} &= \frac{3\tan^2 x - 3}{\tan x + 1}\\ &= \frac{3\left(\tan^2 x - 1\right)}{\tan x + 1}\\ &= \frac{3\cancel{(\tan x + 1)}(\tan x - 1)}{\cancel{\tan x + 1}}\\ &= 3\,(\tan x - 1)\end{aligned}$$

4. Solving an Equation for a Trigonometric Function

- If $3\cos x + 7 = 5 - \cos x$, then

$$\begin{aligned}3\cos x + \cos x &= 5 - 7\\ 4\cos x &= -2\\ \cos x &= -\frac{1}{2}\end{aligned}$$

- If $\cot^2 x - 5\cot x - 6 = 0$, then $(\cot x - 6)(\cot x + 1) = 0$, so $\cot x = 6$ *or* $\cot x = -1$.

Check Your Understanding of Section 13.4

In each case, show how you arrived at your answer by clearly indicating all of the necessary steps, formula substitutions, diagrams, graphs, charts, etc.

1–12. Write each of the following as an equivalent function of a positive acute angle.

1. $\cos 250°$ **3.** $\sin 330°$ **5.** $\tan 225°$

2. $\tan 510°$ **4.** $\cos(-20°)$ **6.** $\csc 150°$

7. $\sec(-130°)$

8. $\cot 470°$

9. $\cos \frac{4}{3}\pi$

10. $\sin \frac{11}{6}\pi$

11. $\tan \frac{7}{12}\pi$

12. $\csc\left(-\frac{8}{9}\pi\right)$

13. Write tan 110° as a function of a positive acute angle less than 45°.

14. Write csc 285° as a function of a positive acute angle less than 45°.

15. Write cos 225° as a function of a positive acute angle less than 45°.

16–18. Factor completely.

16. $3\tan^3 x - 192\tan x$ **17.** $2\sin^2\theta - 18\cos^2\theta$ **18.** $6\sin^2 x + 3\sin x - 3$

19–20. Express in simplest form.

19. $\frac{3\sin y + 6}{4 - \sin^2 y} + \frac{4}{\sin y - 2}$

20. $\frac{1 - \tan^2 x}{6\tan x + 6} \div \frac{\tan^4 x - 1}{6\tan^2 x + 6}$

21–22. Write each complex fraction in simplest form.

21. $\dfrac{\frac{3}{\cos^2 x} + \frac{1}{\cos x}}{1 - \frac{9}{\cos^2 x}}$

22. $\dfrac{\frac{\sin x}{\cos x} - \frac{\cos x}{\sin x}}{\frac{1}{\cos x} - \frac{1}{\sin x}}$

23. Solve for sec x: $2 - 3\sec x = 5 - \sec x$.

24. Solve for sin x: $4\sin^2 x + 5\sin x + 1 = 0$.

25. Solve for tan x: $\frac{3}{\tan x + 3} + \frac{2}{\tan x - 4} = \frac{4}{3}$.

13.5 TRIGONOMETRIC FUNCTIONS OF SPECIAL ANGLES

KEY IDEAS

You should know the *exact* values of trigonometric functions of 30°, 45°, and 60°, as well as the values of trigonometric functions of the *quadrantal angles*: 0°, 90°, 180°, 270°, and 360°.

Quotient Formulas

In a unit circle, $\sin\theta = y$ and $\cos\theta = x$. Since $\frac{\sin\theta}{\cos\theta} = \frac{y}{x}$ and $\tan\theta = \frac{y}{x}$, the quotient of $\sin\theta$ and $\cos\theta$ is $\tan\theta$.

MATH FACTS

$$\tan\theta = \frac{\sin\theta}{\cos\theta} \quad \text{and} \quad \cot\theta = \frac{\cos\theta}{\sin\theta}$$

for all values of θ that do not make the denominators evaluate to 0.

Example 1

Simplify: $\csc\theta \cdot \tan\theta \cdot \cos^2\theta$.

Solution: $\mathbf{\cos\pi}$

Change to sines and cosines using the reciprocal and quotient function relationships. Then simplify.

$$\csc\theta \cdot \tan\theta \cdot \cos^2\theta = \frac{1}{\sin\theta} \cdot \frac{\sin\theta}{\cos\theta} \cdot \cos^2\theta$$

$$= \frac{1}{\cancel{\sin\theta}} \cdot \frac{\cancel{\sin\theta}}{\cancel{\cos\theta}} \cdot \cos^{\cancel{2}}\theta$$

$$= \cos\theta$$

Trigonometric Functions of 30°, 45°, and 60°

The exact values of the three basic trigonometric functions of 30°, 45°, and 60° are summarized in Table 13.2. These values are based on the special right triangle relationships discussed on pages 78 to 80.

TABLE 13.2 TRIGONOMETRIC FUNCTION VALUES OF 30°, 45°, AND 60°

x	$\sin x$	$\cos x$	$\tan x$
30°	$\frac{1}{2}$	$\frac{\sqrt{3}}{2}$	$\frac{1}{\sqrt{3}}$ or $\frac{\sqrt{3}}{3}$
45°	$\frac{\sqrt{2}}{2}$	$\frac{\sqrt{2}}{2}$	1
60°	$\frac{\sqrt{3}}{2}$	$\frac{1}{2}$	$\sqrt{3}$

TIP

To help memorize this table of values, remember that $\sin 30° = \cos 60° = \frac{1}{2}$, $\cos 30° = \sin 60° = \frac{\sqrt{3}}{2}$, and $\sin 45° = \cos 45° = \frac{\sqrt{2}}{2}$. The quotient formulas can be used to obtain the values for $\tan\theta$ using the values of $\sin\theta$ and $\cos\theta$. For example,

$$\tan 30° = \sin 30° \div \cos 30° = \frac{1}{2} \div \frac{\sqrt{3}}{2} = \frac{1}{\sqrt{3}}.$$

Use the reciprocal function relationships to find the values of cosecant, secant, and cotangent of 30°, 45°, and 60°.

Angles greater than 90° or less than 0° may have reference angles of 30°, 45°, or 60°. You should verify that:

- $\cos 120° = -\cos 60° = -\frac{1}{2}$
- $\sin 300° = -\sin 60° = -\frac{\sqrt{3}}{2}$
- $\tan 225° = \tan 45° = 1$
- $\cos(-30)° = \cos 30° = \frac{\sqrt{3}}{2}$
- $\csc 240° = -\frac{1}{\sin 60°} = -\frac{2}{\sqrt{3}}$
- $\sec 405° = \frac{1}{\cos 45°} = \sqrt{2}$

Quadrantal Angles

An angle whose terminal side coincides with a coordinate axis is called a **quadrantal angle**. An angle of rotation of 0°, 90°, 180°, 270°, or 360° is a quadrantal angle. The values of trigonometric functions of quadrantal angles may be ±1, 0, or undefined, as summarized in Table 13.3.

TABLE 13.3 EVALUATING TRIGONOMETRIC FUNCTIONS OF QUADRANTAL ANGLES

Trigonometric Function	0°	$90°\left(\frac{\pi}{2}\right)$	180° (π)	$270°\left(\frac{3\pi}{2}\right)$	360° (2π)
$\sin x$	0	1	0	−1	0
$\cos x$	1	0	−1	0	1
$\tan x$	0	Undefined	0	Undefined	0

TIP

The function values of 0° and 360° are the same because these are coterminal angles. Since division by 0 is undefined, $\tan x$ is undefined whenever $\cos x = 0$. Use the reciprocal function relationships to find the values of cosecant, secant, and cotangent of the quadrantal angles.

Example 2

If $f(x) = 2\sin x + \cos 2x$, what is $f\left(\frac{\pi}{2}\right)$?

Solution: **1**

Let $x = \frac{\pi}{2}$. Then:

$$\begin{aligned} f\left(\frac{\pi}{2}\right) &= 2\sin\frac{\pi}{2} + \cos 2\left(\frac{\pi}{2}\right) \\ &= 2\sin 90° + \cos 2(90°) \\ &= 2(1) + \cos 180° \\ &= 2 + (-1) \\ &= 1 \end{aligned}$$

Example 3

If $f(x) = \csc\frac{x}{2} + \cot x + \sec 3x$, what is the exact value of $f\left(\frac{\pi}{3}\right)$?

Solution: $\mathbf{1 + \frac{1}{\sqrt{3}}}$

Let $x = \frac{\pi}{3}$. Then:

$$\begin{aligned} f\left(\frac{\pi}{3}\right) &= \csc\left(\frac{1}{2}\cdot\frac{\pi}{3}\right) + \cot\frac{\pi}{3} + \sec 3\left(\frac{\pi}{3}\right) \\ &= \csc 30° + \cot 60° + \sec 180° \\ &= 2 + \frac{1}{\sqrt{3}} + (-1) \\ &= 1 + \frac{1}{\sqrt{3}} \end{aligned}$$

Example 4

For what value of x is $\frac{x-1}{1+\sin x}$, not defined in the interval $0 < x \leq 2\pi$?

Solution: $\mathbf{\frac{3\pi}{2}}$ or **270°**

A fraction with a variable expression in the denominator is not defined for any value of the variable that makes the denominator evaluate to 0. Hence, the fraction is not defined when $1 + \sin x = 0$ or $\sin x = -1$. If $\sin x = -1$, then $x = \frac{3\pi}{2}$ or 270°.

Check Your Understanding of Section 13.5

A. Multiple Choice

1. If $\sin x = -\frac{\sqrt{2}}{2}$ and $\cos x = \frac{\sqrt{2}}{2}$, the measure of $\angle x$ is
(1) 45° (2) 135° (3) 225° (4) 315°

2. If $\tan x = -\frac{\sqrt{3}}{3}$ and $\cos x = -\frac{\sqrt{3}}{2}$, the measure of $\angle x$ is
(1) 120° (2) 150° (3) 210° (4) 300°

3. What is the value of $(\sin x - 2\cos y)^2$ when $x = \frac{3}{2}\pi$ and $y = \frac{5\pi}{6}$?
(1) $6\sqrt{3}$ (2) $1+\frac{\sqrt{3}}{4}$ (3) $4-2\sqrt{3}$ (4) 4

4. What is the value of $\sin^2 150° + \cos^2 150°$?
(1) 1 (2) $\frac{1}{2}$ (3) $1\frac{1}{2}$ (4) 0

5. Which expression has the *greatest* numerical value?
(1) cot 210° (2) $\tan^2 135°$ (3) 1 − cos 240° (4) sin 30° + sin 60°

6. Which expression has the *least* numerical value?
(1) sin 0 (2) $\cos\frac{7}{6}\pi$ (3) $\tan\frac{7}{4}\pi$ (4) $\csc\frac{2}{3}\pi$

7. If $f(x) = \sin 2x + \tan x$, what is $f\left(\frac{5}{4}\pi\right)$?
(1) 0 (2) 2 (3) −1 (4) −2

8. If $f(x) = \cos 2x - \tan \frac{x}{2}$, what is $f\left(\frac{\pi}{2}\right)$?

(1) -2 (2) 2 (3) 0 (4) undefined

9. If $\frac{1}{2\sin x - 1}$ is undefined when $x = a°$, what is the value of $\cos(180 + a)°$?

(1) $-\frac{1}{2}$ (2) $\frac{1}{2}$ (3) $-\frac{\sqrt{3}}{2}$ (4) $-\frac{1}{3}$

10. The expression $\csc\left(-\frac{\pi}{2}\right)$ is equivalent to

(1) $2\cos\frac{\pi}{3}$ (2) $\tan \pi$ (3) $2\sin\frac{5}{6}\pi$ (4) $\cos \pi$

11. If θ is an angle in standard position and its terminal side passes through point $\left(-\frac{1}{2}, \frac{\sqrt{3}}{2}\right)$ on a unit circle, then a possible value of θ is

(1) 120° (2) 150° (3) 210° (4) 240°

12. If θ is an angle in standard position and its terminal side passes through point $\left(\frac{\sqrt{3}}{2}, -\frac{1}{2}\right)$ on a unit circle, then a possible value of θ is

(1) 210° (2) 240° (3) 300° (4) 330°

13. What is the image of point (1,0) after a clockwise rotation of 30°?

(1) $\left(\frac{\sqrt{3}}{2}, -\frac{1}{2}\right)$ (2) $\left(\frac{1}{2}, -\frac{\sqrt{3}}{2}\right)$ (3) $\left(-\frac{\sqrt{3}}{2}, 1\right)$ (4) $\left(\frac{1}{2}, -\frac{1}{2}\right)$

14. What is the image of point (1,0) after a counterclockwise rotation of 135°?

(1) $\left(\frac{\sqrt{2}}{2}, -\frac{\sqrt{2}}{2}\right)$ (2) $\left(-\frac{\sqrt{2}}{2}, \frac{\sqrt{2}}{2}\right)$ (3) $(-\sqrt{2}, 1)$ (4) $\left(-\frac{1}{2}, \frac{1}{2}\right)$

15. If $\sin A = b$, then the value of $\sin A \cdot \cos A \cdot \tan A$ is equal to

(1) 1 (2) $\frac{1}{b}$ (3) b (4) b^2

16. The expression $\tan\theta(\cos\theta + \csc\theta)$ is equivalent to

(1) $1 + \sin\theta$ (2) $1 + \cos\theta$ (3) $\frac{1}{\cos\theta\sin\theta}$ (4) $\frac{1 + \cos\theta\sin\theta}{\cos\theta}$

17. In the accompanying diagram of a unit circle, the ordered pair (x, y) represents a point where the terminal side intersects the unit circle. If $\theta = -\frac{\pi}{3}$, what is the value of y?

(1) $-\frac{\sqrt{3}}{2}$ (3) $-\sqrt{3}$

(2) $-\frac{\sqrt{2}}{2}$ (4) $-\frac{1}{2}$

B. Show how you arrived at your answer by clearly indicating all of the necessary steps, formula substitutions, diagrams, graphs, charts, etc.

18. Find all values of x in the interval $0 \le x \le \pi$ that make the fraction $\frac{1}{\sin 2x}$ undefined.

13.6 GRAPHS OF PERIODIC FUNCTIONS

KEY IDEAS

When x varies from 0° to 360°, the y-values for $y = \sin x$ and $y = \cos x$ range from −1 to +1. The same pattern of y-values is repeated every 360°. For example:

$$\cos 60° = \underbrace{\cos 420°}_{\cos(60+360)°} = \underbrace{\cos 780°}_{\cos(60+720)°} = \ldots = \frac{1}{2}.$$

Because of this property, sine and cosine are *periodic functions* with periods of 360°.

Amplitude and Period

A **periodic function** is a function for which there is a repeating pattern of y-values over equal intervals of x. Each complete pattern is called a **cycle**. For any periodic function:

- The **amplitude** is one-half of the positive difference between the maximum and minimum values of the function. For the periodic function in Figure 13.10, the amplitude is $\frac{3-(-2)}{2} = 2.5$.
- The **period** is the length of the interval of x-values over which the function completes one cycle. For the periodic function in Figure 13.10, the period is $8 - 3 = 5$.

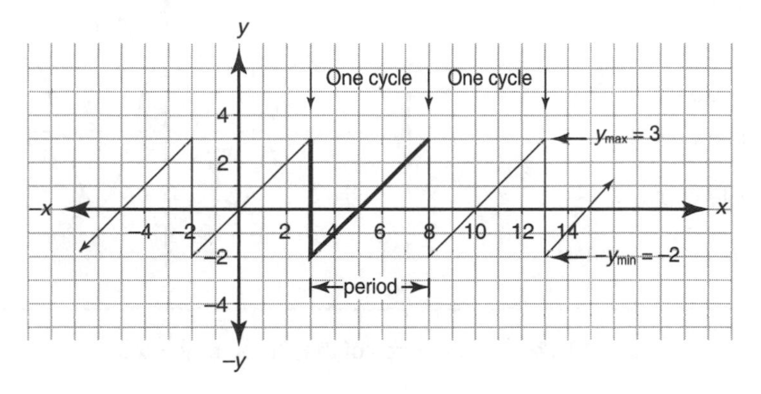

Figure 13.10 Amplitude and Period

Basic Sine and Cosine Graphs

The function $y = \sin x$ has an amplitude of 1 and a period of 2π, as shown in Figure 13.11. The **frequency** of a trigonometric function is the number of cycles that its graph completes in an interval of 2π radians. The frequency of $y = \sin x$ is 1 since the sine curve completes one cycle in each interval of 2π radians.

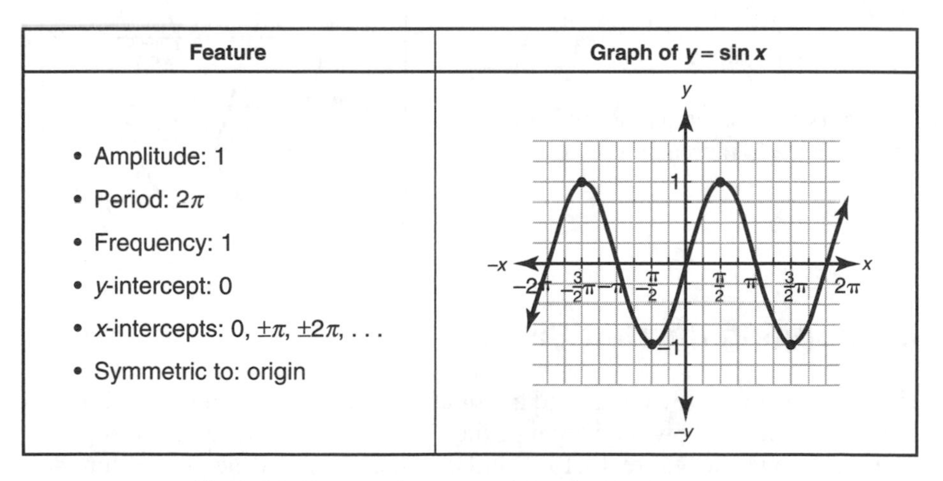

Feature	Graph of $y = \sin x$
• Amplitude: 1 • Period: 2π • Frequency: 1 • y-intercept: 0 • x-intercepts: $0, \pm\pi, \pm 2\pi, \ldots$ • Symmetric to: origin	

Figure 13.11 Key Features of the Graph of $y = \sin x$

The function $y = \cos x$ has an amplitude of 1, a period of 2π, and a frequency of 1, as shown in Figure 13.12. Its graph has the same basic shape as the sine curve but is 90° out of phase with it.

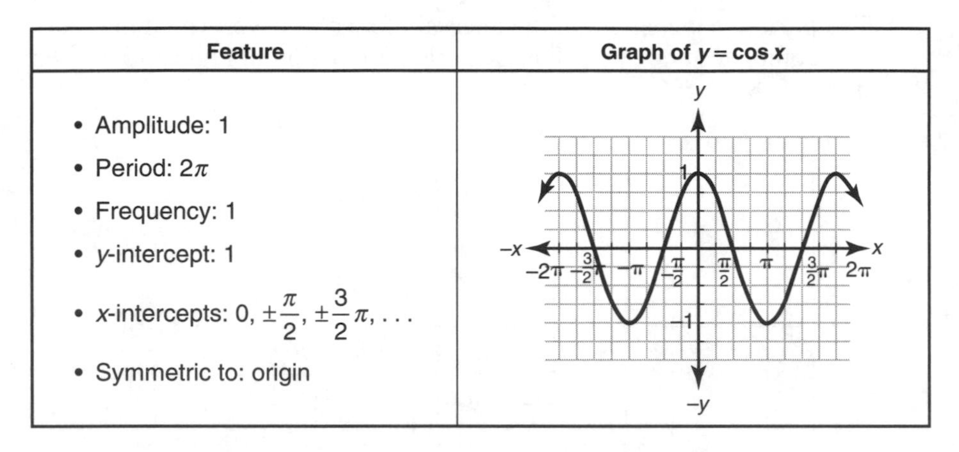

Figure 13.12 Key Features of the Graph of $y = \cos x$

Example 1

In which quadrant is $y = \sin x$ decreasing and $y = \cos x$ increasing?

Solution: **III**

Superimpose the graph of $y = \cos x$ on the graph of $y = \sin x$, as shown in the accompanying figure. In Quadrant III, where $\pi < x < \frac{3\pi}{2}$, the sine curve is decreasing and the cosine curve is increasing.

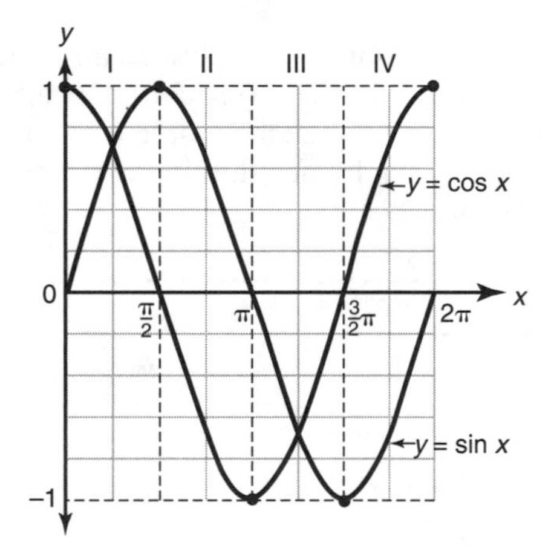

Unwrapping the Unit Circle

If a point (x, y) is moving around a unit circle, the graph of the sine function $y = \sin x$ indicates how the height of the point changes as x varies from 0° to 360°, as shown in Figure 13.13. Similarly, the graph of the cosine function $y = \cos x$ shows how the horizontal position of the point changes.

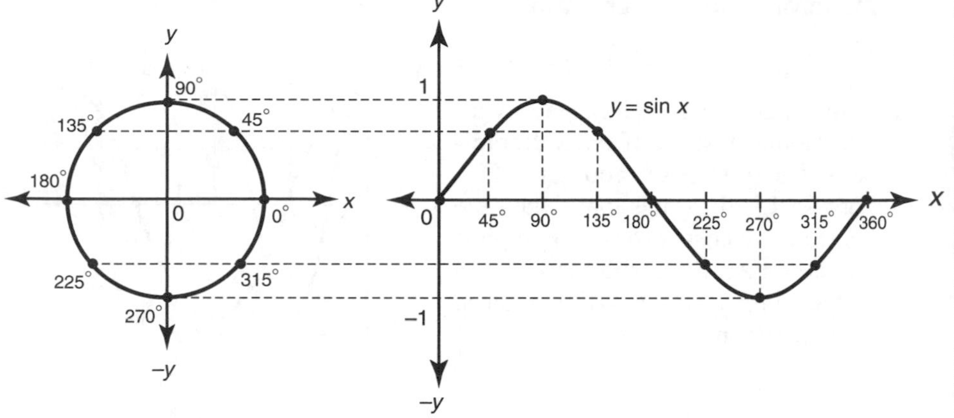

Figure 13.13 Mapping Points of the Unit Circle onto the Sine Function

Amplitude and Period of *y* = *a* sin *bx* and *y* = *a* cos *bx*

In the equations $y = a \sin bx$ and $y = a \cos bx$, the number a affects the amplitude and the number b determines the period. For each of these functions:

- The *range* is $-a \leq y \leq a$ when a is a positive number.
- The *amplitude* is $|a|$ since the amplitude is half the distance between the maximum and minimum values of the function. For example, the amplitude of $y = 2 \sin x$ is 2, as shown in Figure 13.14. The graph of $y = 2 \sin x$ represents a *vertical stretching* of the graph of $y = \sin x$ using a scale factor of 2. The graph of $y = \dfrac{1}{2} \sin x$ represents a *vertical shrinking* of the graph of $y = \sin x$ using a scale factor of $\frac{1}{2}$.

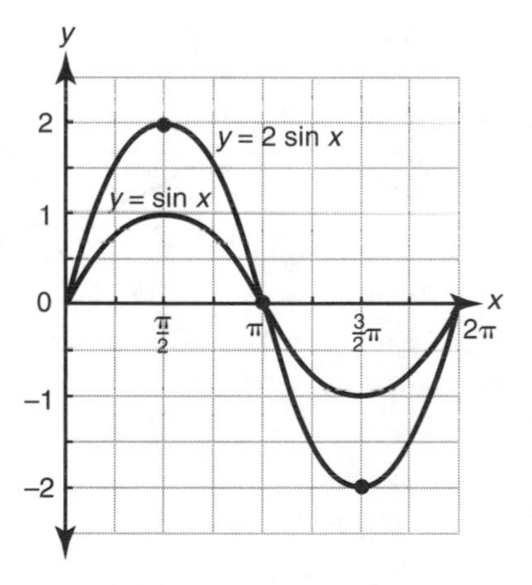

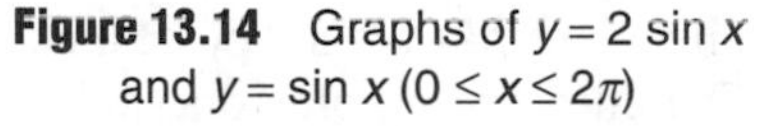

Figure 13.14 Graphs of $y = 2 \sin x$ and $y = \sin x$ $(0 \leq x \leq 2\pi)$

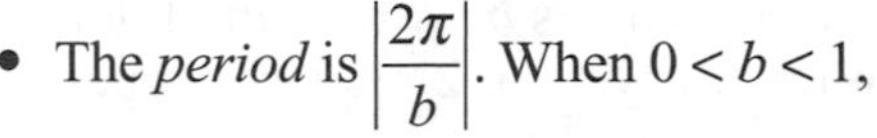

- The *period* is $\left|\dfrac{2\pi}{b}\right|$. When $0 < b < 1$, the period of each function is *greater than* 2π, so each graph represents a *horizontal stretching* of the basic sine or cosine curve. For example, if $y = \cos \dfrac{1}{2}x$, then $b = \dfrac{1}{2}$, so the period is $\dfrac{2\pi}{1/2} = 4\pi$.

Compared to the graph of $y = \cos x$, the graph of $y = \cos \frac{1}{2}x$ is "stretched" because it repeats every 4π radians instead of every 2π radians. The graph of $y = \cos 2x$ in Figure 13.15 illustrates that, when the period of a cosine (or sine) graph is *less than* 2π, the graph represents a *horizontal shrinking* of the basic cosine (or sine) curve.

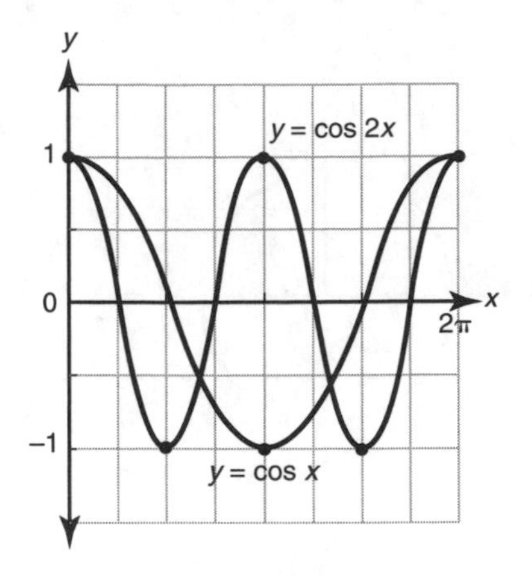

Figure 13.15 Graphs of $y = \cos x$ and $y = \cos 2x$ $(0 \le x \le 2\pi)$

Example 2

Determine an equation of each graph:

a.

b.

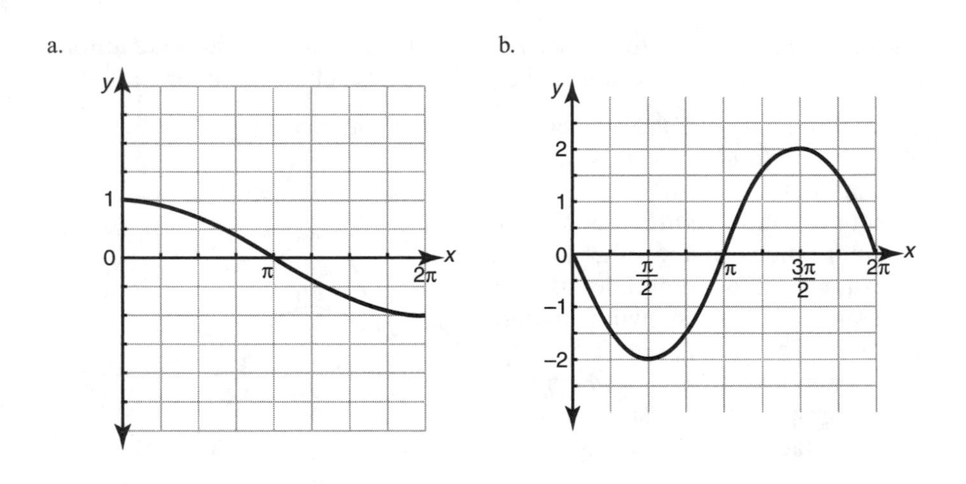

Solution: a. $\mathbf{y = \cos \frac{1}{2}x}$

The curve has the shape of a cosine curve $y = a \cos bx$ whose amplitude is 1. The curve completes one-half of a complete cycle in 2π rad, so its period is 4π rad. Hence, $a = 1$ and $\frac{2\pi}{b} = 4\pi$, so $b = \frac{2\pi}{4\pi} = \frac{1}{2}$. Therefore, an equation of the curve is $y = \cos \frac{1}{2}x$.

b. $y = -2 \sin x$

The curve has the basic shape of a sine curve $y = a \sin x$ whose amplitude is 2 and whose period is 2π, except that it has been reflected in the x-axis. Since the curve reaches a *minimum* of -2 at $x = \frac{\pi}{2}$, $a < 0$, so an equation of the curve is $y = -2 \sin x$.

Check Your Understanding of Section 13.6

A. Multiple Choice

1. In which quadrant is the terminal side of $\angle x$ located if the graphs of $y = \sin x$ and $y = \cos x$ are both decreasing when $\angle x$ is increasing?
(1) I (2) II (3) III (4) IV

2. What is the range of the function $y = 2 \sin 3x$?
(1) all real numbers
(2) $-2\pi \le x \le 2\pi$
(3) $-2 \le y \le 2$
(4) $-3 \le y \le 3$

3. The graph of which equation has an amplitude of $\frac{1}{2}$ and a period of π?
(1) $y = 2 \sin \frac{1}{2}x$
(2) $y = \frac{1}{2} \sin \frac{1}{2}x$
(3) $y = 2 \sin 2x$
(4) $y = \frac{1}{2} \sin 2x$

4. The graph of which equation has a period of π radians and passes through the origin?
(1) $y = \cos 2x$ (2) $y = \sin 2x$ (3) $y = \cos \frac{1}{2}x$ (4) $y = \sin \frac{1}{2}x$

5. The graph of the equation $y = 2 \cos 2x$, in the interval $0 \le x \le 2\pi$, has a line of symmetry at
(1) $x = \pi$ (2) $x = \frac{\pi}{4}$ (3) $y = 2$ (4) x-axis

6. In the interval $0 < \theta \le 2\pi$, the number of solutions of the equation $\sin \theta - \cos \theta = 0$ is
(1) 1 (2) 2 (3) 3 (4) 4

7. Which equation is represented by the graph in the accompanying figure?
(1) $y = 3 \sin 2x$ (2) $y = 2 \sin 3x$ (3) $y = 3 \sin x$ (4) $y = 2 \sin 4x$

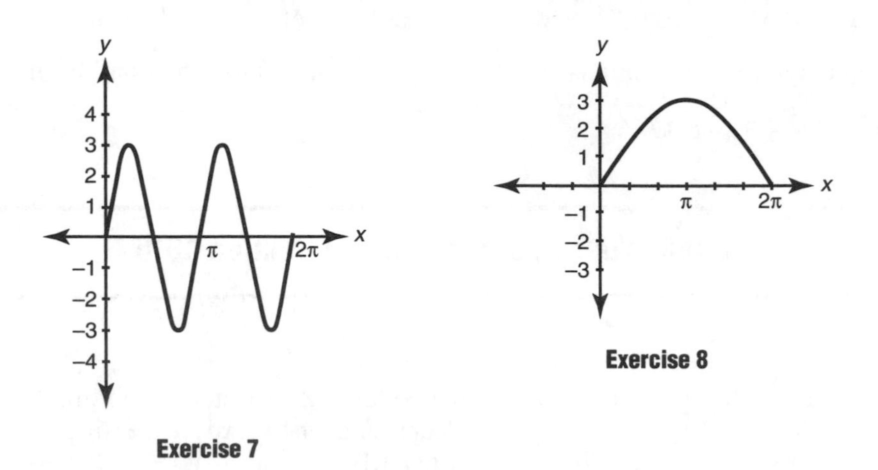

Exercise 7

Exercise 8

8. Which equation is represented by the graph in the accompanying figure?

(1) $y = 3 \sin 2x$ (2) $y = 3 \sin \frac{1}{2}x$ (3) $y = 2 \sin 3x$ (4) $y = \frac{1}{2} \sin 3x$

9. Which equation is represented by the graph in the accompanying figure?

(1) $y = 2 \sin \frac{1}{2}x$

(2) $y = \frac{1}{2} \sin 2x$

(3) $y = 2 \cos \frac{1}{2}x$

(4) $y = \frac{1}{2} \cos 2x$

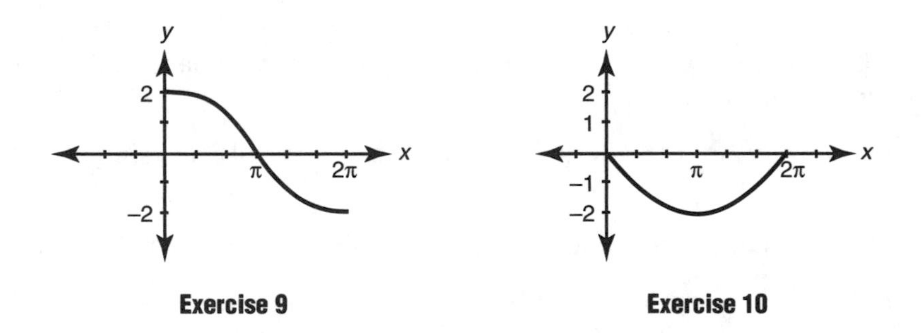

Exercise 9

Exercise 10

10. Which equation is represented by the graph in the accompanying figure?

(1) $y = -2 \sin \frac{1}{2}x$

(2) $y = -\frac{1}{2} \sin 2x$

(3) $y = \frac{1}{2} \sin 2x$

(4) $y = 2 \sin \frac{1}{2}x$

B. *In each case, show how you arrived at your answer by clearly indicating all of the necessary steps, formula substitutions, diagrams, graphs, charts, etc.*

11. A normal breathing cycle consists of inhaling followed by exhaling. Assume that one complete breathing cycle occurs every 5 seconds with a maximum airflow rate of 0.6 liter per second. If the function $f(t) = A \sin Bt$ is used to model this process, where $f(t)$ gives the airflow at time t, what are the values of A and B?

12. The peak demand for water in a certain town occurs during the summer months and can be modeled by the function $W(t) = 1600 \sin \frac{\pi}{90} t + 5000$, where $W(t)$ represents the number of cubic feet of water needed for day number t ($0 \le t \le 90$). June has 30 days, and the months of July and August each have 31 days. Assume $t = 1$ corresponds to June 1.
a. On what date is the demand for water the greatest?
b. How many cubic feet of water are needed on that day?

13.7 GRAPHING $y = a \sin bx$ and $y = a \cos bx$

KEY IDEAS

The graphs of $y = a \sin bx$ and $y = a \cos bx$ can be drawn by hand on graph paper, or a graphing calculator can be used.

Graphing Sine and Cosine Curves by Hand

When sketching the graphs of $y = \sin x$ or $y = \cos x$ from 0 to 2π radians:

- Set up the coordinate axes on graph paper, using a convenient scale with quarter-period marks at $x = \frac{\pi}{2}, \pi, \frac{3}{2}\pi$, and 2π. The quarter-period marks correspond to the division points of the four quadrants.
- "Frame" the graph by drawing broken horizontal lines through the maximum and minimum y-values of 1 and -1, and broken vertical lines through the endpoints of the interval on which the graph is being drawn.
- Plot key points such as the maximum point on the graph, the minimum point on the graph, and intercepts.
- Use your knowledge of the basic shape of the curve to sketch the graph, as illustrated in Figure 13.16.

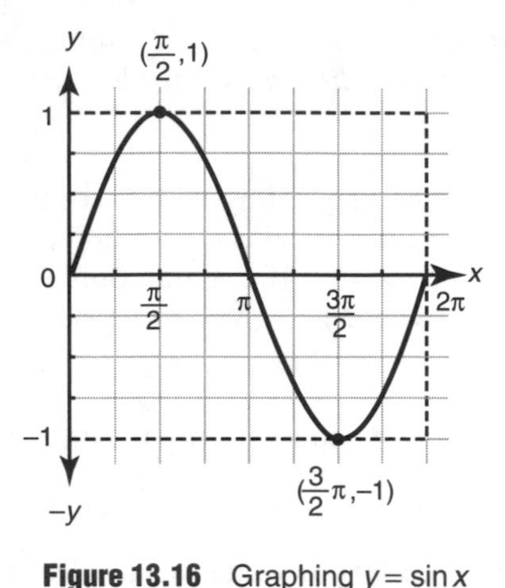

Figure 13.16 Graphing $y = \sin x$

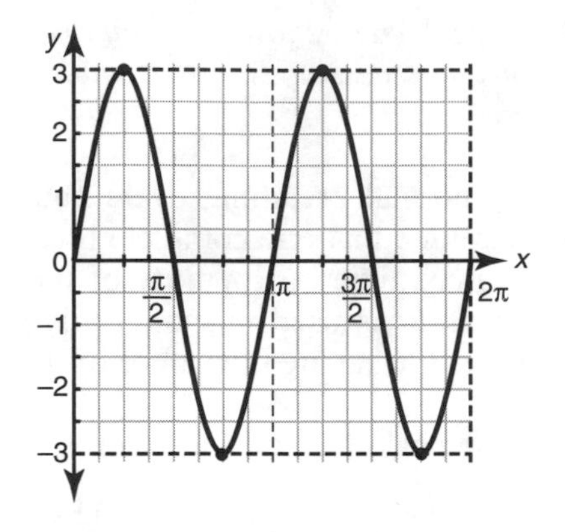

Figure 13.17 Graphing $y = 3\sin 2x$

Functions of the form $y = a \sin bx$ or $y = a \cos bx$ can be graphed in a similar way by dividing the interval from $x = 0$ to $x = \frac{2\pi}{b}$ into four equal parts. For example, to sketch the graph of $y = 3 \sin 2x$ from 0 to 2π radians:

- Find the amplitude and period. The amplitude of $y = 3 \sin 2x$ is 3, and the period is π radians. Hence, the graph completes two cycles from 0 to 2π radians.
- Since the graph completes one cycle in the interval from 0 to π radians, set up the coordinate axes, using a convenient scale with quarter-period marks from 0 to π radians. Also, divide the interval from π to 2π radians into four equal parts.
- Draw broken horizontal lines through the maximum and minimum y-values of +3 and –3, and broken vertical lines at $x = \pi$ and $x = 2\pi$, the right endpoint of each cycle.
- Use your knowledge of the basic shape of sine curves to sketch one cycle of the graph from 0 to π and another cycle of the graph from π to 2π radians, as shown in Figure 13.17.

Example

a. On the same set of axes on graph paper, sketch and label the graphs of the equations $y = 2 \sin \frac{1}{2}x$ and $y = \cos 2x$ in the interval $0 \leq x \leq 2\pi$.

b. What is the value of x in the interval $0 \leq x \leq 2\pi$ for which $2 \sin \frac{1}{2}x - \cos 2x = 1$?

Solution: a. See the accompanying figure.

- The equation $y = 2\sin\frac{1}{2}x$ has the form $y = a\sin bx$, where $a = 2$ and $b = \frac{1}{2}$. Hence the amplitude of this graph is 2, and its period is $\frac{2\pi}{1/2} = 4\pi$. Since the amplitude is 2, the graph reaches a maximum height of 2 at $x = \pi$ and a minimum height of -2 at $x = 3\pi$. A period of 4π rad means that in the interval $0 \leq x \leq 2\pi$ the curve will complete one-half cycle. This represents one arch of the sine curve, as shown in the accompanying figure.

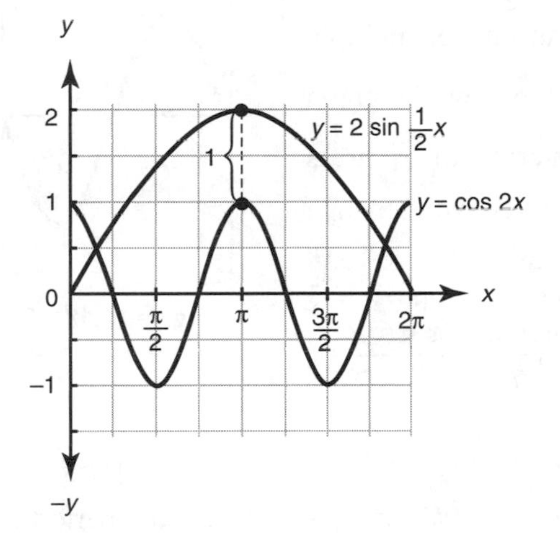

- The equation $y = \cos 2x$ has the form $y = a\cos bx$, where $a = 1$ and $b = 2$. The amplitude of this graph is 1, and its frequency is 2. The curve reaches a maximum height of 1 and a minimum height of -1. Since the frequency of this graph is 2, the graph will complete two cycles in the interval $0 \leq x \leq 2\pi$. Mark off on the x-axis four quarter-period marks from 0 to π rad, and another four quarter-period marks from π to 2π rad. Using the quarter-period marks and amplitude as guideposts, sketch one complete cycle of the cosine curve from 0 to π rad and another complete cycle from π to 2π rad, as shown in the accompanying figure.

b. $\boldsymbol{x = \pi}$

According to the graph drawn in part a, at $x = \pi$ the height of the graph of $y = 2\sin\frac{1}{2}x$ is 2 and the height of the graph of $y = \cos 2x$ is 1. Therefore, $2\sin\frac{1}{2}x - \cos 2x = 1$ at $x = \pi$.

Graphing Using a Calculator

To graph a trigonometric function using your graphing calculator, enter the function and then press ZOOM 7. If the angular mode is set to radians, the graph will be displayed in the interval $-2\pi < x < 2\pi$, using the following preset values:

$$\text{Xmin} = -\left(\frac{47}{24}\right)\pi, \qquad \text{Xmax} = \left(\frac{47}{24}\right)\pi, \qquad \text{Xscl} = \frac{\pi}{2}, \qquad \text{Yscl} = 1.$$

When Xscl is set to $\frac{\pi}{2}$, each tic mark on the x-axis is an integer multiple of $\frac{\pi}{2}$. If you need to view the graph in a different interval of x, press WINDOW and selectively change the values of Xmin, Xmax, and Xscl. For example, to graph $y = 3 \sin 2x$ over the interval $-\pi < x < \pi$, press ZOOM 7 WINDOW.

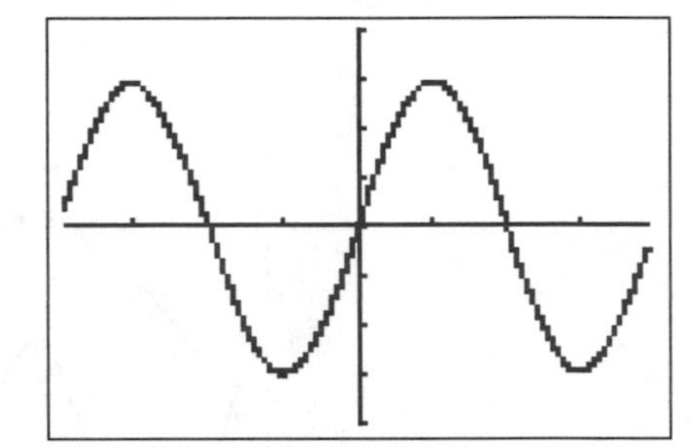

Figure 13.18 Graph of $y = 3\sin 2x$ $(-\pi < x < \pi)$

Divide Xmin, Xmax, and Xscl by 2 so that consecutive tic marks are $\frac{\pi}{4}$ radians apart in the interval $-\pi < x < \pi$. Since the period of $y = 3 \sin 2x$ is π radians, one full cycle will be displayed on either side of the y-axis, as shown in Figure 13.18.

Check Your Understanding of Section 13.7

In each case, show how you arrived at your answer by clearly indicating all of the necessary steps, formula substitutions, diagrams, graphs, charts, etc.

1. a. On the same set of axes on graph paper, sketch and label the graphs of $y = 2 \sin x$ and $y = \cos \frac{1}{2}x$ for all values of x in the interval $0 \le x \le 2\pi$.

b. Each graph drawn in part a is symmetric about which of the following?
(1) line $x = \pi$ (2) x-axis (3) $(\pi, 0)$ (4) $(0, 0)$

2. a. On the same set of axes on graph paper, sketch and label the graphs of the equations $y = -2 \sin x$ and $y = 3 \cos \frac{1}{2}x$ in the interval $0 \le x \le 2\pi$.

b. In the interval $0 \le x \le 2\pi$, which value of x satisfies the equation $3 \cos \frac{1}{2}x + 2 \sin x = 0$?

3. a. On the same set of axes on graph paper, sketch and label the graphs of the equations $y = \sin 2x$ and $y = 3 \cos x$ in the interval $-\pi \le x \le \pi$.
b. On the basis of the graphs drawn in part a, find all values of x in the interval $-\pi \le x \le \pi$ that satisfy the equation $\sin 2x = 3 \cos x$.

4. A person who has just finished exercising has a respiratory cycle modeled by the equation $v = 1.8 \sin\left(\frac{2}{3}\pi t\right)$, where v is the velocity of airflow after t seconds. If one respiratory cycle consists of an inhalation ($v > 0$) and an exhalation ($v < 0$):
a. Determine the number of seconds to the *nearest tenth* for one respiratory cycle.
b. Determine the number of full respiratory cycles completed in 1 minute.
c. Sketch the graph of $v = 1.8 \sin\left(\frac{2}{3}\pi t\right)$ for one complete respiratory cycle.

5. A ball attached to the end of a spring is bobbing up and down according to the equation $d = 20 \sin \frac{1}{2}\pi t$, where d is the distance, in centimeters, from the position of the ball at rest at time t, which is measured in seconds. Assume that the motion of the ball is not affected by friction or air resistance, and that the ball continues to move up and down in a uniform manner. Oscillating motion of this type is called *simple harmonic motion*.
a. What is the distance, in centimeters, between the lowest and highest positions of the bobbing ball?
b. What is the least number of seconds that the ball takes to travel from its lowest to its highest position?
c. Sketch the graph of $d = 20 \sin \frac{1}{2}\pi t$ from $t = 0$ second to $t = 6$ seconds.

6. a. Write an equation of the form $d = a \sin bt$ that describes the simple harmonic motion of a ball attached to a spring, if the motion of the ball starts at its highest position of 8 inches above its rest point, bounces down to its lowest position of 8 inches below its rest point, and then bounces back to its highest position in a total of 6 seconds.
b. Sketch the graph of the equation obtained in part a from $t = 0$ second to $t = 12$ seconds.

13.8 TRANSFORMATIONS OF TRIGONOMETRIC FUNCTIONS

KEY IDEAS

Transformations can be applied to trigonometric functions in the same way they are applied to algebraic functions.

Reflecting Trigonometric Functions

The graphs of $y = -\sin x$ and $y = -\cos x$ are the reflections of the graphs of the basic sine and cosine curves in the x-axis, as shown in Figure 13.19

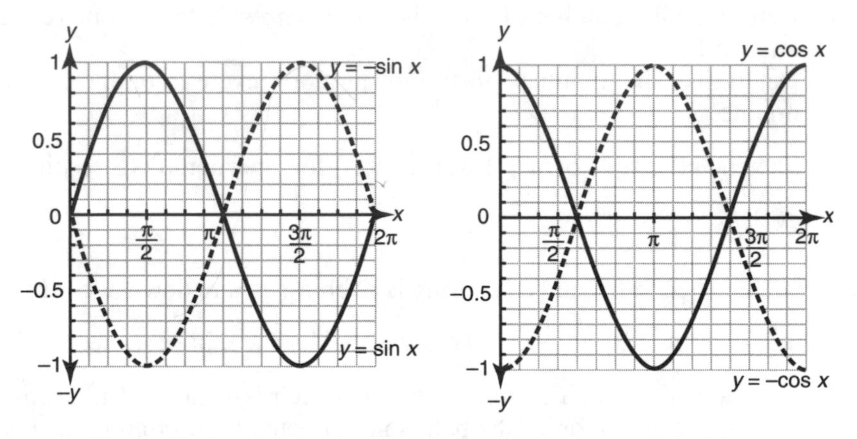

Figure 13.19 Reflecting Sine and Cosine Curves in the x-Axis

You can use your graphing calculator to verify that:

- The graph of $y = \sin(-x)$ is the reflection of the graph of $y = \sin x$ in the y-axis.
- The graph of $y = \cos x$ and its reflection in the y-axis, $y = \cos(-x)$, coincide.

Translating Trigonometric Functions

Trigonometric functions, like algebraic functions, can be shifted horizontally, vertically, or both horizontally and vertically. For example:

- The function $y = \sin x + 3$ shifts the graph of $y = \sin x$ up 3 units, as shown in Figure 13.20. Notice that the line $y = 3$ is a horizontal line of symmetry. In general, the constant k is a horizontal line of symmetry of $y = a \sin bk + k$ and $y = a \cos bk + k$.

- The function $y = \cos\left(x - \frac{\pi}{2}\right)$ shifts the graph of $y = \cos x$ to the right $\frac{\pi}{2}$ units, shown as the solid curve in Figure 13.21. A horizontal translation of a periodic function is called a **phase shift**.

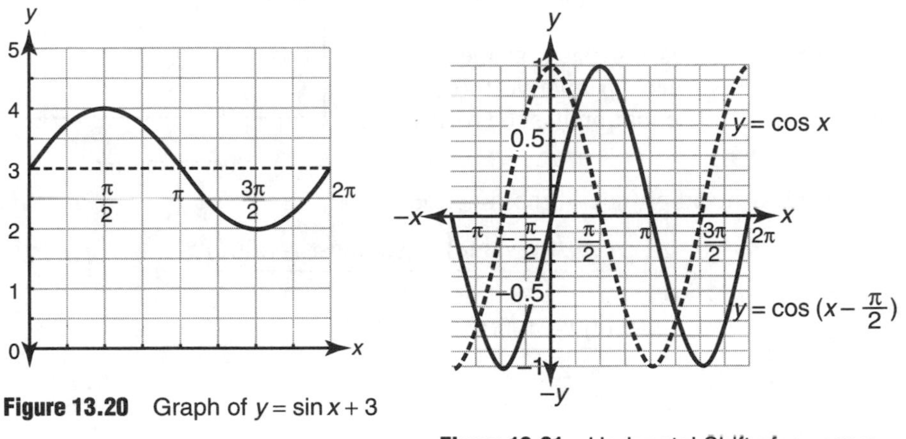

Figure 13.20 Graph of $y = \sin x + 3$

Figure 13.21 Horizontal Shift of $y = \cos x$ by $\frac{\pi}{2}$

Example 1

On the interval $-\pi \leq x \leq \pi$, sketch $y = \sin x$ under the transformation $T_{\pi,0}$.

Solution: See the accompanying figure. Under the transformation $T_{\pi,0}$, the graph of $y = \sin x$ is shifted π radians to the right. Then the equation of the translated graph is $y = \sin\left(x - \frac{\pi}{2}\right)$, as shown by the broken curve in the accompanying diagram.

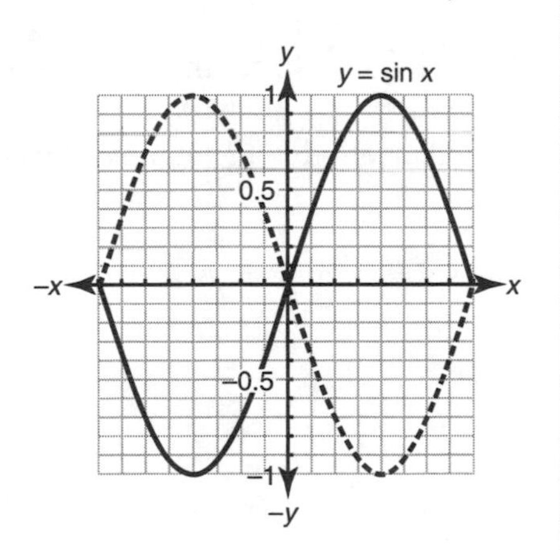

Example 2

The number of visitors at a resort rises and falls during the year according to the graph in the accompanying diagram. Determine an equation of this graph in terms of the month number, t.

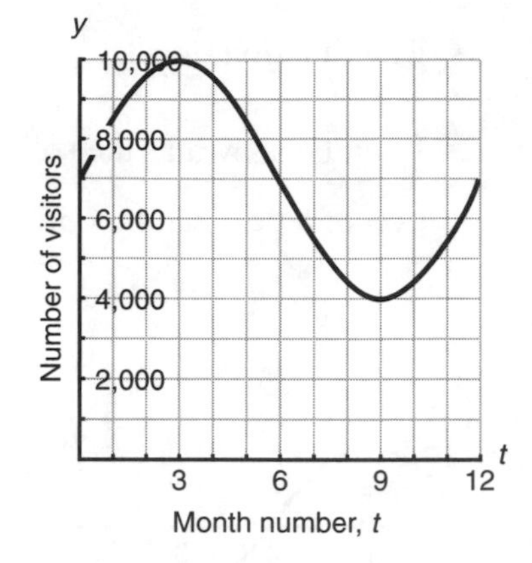

Solution: $\mathbf{y = 3000 \sin\left(\frac{\pi}{6}t\right) + 7000}$

Since the curve has the basic shape of the sine curve shifted up relative to the horizontal axis, its equation has the form $y = a \sin(bt) + k$.

- Find a, the amplitude of the function:

$$\begin{aligned} a &= \frac{(\text{greatest } y\text{-value}) - (\text{smallest } y\text{-value})}{2} \\ &= \frac{10{,}000 - 4000}{2} \\ &= 3000 \end{aligned}$$

- Find k, the vertical shift of the function:

$$\begin{aligned} k &= \frac{(\text{greatest } y\text{-value}) - (\text{smallest } y\text{-value})}{2} \\ &= \frac{10{,}000 + 4000}{2} \\ &= 7000 \end{aligned}$$

- Find b. Since the graph completes one full cycle in 12 months, the period is 12 months. Hence, $\frac{2\pi}{b} = 12$, so $b = \frac{\pi}{6}$.

Since $a = 3000$, $b = \frac{\pi}{6}$, and $k = 7000$, an equation of the curve is

$$y = 3000 \sin\left(\frac{\pi}{6}t\right) + 7000.$$

Dilating the Graphs of Trigonometric Functions

When the graph of a trigonometric function is dilated, both the amplitude and the period are affected. For example, if a dilation with a scale factor of $\frac{1}{2}$ is applied to $y = 2\sin\frac{1}{2}x$, the coordinates of each point of the original graph are twice the corresponding coordinates of its image point. Replacing x by $2x$ and y by $2y$ in the original equation gives $y = \sin x$ as an equation of the dilated graph:

$$y = 2\sin\frac{1}{2}x \xrightarrow{D_{\frac{1}{2}}} 2y = 2\sin\frac{1}{2}(2x) \text{ or } y = \sin x.$$

Check Your Understanding of Section 13.8

A. Multiple Choice

1. What phase shift of the cosine function will translate it onto the sine function?

(1) 1 (2) $\frac{\pi}{2}$ (3) $-\frac{\pi}{2}$ (4) -1

2. If $y = 3\sin 2x + 4$, what is the *maximum* value of y?
(1) 3 (2) 24 (3) 7 (4) 10

3. If $y = 2\cos\frac{1}{2}x - 3$, what is the *minimum* valuc of y?

(1) -3 (2) -2 (3) -1 (4) -5

4. Which is an equation of the image of the graph of $y = 3\sin\frac{1}{2}x$ under a dilation with a scale factor of 2?

(1) $y = 6\sin x$ (2) $y = 6\sin\frac{1}{4}x$ (3) $y = 1.5\sin\frac{1}{4}x$ (4) $y = 1.5\sin x$

B. *In each case, show how you arrived at your answer by clearly indicating all of the necessary steps, formula substitutions, diagrams, graphs, charts, etc.*

5 and 6. If each of the following graphs represents a vertical translation of a basic sine curve, write an equation of the translated graph shown.

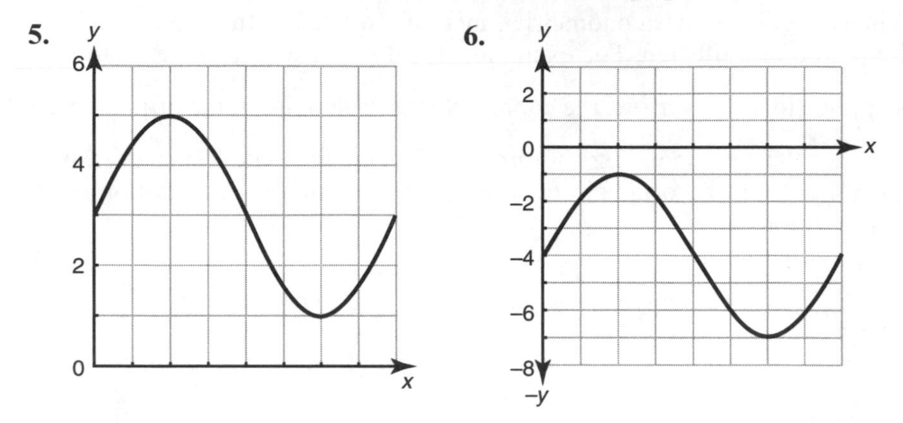

7 and 8. If each of the following graphs represents a vertical translation of a basic cosine curve, write an equation of the translated graph shown.

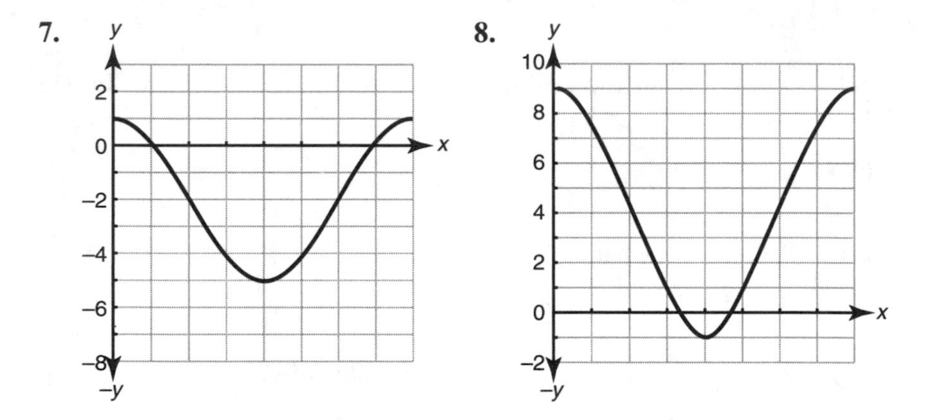

9. Given the function $y = 2\sin\left(x - \dfrac{\pi}{3}\right) + 3$.

a. Determine the amplitude of the function.
b. If the graph completes one full cycle from $x = a$ to $x = b$, find the *smallest* possible values of a and b.
c. Using your answers from part b, sketch the graph, using graph paper in the interval $a \le x \le b$.

10. The populations of a certain species of fish in a river, as shown in the accompanying graph, over a 12-month interval can be modeled by the equation $y = A\cos(Bx) + D$. Determine the values of A, B, and D, and explain how you arrived at these values.

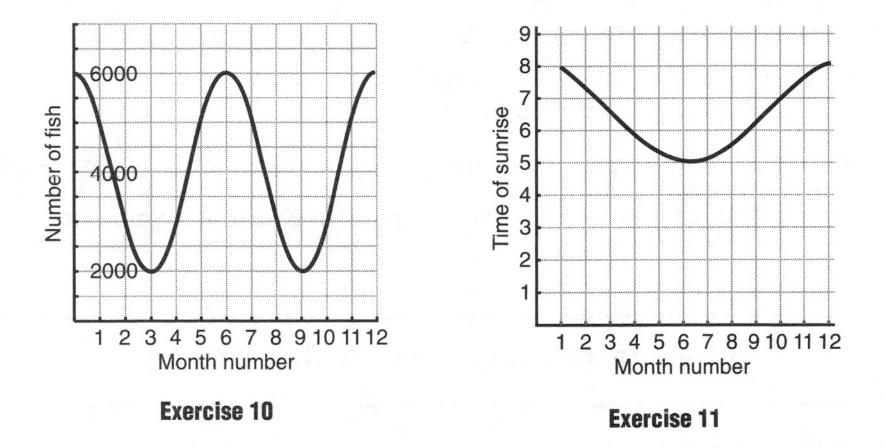

11. The times of average monthly sunrise, as shown in the accompanying diagram, over a 12-month interval can be modeled by the equation $y = A\cos(Bx) + D$. Determine the values of A, B, and D, and explain how you arrived at these values.

13.9 GRAPHING THE TANGENT FUNCTION

KEY IDEAS

The function $y = \tan x$ is a periodic function with a period of 180°.

Graph of $y = \tan x$

The function $y = \tan x$ is not defined at $x = \frac{\pi}{2} (= 90°)$ or at any odd-integer multiple of $\frac{\pi}{2}$. As the graph of $y = \tan x$ approaches these x-values, the corresponding y-values become unbounded. Therefore, there are vertical asymptotes through these x-values, as shown in Figure 13.22.

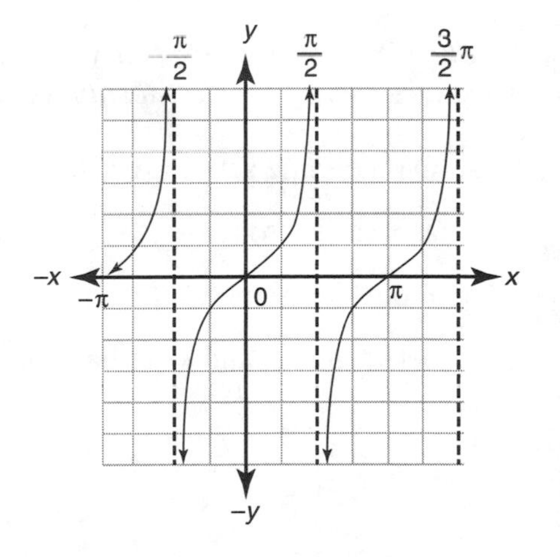

Figure 13.22 Graph of $y = \tan x$

The graph of $y = \tan x$ has:

- A period of π radians and no amplitude.
- x-intercepts at integer multiples of π.
- Vertical asymptotes at lines that are odd multiples of $\pm\frac{\pi}{2}$. The graph completes one full cycle between consecutive vertical asymptotes.

Check Your Understanding of Section 13.9

A. Multiple Choice

1. At which value of x does the graph of $y = \tan\left(x + \frac{\pi}{4}\right)$ have a vertical asymptote?

(1) $-\frac{\pi}{4}$ (2) 0 (3) $\frac{\pi}{4}$ (4) $\frac{\pi}{2}$

2. The graph of the equation $y = \tan x$ is symmetric to the

(1) line $x = \frac{\pi}{2}$ (2) x-axis (3) y-axis (4) origin

3. In the interval $0 \leq \theta \leq 2\pi$, the number of solutions of the equation $\sin 2\theta - \tan\theta = 0$ is

(1) 6 (2) 2 (3) 3 (4) 4

B. *Show how you arrived at your answer by clearly indicating all of the necessary steps, formula substitutions, diagrams, graphs, charts, etc.*

4. a. On the same set of axes, sketch the graphs of $y = \cos 2x$ and $y = \tan x$ as x varies from $-\frac{\pi}{2}$ to $\frac{\pi}{2}$ radians.

b. From the graph sketched in part a, determine the number of points between $-\frac{\pi}{2}$ and $\frac{\pi}{2}$ radians for which $\tan x - \cos 2x = 0$.

13.10 INVERSE TRIGONOMETRIC FUNCTIONS

KEY IDEAS

Since the graphs of the sine, cosine, and tangent functions do not pass the horizontal-line test, these functions do not have inverse functions. Inverse trigonometric functions can be formed, however, by restricting the domains of the original trigonometric functions so that their graphs pass the horizontal-line test.

Forming the Inverse Sine Function

Figure 13.23 shows that, in the interval $-\frac{\pi}{2} \leq x \leq \frac{\pi}{2}$, the graph of $y = \sin x$ passes the horizontal-line test while $\sin x$ takes on its full range of values from -1 to $+1$. If the domain of $y = \sin x$ is restricted to $-\frac{\pi}{2} \leq x \leq \frac{\pi}{2}$, the inverse sine function is formed by interchanging x and y, so that $x = \sin y$. The equation $x = \sin y$ is solved for y by writing $y = \sin^{-1} x$ or $y = \text{Arcsin}\, x$, read as "inverse sine of x." The range of $y = \text{Arcsin}\, x$ is $-90° \leq y \leq 90°$, and the domain is $-1 \leq x \leq 1$, as shown in Figure 13.24.

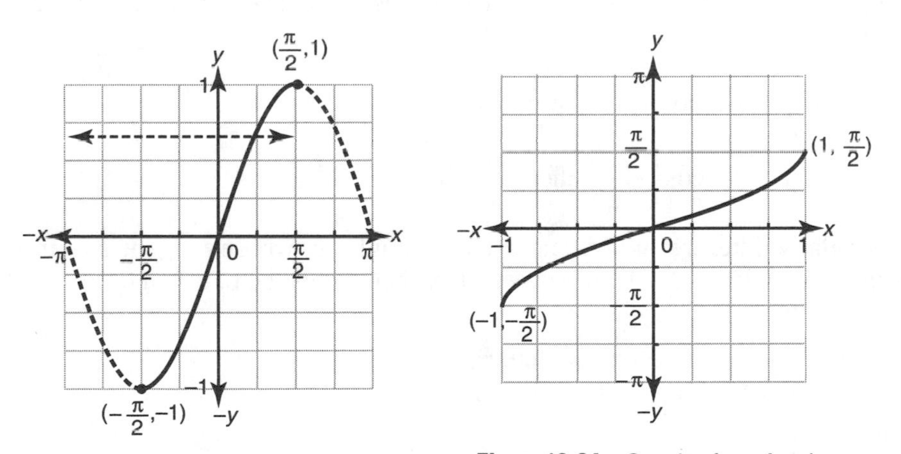

Figure 13.23 Graph of $y = \sin x$

Figure 13.24 Graph of $y = \text{Arcsin}\, x$

Think of $y = \text{Arcsin}\, x$ as the $\angle y$ whose sine is x in the interval $-90° \leq y \leq 90°$. If x is negative, then y must be between 0° and −90°. For example:

- If $y = \text{Arcsin}\,\frac{1}{2}$, then $y = 30°$ since $\sin 30° = \frac{1}{2}$.

- If $y = \text{Arcsin}\left(-\frac{\sqrt{2}}{2}\right)$, then $y = -45°$ since $\sin(-45)° = -\frac{\sqrt{2}}{2}$.

Inverse Cosine and Tangent Functions

The same approach as was used in forming the inverse sine function can be applied also to define the inverse functions of cosine and tangent, as shown in Table 13.4.

TABLE 13.4 INVERSE TRIGONOMETRIC FUNCTIONS

Original Function	Restricted Domain	Inverse Function	Principal Values (Range)
$y = \sin x$	$-90° \le x \le 90°$	$y = \text{Arcsin}\,x$	$-90° \le y \le 90°$
$y = \cos x$	$0 \le x \le 180°$	$y = \text{Arccos}\,x$	$0 \le y \le 180°$
$y = \tan x$	$-90° < x < 90°$	$y = \text{Arctan}\,x$	$-90° < y < 90°$

Think of $y = \text{Arccos}\,x$ as the angle, y, whose cosine is x in the interval $0 \le y \le 180°$. If x is negative, then y must be between 90° and 180°. For example:

- If $y = \text{Arccos}\,\frac{\sqrt{3}}{2}$, then $y = 30°$.
- if $y = \text{Arccos}\left(-\frac{1}{2}\right)$, then $y = 120°$.

Similarly, interpret $y = \text{Arctan}\,x$ as the angle, y, whose tangent is x in the interval $-90° \le y \le 90°$. If x is negative, then y must be between 0° and −90°. For example:

- if $y = \text{Arctan}\,\sqrt{3}$, then $y = 60°$.
- if $y = \text{Arctan}\,(-1)$, then $y = -45$.

Example 1

If $f(x) = \sin(\text{Arctan}\,x)$, what is the value of $f(1)$?

Solution: $\mathbf{\frac{\sqrt{2}}{2}}$

- Write the expression: $f(x) = \sin(\text{Arctan}\,x)$
- Let $x = 1$: $f(1) = \sin(\text{Arctan}\,1)$
- Recognize that, since $\tan 45° = 1$, $\text{Arctan}\,1 = 45°$: $= \sin 45° = \frac{\sqrt{2}}{2}$

Example 2

If $y = \tan\left(\text{Arccos}\left(-\frac{\sqrt{3}}{2}\right)\right)$, what is the value of y?

Solution: $\mathbf{-\frac{\sqrt{3}}{3}}$

Since $\text{Arccos}\left(-\frac{\sqrt{3}}{2}\right) = 150°$:

$$y = \tan\left(\text{Arccos}\left(-\frac{\sqrt{3}}{2}\right)\right) = \tan 150° = -\frac{\sqrt{3}}{3}.$$

Example 3

What is the value of $\cot\left[\text{Arcsin}\left(-\frac{12}{13}\right)\right]$?

Solution: $\mathbf{-\frac{5}{12}}$

- Let $\theta = \text{Arcsin}\left(-\frac{12}{13}\right)$. Then

$$\sin\theta = \frac{y}{r} = \frac{-12}{13}$$

Since $-90° \leq \theta \leq 90°$, θ terminates in Quadrant IV, where the sine function takes on negative values. Sketch the angle, as shown in the accompanying figure.

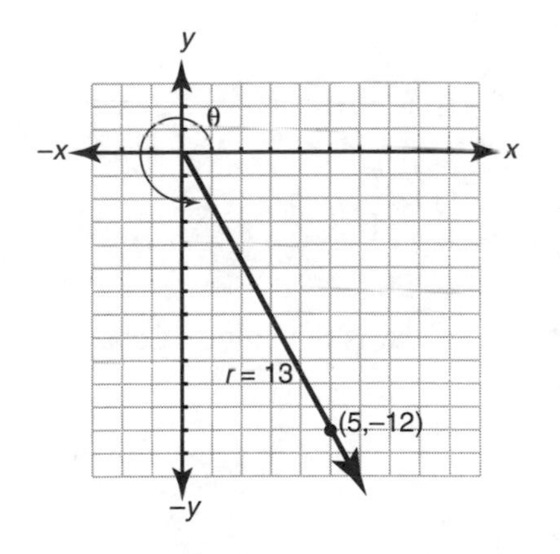

- If $y = -12$ and $r = 13$, then $x = 5$ since $5^2 + (-12)^2 = 13^2$.
- Hence:

$$\cot\left[\text{Arcsin}\left(\frac{-12}{13}\right)\right] = \cot\theta = \frac{x}{y} = -\frac{5}{12}.$$

Check Your Understanding of Section 13.10

Multiple Choice

1. If $\theta = \text{Arccos}\left(-\frac{\sqrt{2}}{2}\right)$, what is the value of $\tan\theta$?

(1) 1 (2) -1 (3) $-\frac{1}{\sqrt{3}}$ (4) $\sqrt{3}$

2. If $f(x) = \sin(\text{Arccos}\, x)$, what is the value of $f\left(\frac{1}{2}\right)$?

(1) 1 (2) $\frac{1}{2}$ (3) $\frac{\sqrt{3}}{2}$ (4) $\frac{\sqrt{2}}{2}$

3. If $y = \sec\left(\text{Arcsin}\,\frac{1}{2}\right)$, what is the value of y?

(1) 30° (2) 2 (3) $\frac{2}{\sqrt{3}}$ (4) $\frac{1}{2}$

4. What is the value of $\csc\left[\text{Arctan}\left(\frac{4}{3}\right)\right]$?

(1) $\frac{5}{3}$ (2) $\frac{5}{4}$ (3) $\frac{3}{5}$ (4) $\frac{4}{5}$

5. If $y = \text{Arcsin}\,\frac{\sqrt{2}}{2} + \text{Arccos}\,\frac{\sqrt{2}}{2}$, then $y =$

(1) $\frac{\pi}{4}$ (2) $\frac{\pi}{2}$ (3) $\frac{2\pi}{3}$ (4) π

6. What is the value of $\cos\left[\cos^{-1}\left(-\frac{1}{2}\right) + \sin^{-1}\left(\frac{1}{2}\right)\right]$?

(1) $-\frac{\sqrt{3}}{2}$ (2) $-\frac{1}{2}$ (3) $\frac{1}{2}$ (4) $\frac{\sqrt{3}}{2}$

7. What is the value of y if $y = \cos\left(\text{Arctan}\,\frac{24}{7}\right)$?

(1) $\frac{7}{25}$ (2) $\frac{24}{25}$ (3) $\frac{25}{24}$ (4) $\frac{25}{7}$

8. What is the value of $\sin\left[\text{Arcsec}\left(-\frac{41}{9}\right)\right]$?

(1) $-\frac{41}{40}$ (2) $\frac{40}{9}$ (3) $-\frac{9}{40}$ (4) $\frac{40}{41}$

9. What is the value of $\cot\left[\text{Arcsin}\left(-\frac{3}{\sqrt{13}}\right)\right]$?

(1) $\frac{2}{3}$ (2) $-\frac{3}{2}$ (3) $\frac{3}{2}$ (4) $-\frac{2}{3}$

10. What is the value of $\sin\left(\text{Arccos}\ \frac{1}{x}\right)$ where $x \neq 0$?

(1) $\frac{\sqrt{1-x^2}}{x}$ (2) $\frac{\sqrt{1+x^2}}{x}$ (3) $\frac{\sqrt{x^2-1}}{x}$ (4) $\frac{x}{\sqrt{1+x^2}}$

CHAPTER 14
TRIGONOMETRIC IDENTITIES AND EQUATIONS

14.1 TRIGONOMETRIC IDENTITIES

KEY IDEAS

An **identity** is an equation that is true for all possible replacement values of the variable. The equation $\tan\theta = \dfrac{\sin\theta}{\cos\theta}$ is an example of a trigonometric identity. Additional trigonometric identities can be derived using the unit circle.

Conditional Equations Versus Identities

A **conditional equation** is true only for particular values of the variable, whereas an identity is true for each value in the domain of the variable. For example:

- The equation $2x = x + 1$ is a *conditional equation*. The graphs of $y = 2x$ and $y = x + 1$ intersect in exactly one point. Therefore, the left and right members of the equation $2x = x + 1$ agree for only one value of x.
- The equation $x^2 + 1 = (x + 1)(x - 1) + 2$ is an *identity*. The graphs of $y = x^2 + 1$ and $y = (x + 1)(x - 1) + 2$ exactly coincide. Therefore, the left and right members of the equation $x^2 + 1 = (x + 1)(x - 1) + 2$ agree for all possible replacement values of x.

Pythagorean Trigonometric Identities

In a unit circle, $x^2 + y^2 = 1$. Replacing x with $\cos\theta$ and y with $\sin\theta$ produces the Pythagorean trigonometric identity

$$\sin^2\theta + \cos^2\theta = 1.$$

Dividing each term of $\sin^2\theta + \cos^2\theta = 1$ by either $\sin^2\theta$ or $\cos^2\theta$ forms two additional Pythagorean identities. The three Pythagorean identities are given in the following table.

PYTHAGOREAN TRIGONOMETRIC IDENTITIES

Basic Pythagorean Trigonometric Identities	Equivalent Forms
$\sin^2\theta + \cos^2\theta = 1$	$\sin^2\theta = 1 - \cos^2\theta$ or $\cos^2\theta = 1 - \sin^2\theta$
$\tan^2\theta + 1 = \sec^2\theta$	$\tan^2\theta = \sec^2\theta - 1$ or $\sec^2\theta - \tan^2\theta = 1$
$\cot^2\theta + 1 = \csc^2\theta$	$\cot^2\theta = \csc^2\theta - 1$ or $\csc^2\theta - \cot^2\theta = 1$

Example 1

Multiple Choice:

If $y = \cos A(\sec A - \cos A)$, then $y =$
(1) $\cos^2 A$ (2) $\cos A - \sin A$ (3) $\sin^2 A$ (4) $\cot A - 1$

Solution: **(3)**

Change to sines and cosines and simplify:

$$\begin{aligned} y &= \cos A(\sec A - \cos A) \\ &= \cos A\left(\frac{1}{\cos A} - \cos A\right) \\ &= 1 - \cos^2 A \\ &= \sin^2 A \end{aligned}$$

Example 2

If $\angle A$ is obtuse and $\sin A = \frac{\sqrt{5}}{3}$, what is the exact value of $\cos A$?

Solution: $\mathbf{-\frac{2}{3}}$

$$\begin{aligned} \cos^2 A &= 1 - \sin^2 A \\ &= 1 - \left(\frac{\sqrt{5}}{3}\right)^2 \\ &= 1 - \frac{5}{9} \\ \cos A &= \pm\sqrt{\frac{4}{9}} = \pm\frac{2}{3} \end{aligned}$$

Since $\angle A$ is obtuse and cosine is negative in Quadrant II, $\cos A = -\frac{2}{3}$.

Example 3

Multiple Choice:

If $\csc x \neq -1$, then $\dfrac{\cot^2 x}{1+\csc x}$ is equivalent to

(1) $1 - \csc x$ (2) $\csc x - 1$ (3) $-\csc x$ (4) $\csc x - \cot x$

Solution: **(2)**

- Let $\cot^2 x = \csc^2 x - 1$: $\dfrac{\cot^2 x}{1+\csc x} = \dfrac{\csc^2 x - 1}{1+\csc x}$
- Factor: $= \dfrac{(\csc x + 1)(\csc x - 1)}{1+\csc x}$
- Divide: $= \dfrac{\cancel{(\csc x + 1)}(\csc x - 1)}{\cancel{1+\csc x}} = \csc x - 1$

Example 4

Express $\dfrac{\sec x - \cos x}{\tan x}$ as a single trigonometric function for all values of x for which the fraction is defined.

Solution: $\mathbf{\sin x}$

- Change to sines and cosines: $\dfrac{\sec x - \cos x}{\tan x} = \dfrac{\dfrac{1}{\cos x} - \cos x}{\dfrac{\sin x}{\cos x}}$
- Simplify the complex fraction: $= \dfrac{1 - \cos^2 x}{\cos x} \div \dfrac{\sin x}{\cos x}$
- Let $\sin^2 x = 1 - \cos^2 x$: $= \dfrac{\cancel{\sin^2} x}{\cancel{\cos x}} \cdot \dfrac{\cancel{\cos x}}{\cancel{\sin x}}$

$= \sin x$

Check Your Understanding of Section 14.1

A. Multiple Choice

1. The expression $k - \dfrac{1}{\sec^2 x} = \sin^2 x$ is an identity if k is equal to

(1) 1 (2) 0 (3) $\cos^2 x$ (4) $\csc^2 x$

2. If $\cos x \neq \pm 1$, then $\dfrac{\cos x}{1 - \cos x} \cdot \dfrac{\cos x}{1 + \cos x}$ is equivalent to

(1) $\tan^2 x$ (2) $\sec^2 x$ (3) $\cot^2 x$ (4) $\sec x \cdot \csc x$

3. If $\sec x \neq 1$, then $\dfrac{\tan^2 x}{1-\sec x}$ is equivalent to
(1) $1 + \sec x$ (2) $\sec x - 1$ (3) $-1 - \sec x$ (4) $-\sec x$

4. The expression $\sec x - \sin x \tan x$ is equivalent to
(1) $\csc x$ (2) $\sec x$ (3) $\sin x$ (4) $\cos x$

5. If $\log \sin x = a$ and $\log \cos x = b$, then $\log(\cot^2 x + 1)$ is equivalent to
(1) $2(a - b)$ (2) $\dfrac{1}{a^2}$ (3) $2b + 1$ (4) $-2a$

6. The expression $\sin^2 x - b^2 + \cos^2 x$ is equivalent to
(1) 1
(2) b^2
(3) $(1 + b)(1 - b)$
(4) $(\sin x - b)(\cos x + b)$

7. The expression $\dfrac{\sec x}{\cot x + \tan x}$ is equivalent to
(1) $\tan x$ (2) $\sin x$ (3) $\cos x$ (4) $\cot x$

B. *In each case, show how you arrived at your answer by clearly indicating all of the necessary steps, formula substitutions, diagrams, graphs, charts, etc.*

8. Express $\sin^2 \theta(1 + \tan^2 \theta)$ as a single trigonometric function for all values of θ for which the expression is defined.

9. Express $(\tan x + \cot x)^2 - \csc^2 x$ as a single trigonometric function for all values of x for which the expression is defined.

14.2 SOLVING TRIGONOMETRIC EQUATIONS

KEY IDEAS

Trigonometric equations that are not identities can be solved in much the same way as algebraic equations.

Solving Linear Trigonometric Equations

Trigonometric equations are solved by first solving for the trigonometric function.

Example 1

Solve $2 \sin x + 1 = 0$ for x in the interval $0° \le x < 360°$.

Solution: $\mathbf{x = 210°}$ or $\mathbf{330°}$

If $2 \sin x + 1 = 0$, then $\sin x = -\frac{1}{2}$ and $x = \sin^{-1}\left(-\frac{1}{2}\right)$ in the interval $0° \le x < 360°$. The reference angle is 30°. Since sine is negative in Quadrants III and IV, there are two possible solutions:

$$Q_{III}: x_1 = 210° \qquad \text{or} \qquad Q_{IV}: x_2 = 330°.$$

Example 2

Given the system of equations:

$$0.4 \cos A + 0.3 \sin B = 0.35$$
$$0.3 \cos A - 0.4 \sin B = 0.20$$

a. If $0° \le A < 360°$, solve for $\angle A$ to the *nearest hundredth of a degree.*
b. If $\angle B$ is acute, solve for $\angle B$ to the *nearest hundredth of a degree.*

Solution: a. $\mathbf{\angle A = 36.87}$ or **323.13**

- Multiply both sides of the first equation by 4, and multiply both sides of the second equation by 3. Then add the resulting equations to eliminate $\sin A$:

$$4\{0.4 \cos A + 0.3 \sin B = 0.35\} \Rightarrow 1.6 \cos A + 1.2 \sin B = 1.4$$
$$3\{0.3 \cos A - 0.4 \sin B = 0.20\} \Rightarrow \underline{0.9 \cos A = 1.2 \sin B = 0.6}$$
$$2.5 \cos A + \quad 0 \qquad = 2.0$$

- Thus, $2.5 \cos A = 2.0$, $\cos A = \frac{2.0}{2.5} = 0.8$. The reference angle is $A_R = \cos^{-1} 0.8 = 36.87°$.
- Since cosine is positive in Quadrants I and IV, there are two possible solutions:

$$Q_I: A_1 = 36.87° \qquad \text{or} \qquad Q_{IV}: A_2 = 360° - 36.87° = 323.13°.$$

b. $\mathbf{\angle B = 5.74}$

Since $\cos A = 0.8$, replace $\cos A$ with 0.8 in either of the original equations and solve the resulting equation for $\sin B$:

$$0.4 \cos A + 0.3 \sin B = 0.35$$
$$0.4(0.8) + 0.3 \sin B = 0.35$$
$$0.3 \sin B = 0.35 - 0.32$$
$$\sin B = \frac{0.03}{0.3} = 0.1$$
$$B = \sin^{-1} 0.1 \approx 5.74\,°$$

Solving Quadratic Trigonometric Equations

To solve a quadratic trigonometric equation, use factoring or the quadratic formula to solve for the trigonometric function. Then find all of the possible values of the angle.

Example 3

Solve for x to the *nearest tenth of a degree* in the interval $0° \le x < 360°$: $\tan^2 x = \tan x + 2$.

Solution: $\mathbf{x = 63.4}$, **135**, **243.4**, or **315**

Write the quadratic equation in standard form and solve for $\tan x$ by factoring:

$$\tan^2 x - \tan x - 2 = 0$$
$$(\tan x + 1)(\tan x - 2) = 0$$
$$\tan x + 1 = 0 \quad \text{or} \quad \tan x - 2 = 0$$

$\tan x = -1$	$\tan x = 2$
$Q_{II}: x_1 = 135°$	$x = \tan^{-1} 2$
$Q_{IV}: x_1 = 315°$	Use a calculator to find x to the *nearest tenth of a degree.*
	$Q_I: \ x_3 = 63.4°$
	$Q_{III}: x_4 = 180° + 63.4° = 243.4°$

Trigonometric Equations Requiring Substitutions

If a trigonometric equation contains two different trigonometric functions, use a trigonometric identity or reciprocal function relationship to transform the equation into an equivalent equation that contains the same trigonometric function.

Example 4

Solve for x to the *nearest tenth of a degree* in the interval $0° \le x < 360°$: $3 \cos^2 x + 5 \sin x = 4$.

Solution: $\mathbf{x = 13.5}$ or **166.5**

- Transform the original equation into an equivalent equation that contains only the sine function by replacing $\cos^2 x$ with $1 - \sin^2 x$:

$$3\cos^2 x + 5\sin x = 4$$
$$3\left(1 - \sin^2 x\right) + 5\sin x = 4$$
$$3 - 3\sin^2 x + 5\sin x = 4$$
$$3\sin^2 x - 5\sin x + 1 = 0$$

- Use the quadratic formula to solve for $\sin x$:

Write the formula: $\sin x = \dfrac{-b \pm \sqrt{b^2 - 4ac}}{2a}$

Let $a = 3$, $b = -5$, $c = 1$: $= \dfrac{-(-5) \pm \sqrt{(-5)^2 - 4(3)(1)}}{2(3)}$

$$= \frac{5 \pm \sqrt{13}}{6}$$

$$\approx \frac{5 \pm 3.60555}{6}$$

- Solve for x:

$\sin x \approx \dfrac{5 - 3.60555}{6}$

≈ 0.2324

$x \approx \sin^{-1} 0.2324$

Q$_{\text{I}}$: $x_1 \approx 13.5°$

Q$_{\text{II}}$: $x_2 \approx 166.5°$

or

$\sin x = \dfrac{5 + 3.60555}{6}$

≈ 1.39343

Reject since the maximum value of $\sin x$ is 1.

Check Your Understanding of Section 14.2

A. Multiple Choice

1. Which angle is a solution of the equation $\cos^2 x - \cos x - 2 = 0$?
(1) $\dfrac{11\pi}{6}$ (2) $\dfrac{5\pi}{3}$ (3) π (4) 0

2. Which value of θ satisfies the equation $2\sin^2\theta - 5\sin\theta - 3 = 0$?
(1) 300° (2) 210° (3) 150° (4) 30°

3. If θ is an angle in Quadrant I and $\cot^2\theta - 4 = 0$, what is the value of θ to the *nearest degree*?
(1) 45 (2) 2 (3) 27 (4) 63

4. In the interval $0 \le x < 2\pi$, the solutions of the equation $\sin^2 x = \sin x$ are

(1) $0, \frac{\pi}{2}, \pi$ (2) $\frac{\pi}{2}, \frac{3\pi}{2}$ (3) $0, \frac{\pi}{2}, \frac{3\pi}{2}$ (4) $\frac{\pi}{2}, \pi, \frac{3\pi}{2}$

5. Given this system of equations:

$$0.4 \sin x + \cos y = -0.2$$
$$0.8 \sin x - \cos y = 1.1$$

Which value of y can be a solution?

(1) 90° (2) 180° (3) 240° (4) 330°

B. *In each case, show how you arrived at your answer by clearly indicating all of the necessary steps, formula substitutions, diagrams, graphs, charts, etc.*

6–10. Find all values of x in the interval $0° \le x < 360°$ that satisfy each equation.

6. $\sqrt{4\cos^2 x - 1} - 1 = 0$

7. $4\cos^2 x - 3 = 0$

8. $3\tan^2 x + \sqrt{3}\tan x = 0$

9. $\csc^2 x + \cot x = 1$

10. $2\cos x - 3 = 2\sec x$

11. A particle moves along a straight line. The distance, s, of this particle from the origin t seconds after the particle begins to move can be modeled by the function $s(t) = 2\sin\left(\frac{t}{2}\right) - \cos t$, where $t \ge 0$. Find the smallest value of t for which the particle is 3 units from the origin.

12. Given this system of equations:

$$0.8\cos x + 0.3\sin y = -0.10$$
$$-0.7\cos x + 0.4\sin y = 0.75$$

Find x and y where each angle can vary from 0° to 360°.

13. Given this system of equations:

$$0.5\sin\theta + 0.4\cos\theta = 0.4$$
$$0.4\sin\theta - 0.5\cos\theta = 0.5$$

Find θ correct to the *nearest hundredth of a degree.*

14. Given the system of equations:

$$0.9 \sin A + 0.4 \cos B = 0.7$$
$$0.8 \sin A + 0.3 \cos B = 0.6$$

Find, to the *nearest hundredth of a degree*, A and B if $\angle A$ is obtuse and $\angle B$ is acute.

15–20. Find, correct to the *nearest tenth of a degree*, all values of x in the interval $0° \leq x < 360°$ that satisfy the given equation.

15. $4 \sin^2 x = 5 \sin x + 6$

16. $4 \cos^2 x = 5 (\sin x + 1)$

17. $\sec^2 x + 2 \tan x = 4$

18. $2 \sin^2 x + 2 \cos x = 1$

19. $2 \tan x (\tan x - 1) = 3$

20. $\dfrac{3 \sin x}{6 \sin x + 1} = \dfrac{\csc x}{3}$

14.3 SUM AND DIFFERENCE IDENTITIES

KEY IDEAS

You can easily verify that $\sin (30 + 60)° \neq \sin 30° + \sin 60°$. However, although $\sin (A + B) \neq \sin A + \sin B$, there are identities that allow trigonometric functions of the sum or difference of two angles to be expressed in terms of combinations of trigonometric functions of the individual angles.

Functions of the Sum of Two Angles

If A and B represent angle measures, then:

- $\sin (A + B) = \sin A \cos B + \cos A \sin B$.
- $\cos (A + B) = \cos A \cos B - \sin A \sin B$.
- $\tan (A + B) = \dfrac{\tan A + \tan B}{1 - \tan A \cdot \tan B}$.

Example 1

What is the exact value of $\sin 17° \cos 13° + \cos 17° \sin 13°$?

Solution: $\mathbf{\frac{1}{2}}$

The given expression has the same form as the right side of the identity for the sum of the sines of angles A and B, where $A = 17°$ and $B = 13°$. Hence:

$$\sin 17° \cos 13° + \cos 17° \sin 13° = \sin(17+13)° = \sin 30° = \frac{1}{2}.$$

Example 2

If $\sin A = \frac{4}{5}$, $\cos B = \frac{12}{13}$, $\angle A$ is obtuse, and $\angle B$ is acute, what is the value of $\cos(A+B)$?

Solution: $-\frac{\mathbf{56}}{\mathbf{65}}$

Before the identity for the cosine of the sum of two angles can be used, the values of $\cos A$ and $\sin B$ need to be determined by first locating the reference triangles, as shown in the accompanying diagrams.

- $\sin A = \frac{4}{5} = \frac{y}{r}$. The reference triangle is a $3-4-5$ right triangle. Since $\angle A$ lies in Quadrant II, where $x < 0$, then $x = -3$ and $\cos A = \frac{x}{r} = \frac{-3}{5}$.

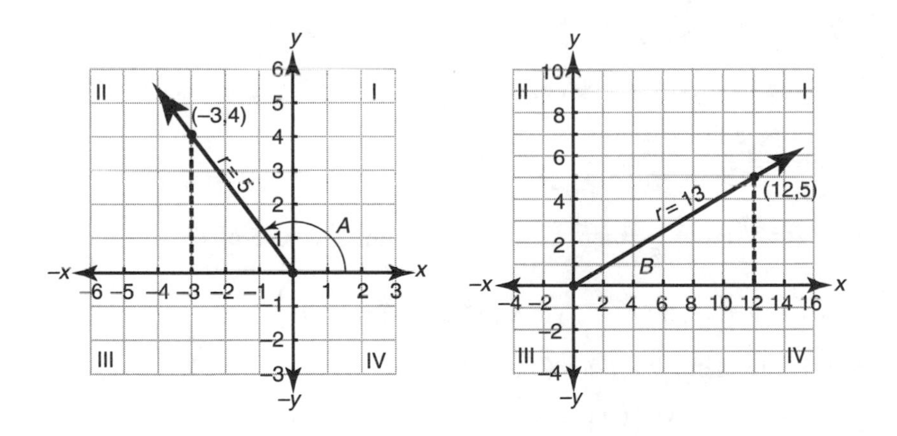

- $\sin B = \frac{y}{r} = \frac{5}{13}$. Since the reference triangle is a $5-12-13$ right triangle in Quadrant I, $x = 12$ and $\cos B = \frac{x}{r} = \frac{12}{13}$.
- Evaluate the identity for $\cos(A+B)$:

$$
\begin{aligned}
\cos(A+B) &= \cos A\cos B - \sin A\sin B \\
&= \left(\frac{-3}{5}\right)\left(\frac{12}{13}\right) - \left(\frac{4}{5}\right)\left(\frac{5}{13}\right) \\
&= \left(-\frac{36}{65}\right) - \left(\frac{20}{65}\right) \\
&= -\frac{56}{65}
\end{aligned}
$$

Functions of the Difference of Two Angles

If A and B represent angle measures, then:

- $\sin(A-B) = \sin A\cos B - \cos A\sin B$
- $\cos(A-B) = \cos A\cos B + \sin A\sin B$
- $\tan(A-B) = \dfrac{\tan A - \tan B}{1+\tan A\cdot\tan B}$

Example 3

If $A = 30°$ and $B = \text{Arccos}\,\dfrac{3}{5}$, what is the exact value of $\sec(A-B)$?

Solution: $\mathbf{\dfrac{10}{3\sqrt{3}+4}}$

To find $\sec(A-B)$, first calculate its reciprocal function, $\cos(A-B)$.

- Since $A = 30°$, $\sin A = \dfrac{1}{2}$ and $\cos A = \dfrac{\sqrt{3}}{2}$.
- If $B = \text{Arccos}\,\dfrac{3}{5}$, then $\angle B$ is in Quadrant I, where $\cos B = \dfrac{3}{5} = \dfrac{x}{r}$, and $x = 3$, $y = 4$, and $r = 5$. Therefore, $\sin B = \dfrac{y}{r} = \dfrac{4}{5}$.
- Use the identity for $\cos(A-B)$:

$$
\begin{aligned}
\cos(A-B) &= \cos A\cos B + \sin A\sin B \\
&= \left(\frac{\sqrt{3}}{2}\right)\left(\frac{3}{5}\right) + \left(\frac{1}{2}\right)\left(\frac{4}{5}\right) \\
&= \frac{3\sqrt{3}+4}{10}
\end{aligned}
$$

- Since $\cos(A-B)=\dfrac{3\sqrt{3}+4}{10}$, $\sec(A-B)=\dfrac{10}{3\sqrt{3}+4}$.

Example 4

If $x > 0$, show that $\tan(-x) = -\tan x$.

Solution:

Use the identity for $\tan(A - B)$, where $A = 0$ and $B = x$:

$$\begin{aligned}\tan(0-x) &= \frac{\tan 0 - \tan x}{1+\tan 0 \cdot \tan x} \\ &= \frac{0-\tan x}{1+0\cdot \tan x} \\ &= -\tan x\end{aligned}$$

Example 4 proves that $\tan(-x) = -\tan x$. Using a similar approach, you should verify that $\sin(-x) = -\sin x$ and $\cos(-x) = \cos x$.

MATH FACTS

- $\sin(A \pm B) = \sin A \cos B \pm \cos A \sin B$
- $\cos(A \pm B) = \cos A \cos B \mp \sin A \sin B$
- $\tan(A \pm B) = \dfrac{\tan A + \tan B}{1 \mp \tan A \cdot \tan B}$
- $\sin(-A) = -\sin A$
- $\cos(-A) = \cos A$
- $\tan(-A) = -\tan A$

Check Your Understanding of Section 14.3

A. Multiple Choice

1. The expression $\cos(270° + x)$ is equivalent to
(1) $\sin x$ (2) $-\sin x$ (3) $\cos x$ (4) $-\cos x$

2. The expression $\cos 80° \cos 20° - \sin 80° \sin 20°$ is equivalent to
(1) $\cos 60°$ (2) $\cos 100°$ (3) $\sin 100°$ (4) $\sin 60°$

3. If $A = \text{Arctan}\,\dfrac{2}{3}$ and $B = \text{Arctan}\,\dfrac{1}{2}$, what is the value of $\tan(A + B)$?
(1) $\dfrac{1}{8}$ (2) $\dfrac{7}{8}$ (3) $\dfrac{1}{4}$ (4) $\dfrac{7}{4}$

4. The expression sin $(270° - A)$ is equivalent to
(1) $\cos A$ (2) $-\cos A$ (3) $\sin A$ (4) $-\sin A$

5. The expression $\dfrac{\sin(90° + x)}{\sin(-x)}$ is equivalent to
(1) -1 (2) 1 (3) $-\cot x$ (4) $\cot x$

6. If $\tan\theta = a$, the expression $\tan(45° - \theta)$ is equivalent to
(1) $\dfrac{1-a}{1+a}$ (2) $\dfrac{a-1}{a+1}$ (3) $\dfrac{1}{a}$ (4) a

7. If $\cos x = \dfrac{12}{13}$ and $\sin y = \dfrac{4}{5}$, and x and y are acute angles, what is the value of $\sin(x - y)$?
(1) $\dfrac{72}{65}$ (2) $\dfrac{56}{65}$ (3) $-\dfrac{16}{65}$ (4) $-\dfrac{33}{65}$

8. In $\triangle ABC$, $\sin(A + B) = \dfrac{3}{5}$. What is the value of $\sin C$?
(1) $\dfrac{2}{5}$ (2) $\dfrac{2}{3}$ (3) $\dfrac{3}{5}$ (4) $\dfrac{11}{18}$

9. If $\sin A = \dfrac{3}{5}$ and $\sin B = \dfrac{2}{3}$, and A and B are acute angles, what is the value of $\cos(A - B)$?
(1) $-\dfrac{2}{3}$ (2) $\dfrac{4\sqrt{5}-6}{15}$ (3) $\dfrac{4\sqrt{5}+2}{5}$ (4) $\dfrac{4\sqrt{5}+6}{15}$

B. *In each case, show how you arrived at your answer by clearly indicating all of the necessary steps, formula substitutions, diagrams, graphs, charts, etc.*

10. Angle x is acute, angle y is obtuse, $\cos x = \dfrac{15}{17}$, and $\sin y = \dfrac{3}{5}$. What is the exact value of $\sin(x + y)$?

11. When $\sin x = -\dfrac{8}{17}$ and x terminates in Quadrant III, and $\cos y = -\dfrac{4}{5}$ and y terminates in Quadrant II, what is the exact value of $\cos(x - y)$?

12. If $\text{Arcsin}(-0.6) = x$ and $\text{Arccos}\ 0.5 = y$, what is the exact value of $\cot(x - y)$?

13. If $A = 45°$ and $B = \text{Arcsin}\,\frac{3}{5}$, what is the exact value of csc $(A - B)$?

14. Find all values of θ in the interval $0° \leq x < 360°$ that satisfy the equation

$\sin\theta\cos\theta + \frac{1}{2}\cos(90° - \theta) = 0.$

15. Given $\tan A = 8$ and $\tan(A + B) = -\frac{17}{6}$.

a. What is the exact value of tan B?
b. What is the exact value of $\sec^2 A + \sec^2 B$?

14.4 DOUBLE-ANGLE IDENTITIES

KEY IDEAS

Because the graphs of $y = \sin 2x$ and $y = 2\sin x$ do not coincide, we know that $\sin 2x = 2\sin x$ is *not* an identity. For example, $\sin 2(30°) \neq 2\sin 30°$. There are identities, however, that allow a trigonometric function of a double angle to be expressed in terms of a trigonometric function of a single angle.

Functions of the Double Angle

The double-angle trigonometric identities can be derived from the identities for $\sin(A + B)$, $\cos(A + B)$, and $\tan(A + B)$ by letting $B = A$.

$\sin 2A = 2\sin A\cos A$	$\cos 2A = \cos^2 A - \sin^2 A$ or $\cos 2A = 2\cos^2 A - 1$ or $\cos 2A = 1 - 2\sin^2 A$	$\tan 2A = \frac{2\tan A}{1 - \tan^2 A}$

Example 1

Find the value of sin $2A$ if $\sin A = \frac{3}{5}$ and $\angle A$ is obtuse.

Solution: $-\frac{\mathbf{24}}{\mathbf{25}}$

- Since $\sin A = \frac{3}{5} = \frac{y}{r}$ and $\angle A$ is obtuse:

$$\cos A = \frac{x}{r} = -\frac{4}{5}.$$

- Then: $\sin 2A = 2 \sin A \cos A$

$$= 2\left(\frac{3}{5}\right)\left(\frac{-4}{5}\right)$$

$$= -\frac{24}{25}$$

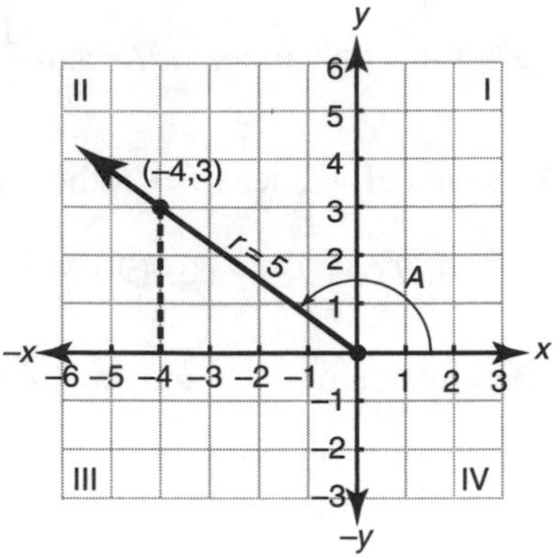

Example 2

If $\sin x = \frac{3}{4}$, what is the value of $\cos 2x$?

Solution: $\mathbf{-\frac{1}{8}}$

Since the value of $\sin x$ is given, choose the form of the identity for $\cos 2x$ that involves only sine:

$$\cos 2x = 1 - 2\sin^2 x$$

$$= 1 - 2\left(\frac{3}{4}\right)^2$$

$$= -\frac{1}{8}$$

Example 3

Find all values of x in the interval $0° \leq x < 360°$ that satisfy the equation $\cos 2x - \cos x = 0$.

Solution: $\mathbf{x = 0°, 120°, 240°}$

- Transform the given equation into an equation that does not contain a double angle. Since the equation also contains $\cos x$, choose the form of the identity for $\cos 2x$ that is expressed only in terms of $\cos x$:

$$\cos 2x - \cos x = 0$$

- Replace $\cos 2x$ with $2\cos^2 x - 1$: $(2\cos^2 x - 1) - \cos x = 0$

- Factor: $(2\cos x + 1)(\cos x - 1) = 0$

- If $2\cos x + 1 = 0$, then $\cos x = -\frac{1}{2}$. The reference angle is 60°. Since cosine is negative in Quadrants II and III:

$$Q_{II}: x_1 = 120° \quad \text{or} \quad Q_{III}: x_2 = 240°.$$

- If $\cos x - 1 = 0$, then $\cos x = 1$, so $x_3 = 0°$.

Example 4

Express $\frac{1-\cos 2A}{\sin 2A}$ as a *single* trigonometric function of $\angle A$ for all values of A for which the fraction is defined.

Solution: **tan *A***

Eliminate 1 in the numerator by substituting $1 - 2\sin^2 A$ for $\cos 2A$, and substitute $2\sin A\cos A$ for $\sin 2A$ in the denominator:

$$\frac{1-\cos 2A}{\sin 2A} = \frac{1-(1-2\sin^2 A)}{2\sin A\cos A} = \frac{\cancel{2\sin^2} A}{\cancel{2\sin} A\cos A}$$

$$= \frac{\sin A}{\cos A}$$

$$= \tan A$$

Check Your Understanding of Section 14.4

A. Multiple Choice

1. The expression $\frac{\cos 2x}{\cos x + \sin x}$ is equivalent to

(1) $\cos x - \sin x$ (2) $\sin x - \cos x$ (3) $1 - \sin x$ (4) $1 - \cos x$

2. If $\tan A = \frac{1}{2}$, what is the value of $\tan 2A$?

(1) 1 (2) $\frac{1}{4}$ (3) $\frac{3}{4}$ (4) $\frac{4}{3}$

3. The expression $\frac{\sin 2A}{2\tan A}$ is equivalent to

(1) $\cot^2 A$ (2) $\sin^2 A$ (3) $\cos^2 A$ (4) $\sec^2 A$

4. The expression $(\sin x + \cos x)^2 - 1$ is equivalent to

(1) $\sin 2x$ (2) $\cos 2x$ (3) $\tan^2 x$ (4) 0

5. If x is an acute angle, which statement is *not* always true?
(1) $\log(\sin^2 x + \cos^2 x) = 0$
(2) $\log \cos 2x = 2(\log \cos x - \log \sin x)$
(3) $\log \sin 2x = \log \sin x + \log \cos x + \log 2$
(4) $\log(\tan^2 x + 1) + 2 \log \cos x = 0$

***B.** In each case, show how you arrived at your answer by clearly indicating all of the necessary steps, formula substitutions, diagrams, graphs, charts, etc.*

6. Express $\dfrac{\sin^2 A}{\cos 2A + \sin^2 A}$ as a single trigonometric function for all values of A for which the fraction is defined.

7. What are the possible values of θ if $\sin 2\theta - \sin \theta = 0$ and $0° \le \theta < 360°$?

8–10. Find to the *nearest tenth of a degree* all nonnegative values of θ less than 360° that satisfy the equation.

8. $2 \sin 2x + \cos x = 0$ **9.** $2 \cos 2\theta + \cos \theta = 1$ **10.** $\sin \theta + 3 \cos 2\theta = 1$

14.5 HALF-ANGLE IDENTITIES

KEY IDEAS

Although $\sin \frac{1}{2} x \neq \frac{1}{2} \sin x$, there are identities that allow a trigonometric function of a half angle to be expressed in terms of trigonometric function of a single angle.

Functions of the Half Angle

The half-angle trigonometric identities can be derived from the identities for $\sin 2A$, $\cos 2A$, and $\tan 2A$ by letting $2A = x$, $A = \frac{1}{2}x$, and then solving for the trigonometric function of $\angle \frac{1}{2}x$ to obtain these formulas:

$\sin \frac{1}{2}x = \pm\sqrt{\dfrac{1-\cos x}{2}}$	$\cos \frac{1}{2}x = \pm\sqrt{\dfrac{1+\cos x}{2}}$	$\tan \frac{1}{2}x = \pm\sqrt{\dfrac{1-\cos x}{1+\cos x}}$

The choice of a positive or negative sign in front of each radical depends on the sign of the trigonometric function in the quadrant in which $\angle\frac{1}{2}x$ lies.

Example 1

If $\cos x = \frac{7}{8}$ and $\frac{3\pi}{2} \le x < 2\pi$, find the value of: a. $\sin\frac{1}{2}x$ b. $\cos\frac{1}{2}x$

Solution: a. $\mathbf{\frac{1}{4}}$

- It is given that $\frac{3\pi}{2} \le x < 2\pi$ or, equivalently, $270° \le x < 360°$.
- To determine the quadrant in which $\frac{1}{2}x$ lies, divide each member of the inequality by 2. The result is $135° \le \frac{1}{2}x < 180°$.
- Since $\angle\frac{1}{2}x$ lies in Quadrant II, where sine is positive, use the positive value of the radical in the identity for $\sin\angle\frac{1}{2}x$:

$$\sin\frac{1}{2}x = +\sqrt{\frac{1-\cos x}{2}}$$

- Let $\cos x = \frac{7}{8}$:

$$= \sqrt{\frac{1-\left(\frac{7}{8}\right)}{2}} = \sqrt{\frac{1}{16}} = \frac{1}{4}$$

b. $\mathbf{-\frac{\sqrt{15}}{4}}$

Since $\angle\frac{1}{2}x$ lies in Quadrant II, where cosine is negative, use the negative value of the radical in the identity for $\cos\angle\frac{1}{2}x$:

$$\cos\frac{1}{2}x = -\sqrt{\frac{1+\cos x}{2}} = -\sqrt{\frac{1+\frac{7}{8}}{2}} = -\sqrt{\frac{15}{16}} = -\frac{\sqrt{15}}{4}$$

Example 2

If $\angle A$ is obtuse and $\sin A = \frac{\sqrt{5}}{3}$, what is the exact value of $\cos \frac{1}{2}A$?

Solution: $\mathbf{\frac{1}{\sqrt{6}}}$

From Example 2 on page 366, $\cos A = -\frac{2}{3}$. Since $\angle A$ is obtuse, $\frac{1}{2}A$ is a Quadrant I angle. Hence:

$$\cos \frac{1}{2}A = \sqrt{\frac{1+\cos A}{2}} = \sqrt{\frac{1-\frac{2}{3}}{2}} = \frac{1}{\sqrt{6}}.$$

Check Your Understanding of Section 14.5

A. *Multiple Choice*

1. If $270° < x < 360°$ and $\cos x = 0.28$, what is the value of $\sin \frac{1}{2}x$?

(1) 0.56 (2) 0.6 (3) –0.56 (4) –0.6

2. If $\cos \frac{1}{2}x = \frac{\sqrt{5}}{4}$, what is the value of $\cos^2 x$?

(1) $\frac{1}{64}$ (2) $\frac{1}{4}$ (3) $\frac{9}{64}$ (4) $\frac{16}{25}$

3. If $1 - \cos y = 0.2$ and $\angle y$ is acute, what is the value of $\tan \frac{1}{2}y$?

(1) $\frac{1}{3}$ (2) $\frac{\sqrt{3}}{3}$ (3) $\frac{2}{3}$ (4) $\frac{1}{2}$

4. If $1 + \cos A = 0.38$ and $\angle A$ is obtuse, what is the value of $\sin \frac{1}{2}A$?

(1) 0.44 (2) 0.1 (3) $\frac{\sqrt{19}}{10}$ (4) 0.9

B. *In each case, show how you arrived at your answer by clearly indicating all of the necessary steps, formula substitutions, diagrams, graphs, charts, etc.*

5. If $\tan A = \frac{24}{7}$ and $180° < A < 270°$, determine the exact value of $\cos \frac{1}{2}A$.

6. If $\sin A = \frac{\sqrt{15}}{8}$ and $90° < A < 180°$, determine the exact value of $\cot \frac{1}{2}A$.

7. If $\sin A = -\frac{9\sqrt{19}}{50}$ and $180° < A < 270°$, determine the exact value of $\csc \frac{1}{2}A$.

8. If $\tan \frac{1}{2}\theta = \frac{2}{5}$ and $\angle\theta$ is acute, determine the exact value of $\sin\theta$.

CHAPTER 15
LAWS OF SINES AND COSINES

15.1 AREA OF A TRIANGLE

KEY IDEAS

The area of a triangle can be expressed in terms of the lengths of two sides and the sine of the included angle.

SAS Formula for the Area of a Triangle

In Figure 15.1, base = $AB = c$ and height = h, where $A = \frac{h}{b}$, so $h = b \times \sin A$. Hence, the area of $\triangle ABC = \frac{1}{2}(AB)(h) = \frac{1}{2}bc \sin A$. Thus, the area of a triangle is equal to one-half the product of the lengths of *any* two sides and the sine of the included angle.

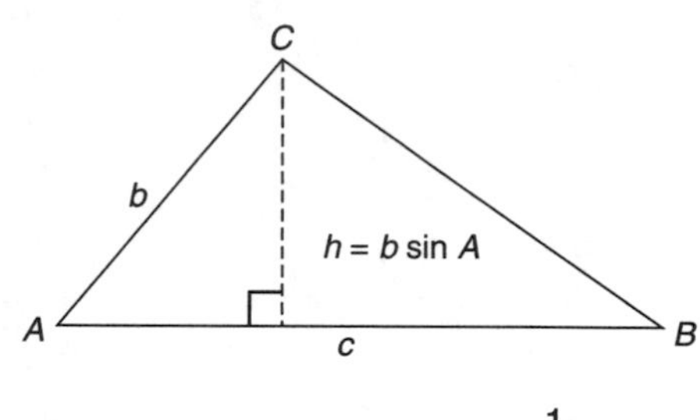

Figure 15.1 Area of $\triangle ABC = \frac{1}{2}bc \sin A$

Example 1

If m$\angle A = 150$, $b = 8$ centimeters, and $c = 10$ centimeters, find the area of $\triangle ABC$.

Solution: **20 cm²**

$$\begin{aligned} \text{Area } \Delta ABC &= \frac{1}{2}bc \sin A \\ &= \frac{1}{2}(8)(10)\sin 150 \\ &= 40(0.5) \\ &= 20 \end{aligned}$$

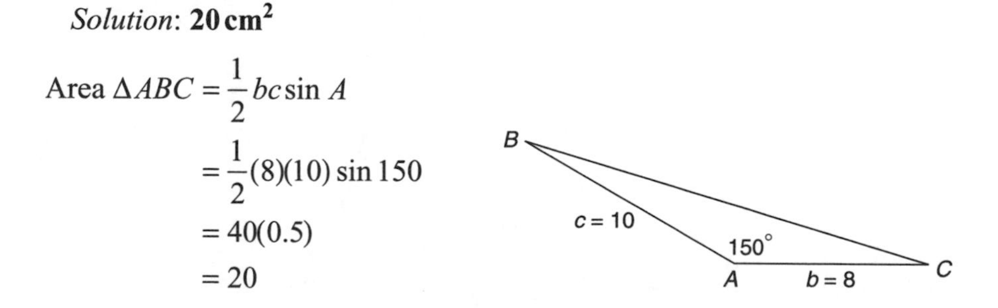

Example 2

In parallelogram $ABCD$, $AB = 20$, $AD = 10$, and $m\angle A = 45$. What is the exact area of parallelogram $ABCD$?

Solution: $\mathbf{100\sqrt{2}}$ **square units**

- Draw diagonal $\overline{DB}$, as shown in the accompanying diagram. Then find the area of $\triangle DAB$:

$$\text{Area } \triangle DAB = \frac{1}{2}(AD)(AB)\sin 45^\circ$$
$$= \frac{1}{2}(10)(20)\frac{\sqrt{2}}{2}$$
$$= 50\sqrt{2}$$

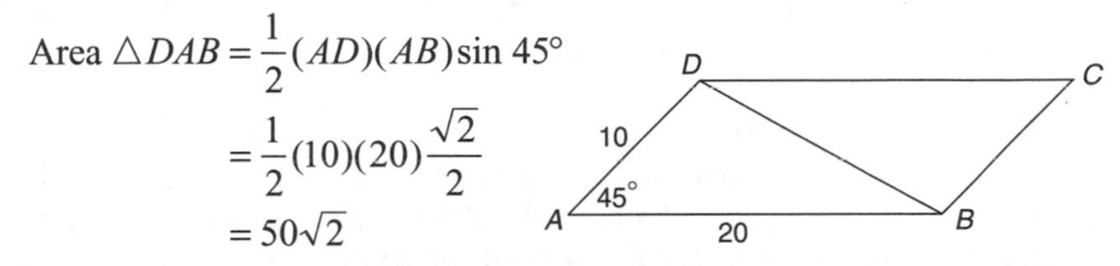

- Since a diagonal of a parallelogram divides the parallelogram into two congruent triangles, area $\Delta DBC = 50\sqrt{2}$.
- Hence, area of parallelogram $ABCD = 50\sqrt{2} + 50\sqrt{2} = 100\sqrt{2}$ square units.

Check Your Understanding of Section 15.1

A. Multiple Choice

1. In $\triangle ABC$, $a = 8$, $b = 9$, and $m\angle C = 135$. What is the area, in square units, of $\triangle ABC$?
(1) 18 (2) 36 (3) $18\sqrt{2}$ (4) $36\sqrt{2}$

2. An angle of a parallelogram has a measure of 150°. If the sides of the parallelogram measure 10 and 12 centimeters, what is the area of the parallelogram?
(1) $30\,\text{cm}^2$ (2) $60\,\text{cm}^2$ (3) $60\sqrt{2}\,\text{cm}^2$ (4) $60\sqrt{3}\,\text{cm}^2$

3. In $\triangle ABC$, side a is twice as long as side b and $m\angle C = 30$. In terms of b, the area, in square units, of $\triangle ABC$ is
(1) $0.25b^2$ (2) $0.5b^2$ (3) $0.866b^2$ (4) b^2

4. If the area of equilateral triangle ABC is $36\sqrt{3}$ square units, what is the length of a side?
(1) 12 (2) 9 (3) $6\sqrt{3}$ (4) $12\sqrt{3}$

5. In isosceles triangle ABC, $\overline{AB} \cong \overline{BC}$ $m\angle B = 45$, and $AB = 3\sqrt{2}$. The area of the triangle is
(1) $\frac{9}{2}$ (2) $9\sqrt{2}$ (3) $\frac{9\sqrt{2}}{2}$ (4) 9

B. *In each case, show how you arrived at your answer by clearly indicating all of the necessary steps, formula substitutions, diagrams, graphs, charts, etc.*

6. The coordinates of the vertices of $\triangle ABC$ are $A(-6, 8)$, $B(8, 0)$, and $C(0, 0)$. If the area of $\triangle ABC$ is 20, find the number of degrees in the measure of $\angle C$.

7. In isosceles triangle ABC, $m\angle C = 30$ and $BC = 10$. What is the least possible area, in square units, of $\triangle ABC$?

8. Gregory wants to build a garden in the shape of an isosceles triangle with one of the congruent sides equal to 12 yards. If the area of the garden will be 55 square yards, find, to the *nearest tenth of a degree*, the measures of the three acute angles of the triangle.

9. A forest preserve has the shape of quadrilateral $ABCD$, shown in the accompanying figure, where $AB = 3.6$ kilometers, $AD = 4.8$ kilometers, $m\angle DAB = 90$, $m\angle DBC = 50$, and $BC = 2.8$ kilometers. Find to the *nearest tenth* the number of square kilometers in the area of the forest preserve.

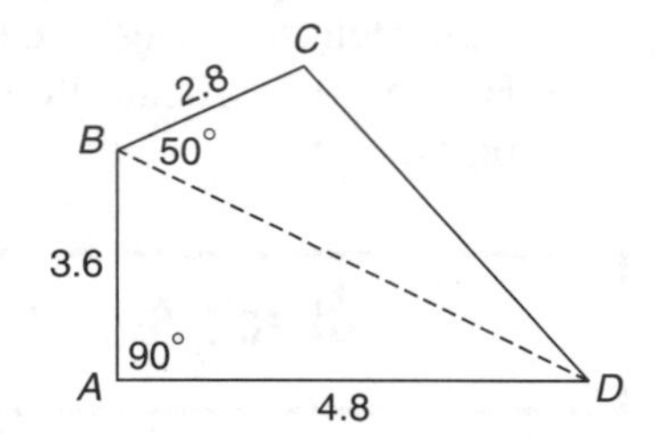

10. In the accompanying diagram, $\overline{PB}$ is tangent to circle O at B, chord $\overline{AB}$ intersects secant $\overline{PCD}$ at E, A is the midpoint of $\overset{\frown}{CD}$. If $PB = 20$ cm and $m\overset{\frown}{AC} : m\overset{\frown}{CB} : m\overset{\frown}{BD} = 3:10:14$, find to the *nearest tenth* the number of square centimeters in the area of $\triangle PEB$.

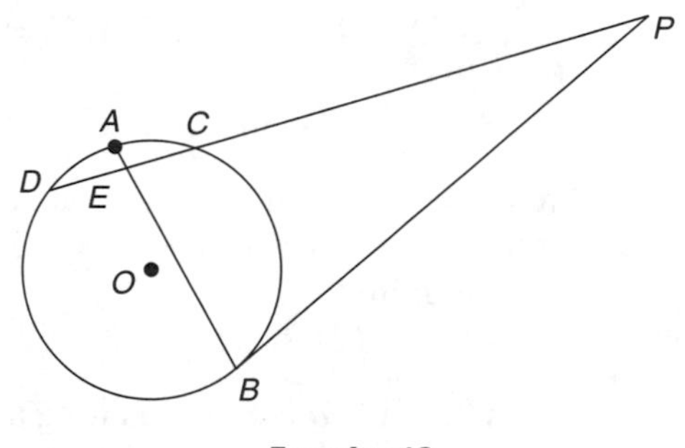

Exercise 10

15.2 THE LAW OF SINES

KEY IDEAS

The Law of Sines is a proportion that relates the lengths of any two sides of a triangle to the sines of the angles opposite these sides.

Deriving the Law of Sines

The Law of Sines can be obtained using the formula for the area of a triangle. Since area of $\triangle ABC = \frac{1}{2}ac \sin B$ and area of $\triangle ABC = \frac{1}{2}bc \sin A$:

$$\cancel{\tfrac{1}{2}}\, a \cancel{c} \sin B = \cancel{\tfrac{1}{2}}\, b \cancel{c} \sin A$$

$$a \sin B = b \sin A$$

$$\frac{a}{\sin A} = \frac{b}{\sin B}$$

Similarly, it can be shown that $\frac{a}{\sin A} = \frac{c}{\sin C}$. These proportions are collectively referred to as the **Law of Sines**.

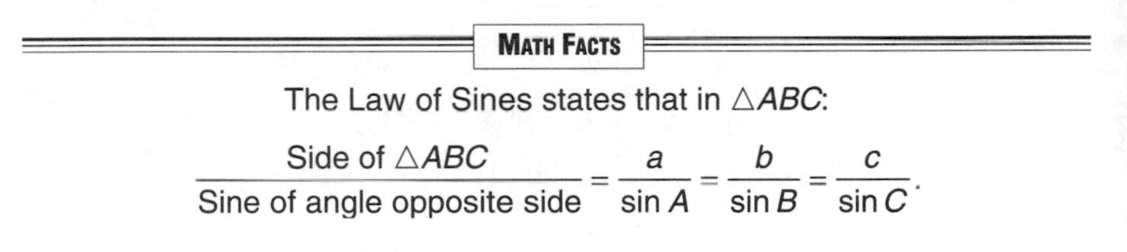

MATH FACTS

The Law of Sines states that in $\triangle ABC$:

$$\frac{\text{Side of } \triangle ABC}{\text{Sine of angle opposite side}} = \frac{a}{\sin A} = \frac{b}{\sin B} = \frac{c}{\sin C}.$$

Using the Law of Sines Given AAS

If *A*ngle-*A*ngle-*S*ide measurements of a triangle are given, the Law of Sines can be used to find the length of a side that is not given.

Example 1

In $\triangle ABC$, $a = 12$, $\sin A = 0.6$, and $\sin B = 0.4$. What is the length of side b?

Solution: **8**

Since AAS is given, solve for b $(= AC)$ by using the Law of Sines:

$$\frac{a}{\sin A} = \frac{b}{\sin B}$$

$$\frac{12}{0.6} = \frac{b}{0.4}$$

$$0.6b = 0.4(12)$$

$$b = \frac{4.8}{0.6} = 8$$

Before using the Law of Sines, you may need to find the measure of the third angle of the triangle.

Example 2

In $\triangle ABC$, m$\angle A$ = 59.0, m$\angle B$ = 74.0, and b = 100.0 meters. Find c to the *nearest tenth of a meter*.

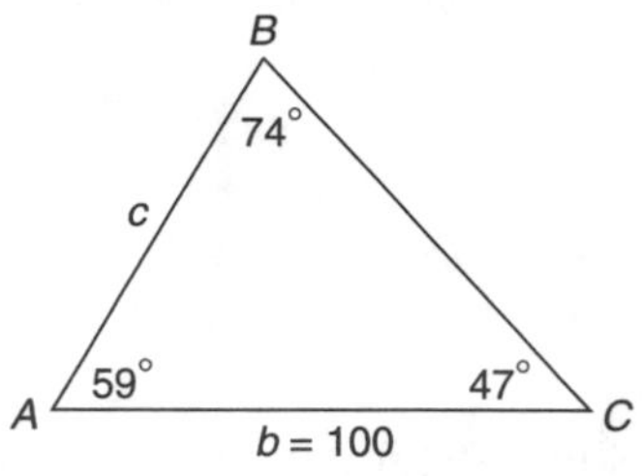

Solution: **76.1**

Since AAS is given, first find the measure of the third angle of the triangle:

$m\angle C = 180 - (59 + 74) = 47,$

as shown in the accompanying diagram.

Find c using the Law of Sines:

$$\frac{b}{\sin B} = \frac{c}{\sin C}$$
$$\frac{100}{\sin 74^\circ} = \frac{c}{\sin 47^\circ}$$
$$c \sin 74^\circ = 100 \sin 47^\circ$$
$$c = \frac{100 \sin 47^\circ}{\sin 74^\circ} \approx 76.1$$

TIP

When calculating c using the proportion in Example 2, first solve for c without making any substitutions or performing intermediate calculations. Then perform all the required calculator operations at the same time. Store intermediate results in memory until the final answer is displayed. Then round the answer to the required number of digits.

Using the Law of Sines Given SSA

If *S*ide-*S*ide-*A*ngle measurements of a triangle are given, the Law of Sines can be used to find the measure of an angle that is not given.

Example 3

In $\triangle JKL$, $j = 88.2$, $k = 100$, $\angle J$ measures $26^\circ 10'$, and $\angle K$ is acute, as shown in the accompanying diagram. Find the measure of $\angle L$ correct to the *nearest 10 minutes* or *hundredth of a degree*.

Solution: **123°50′** or **123.83**

Since SSA is given, first find m∠*K* by using the Law of Sines. Then use m∠*J* and m∠*J* to find m∠*L*.

$$\frac{j}{\sin J} = \frac{k}{\sin K}$$

$$\frac{88.2}{\sin 26°10'} = \frac{100}{\sin K}$$

$$88.2(\sin K) = 100(\sin 26.17°)$$

$$\sin K = \frac{100 \sin 26.17°}{88.2} \approx 0.5$$

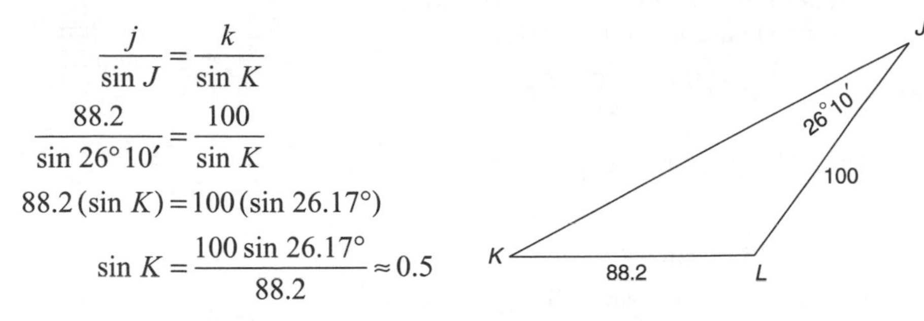

If $\sin K = 0.5$, then ∠*K* measures 30° or 150°. It is given that ∠*K* is acute, so ∠*K* measures 30°. Since the sum of the degree measures of the three angles of a triangle is 180:

$$\begin{aligned} \angle L &= 180° - (26°10' + 30°) \\ &= 180° - 56°10' \\ &= 179°60' - 56°10' \\ &= 123°50' \\ &= \left(123 + \frac{50}{60}\right)^{\circ} = 123.83° \end{aligned}$$

Solving "Double-Triangle" Problems

Two triangles may overlap so that they share an angle or a side. To use a trigonometric relationship in one triangle, it may be necessary to first find the measure of a shared angle or side by working in the other triangle.

Example 4

The angle of elevation from a ship at point *A* to the top of a lighthouse at point *B* is 43°. When the ship reaches point *C*, 300 meters closer to the lighthouse, the angle of elevation is 56°. Find, to the *nearest meter*, the height of the lighthouse.

Solution: **754**

Reason backwards: in the accompanying diagram of right triangle *BDC*, if the length of hypotenuse $\overline{BC}$ were known, *BD*, the height of the lighthouse, could be determined by using the sine ratio. To solve for *BC*, use the Law of Sines in $\triangle ABC$.

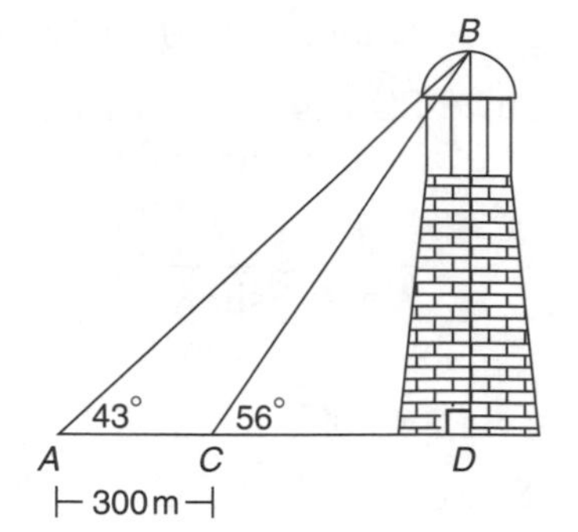

- Before the Law of Sines can be used, you need to find $\angle ABC$. Since the measure of an exterior angle of a triangle is equal to the sum of the measures of the two nonadjacent interior angles:

$$\angle ABC + 43° = 56°$$
$$\angle ABC = 13°$$

- In $\triangle ABC$, use the Law of Sines:

$$\frac{300}{\sin 13°} = \frac{BC}{\sin 43°}$$
$$BC(\sin 13°) = 300(\sin 43°)$$
$$BC = \frac{300 \sin 43°}{\sin 13°} \approx 909.33$$

- In $\triangle BDC$, calculate BD using the sine ratio:

$$\sin 56° = \frac{BD}{909.33}$$
$$BD = 909.33 \sin 56° \approx 754$$

Solving Triangles That Intersect Circles

It may be necessary to use circle measurement relationships before the Law of Sines can be applied.

Example 5

In the accompanying diagram, $\overline{AB}$ is tangent to circle O at B, $\overline{AB}$ and $\overline{ADC}$ intersect at A, and chord $\overline{BC} \cong$ chord $\overline{CD}$. If m$\overset{\frown}{BD} = 60$ and $BC = 8$, find the length of tangent segment $\overline{AB}$.

Solution: $\frac{8}{\sqrt{2}}$

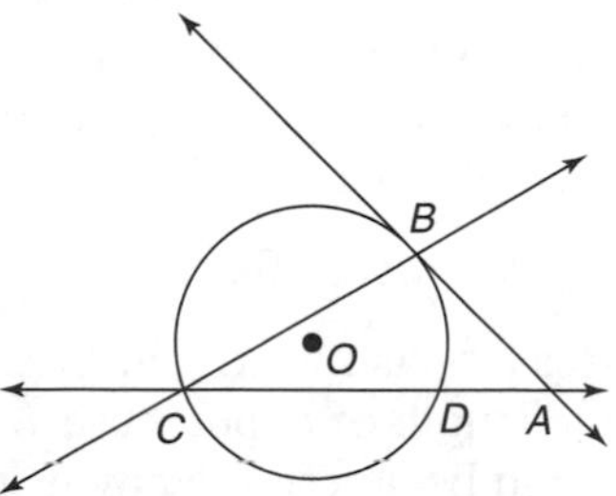

- $BC = 8$ and $\text{m}\angle C = \frac{1}{2}\text{m}\overset{\frown}{BD} = \frac{1}{2}(60) = 30$.
- In a circle, congruent chords intercept congruent arcs. If $\text{m}\overset{\frown}{BC} = \text{m}\overset{\frown}{CD} = x$, then $x + x + 60 = 360$, so $x = 150$. Since $\angle A$ is formed by a tangent and a secant, it is equal in measure to one-half the difference of the measures of its intercepted arcs. Hence:

$$
\begin{aligned}
m\angle A &= \frac{1}{2}\left(m\widehat{BC} - m\widehat{BD}\right)\\
&= \frac{1}{2}(150 \quad - 60)\\
&= 45
\end{aligned}
$$

- Apply the Law of Sines:

$$
\begin{aligned}
\frac{BC}{\sin A} &= \frac{AB}{\sin C}\\
\frac{8}{\sin 45^\circ} &= \frac{AB}{\sin 30^\circ}\\
AB(\sin 45^\circ) &= 8(\sin 30^\circ)\\
AB\left(\frac{\sqrt{2}}{2}\right) &= 8(0.5)\\
AB &= \frac{8}{\sqrt{2}}
\end{aligned}
$$

Check Your Understanding of Section 15.2

A. *Multiple Choice*

1. In $\triangle ABC$, $\sin A = \frac{1}{2}$ and $\sin B = \frac{1}{2}\sqrt{2}$. The value of $\frac{b}{a}$ is

(1) $\frac{1}{2}$ (2) 2 (3) $\sqrt{2}$ (4) $\frac{1}{2}\sqrt{2}$

2. In $\triangle ABC$, $m\angle A = 75$, $m\angle B = 40$, and $b = 35$. What is the length of side c?

(1) $\frac{35 \sin 40^\circ}{\sin 65^\circ}$ (2) $\frac{35 \sin 75^\circ}{\sin 40^\circ}$ (3) $\frac{35 \sin 40^\circ}{\sin 75^\circ}$ (4) $\frac{35 \sin 65^\circ}{\sin 40^\circ}$

3. If $a = 4$, $b = 6$, and $\sin A = \frac{3}{5}$ in $\triangle ABC$, then $\sin B$ equals

(1) $\frac{3}{20}$ (2) $\frac{6}{10}$ (3) $\frac{8}{10}$ (4) $\frac{9}{10}$

B. *In each case, show how you arrived at your answer by clearly indicating all of the necessary steps, formula substitutions, diagrams, graphs, charts, etc.*

4. In $\triangle ABC$, $m\angle A = 65$, $m\angle B = 70$, and the length of the side opposite vertex B is 7.
  a. Find the length of the side opposite vertex A.
  b. Find the area, in square units, of $\triangle ABC$.

5. In parallelogram $ABCD$, $AD = 11$, diagonal $AC = 15$, and $m\angle BAD = 63.8$.
  a. Find $m\angle ACD$ to the *nearest tenth of a degree.*
  b. Find, *to the nearest tenth of a square unit*, the area of parallelogram $ABCD$.

6. A 54-foot entrance ramp makes an angle of 4.3° with the level ground. To comply with the most recent wheelchair-accessibility guidelines, the ramp must be extended, as indicated in the accompanying diagram, so that it makes an angle of, at most, 3° with the level ground. What is the *minimum* distance from the point on the ground at which the incline of the old ramp begins to the point on the ground where the incline of the new ramp must begin? Approximate your answer correct to the *nearest inch*.

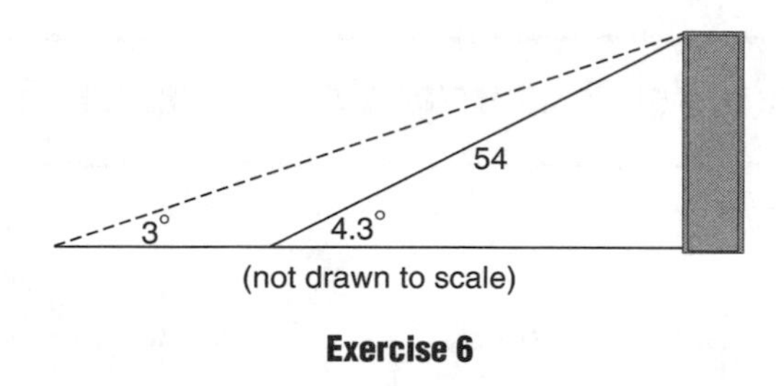

Exercise 6

7. In the accompanying diagram, an airplane traveling at a level altitude of 2050 feet sights the top of a 50-foot tower at an angle of depression of 28° from point A. After the airplane continues in level flight to point B, the angle of depression to the same tower is 34°. Find, to the *nearest foot*, the distance that the plane has traveled from point A to point B.

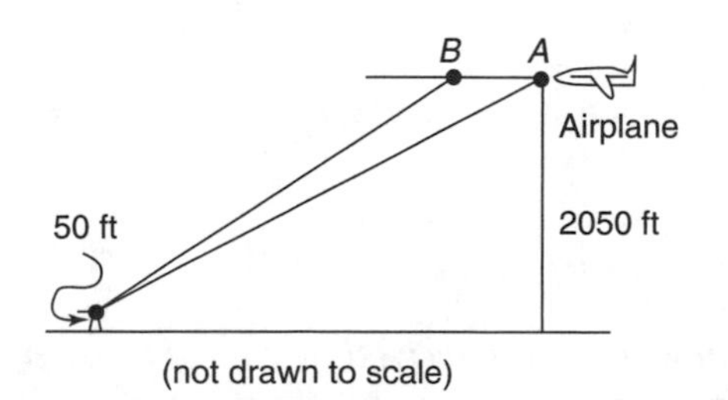

Exercise 7

8. In the accompanying diagram of circle O, $\triangle ABC$ is formed by tangent $\overline{AB}$, secant $\overline{BDC}$, and chord $\overline{AC}$; $\overline{CA} \cong \overline{CD}$; $m\overset{\frown}{AC} = 140$; and $AC = 10$. Find the measures of:

a. $\overset{\frown}{AD}$ b. $\angle B$ c. AB to the *nearest tenth*

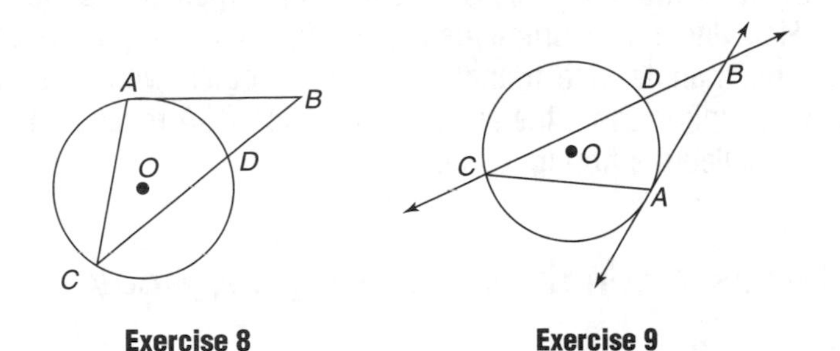

Exercise 8 **Exercise 9**

9. In the accompanying diagram of circle O, $\overleftrightarrow{CDB}$ is a secant and $\overleftrightarrow{AB}$ is tangent at A. If $BD = 12$, $CD = 15$, and $m\overset{\frown}{AD} = 60$, determine:

a. the length of $\overline{AB}$

b. the measure of obtuse angle BAC correct to the *nearest tenth of a degree*

10. Engineers are designing a straight tunnel through a hill from point A to point B, which are on opposite sides of the hill but at the same level. Point C is chosen on the top of the hill in such a way that points A, B, and C lie in the same vertical plane. From points A and B, the angles of elevation to C are $26°40'$ and $38°10'$, respectively. If the distance from A to C is 400 meters, find the length of tunnel $\overline{AB}$ to the *nearest tenth of a meter*.

11. To determine the distance across a river, a surveyor marked three points H, G, and F, along a riverbank, as shown in the accompanying diagram. She also marked one point, K, on the opposite bank in such a way that $\overline{KH} \perp \overline{HGF}$, $m\angle KGH = 41$, and $m\angle KFH = 37$. The distance between G and F is 45 meters. Find KH, the width of the river, to the *nearest tenth of a meter*.

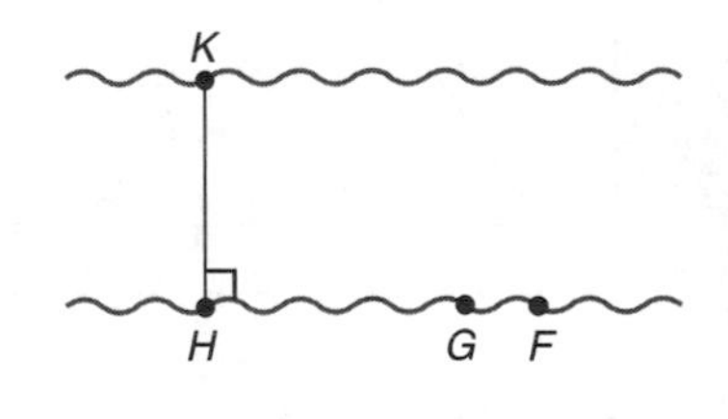

15.3 THE AMBIGUOUS CASE

KEY IDEAS

Two triangles are not necessarily congruent when $SSA \cong SSA$. When *S*ide-*S*ide-*A*ngle measurements are given, it may be possible to construct no triangle, one triangle, or two triangles with these dimensions. For this reason, the situation in which SSA measurements are given is called the **Ambiguous Case**.

Four Cases When the Given Angle is Acute

When SSA measurements are known and the given angle is acute, four situations are possible. See Figure 15.2.

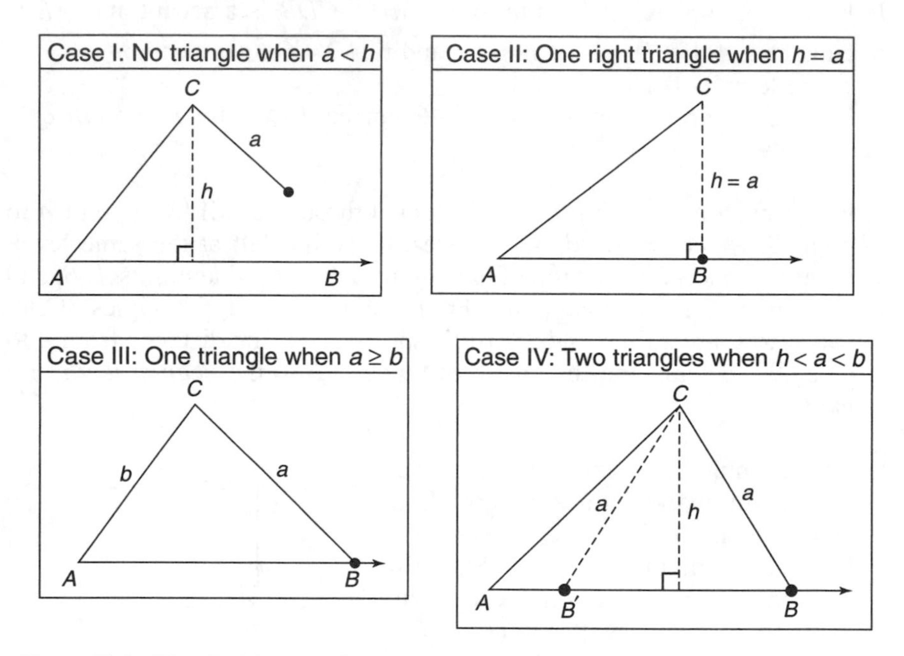

Figure 15.2 The Ambiguous Case Where $\angle A$ is Acute and $h = b\sin A$. In Case III, $\angle C$ can be acute, right, or obtuse

Counting Triangles Given SSA

When *S*ide-*S*ide-*A*cute *A*ngle measurements are given, you can figure out the possible number of triangles, if any, that have these measurements. Suppose $a = 7$, $b = 10$, and $m\angle A = 37$. See Figure 15.3.

- <u>Method 1: Compare side a with height h.</u>
  Find the height, h, of a possible triangle:

$$h = b \sin A$$
$$= 10 \sin 37^\circ$$
$$\approx 6.02$$

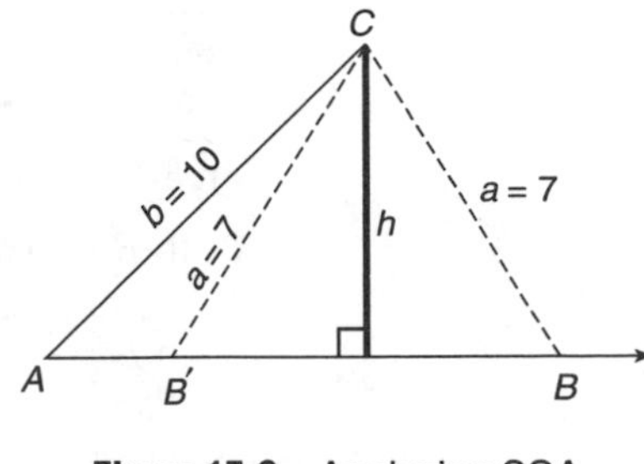

Figure 15.3 Analyzing SSA Measurements

  Compare a, h, and b. Since $a = 7$, $h = 6.02$, and $b = 10$, then $h < a < b$. Hence, *two* triangles are possible: $\triangle ABC$ and $\triangle AB'C$.
- <u>Method 2: Use the Law of Sines.</u>
  Assume a triangle is possible. Use the Law of Sines to find m$\angle B$. If $\dfrac{7}{\sin 37^\circ} = \dfrac{10}{\sin B'}$ then

$$7 \sin B = 10 \sin 37^\circ$$
$$\sin B = \frac{10 \sin 37^\circ}{7} \approx 0.8597$$

  and $\angle B = \sin^{-1} 0.8597 \approx 59^\circ$ or, if $\angle B$ is obtuse, $\angle B = 180^\circ - 59^\circ = 121^\circ$.
  Check whether it is possible for $\angle B$ to be obtuse. Since m$\angle A$ + m$\angle B$ = 37 + 121 = 158 and 158 is less than 180, an obtuse triangle is possible in which m$\angle C$ = 180 − 158 = 22. Hence, two triangles are possible.

Example 1

Find the maximum number of triangles that can be constructed if $a = 2$, $b = 5$, and $\angle A = 30^\circ$.

Solution: **No triangle**

- <u>Method 1: Compare side a with altitude h.</u>
  The *shortest* distance from C to the base of a possible triangle is $h = 5 \sin 30^\circ = 2.5$, as shown in the accompanying diagram. Thus, a must be *greater than* 2.5. But this contradicts the given statement that $a = 2$.

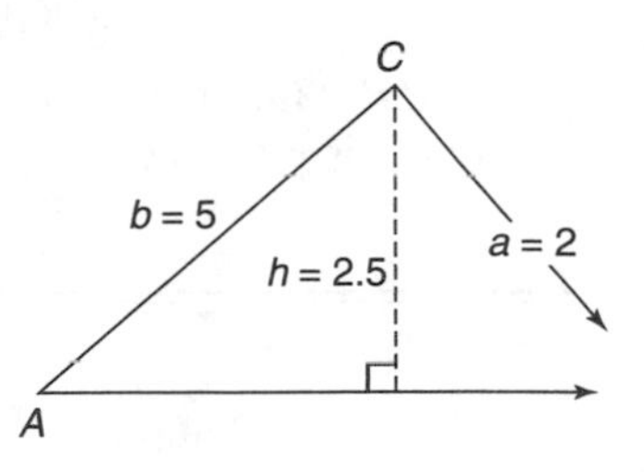

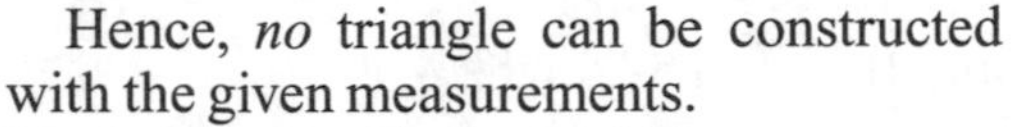

  Hence, *no* triangle can be constructed with the given measurements.
- <u>Method 2: Use the Law of Sines.</u>
  Assume $\triangle ABC$ is possible. Then:

$$\frac{a}{\sin A} = \frac{b}{\sin B}$$
$$\frac{2}{\sin 30°} = \frac{5}{\sin B}$$
$$2 \sin B = 5(0.5)$$
$$\sin B = \frac{2.5}{2} = 1.25 \quad \leftarrow \text{Impossible!}$$

Since $\sin A$ cannot be greater than 1, the assumption that a triangle is possible is *not* correct.

Example 2

If $r = 7$, $s = 10$, and $m\angle S = 50$, how many distinct triangles can be constructed?

Solution: **One**

- Method 1: Compare r and s.
  Here the given angle is $\angle S$. As shown in the accompanying figure, s, the side opposite the given angle, is longer than side r. Hence, $\triangle RST$ is possible.

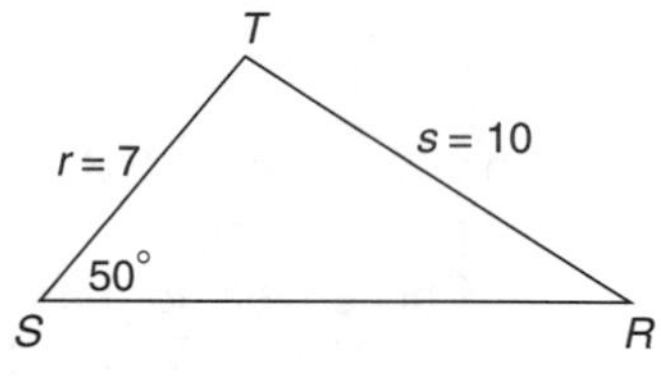

- Method 2: Use the Law of Sines.
  Assume a triangle is possible. If $\frac{7}{\sin R} = \frac{10}{\sin 50°}$, then

$$\sin R = \frac{7 \sin 50°}{10} \approx 0.5362$$

and $\angle R = \sin^{-1} 0.5362 \approx 32.4°$. Hence, at least one triangle is possible. Sine is also positive in Quadrant II. If $\angle R$ is obtuse, then $\angle R = 180° - 32.4° = 147.6°$. But

$$m\angle R + m\angle S = 147.6 + 50 > 180,$$

so $\angle R$ cannot be obtuse. Hence, one triangle, in which $\angle R$ measures $32.4°$, is possible.

The remaining angle, $\angle T$, of the triangle may be an obtuse angle. Since

$$m\angle S + m\angle R = 50 + 32.4 = 82.4,$$

$m\angle T = 180 - 82.4 = 97.6$. Hence, the one triangle that can be formed is an obtuse triangle.

Two Cases When the Given Angle Is Obtuse

Figure 15.4 summarizes the two possible situations when the given angle is obtuse.

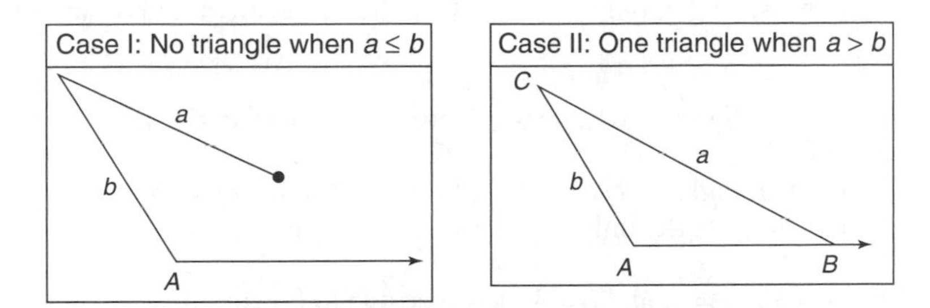

Figure 15.4 The Ambiguous Case When $\angle A$ Is Obtuse

Check Your Understanding of Section 15.3

Multiple Choice

1. If $a = 5$, $c = 12$, and $m\angle A = 30$, what is the total number of distinct triangles that can be constructed?
(1) 1 (2) 2 (3) 3 (4) 0

2. For which set of measurements can more than one triangle be constructed?
(1) $r = 6, s = 5, m\angle R = 100$
(2) $r = 6, s = 5, m\angle R = 30$
(3) $r = 3, s = 5, m\angle R = 30$
(4) $r = 3, s = 6, m\angle R = 30$

3. If $m\angle A = 30$, $BC = 10$, and $AC = 12$, then $\angle C$ in $\triangle ABC$ can be
(1) an acute angle, only
(2) an obtuse angle, only
(3) a right angle
(4) either an acute angle or an obtuse angle

4. If $a = 5\sqrt{2}$, $b = 8$, and $m\angle A = 45$, how many distinct triangles can be constructed?
(1) 1 (2) 2 (3) 3 (4) 0

5. If $\sin A = 0.75$ and $b = 8$, for which value of a is it possible to construct two distinct triangles?
(1) 5 (2) 6 (3) 7 (4) 9

6. If $a = 5$ and $b = 10$, for which measure of $\angle A$ can two distinct triangles be constructed?
(1) 20 (2) 30 (3) 45 (4) 120

7. If m$\angle A = 48$, $BC = 7$, and $AC = 9$, then $\angle C$ in $\triangle ABC$ can be
(1) an acute angle, only
(2) an obtuse angle, only
(3) a right angle
(4) either an acute angle or an obtuse angle

8. If $a = 9$, $c = 8$, and m$\angle A = 60$, which type of triangle, if any, can be constructed?
(1) a right triangle, only
(2) an acute triangle, only
(3) an obtuse triangle, only
(4) no triangle

9. If m$\angle A = 32$, $a = 5$, and $b = 3$, it is possible to construct
(1) an obtuse triangle
(2) two distinct triangles
(3) no triangle
(4) a right triangle

10. In $\triangle ABC$, m$\angle B = 35$, and $c = 10$. Which value of b will make it possible for two different triangles to be formed?
(1) $b = 5$ (2) $b = 8$ (3) $b = 10$ (4) $b = 13$

11. Avenue A and Beach Boulevard are straight streets that intersect at an angle of 30°. A local post office is located on Avenue A, 3.5 miles from the point at which Avenue A and Beach Boulevard intersect. A new mailbox is to be placed on Beach Boulevard at a distance of 1.75 miles from the post office. Which conclusion is valid?
(1) No such location is possible.
(2) The mailbox can be placed at only one location on Beach Boulevard.
(3) The mailbox can be placed at one of two locations on Beach Boulevard.
(4) The mailbox can be placed at one of three locations on Beach Boulevard.

12. Main Street and Central Avenue intersect, making an angle of 43°. Angela lives at the intersection of the two roads, and Caitlin lives on Central Avenue 10 miles from the intersection. If Leticia lives 7 miles from Caitlin, which conclusion is valid?
(1) Leticia cannot live on Main Street.
(2) Leticia can live at only one location on Main Street.
(3) Leticia can live at one of two locations on Main Street.
(4) Leticia can live at one of three locations on Main Street.

15.4 THE LAW OF COSINES

KEY IDEAS

Solving a triangle means finding the missing measures of its sides and angles. To solve a triangle that does not contain a right angle, you need to know the measure of at least one side and the measures of any two other parts of the triangle. The Law of Cosines relates the cosine of any angle of a triangle to the lengths of the three sides of the triangle.

- Use the Law of Sines when AAS, SSA, or ASA is given.
- Use the Law of Cosines when SAS or SSS is given.

Law of Cosines Given SAS

The accompanying table summarizes the formulas for finding the length of a side of a triangle when the measures of the other two sides and their included angle are given. In the formula $c^2 = a^2 + b^2 - 2ab\cos C$, when $\angle C$ is a right angle, $\cos C = 0$, so the Law of Cosines reduces to $c^2 = a^2 + b^2$. Hence, the Law of Cosines can be considered to be a generalization of the Pythagorean theorem since it works in any triangle.

LAW OF COSINES GIVEN SAS

Figure	Given SAS	To Find	Law of Cosines
	$b, \angle A, c$	a	$a^2 = b^2 + c^2 - 2bc\cos A$
	$a, \angle B, c$	b	$b^2 = a^2 + c^2 - 2ac\cos B$
	$a, \angle C, b$	c	$c^2 = a^2 + b^2 - 2ab\cos C$

Example 1

In $\triangle ABC$, $a = 6$, $b = 10$, and $m\angle C = 120$, as shown in the accompanying diagram. What is the length of side c?

Solution: **14**

Since SAS are given, use the Law of Cosines to find the length of side c.

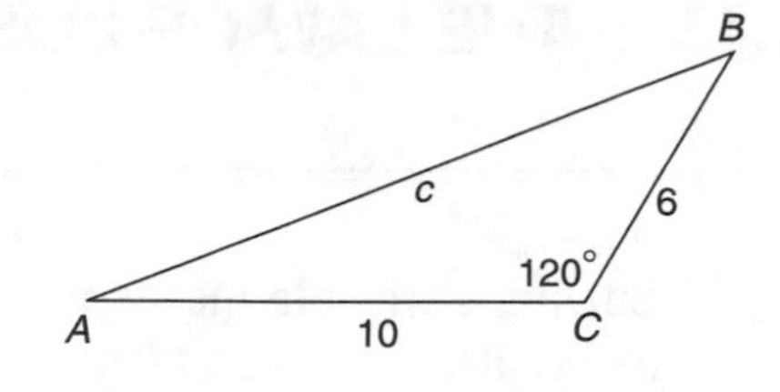

$$
\begin{aligned}
c^2 &= a^2 + b^2 - 2ab\cos C \\
&= 6^2 + 10^2 - 2(6)(10)\cos 120° \\
&= 36 + 100 - 120(-0.5) \\
&= 136 + 60 \\
&= 196 \\
c &= \sqrt{196} = 14
\end{aligned}
$$

The length of side c is **14**.

Law of Cosines Given SSS

If the lengths of the three sides of $\triangle ABC$ are known (SSS), then the measure of any angle of the triangle can be solved for by using the form of the Law of Cosines that involves that angle.

Example 2

Peter (P) and Jamie (J) have computer factories that are 132 miles apart. They both ship computer parts to Diane (D). Diane is 72 miles from Peter and 84 miles from Jamie, as shown in the accompanying diagram. If points P, D, and J are located on a map drawn to scale, find, to the *nearest tenth of a degree*, the measure of the largest angle of $\triangle PDJ$.

Solution: **115.4**

The largest angle of a triangle lies opposite the longest side of the triangle. The longest side of $\triangle PDJ$ is $\overline{PJ}$. Since SSS are given, use the Law of Cosines to find $m\angle D$.

$$
\begin{aligned}
(PJ)^2 &= (PD)^2 + (JD)^2 - 2(PD)(JD)\cos D \\
(132)^2 &= (72)^2 + (84)^2 - 2(72)(84)\cos D \\
17{,}424 &= 5184 + 7056 - 12{,}096\cos D \\
17{,}424 &= 12{,}240 - 12{,}096\cos D \\
17{,}424 - 12{,}240 &= -12{,}096\cos D \\
5184 &= -12{,}096\cos D \\
\text{or } \cos D &= -\frac{5184}{12{,}096} \approx -0.42857 \\
D &= \cos^{-1}(-0.42857) \approx 115.4°
\end{aligned}
$$

Check Your Understanding of Section 15.4

A. Multiple Choice

1. In $\triangle ABC$, $a = 8$, $b = 2$, and $c = 7$. What is the value of $\cos C$?
(1) $-\frac{19}{32}$ (2) $-\frac{11}{28}$ (3) $\frac{109}{112}$ (4) $\frac{19}{32}$

2. In $\triangle DEF$, if $d = \sqrt{3}$, $e = 4$, and $m\angle F = 30$, the length of side f is
(1) 7 (2) $\sqrt{17}$ (3) $\sqrt{7}$ (4) $\sqrt{3}$

3. In $\triangle ABC$, $a = 4$, $b = 3$, and $c = \sqrt{37}$. What is the degree measure of the largest angle of the triangle?
(1) 60 (2) 120 (3) 135 (4) 150

4. The sides of a triangle measure 6, 7, and 9. What is the value of the cosine of the largest angle?
(1) $-\frac{4}{84}$ (2) $\frac{2}{21}$ (3) $\frac{4}{84}$ (4) $-\frac{1}{81}$

5. In isosceles triangle ABC, $BC = 1$ and $m\angle C = 120$. The length of $\overline{AB}$ is
(1) 1 (2) $\sqrt{2}$ (3) $\sqrt{2.5}$ (4) $\sqrt{3}$

B. *In each case, show how you arrived at your answer by clearly indicating all of the necessary steps, formula substitutions, diagrams, graphs, charts, etc.*

6. Main Street and Park Avenue intersect at an angle of 74°. Michael lives on Main Street, 50 meters from the intersection, and Paul lives on Park Avenue, 40 meters from the intersection. The triangle formed by the intersection and the houses is an acute triangle. Find, to the *nearest tenth of a meter*, the distance between Michael's house and Paul's house.

7. Patricia and Quentin are separated by a wall. To calculate the distance between them, Akim positions himself 50 meters from Patricia and 75 meters from Quentin. If Akim's horizontal line of sight must change by 120°40′ when he switches his view from Patricia to Quentin, find, to the *nearest tenth of a meter*, the distance between Patricia and Quentin.

8. In $\triangle ABC$, $AC = 8$, $BC = 17$, and $AB = 20$.
a. Find the measure of the largest angle of the triangle, correct to the *nearest degree*.
b. Find the area in square units of $\triangle ABC$ to the *nearest integer*.

9. A metal frame is constructed in the form of an isosceles trapezoid, with diagonals acting as braces to strengthen the frame. Each base angle measures $73°30'$, the length of the shorter base is 8.0 feet, and the length of each of the nonparallel sides is 5.0 feet. Find, to the *nearest tenth of a foot*, the length of a diagonal brace of this frame.

10. Home plate and the three bases on a Little League baseball field are at the vertices of a square in which each side has a regulation length of 60 feet. If the pitcher's mound is 46 feet from home plate, what is the number of feet to the *nearest tenth* from the pitcher's mound to first base?

11. Michael and his friends are plotting a course for a race. They decide to make the course in the shape of a triangle, PQR. Beginning at point P, participants will run 1.4 miles to Q, then from Q to R, and finally 2.6 miles from R back to P. Angle QPR measures $38°30'$.

a. Find, to the *nearest tenth of a mile*, the total distance for the entire race.

b. Find, to the *nearest tenth of a square mile*, the area of $\triangle PQR$.

12. In $\triangle ABC$, the lengths of sides a, b, and c are in the ratio $4:6:8$. Find the ratio of the cosine of $\angle C$ to the cosine of $\angle A$.

13. In the accompanying diagram of $\triangle ABC$, $AB = 12$ feet, $DC = 17$ feet, $m\angle ABD = 40$, and $m\angle ADB = 110$. Find AC to the *nearest foot*.

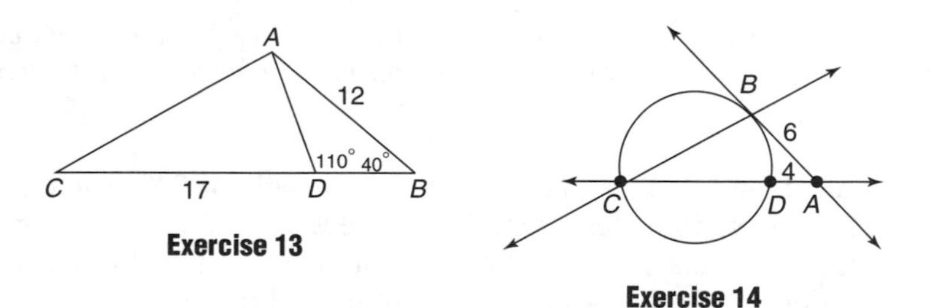

Exercise 13

Exercise 14

14. In the accompanying diagram, $\overleftrightarrow{AB}$ is tangent to circle O at B, $\widehat{BC} \cong \widehat{CD}$, $AB = 6$, and $AD = 4$. Determine the measures of:

a. $\overline{ADC}$ b. $\angle A$ correct to the *nearest tenth of a degree*

15. A parcel of land is shaped like an isosceles triangle in which the angle included between the congruent sides measures $53°10'$. The area of the parcel of land is 1 acre, which is equal to 43,560 square feet. Find the perimeter of the parcel of land, correct to the *nearest foot*.

15.5 PARALLELOGRAM OF FORCES

KEY IDEAS

If two forces act simultaneously on a body, the single force that has the same effect as these two forces, called the **resultant**, is the diagonal of a parallelogram whose adjacent sides represent the two forces.

Finding a Resultant Force

If a 20-pound force and a 30-pound force act on a body at an angle of 60° with each other, then the magnitude of the resultant is the length of the diagonal of the parallelogram whose adjacent sides are the two forces, as shown in Figure 15.5. To calculate the magnitude of the resultant to the *nearest tenth of a pound*, find AC using the Law of Cosines.

- Since consecutive angles of a parallelogram are supplementary, $m\angle B = 180 - 60 = 120$.
- Since opposite sides of a parallelogram have the same length, $DC = AB = 20$.
- In $\triangle ABC$, use the Law of Cosines:

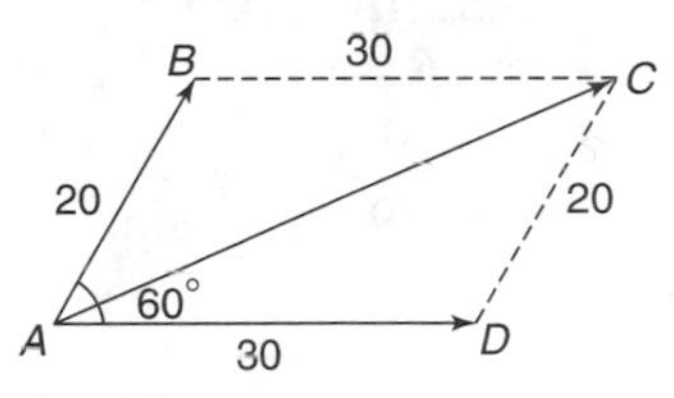

Figure 15.5 Resultant Force

$$\begin{aligned}(AC)^2 &= (BC)^2 + (AB)^2 - 2(BC)(AB)\cos\angle B\\ &= (30)^2 + (20)^2 - 2(30)(20)\cos 120°\\ &= 900 + 400 - 1200(-0.5)\\ &= 1900\\ AC &= \sqrt{1900} \approx 43.6\end{aligned}$$

Example 1

Two forces of 437 pounds and 876 pounds act on a body at an acute angle with each other. The angle between the resultant and the 437-pound force is 41°10′. Find, to the *nearest 10 minutes* or *nearest tenth of a degree*, the measure of the angle between the original two forces.

Solution: **60°20′** or **60.3°**

- Draw parallelogram $ABCD$ with $AB = 437$, $AD = 876$, and $\angle CAB = 41°10'$, as shown in the accompanying diagram. Find the measure of $\angle DAB$ by first finding the measure of $\angle DAC$. Let $x = m\angle DAC = m\angle BCA$.

- In $\triangle ABC$, use the Law of Sines:

$$\frac{876}{\sin 41^\circ 10'} = \frac{437}{\sin x}$$

$$\sin x = \frac{437 \times \sin 41.167^\circ}{876}$$

$$x \approx 19.17^\circ \text{ or } 19^\circ 10'$$

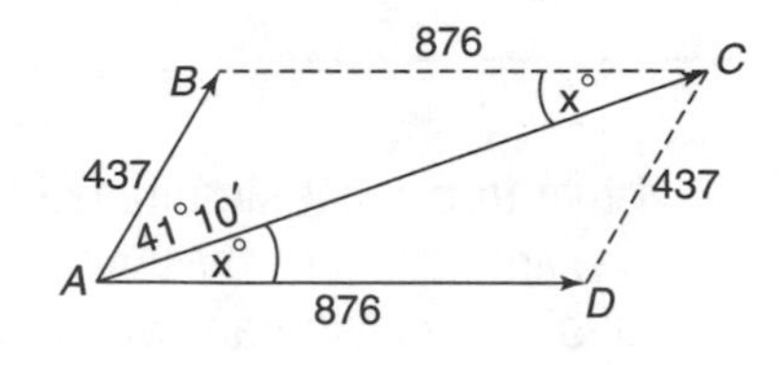

- $m\angle DAB = 41^\circ 10' + 19^\circ 10' = 60^\circ 20'$ or, equivalently, 60.3°.

Example 2

For the body in Example 1, find the magnitude of the resultant force to the *nearest pound.*

Solution: **1156**

- In $\triangle ABC$, find AC by using the Law of Sines.
- $m\angle B = 180 - m\angle DAB = 180 - 60^\circ 20' = 119^\circ 40'$.
- $\dfrac{AC}{\sin 119^\circ 40'} = \dfrac{876}{\sin 41^\circ 10'}$, so $AC = \dfrac{\sin 119.667^\circ \times 876}{\sin 41.167^\circ} \approx 1156.$

TIP

In $\triangle ABC$, the Law of Cosines can be used to find AC since

$$(AC)^2 = (437)^2 + (876)^2 - 2(437)(876)\cos 119^\circ 40'.$$

Check Your Understanding of Section 15.5

In each case, show how you arrived at your answer by clearly indicating all of the necessary steps, formula substitutions, diagrams, graphs, charts, etc.

1. Two forces of 50 and 68 pounds act on a body to produce a resultant force of 70 pounds. Find, to the *nearest ten minutes* or *nearest tenth of a degree*, the angle formed between the resultant force and the smaller force.

2. Two forces of 130 and 150 pounds yield a resultant force of 170 pounds. Find, to the *nearest 10 minutes* or *nearest tenth of a degree*, the measure of the angle between the original two forces.

3. Two forces of 30 pounds and 40 pounds act on a body, forming an acute angle with each other. The angle between the resultant force and the

30-pound force is 35.2°. Find, to
between the two original forces.

4. Two forces acting on a body make an an
magnitude of the first force is 390 pounds.
of 47° with the first force, what is the magni
nearest tenth of a pound?

5. Two people are needed to carry a crate. One person
pounds, and the other person exerts a force of 117 poun
the accompanying figure. If the angle between the two f
56.8°, find the weight of the crate to the *nearest tenth of a p*

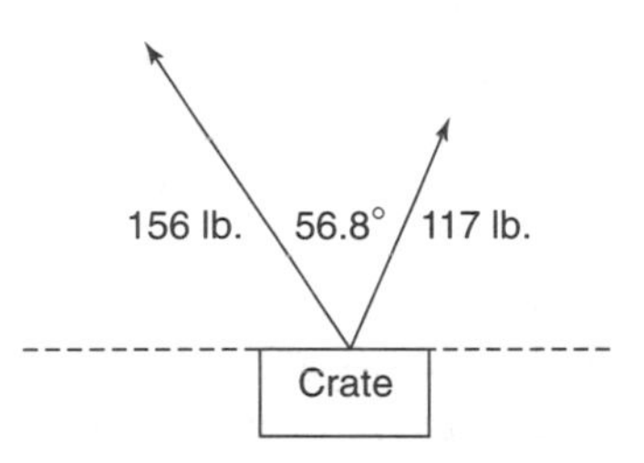

6. Two forces are applied to an object. The measure of the angle between the 30.2-pound applied force and the 50.1-pound resultant is 25°.

a. Find the magnitude of the second applied force to the *nearest tenth of a pound.*

b. Using the answer found in part a, find the measure of the angle between the second applied force and the resultant to the *nearest degree.*

LAWS

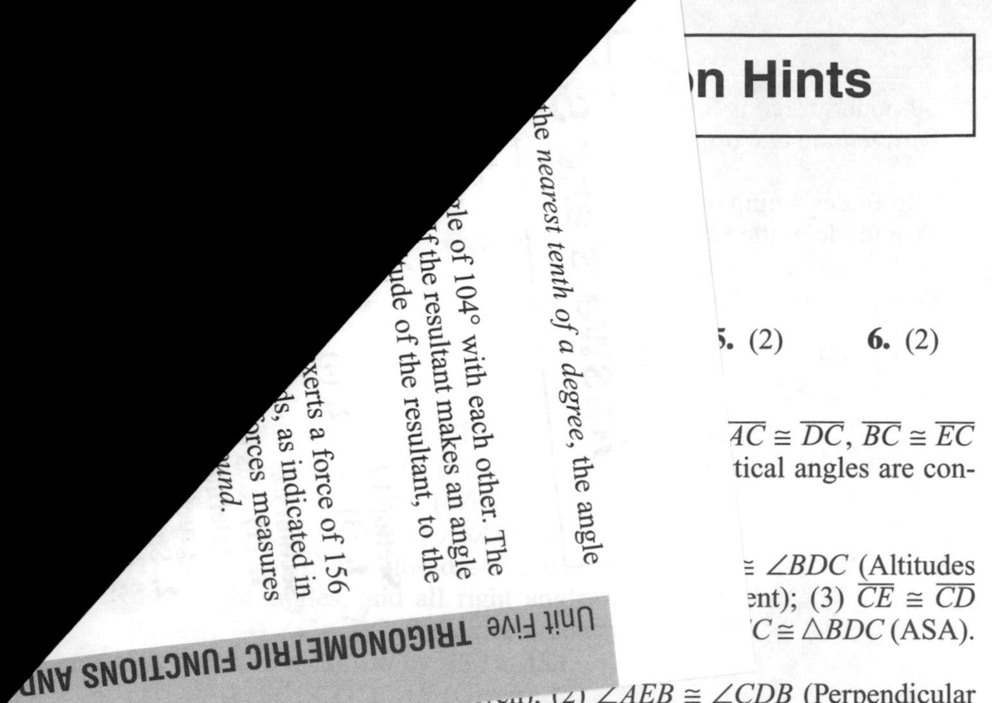

(2) $\angle AEB \cong \angle CDB$ (Perpendicular lines form right angles, and all right angles are congruent); (3) $\overline{BD} \cong \overline{BE}$ (Given); (4) $\angle B \cong \angle B$ (Reflexive property); (5) $\triangle ABE \cong \triangle CBD$ (ASA).

6. Show $\triangle DBC \cong \triangle ACB$ by $SAS \cong SAS$.
7. (1) $\overline{AB} \cong \overline{DC}$, $\overline{AB} \perp \overline{BE}$, $\overline{DC} \perp \overline{CE}$ m$\angle 1$ = m$\angle 2$ (Given); (2) m$\angle ABE$ = m$\angle DCE$ (Perpendicular lines form right angles, and all right angles are equal in measure); (3) m$\angle ABE$ + m$\angle 1$ = m$\angle DCE$ + m$\angle 2$ (Addition property); (4) $\angle ABC \cong \angle DCB$ (Substitution); (5) $\overline{BC} \cong \overline{BC}$ (Reflexive property); (6) $\triangle DBC \cong \triangle ACB$ (SAS).

Section 1.3

1. (3) **2.** (1) **3.** (2)
4. $\triangle ABD \cong \triangle CBD$ by *Hy-Leg* $\cong$ *Hy-Leg*, so $\angle DAB \cong \angle DCB$ by CPCTC. Hence, $\angle 1 \cong \angle 2$ since supplements of congruent angles are congruent.
5. (1) $\angle 1 \cong \angle 2$, $\overline{DB} \perp \overline{AC}$ (Given); (2) $\angle DAB \cong \angle DCB$ (Supplements of congruent angles are congruent); (3) $\angle DBA \cong \angle DBC$ (Right angles are congruent); (4) $\overline{DB} \cong \overline{DB}$ (Reflexive property); (5) $\triangle DBA \cong \triangle DBC$ (AAS); (6) $\overline{AB} \cong \overline{CB}$ (CPCTC); (7) $\overline{DB}$ is a median (Definition of median).
6. $\triangle ADB \cong \triangle CDB$ by $AAS \cong AAS$, so $\overline{AD} \cong \overline{DB}$ by CPCTC.
7. Using the subtraction property, $\overline{AE} \cong \overline{CF}$ (leg). It is given that $\overline{AB} \cong \overline{CD}$ (hypotenuse). $\triangle AEB \cong \triangle CFD$ by *Hy-Leg* $\cong$ *Hy-Leg*, so $\overline{BE} \cong \overline{DF}$ by CPCTC.
8. $\triangle TLS \cong \triangle SWT$ by *Hy-Leg* $\cong$ *Hy-Leg*, since $\overline{ST} \cong \overline{ST}$ (hypotenuse) and, from the given, $\overline{TL} \cong \overline{SW}$ (leg). Hence, $\overline{SL} \cong \overline{TW}$ by CPCTC.
9. (1) $\overline{TL}$ and $\overline{SW}$ are altitudes (Given); (2) $\angle SWR \cong \angle TLR$ (Altitudes form right angles, and all right angles are congruent); (3) $\angle R \cong \angle R$ (Reflexive

property); (4) $\overline{RS} \cong \overline{RT}$ (Given); (5) $\triangle SWR \cong \triangle TLR$ (AAS); (6) $\overline{RW} \cong \overline{RL}$ (CPCTC).

10. Given: Isosceles triangle ABC with $\overline{AB} \cong \overline{BC}$, $\overline{BM}$ median. Prove: $\overline{BM} \perp \overline{AC}$. Since $\overline{AM} \cong \overline{CM}$, $\triangle ABM \cong \triangle CBM$ by $SSS \cong SSS$. Adjacent angles BMA and BMC are congruent by CPCTC, so $\overline{BM} \perp \overline{AC}$.

11. a. $\triangle ADF \cong \triangle CEF$ by *Hy-Leg* $\cong$ *Hy-Leg*. b. By CPCTC, $\overline{DF} \cong \overline{EF}$, so $\triangle BDF$ and $\triangle BEF$ by *Hy-Leg* $\cong$ *Hy-Leg*. Since $\angle DBE \cong \angle EBF$ by CPCTC, $\overline{BF}$ bisects $\angle DFE$.

Section 1.4

1. (1) $\angle 2 \cong \angle 4$ (Given); (2) $\overline{AD} \cong \overline{CD}$ (Converse of the Base Angles Theorem); (3) $\angle BDA \cong \angle BDC$ (Given); (4) $BD \cong BD$ (Reflexive property); (5) $\triangle ADB \cong \triangle CDB$ (SAS); (6) $\overline{AB} \cong \overline{CB}$ (CPCTC); (7) $\triangle ABC$ is isosceles (A triangle that has two congruent sides is isosceles).

2. Since $\overline{AB} \cong \overline{BC}$ (Given), $\angle BAC \cong \angle BCA$. It is also given that $\angle 1 \cong \angle 3$. By the subtraction property, $\angle 2 \cong \angle 4$ so $\overline{AD} \cong \overline{DC}$. Therefore, $\triangle ADC$ is isosceles.

3. (1) $\angle 1 \cong \angle 2$ (Given); (2) $\angle AFD \cong \angle CGE$ (Supplements of congruent angles are congruent); (3) $\overline{AB} \cong \overline{BC}$, and F and G are midpoints (Given); (4) $AF = \frac{1}{2}AB,\ CG = \frac{1}{2}BC$ (Definition of midpoint); (5) $\overline{AF} \cong \overline{CG}$ (Halves of congruent segments are congruent); (6) $\angle A \cong \angle C$ (Base Angles Theorem); (7) $\triangle AFD \cong \triangle CGE$ (ASA); (8) $\overline{FD} \cong \overline{GE}$ (CPCTC).

4. Right angles ADF and CEG are congruent. $\angle AFD \cong \angle CGD$ since supplements of congruent angles ($\angle 1 \cong \angle 2$) are congruent. Since $\overline{AE} \cong \overline{CD}$, $\overline{AD} \cong \overline{CE}$ by the subtraction property. $\triangle ADF \cong \triangle CEG$ by $AAS \cong AAS$, so $\angle A \cong \angle C$, making $\overline{AB} \cong \overline{BC}$. Therefore, $\triangle ADC$ is isosceles.

5. (1) $\angle B \cong \angle C$ (Given); (2) $AB = AC$ (Converse of the Base Angles Theorem); (3) $DB = EC$ (Given); (4) $AD = AE$ (Subtraction property); (5) $\angle ADE \cong \angle AED$ (Base Angles Theorem); (6) $\angle 1 \cong \angle 2$ (Supplements of congruent angles are congruent).

6. $\triangle ABD \cong \triangle CBD$ by $ASA \cong ASA$ since $\angle ADB \cong \angle CEB$ (Supplements of congruent angles), $\overline{BD} \cong \overline{BE}$ (Converse of Base Angles Theorem), and $\angle ABD \cong \angle CBE$ (Subtraction property). Hence, $\angle A \cong \angle C$, so $\overline{AB} \cong \overline{BC}$. Therefore, $\triangle ADC$ is isosceles.

7. (1) $\overline{DF} \cong \overline{CF}$, $\overline{AD} \cong \overline{FC}$, and $\overline{BC} \cong \overline{ED}$ (Given); (2) $\angle ACB \cong \angle FDE$ (Base Angles Theorem); (3) $\overline{AC} \cong \overline{FD}$ (Addition property); (4) $\triangle ABC \cong \triangle FED$ ($SAS \cong SAS$); (5) $\angle B \cong \angle E$ (CPCTC).

8. Given: Isosceles triangle ABC with $\overline{AB} \cong \overline{BC}$, $\overline{CX}$ and $\overline{AY}$ are medians. Prove: $\overline{CX} \cong \overline{AY}$. Show $\triangle AXC \cong \triangle CYA$ by $SAS \cong SAS$. Then $\overline{CX} \cong \overline{AY}$ by CPCTC.

9. (1) $\overline{AF} \cong \overline{DE}$, $\overline{AF}$ and $\overline{DE}$ bisect each other at G (Given); (2) $EG = \frac{1}{2}DE$, $FG = \frac{1}{2}AF$ (Definition of bisector); (3) $\overline{EG} \cong \overline{FG}$ (Halves of congruent segments are congruent); (4) $\angle AFG \cong \angle DEF$ (Base Angles Theorem); (5) $BE = FC$ (Given); (6) $EF = EF$ (Reflexive property); (7) $BE + EF = FC + EC$ (Addition property); (8) $\overline{BF} \cong \overline{CE}$ (Substitution); (9) $\triangle AFB \cong \triangle DEC$ (SAS); (10) $\overline{AB} \cong \overline{DC}$ (CPCTC).

10. Given: Isosceles $\triangle ABC$, $\overline{AB} \cong \overline{AC}$, altitudes $\overline{CX}$ and $\overline{AY}$. Prove: $\overline{CX} \cong \overline{AY}$. Show $\triangle AXC \cong \triangle CYA$ by $AAS \cong AAS$. Then $\overline{CX} \cong \overline{AY}$ by CPCTC.

Section 1.5

1. a. $\triangle RLS \cong \triangle RLT$ (Hy-Leg). b. (1) $\overline{SL} \cong \overline{TL}$ and $\angle SLR \cong \angle TLR$ (CPCTC); (2) $\angle SLW \cong \angle TLW$ (Supplements of congruent angles are congruent); (3) $\overline{LW} \cong \overline{LW}$ (Reflexive property); (4) $\triangle SLW \cong \triangle TLW$ (SAS); (5) $\angle SWL \cong \angle TWL$ (CPCTC); (6) $\overline{WL}$ bisects $\angle SWT$ (Definition of bisector).

2. $\triangle KQM \cong \triangle LQN$ by $SAS \cong SAS$. By CPCTC, $\angle K \cong \angle L$ and $\overline{KM} \cong \overline{LN}$. Then $\triangle KPM \cong \triangle LRN$ by $ASA \cong ASA$.

3. (1) $\triangle ABC \cong \triangle ADC$ (SAS): (2) $\overline{BC} \cong \overline{DC}$ and $\angle ACB \cong \angle ACD$ (CPCTC); (3) $\angle BCE \cong \angle DCE$ (Supplements of congruent angles are congruent): (4) $\overline{CE} \cong \overline{CE}$ (Reflexive property); $\triangle BCE \cong \triangle DCE$ (SAS).

4. Show $\triangle BEC \cong \triangle DEC$ by $SSS \cong SSS$. By CPCTC, $\overline{BC} \cong \overline{DC}$ and $\angle BCE \cong \angle DCE$. Then $\angle BCA \cong \angle DCA$ (Supplements of congruent angles are congruent). Hence, $\triangle ABC \cong \triangle ADC$ by $SAS \cong SAS$.

5. (1) $\triangle AFD \cong \triangle DFE$ (AAS); (2) $\overline{FD} \cong \overline{FE}$ (CPCTC); (3) $\overline{BF} \cong \overline{BF}$ (Reflexive property); (4) right triangle $FDB \cong$ right triangle FEB (Hy-Leg); (5) $\angle FBD \cong \angle FEB$ (CPCTC); (6) $\overline{BF}$ bisects $\angle DBE$ (Definiton of bisector).

6. $\triangle BDF \cong \triangle BEF$ by $SSS \cong SSS$. By CPCTC, $\angle BDF \cong \angle BEF$, so $\angle ADF \cong \angle CEF$ (Supplements of congruent angles are congruent). Also, $\angle AFD \cong \angle CFE$ (Vertical angles are congruent). $\triangle AFD \cong \triangle CFE$ by $ASA \cong ASA$. Then $\overline{AF} \cong \overline{CF}$ by CPCTC, so $\triangle AFC$ is isosceles.

7. Since $AP = PB = AQ = QB$, $\triangle APB \cong \triangle AQB$ by $SSS \cong SSS$, so $\angle APQ \cong \angle BPQ$. $\triangle APC \cong \triangle BPC$ by $SAS \cong SAS$. Hence, $\overline{AC} \cong \overline{BC}$ and angles ACP and BCP are congruent adjacent angles, so $\overline{PQ} \perp \overline{AB}$ at its midpoint C.

Section 1.6

1. (1) $\angle 1 \cong \angle 2$, $\overline{BD}$ does *not* bisect $\angle ABC$ (Given); (2) assume $\overline{AB} \cong \overline{BC}$ (Opposite of "Prove"); (3) $\overline{AD} \cong \overline{CD}$ (Converse of Base Angles Theorem); (4) $\overline{BD} \cong \overline{BD}$ (Reflexive property); (5) $\triangle ABD \cong \triangle CBD$ ($SSS \cong SSS$); (6) $\angle ABD \cong \angle CBD$ (CPCTC); (7) $\overline{BD}$ bisects $\angle ABC$ (Definition of angle bisector); (8) Statement 2 is false, so $\overline{AB} \ncong \overline{BC}$ (Statement 7 contradicts statement 1).

2. Assume $\overline{JK} \cong \overline{ML}$. Then $\triangle JLM \cong \triangle MKJ$ by $Hy\text{-}Leg \cong Hy\text{-}Leg$, so $\angle AJM \cong \angle AMJ$ by CPCTC. Since $\overline{AJ} \cong \overline{AM}$ contradicts the "Given" that $\triangle JAM$ is scalene, $\overline{JK} \ncong \overline{ML}$.

3. Given: Scalene triangle ABC with M the midpoint of $\overline{AB}$, $\overline{MX} \perp \overline{AC}$, and $\overline{MY} \perp \overline{BC}$. Prove: $\overline{MX} \not\cong \overline{MY}$. Assume $\overline{MX} \cong \overline{MY}$. Right triangles AXM and BYM are congruent by *Hy-Leg* $\cong$ *Hy-Leg*, so $\angle A \cong \angle B$. Since $\overline{AC} \cong \overline{BC}$ contradicts the "Given" that $\triangle ABC$ is scalene, $\overline{MX} \not\cong \overline{MY}$.

4. Assume $\overline{BE} \cong \overline{EC}$. Since $\overline{AB} \cong \overline{AC}$ and $\overline{AE} \cong \overline{AE}$, $\triangle AEB \cong \triangle AEC$ by $SSS \cong SSS$. By CPCTC, $\angle AEB \cong \angle AEC$, so $\angle BED \cong \angle CED$ (Supplements of congruent angles are congruent). Because $\overline{ED} \cong \overline{ED}$, $\triangle BED \cong \triangle CED$, so $\overline{BD} \cong \overline{CD}$ by CPCTC. But this contradicts the "Given" that $\overline{BD} \not\cong \overline{CD}$. Hence the assumption that $\overline{BE} \cong \overline{EC}$ is false, so $\overline{BE} \not\cong \overline{EC}$.

Section 1.7

1. (4) **2.** (3) **3.** (2) **4.** (2) **5.** (1) **6.** (3)

7. Since $m\angle 1 > m\angle A$ and $m\angle 1 = m\angle 2$, $m\angle 2 > m\angle A$, so $AD > ED$.

8. By the Base Angles Theorem, $m\angle CAB = m\angle CBA$ and $m\angle 1 = m\angle 2$ so, by subtraction, $m\angle 3 = m\angle 4$. Since $m\angle AED > m\angle 4$, $m\angle AED > m\angle 3$, making $AD > DE$.

9. $m\angle 4 > m\angle 3$, and $m\angle 3 > m\angle AEC$. Hence, $m\angle 4 > m\angle AEC$.

10. Since $AD > BD$, $m\angle 3 > m\angle 2$, $m\angle 1 = m\angle 2$, and, by substitution, $m\angle 3 > m\angle 1$. But $m\angle 4 > m\angle 3$, so $m\angle 4 > m\angle 1$, making $AC > DC$.

CHAPTER 2

Section 2.1

1. (1) $\overline{RP} \parallel \overline{SW}$, $\overline{SP} \parallel \overline{TW}$ (Given); (2) $\angle R \cong \angle TSW$, $\angle RSP \cong \angle T$ (If two lines are parallel, corresponding angles are congruent); (3) $\overline{TW} \cong \overline{SP}$ (Given); (4) $\triangle RSP \cong \triangle STW$ (AAS); (5) $\overline{RS} \cong \overline{TS}$ (CPCTC), so $\overline{PS}$ bisects $\overline{RT}$.

2. $\triangle RSP \cong \triangle TSW$ by $SAS \cong SAS$ since $\overline{RS} \cong \overline{TS}$, $\angle PRS \cong \angle WST$, and $\overline{RP} \cong \overline{SW}$. By CPCTC, corresponding angles RSP and WTS are congruent, making $\overline{SP} \parallel \overline{TW}$.

3. (1) $\overline{QL} \cong \overline{QM}$ (Given); (2) $\angle 1 \cong \angle 2$ (Base Angles Theorem); (3) $\overline{LM} \parallel \overline{PR}$ (Given); (4) $\angle 1 \cong \angle 2$ and $\angle 3 \cong \angle 4$ (If two lines are parallel, then corresponding angles are congruent); (5) $\angle 2 \cong \angle 4$ (Transitivity); (6) $\overline{QR} \cong \overline{QP}$ (Converse of Base Angles Theorem); (7) $\triangle PQR$ is isosceles (A triangle that contains two congruent sides is isosceles).

4. a. (1) $\overline{RT}$ bisects $\angle BRS$ (Given); (2) $\angle BRM \cong \angle SRM$ (Definition of angle bisector); (3) $\overleftrightarrow{ARB} \parallel \overleftrightarrow{CST}$ (Given); (4) $\angle T \cong \angle BRM$ (If two lines are parallel, alternate interior angles are congruent); (5) $\angle SRM \cong \angle T$ (Transitivity); (6) $\overline{RS} \cong \overline{ST}$ (Converse of Base Angles Theorem). b. (7) M is the midpoint of $\overline{RT}$ (Given); (8) $\overline{RM} \cong \overline{TM}$ (Definition of midpoint); (9) $\overline{SM} \cong \overline{SM}$ (Reflexive property); (10) $\triangle SRM \cong \triangle STM$ (SSS); (11) $\angle RSM \cong \angle TSM$ (CPCTC); (12) $\overline{SM}$ bisects $\angle RST$ (Definition of angle bisector).

5. $\triangle AMP \cong \triangle BMC$ by SAS since $\overline{BM} \cong \overline{AM}$ (A median divides a segment into two congruent segments), $\angle BMC \cong \angle AMP$ (Vertical angles are congruent), and $\overline{CM} \cong \overline{MP}$ (Given). Therefore, $\angle P \cong \angle BCM$ by CPCTC. It follows that $\overline{AP} \parallel \overline{CB}$ since two lines are parallel if a pair of alternate interior angles are congruent.

6. Assume $\overline{AB} \parallel \overline{DE}$. Then $\angle B \cong \angle D$. Since $\overline{DE} \cong \overline{CE}$, $\angle D \cong \angle DCE$. By transitivity, $\angle B \cong \angle DCE$. Since $\angle ACB \cong \angle DCE$, $\angle B \cong \angle ACB$, implying $\overline{AC} \cong \overline{AB}$. But this contradicts the "Given" ($AC > AB$). Hence, $\overline{AB}$ is *not* parallel to $\overline{DE}$.

7. Assume $\overline{AC}$ bisects $\angle BAD$. Then $\angle BAC \cong \angle DAC$. Since $\overline{BC} \parallel \overline{AD}$, $\angle DAC \cong \angle BCA$. By transitivity, $\angle BAC \cong \angle BCA$, so $\triangle ABC$ is isosceles. But this contradicts the "Given." Hence, $\overline{AC}$ does *not* bisect $\angle BAD$.

Section 2.2

1. (3) **2.** (4) **3.** (4) **4.** (2) **5.** (3) **6.** (4)

7. $F = 1.8C + 32$ **8.** -2 **9.** -4.5 **10.** $y = -\frac{2}{3}x + 3$

11. $y = -2x + 3$ **12.** $t = -7, k = -25$

13. a. Slope $\overline{BC} \times$ slope $\overline{AC} = \left(\frac{1}{2}\right) \times (-2) = -1$ b. $y = 3x - 11$

14. a. Slope $\overline{BC} \times$ slope $\overline{AC} = (1) \times (1) = -1$ b. $y = 7x + 15$

15. a. $y = -\frac{4}{3}x$ b. $y = \frac{3}{4}x + 3$ c. Yes

Section 2.3

1. (1) $ABCD$ is a parallelogram (Given); (2) $\overline{AB} \parallel \overline{DC}$ (Opposite sides of a parallelogram are parallel); (3) $\angle E \cong \angle CDF$ (If two lines are parallel, alternate interior angles are congruent); (4) $\angle EFB \cong \angle DFC$ (Vertical angles are congruent); (5) B is the midpoint of $\overline{AE}$ (Given); (6) $\overline{EB} \cong \overline{AB}$ (Definition of midpoint); (7) $\overline{AB} \cong \overline{CD}$ (Opposite sides of a parallelogram are congruent); (8) $\overline{EB} \cong \overline{CD}$ (Transitivity); (9) $\triangle BEF \cong \triangle CDF$ (AAS): (10) $\overline{EF} \cong \overline{FD}$ (CPCTC).

2. $\triangle EAF \cong \triangle HGC$ by $AAS \cong AAS$ since $\angle AEF \cong \angle CHG$, $\angle EAF \cong \angle HCG$, and $\overline{EF} \cong \overline{HG}$.

3. a. $\triangle CGF \cong \triangle AGE$ by $SAS \cong SAS$. $\triangle ABC \cong \triangle CDF$ by $AAS \cong AAS$ since $\angle 3 \cong \angle 4$ (CPCTC), $\angle B \cong \angle D$ (Given), and $\overline{AC} \cong \overline{AC}$ (Reflexive property). Therefore, $\overline{BC} \cong \overline{AD}$ by CPCTC. b. Since $\angle 3 \cong \angle 4$, $\overline{BC} \parallel \overline{AD}$. Hence, $ABCD$ is a parallelogram since one pair of sides is congruent and parallel.

4. Since $AD > DC$, $m\angle ACD > m\angle DAC$. Since $\overline{AB} \parallel \overline{CD}$, $m\angle BAC = m\angle ACD$. By substitution, $m\angle BAC > m\angle DAC$.

5. Since $\triangle AED \cong \triangle CFB$ by SAS, $\overline{AD} \cong \overline{BC}$ and $\angle EAD \cong \angle FCB$ by CPCTC. Since alternate interior angles are congruent, $\overline{AD} \parallel \overline{BC}$. Since $\square ABCD$ has two sides that are both parallel and congruent, $ABCD$ is a parallelogram.

6. $\triangle ABL \cong \triangle DCM$ by $SAS \cong SAS$ since $\overline{AL} \cong \overline{CM}$, $\angle ALB \cong \angle CMB$ (Supplements of congruent angles are congruent), and $\overline{BL} \cong \overline{DM}$. By CPCTC, $\overline{AB} \cong \overline{CD}$ and $\angle BAL \cong \angle DCM$, making $\overline{AB} \| \overline{CD}$. Hence, *ABCD* is a parallelogram since one pair of sides is both congruent and parallel.

7. $\overline{AB} \cong \overline{CD}$ and $\overline{AB} \| \overline{CD}$, so $\angle BAL \cong \angle DCM$. $\triangle ALB \cong \triangle CMD$ (ASA). $\overline{BL} \cong \overline{DM}$ and $\angle ALB \cong \angle CMD$ (CPCTC). Taking supplements of congruent angles, $\angle BLM \cong \angle DML$, so that $\overline{BL} \| \overline{DM}$. Hence, *BLDM* is a parallelogram.

8. Show midpoint $\overline{AC}$ = midpoint $\overline{BD}$ = (2.5, 3).

9. a. Show midpoint $\overline{AC} \neq$ midpoint $\overline{BD}$. b. If *P*, *Q*, *R*, and *S* are the midpoints of $\overline{AB}$, $\overline{BC}$, $\overline{CD}$, and $\overline{AD}$, respectively, show midpoint $\overline{PR}$ = midpoint $\overline{QS}$.

10. a. $k = -3$ b. $y = 4x - 3$

Section 2.4

1. (4) **2.** (4) **3.** (2) **4.** (2) **5.** (3)

6. Show slope $\overline{MA}$ = slope $\overline{TH} = \frac{7}{3}$ and slope $\overline{AT}$ = slope $\overline{MH} = -\frac{3}{7}$, so *MATH* is a parallelogram. Since the slopes of adjacent sides are negative reciprocals, *MATH* contains 4 right angles and is, therefore, a rectangle. Use the distance formula to show that a pair of adjacent sides are congruent.

7. a. Show midpoint $\overline{AC}$ = midpoint $\overline{BD}$ = (3, –0.5). b. Show $AB \neq AD$.

8. a. Show midpoint $\overline{RC}$ = midpoint $\overline{ET}$ = (4.5, 6), and diagonal RC = diagonal $ET = \sqrt{125}$. b. Show $RE \neq RT$.

9. a. Show $RH = HO = OM = RM = \sqrt{50}$. b. Show slope of $\overline{RH} \times$ slope of $\overline{RM} \neq -1$.

10. a. Show $LM = \sqrt{5}$ and $BC = 2\sqrt{5}$. b. Show slope $\overline{LM}$ = slope $\overline{BC}$ = 0.5.

11. $\triangle ABL \cong \triangle BCM$ by $SSS \cong SSS$, so $\angle ABL \cong BCM$. Since $\overline{AB} \| \overline{CD}$, angles *ABL* and *BCM* are supplementary and congruent, making each a right angle, so *ABCD* is a square.

12. $\triangle VWT \cong \triangle SXT$ by $ASA \cong ASA$ since $\angle WTV = \angle XTS$ (Vertical angles), $\overline{VT} \cong \overline{ST}$ (Rhombus is equilateral), $\angle WVT \cong \angle XST$ (Subtract corresponding sides of the two equations $m\angle RSX \cong m\angle RVW$ and $m\angle RST = m\angle TVR$). By CPCTC, $\overline{TX} \cong \overline{TW}$.

13. Use an indirect proof. Assume *ABCD* is a rectangle. Then $\triangle BAD \cong \triangle CDA$ (SAS). $\angle 1 \cong \angle 2$ (CPCTC), contradicting the "Given."

14. (1) *ABCD* is a square (Given); (2) $\angle B$ and $\angle D$ are right angles (A square contains four right angles); (3) $\overline{AB} \cong \overline{AD}$ (Adjacent sides of a square are congruent); $\angle 1 \cong \angle 2$ (Given); $\overline{AE} \cong \overline{AF}$ (Converse of the Base Angles Theorem); (6) $\triangle ABE \cong \triangle ADF$ (*Hy-Leg*); (7) $\overline{BE} \cong \overline{DF}$ (CPCTC).

15. $\triangle ABE \cong \triangle ADF$ by $AAS \cong AAS$ since $\angle B \cong \angle D$, $\angle BEA \cong \angle DFA$ (By subtracting corresponding sides of the two equations $m\angle BEF = m\angle DFE$ and $m\angle 1 = m\angle 2$), and $\overline{EA} \cong \overline{FA}$ (Since $\angle 1 \cong \angle 2$). By CPCTC,

$\overline{AB} \cong \overline{AD}$. Since an adjacent pair of sides of rectangle $ABCD$ are congruent, $ABCD$ is a square.

16. a. Since $\angle CEF \cong \angle CFE$, $EC = FC$ and $BE = DF$ (Given). By addition, $BC = CD$ so rectangle $ABCD$ is a square. b. Because $ABCD$ is a square, $\overline{AB} \cong \overline{AD}$, so $\triangle ABE \cong ADF$ by $SAS \cong SAS$. By CPCTC, $\overline{AE} \cong \overline{AF}$, so $\triangle EAF$ is isosceles.

Section 2.5

1. Since parallel lines are everywhere equidistant, $\overline{BE} \cong \overline{CF}$. $\triangle AEB \cong \triangle DCF$ by $SAS \cong SAS$. By CPCTC, $\overline{AB} \cong \overline{CD}$, so trapezoid $ABCD$ is isosceles.

2. $\triangle RSW \cong \triangle WTR$ by SAS. By CPCTC, $\angle TRW \cong \angle SWR$, so $\overline{RP} \cong \overline{WP}$ (Converse of the Base Angles Theorem), and $\triangle RPW$ is isosceles.

3. $\triangle AGB \cong \triangle DCF$ by $SAS \cong SAS$ since $\overline{BG} \cong \overline{CF}$, $\angle BGA \cong \angle CFD$ (Base Angles Theorem), and $\overline{AG} \cong \overline{DF}$ (Addition property). By CPCTC, $\overline{AB} \cong \overline{CD}$, so trapezoid $ABCD$ is isosceles.

4. By the Base Angles Theorem, $\angle PLM \cong \angle PML$. Since $\overline{LM}$ is a median, $\overline{LM} \parallel \overline{AD}$, so $\angle PLM \cong \angle APL$ and $\angle PML \cong \angle DPM$. By the transitive property, $\angle APL \cong \angle DPM$. Show $\triangle LAP \cong \triangle MDP$ by SAS. By CPCTC, $\angle A \cong \angle D$, so trapezoid $ABCD$ is isosceles.

5. Since $\angle BAK \cong \angle BKA$. $\overline{BK} \cong \overline{BA}$. $\overline{BA} \cong \overline{CD}$, so $\overline{BK} \cong \overline{CD}$ by the transitive property. Similarly, $\angle BAK \cong \angle CDA$ and $\angle BKA \cong \angle BAK$, so $\angle BKA \cong \angle CDA$, making $\overline{BK} \parallel \overline{CD}$. Since one pair of sides of $BKDC$ are parallel and congruent, $BKDC$ is a parallelogram.

6. Show slope $\overline{PQ}$ = slope $\overline{RS}$ = .75, slope $\overline{QR} \neq$ slope $\overline{PS}$, and $PR = QS = \sqrt{200}$

7. a. Show slope $\overline{BC}$ = slope $\overline{AD}$ = 1, and slope $\overline{AB} \neq$ slope $\overline{CD}$. b. $h = 3$, $k = 2$. c. Since slope $\overline{AB} \times$ slope $\overline{BE} = -1$, $\angle ABE$ is a right angle, so $ABED$ is a rectangle.

8. Show slope $\overline{JA}$ = slope $\overline{KE}$ = 0, slope $\overline{AK} \neq$ slope $\overline{JE}$, and $JK = AE = 5a$, so $JAKE$ is an isosceles trapezoid.

Section 2.6

1. Show $AH = CK = \sqrt{9a^2 + b^2}$.

2. a. $C(t,t)$ b. $AC = BC = t\sqrt{2}$. c. Slope $\overline{BD} \times$ slope $\overline{AC} = -1 \times 1 = -1$.

3. For rectangle $A(0,0)$, $B(0,b)$, $C(d,b)$, and $D(d,0)$, $AC = BC = \sqrt{b^2 + d^2}$.

4. a. Midpoint $\overline{MT}$ = midpoint $\overline{AH} = \left(\frac{s}{2}, \frac{t}{2}\right)$ b. Show $MT \neq AH$.

5. Slope $\overline{BC}$ = slope $\overline{AD}$ = 0, slope $\overline{AB} \neq$ slope $\overline{CD}$, and $AC = BD = \sqrt{c^2 + b^2}$.

6. a. $y = \sqrt{a^2 - b^2}$.

b. Slope $\overline{AC} \times$ slope $\overline{BD} = \frac{y}{a+b} \times \frac{y}{b-a} = \frac{\left(\sqrt{a^2 - b^2}\right)^2}{-\left(a^2 - b^2\right)} = -1$.

7. a. $h = a - b, k = c$. b. Show $AC = BD = \sqrt{(a-b)^2 + c^2}$.

8. If $L(0, b)$ and $M(a + d, b)$ are the endpoints of the midpoints of the legs, then $LM = a + d$, $BC = 2a$, and $AD = 2d$, so $LM = \frac{1}{2}(BC + CD)$.

9. In $\triangle WST$, $\overline{BC} \parallel \overline{WT}$, so $\overline{AC} \parallel \overline{WT}$. Since segments of parallel lines are parallel, $\overline{AW} \parallel \overline{CT}$, making $WACT$ a parallelogram.

10. Let P, Q, R, and S be the midpoints of the sides $\overline{AB}$, $\overline{BC}$, $\overline{CD}$, and $\overline{AD}$, respectively, of the rectangle. Draw diagonal $\overline{AC}$. Since $\overline{PQ}$ and $\overline{SR}$ are parallel to $\overline{AC}$ and equal in length to one-half of AC, $\overline{PQ} \parallel \overline{SR}$, and $\overline{PQ} \cong \overline{SR}$, so $PQRS$ is a parallelogram. Right triangle $PAS \cong$ right triangle PBQ, so $\overline{PS} \cong \overline{PQ}$ by CPCTC, making $PQRS$ a rhombus.

CHAPTER 3

Section 3.1

1. (4) **2.** (2) **3.** (3) **4.** (1) **5.** (4) **6.** –22
7. 5 **8.** –2, 8 **9.** 45 **10.** 117 lb **11.** 14 **12.** 125
13. $\frac{1}{8}$ **14.** 10 **15.** 50 of 35% solution, 30 of 75% solution
16. 40

Section 3.2

1. The fact that $\angle W \cong \angle Y$ and right triangle $HAW \cong$ right triangle KBY implies $\triangle HWA \sim \triangle KYB$.

2. Since $\overline{AG} \cong \overline{AE}$, $\angle AGH \cong \angle AEH$. $\overline{AC}$ bisects $\angle FAB$, so $\angle GAH \cong \angle EAH$. Because $\overline{AB} \parallel \overline{CD}$, $\angle AEH \cong \angle CDH$ and $\angle EAH \cong \angle DCH$. By the transitive property, $\angle AGH \cong \angle CDH$ and $\angle GAH \cong \angle DCH$, so $\triangle AHG \sim \triangle CHD$.

3. Show $\triangle RMN \sim \triangle RAT$. $\angle RMN \cong \angle A$, and $\angle RNM \cong \angle T$. Write $\frac{MN}{AT} = \frac{RN}{RT}$. By the converse of the Base Angles Theorem, $NT = MN$. Substitute NT for MN in the proportion.

4. Show $\triangle SQP \sim \triangle WRP$. $\angle SPQ \cong \angle WPR$ (Angle). Since $\overline{SR} \cong \overline{SQ}$, $\angle SRQ \cong \angle SQR$. $\angle SRQ \cong \angle WRP$ (Definition of angle bisector). By the transitive property, $\angle SQR \cong \angle WRP$ (Angle).

5. Show $\triangle PMQ \sim \triangle MKC$. Right triangle $MCK \cong$ right triangle PMQ. Since $\overline{TP} \cong \overline{TM}$, $\angle TPM \cong \angle TMP$.

6. Show $\triangle PMT \sim \triangle JKT$. Since $\overline{MP} \cong \overline{MQ}$, $\angle MPQ \cong \angle MQP$. Since $\overline{JK} \parallel \overline{MQ}$, $\angle J \cong \angle MQP$ so, by the transitive property, $\angle J \cong \angle MPQ$ (Angle). $\angle K \cong \angle QMT$ (Congruent alternate interior angles). $\angle QMT \cong \angle PMT$ (since $\triangle MTP \cong \triangle MTQ$ by SSS). By the transitive property, $\angle K \cong \angle PMT$ (Angle). Write the proportion $\frac{PM}{JK} = \frac{PT}{JT}$. Substitute TQ for PT (see the "Given") in the proportion.

Section 3.3

1. Show $\triangle AEH \sim \triangle BEF$. $\angle BEF \cong \angle HEA$ and $\angle EAH \cong \angle EBF$ (Halves of equals are equal).
2. First prove the proportion $\frac{JY}{XZ} = \frac{YX}{ZL}$ by proving $\triangle JYX \sim \triangle XZL$. To prove these triangles similar, show $\angle JYX \cong \angle XZL$ and $\angle J \cong \angle ZXL$. Since $KYXZ$ is a parallelogram, $KZ = YX$, so $\frac{JY}{XZ} = \frac{YX}{ZL} = \frac{KZ}{ZL}$. In the proportion $\frac{JY}{XZ} = \frac{KZ}{ZL}$, cross-multiplying gives the desired product.
3. Show $\triangle EIF \sim \triangle HIG$. Since $\overline{EF}$ is a median, $\overline{EF} \parallel \overline{AD}$, so $\measuredangle FEI \cong \measuredangle GHI$ and $\measuredangle EFI \cong \measuredangle HGI$.
4. Rewrite the product as $\frac{RW}{RV} = \frac{TW}{SV}$. Prove $\triangle RSV \sim \triangle RTW$. Since RVW bisects $\angle SRT$ (Given), $\angle SRV \cong \angle TRW$ (Angle). Since $\overline{TW} \cong \overline{TV}$ (Given), $\angle TVW = \angle W$. Also, $\angle RVS \cong \angle TVW$ (Vertical angles are congruent). By the transitive property, $\angle RVS \cong \angle W$ (Angle). Hence, $\triangle RSV \sim \triangle RTW$ by the AA Theorem.
5. a. $\overline{EF} \parallel \overline{AC}$, $\measuredangle ACF \cong \measuredangle GFE$ since parallel lines form congruent alternate interior angles. Similarly, since $\overline{DE} \parallel \overline{AB}$, $\measuredangle AFC \cong \measuredangle EGF$. Hence, $\triangle CAF \sim \triangle FEG$.
   b. $\triangle CAF \sim \triangle CDG$. By the transitive property, $\triangle CDG \sim \triangle FEG$, so $\frac{DG}{EG} = \frac{GC}{GF}$, implying $DG \times GF = EG \times GC$.
6. Draw right triangle ABC with altitude $\overline{CH}$ drawn to hypotenuse $\overline{AB}$. Prove $AC \times BC = AB \times CH$ by showing $\triangle ABC \sim \triangle ACH$, using $\angle A \cong \angle A$ and right triangle $ACB \cong$ right triangle AHC.

Section 3.4

1. (1) **2.** (3) **3.** (4) **4.** (2) **5.** (1) **6.** $\sqrt{320}$
7. $r = 9, s = 16, t = 15$ **8.** $r = 31.2, s = 28.8, t = 5$ **9.** 18
10. 13 **11.** 25 **12.** $\sqrt{27}$ **13.** 12 **14.** 13.1 **15.** 13
16. $9\sqrt{5} + 15$ **17.** (17, 26) and (−11, 6)
18. (a) 3.8 (b) 7.1

Section 3.5

1. (1) **2.** (3) **3.** (2) **4.** (1) **5.** (1) **6.** (2)
7. $10\sqrt{3}$ **8.** (−2, 13) and (−2, −5) **9.** 73.2 **10.** a. 24 b. $96\sqrt{3}$

CHAPTER 4

Section 4.1

1. (1) **2.** (2) **3.** (2) **4.** (3) **5.** 120 **6.** 95
7. 142 **8.** $x = 10, y = 13$

9. Since $OXEY$ is a square, $\overline{OX} \perp \overline{OT}$, $\overline{OY} \perp \overline{PJ}$, $OX = OY$, so $\overline{QT} \cong \overline{JP}$. Subtracting corresponding sides of $m\widehat{QT} = m\widehat{JP}$ and $m\widehat{PT} = m\widehat{PT}$ gives $\widehat{QP} \cong \widehat{JT}$.

10. Since $\overline{DE} \cong \overline{FG}$, $\overline{PB} \cong \overline{PC}$, right triangle $APB \cong$ right triangle ACP by *Hy-Leg* $\cong$ *Hy-Leg*. By CPCTC, $\angle BAP \cong \angle CAP$, so $\overline{PA}$ bisects $\angle FAD$.

Section 4.2

1. (2) **2.** (4) **3.** (3) **4.** (2) **5.** (3) **6.** (4)
7. (2) **8.** 31.4 **9.** 16:9 **10.** 27π **11.** a. 27 b. 40
12. a. 11.5 b. 5.5 **13.** a. $\frac{196\pi}{3}$ b. $\frac{196\pi}{3} - 49\sqrt{3}$
14. a. 120 b. $24\pi + 36\sqrt{3}$

Section 4.3

1. 64 **2.** 66 **3.** 50 **4.** 22 **5.** 60 **6.** 20
7. 70 **8.** 80 **9.** 35 **10.** a. 90 b. 120 c. 15 d. 135
11. a. 30 b. 75 **12.** 25°

Section 4.4

1. (2) **2.** (4) **3.** (3) **4.** 12 **5.** 10 **6.** 5.5
7. 16 **8.** 6 **9.** 12 **10.** 10

11. $\triangle KLP \sim \triangle KJM$ since right triangle $KLP \cong$ right triangle KJM and $\angle LKP \cong \angle MKJ$ (They intercept congruent arcs).

12. Show $\triangle KLP \sim \triangle KJM$. Right triangle $KLP \cong$ right triangle KJM. Vertical angles are congruent so $\angle KPL \cong \angle JPM$. Since $\overline{JP} \cong \overline{JM}$, $\angle JPM \cong \angle JMK$ so, by the transitive property $\angle KPL \cong \angle JMK$.

13. Show $\triangle HBW \sim \triangle MBL$. Because $m\angle ABW$ and $m\angle H$ are each equal to one-half the meassure of the same arc, $\angle H \cong \angle ABW$. Since $ABLM$ is a parallelogram, $\overline{AB} \parallel \overline{ML}$, so $\angle ABW \cong \angle BML$. By the transitive property of congruence, $\angle H \cong \angle BML$ (Angle). Angle HBW is contained in both triangles. Hence, $\angle HBW \cong \angle MBL$ (Angle). Therefore, $\triangle HBW \sim \triangle MBL$, so $\frac{BL}{BW} = \frac{BM}{BH}$.

14. Show $\triangle BCD \sim \triangle ABE$. Because $m\angle DBC$ and $m\angle A$ are each equal to one-half the measure of the same arc, $\angle DBC \cong \angle A$. Angle ABE is a right angle, since a diameter is perpendicular to a chord at the point of tangency. Since $\overline{AB} \parallel \overline{CD}$ and interior angles on the same side of the transversal are supplementary, $\angle DCB$ is a right angle, which means that $\angle DCB \cong \angle ABE$. Therefore, $\triangle BCD \sim \triangle ABE$, so $\frac{BD}{AE} = \frac{CD}{BE}$.

15. Show $\triangle WTK \sim \triangle JTW$. Since $\angle T$ is contained in both triangles, $\angle WTK \cong \angle JTW$ (Angle). Because $m\angle NTJ$ and $m\angle JWT$ are equal to one-half of the measure of the same arc, $\angle JWT \cong \angle NTJ$. It is given that $\overline{TK}$ bisects

$\angle NTW$, which means that $\angle NTJ \cong \angle JTW$. Hence, $\angle JTW \cong \angle JWT$. It is also given that $\overline{WK} \cong \overline{WT}$, so $\angle K \cong \angle JTW$. Since $\angle JTW \cong \angle JWT$ and $\angle K \cong \angle JTW$, by the transitive property of congruence, $\angle K \cong \angle JWT$ (Angle). Then $\triangle WTK \sim \triangle JTW$, so $\frac{JT}{TW} = \frac{TW}{TK}$.

CHAPTER 5

Section 5.1

1. (2) **2.** (3) **3.** (1) **4.** (2) **5.** (4) **6.** (2)
7. (1) **8.** (3) **9.** (1) **10.** (3) **11.** (3) **12.** (4)
13. 0.5 **14.** 4 **15.** 2 **16.** $x, \sqrt[3]{x}, 1, \frac{1}{\sqrt[3]{x}}, \frac{1}{\sqrt{x}}, \frac{1}{x}$

Section 5.2

1. (2) **2.** (3) **3.** (3) **4.** (1) **5.** (3) **6.** (1)
7. (1) **8.** $4b^2 + 5b - 6$ **9.** $2x^2 + 5x - 63$
10. $25w^2 - 64$ **11.** $0.09y^4 - 1$ **12.** $4x^2 - 12x + 9$
13. $9x^2 + 24xy + 16y^2$ **14.** $(x + 5)(x + 3)$ **15.** $(x - 3)(x - 7)$
16. $(y + 3)(y + 3)$ **17.** $(10a + 7b)(10a - 7b)$
18. $(a - 9)(a + 5)$ **19.** $(b + 8)(b - 5)$ **20.** $\left(\frac{2}{3}c+1\right)\left(\frac{2}{3}c-1\right)$
21. $(w - 6)(w - 7)$ **22.** $(0.9 + 0.5x)(0.9 - 0.5x)$
23. $(4n - 1)(n + 3)$ **24.** $(3x - 7)(x + 3)$ **25.** $(5s + 1)(s - 3)$
26. $2y(y + 5)(y - 5)$ **27.** $-5(t + 1)(t - 1)$
28. $4(m^2 + n^2)(m + n)(m - n)$ **29.** $8xy(y + 3)(y - 3)$
30. $2x(x + 8)(x - 7)$ **31.** $\frac{1}{2}x(x+6)(x-6)$
32. $-2(y + 2)(y - 5)$ **33.** $10y^2(y + 10)(y - 5)$
34. b. (2) c. Show $(2n + 3)^2 - (2n + 1)^2 = \underline{8}(n + 1)$.

Section 5.3

1. (2) **2.** (4) **3.** (1) **4.** (3) **5.** $\frac{-3}{x+2}$ **6.** $\frac{-1}{y-a}$
7. $\frac{3}{2-x}$ **8.** $\frac{x+1}{x-2}$ **9.** $\frac{2(x+3)}{x-3}$ **10.** $\frac{2(x-5)}{2x-5}$
11. $\frac{a(ab+1)}{b}$ **12.** $\frac{x-2y}{x+2y}$ **13.** $\frac{1}{x^2-y^2}$
14. $\frac{5x+15}{x^2+2x-3}$ **15.** $\frac{x^2-4}{x^2-x-6}$

Section 5.4

1. (2) **2.** (3) **3.** (3) **4.** $\frac{x^4}{ab^5}$ **5.** $\frac{6(x-7)}{xy^3}$

6. $-\frac{1}{2}$ **7.** $\frac{5}{x+7}$ **8.** $\frac{x(r+s)}{r(r-s)}$ **9.** $\frac{x^2(x+3)}{y(x-3)}$

10. $\frac{-2(2x+1)}{5}$ **11.** $\frac{-4}{9(t+2)}$ **12.** $4(x-2)$

Section 5.5

1. (1) **2.** (3) **3.** (2) **4.** (1) **5.** (1) **6.** (4)

7. 3 **8.** $a+b$ **9.** $\frac{5(x+1)}{(x^2-4)(x+3)}$ **10.** $\frac{1}{a-1}$

11. $\frac{x^2+2x-10}{(x^2-9)(x-2)}$ **12.** $\frac{x-4}{2(x-1)(x+4)}$

Section 5.6

1. (1) **2.** (4) **3.** (1) **4.** (3) **5.** $\frac{1}{y+3}$ **6.** $-w$

7. $\frac{3}{3n+1}$ **8.** $\frac{x}{y}$ **9.** $\frac{m+1}{m-1}$ **10.** $\frac{b+6}{b}$

Section 5.7

1. (2) **2.** (3) **3.** (4) **4.** (3) **5.** (3)

6. $2y^2\sqrt{10}$ **7.** $4x^4$ **8.** $-2x^2y^{5/3}$ **9.** $\frac{64y^6}{125x^3}$

10. $2a^6b^5c^{5/4}$ **11.** $\frac{9b^4}{4a^6}$ **12.** $24\sqrt{3}$ **13.** $5\sqrt{7x}$

14. $-6x\sqrt{3}$ **15.** $3y\sqrt[3]{2}$ **16.** $y^{4/3}$ **17.** $5x^2\sqrt[4]{2}$

18. 44 **19.** -4 **20.** $30-12\sqrt{6}$ **21.** $4\sqrt{3}$

22. $-1.5\sqrt{5}$ **23.** $5\sqrt{2}$ **24.** $5-2\sqrt{5}$

25. $\frac{11+6\sqrt{2}}{7}$ **26.** $-\frac{23+6\sqrt{10}}{13}$

CHAPTER 6

Section 6.1

1. (3) **2.** (3) **3.** (1) **4.** (1) **5.** (4) **6.** (2)

7. (3) **8.** (1) **9.** (4) **10.** $a=-0.5$, $b=4$ **11.** $\frac{16}{3}$

12. ±4 **13.** $2x+h$ **14.** $\frac{-1}{x(x+h)}$ **15.** 0

Section 6.2

1. $x = 3$ **2.** (2, 2) **3.** (2, 7) **4.** a. $-40°$
5. a. (1) $f = 25 + 1.75x$ (2) $g = 49 + x$ b. 32
6. a. Day: $y = 5x + 10$, Knight: $y = 6x - 1.5(20 - x)$ b. 90

Section 6.3

1. (4) **2.** (1) **3.** b. $(1, -4); x = 3$ **4.** b. $(3, 2); x = 3$
5. b. $(-0.5, 3.75); x = -0.5$
6. a.

x	−4	−3	−2	−1	0	1	2
y	0	5	8	9	8	5	0

Section 6.4

1. (4) **2.** (3) **3.** (1) **4.** (2) **5.** (3) **6.** $\pm\frac{3\sqrt{3}}{2}$
7. $\pm\frac{\sqrt{2}}{2}$ **8.** $\{-1, -2\}$ **9.** $\{0, 6\}$ **10.** $\left\{\frac{1}{3}, -5\right\}$
11. $\left\{-\frac{1}{2}, 3\right\}$ **12.** $\left\{\frac{2}{3}\right\}$ **13.** $\left\{-\frac{1}{3}, \frac{3}{2}\right\}$
14. $\{-3, 7\}$ **15.** $\{1, 6\}$ **16.** $\{0, 5\}$
17. a. $\frac{-x^2+4x+3}{1-x^2}$ b. $-\frac{1}{5}, -3$ **18.** a. 1.5, 7.5 b. 9 **19.** 7 in.
20. 6.16, −0.16 **21.** 2.46, 0.54 **22.** 4.44, 0.56
23. 8.7 **24.** (1, 6), (4, 9) **25.** (2, −2), (5, 1)
26. (2, −3), (6, 5)

Section 6.5

1. (2) **2.** (1) **3.** (3) **4.** $96{,}800 \text{ yd}^2$
5. a. 4900 ft b. 30 **6.** 0.3 **7.** $w = 9.75$ in., $h = 10.25$ in.
8. 17
9. a. $c(x) = 0.06x^2$ b. $w = \sqrt{11.5}$ in., $\ell = 3\sqrt{11.5}$ in., $h = 1.5\sqrt{11.5}$ in.
10. a. 384 ft b. 240 ft
c. 4.29 (Compare the two functions using the table feature of your calculator with the step value adjusted to .001. The function values will be closest when $x \approx 14.29$ or $14.29 - 10 = 4.29$ sec after the *second* rocket is launched.)

CHAPTER 7

Section 7.1

1. (4) **2.** (1) **3.** (2) **4.** (2) **5.** (2) **6.** (2)
7. (4) **8.** (2) **9.** $x < -\frac{7}{3}$ or $x > 5$ **10.** $-3 \le x \le 7$

11. $x \geq -\frac{1}{2}$ **12.** 5 **13.** $2.5 < t < 5.5$

14. a. (3) b. min. = 22.2, max. = 30

Section 7.2

1. (4) **2.** (1) **3.** $x < -5$ or $x > 1$
4. $0 < x < 7$ **5.** $-0.5 < x < 3$ **6.** $t \leq -2$ or $t \geq 3$
7. $-9 \leq x \leq 7$ **8.** all real numbers **9.** $1 \leq t \leq 3$
10. $8.5 \leq t \leq 15$

Section 7.3

1. (4) **2.** (1) **3.** (2) **4.** (3) **5.** $n = 40$ **6.** $x = 4$
7. $x = 7$ **8.** $x = -1$ **9.** $y = 3$ **10.** $n = 7$ **11.** $x = -4$
12. $r = -5, -1$ **13.** $t = 2$ **14.** $x = 125$
15. $x = \frac{16}{81}$ **16.** $y = -4$ **17.** 3.2 **18.** $-3, 0$

Section 7.4

1. $x = 5, -1$ **2.** $b = 5$ **3.** $y = -5, 2$ **4.** $t = \frac{3}{2}$
5. $x = -\frac{4}{3}, 6$ **6.** $x = -2, 1$ **7.** $m = 2, 6$
8. $r = -2$ **9.** a. $\frac{6}{t} + \frac{6}{t+16} = 1$ b. 8
10. 30 **11.** 35 mph **12.** 6 **13.** 3 mph **14.** 200
15. 2 mph **16.** 16 **17.** 28

Chapter 8

Section 8.1

1. (4) **2.** (3) **3.** (2) **4.** (4) **5.** (3) **6.** (1)
7. $21i$ **8.** $-5i$ **9.** -28 **10.** 1 **11.** $-5i\sqrt{3}$ **12.** $-i$
13. $9i$ **14.** 0 **15.** $\frac{1}{i}$

Section 8.2

1. (3) **2.** (1) **3.** (2) **4.** (4) **5.** (2) **6.** (4)
7. (4) **8.** (3) **9.** (1) **10.** (2) **11.** $11 - 2i$
12. $0.8 + 1.6i$ **13.** $45 + 0i$
14. $-0.2 + 1.6i$ **15.** $\frac{\sqrt{2}}{11} + \frac{3}{11}i$

Section 8.3

1. (2) **2.** (1) **3.** (2) **4.** (4) **5.** (3) **6.** 4

7. 4 **8.** $\dfrac{1\pm\sqrt{61}}{6}$ **9.** $\dfrac{5\pm\sqrt{23}}{2}$

10. $3\pm\sqrt{3}i$ **11.** $\dfrac{1}{3}\pm\dfrac{1}{3}i$ **12.** $\dfrac{7}{4}\pm\dfrac{3}{4}i$ **13.** $-6\pm2\sqrt{10}$

14. $\dfrac{4}{5}\pm\dfrac{3}{5}i$ **15.** $\dfrac{3}{2}\pm\dfrac{1}{2}i$ **16.** $-4\pm3i$

17. a. $1\pm\dfrac{\sqrt{3}}{3}i$ b. 2 **18.** a. 0.1, 3.9 **19.** 12.6

20. 11.72, 23.44 **21.** a. 23.5 b. $6.7 < t < 16.8$

Section 8.4

1. (2) **2.** (4) **3.** (2) **4.** (3) **5.** (1)

6. $x^2+x-2=0$ **7.** $3x^2+x-2=0$ **8.** $x^2-4x+1=0$

9. $x^2+2x+5=0$

CHAPTER 9

Section 9.1

1. (1) **2.** (3) **3.** (4) **4.** (2) **5.** (1)

6. $y=-\dfrac{1}{3}x$ **7.** $y=-\dfrac{1}{2}x+1$ **8.** $y=-\dfrac{1}{2}x$

9 and 10.

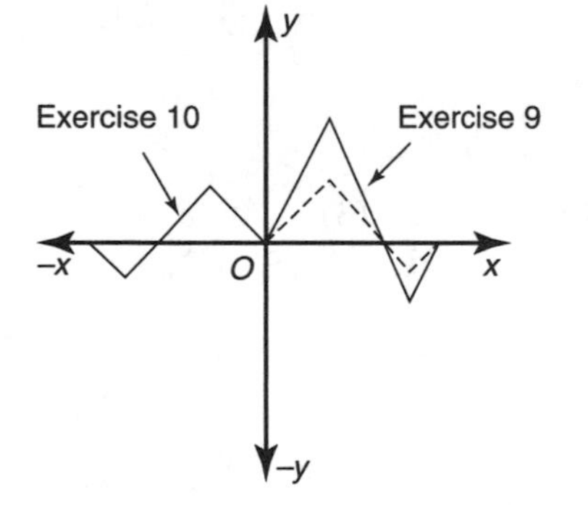

11 and 12.

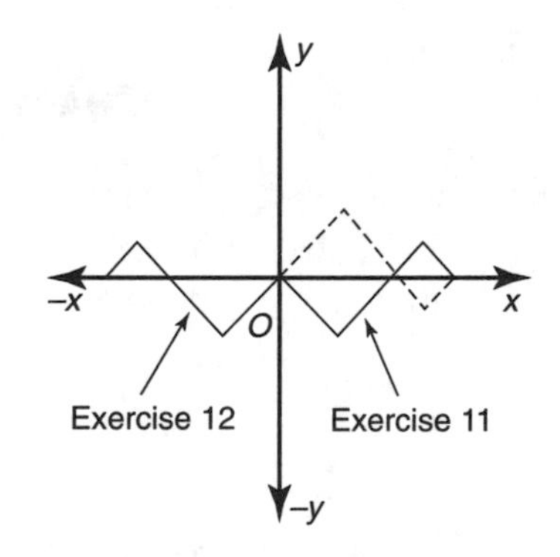

13.

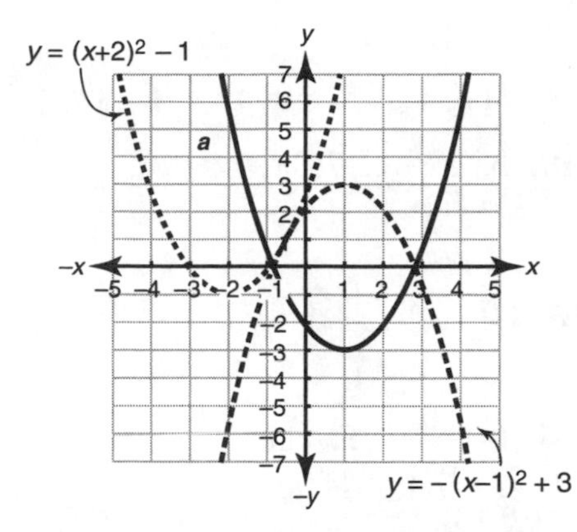

14. a. b. $x = 3\frac{1}{2}$

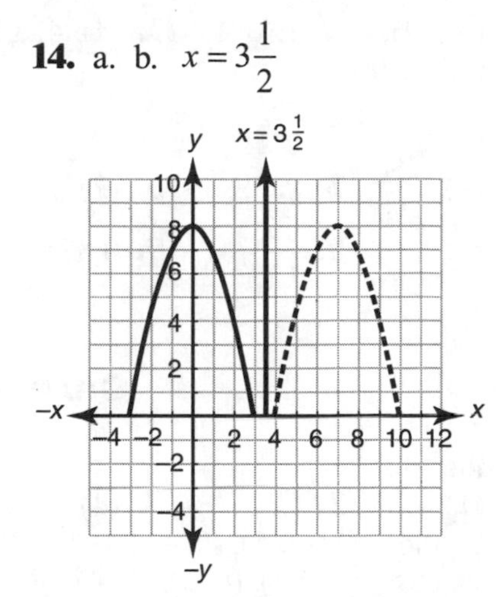

Section 9.2

1. (1) **2.** (3) **3.** (2) **4.** (4)
5. b. L′(7, −1) c. $7 - i$ **6.** a. $g(x) = -x^2 - 2x$ b. 3, −1

Section 9.3

1. (1) **2.** (2) **3.** (1) **4.** (3) **5.** (3) **6.** (3)
7. (3) **8.** (4) **9.** (1) **10.** ±2
11. a. $g(c) = \dfrac{C}{2\pi}$ b. $f(g(C)) = \dfrac{C^2}{4\pi}$ c. 19%
12. a. $0.78\sqrt[3]{7500 + 45t^2} + 15{,}000$ b. 15,020.8

Section 9.4

1. (2) **2.** (2) **3.** (1) **4.** (3) **5.** (2) **6.** (3)
7. (4) **8.** (4) **9.** (1) **10.** (4) **11.** (2) **12.** (1)
13. (1) **14.** (3) **15.** (1, 7)
16. a. $A'(0, 0), B'(-1, 8), C'(-4, 8)$ b. (1)
17. a. $A'(0, -8), B'(8, -7), C'(8, -4)$, b. (4)
18. (a) E (b) D (c) F (d) B

Section 9.5

1. (3) **2.** (4) **3.** (3) **4.** (1) **5.** (2) **6.** (4)
7. (1) **8.** a. $x \geq -3; y \geq 0$ b. $f^{-1}(x) = 4x^2 - 3; x \geq 0, y \geq -3$
9. b. (0, 0), (0.5, 0.5)

Section 9.6

1. (2) **2.** (3) **3.** (3) **4.** (1) **5.** (3)
6. a. $(x + 1)^2 + (y - 2)^2 = 36$ b. $(x + 2)^2 + (y + 1)^2 = 36$
c. $(x + 2)^2 + (y + 1)^2 = 144$
7. 15 **8.** Eight **9.** c. $xy = 10$ **10.** (2, 1), (6, −3)

11. (5, –4), (2.2, 1.6) **12.** (±6.4, ±2.8), (±2.8, ±6.4)
13. (±1.3, 3.8)

Section 9.7

1. (4) **2.** (1) **3.** (2) **4.** (1) **5.** 4000 **6.** 8
7. 8 **8.** $\frac{32}{135}$ **9.** 0.375 **10.** b. 28 c. 75%

Chapter 10

Section 10.1

1. (1) **2.** (3) **3.** (4) **4.** (4) **5.** (1)
6. b. Plot: (1, –0.5), (0, –1), (–1, –2), (–2, –4), (–3, –8).
c. Plot: (5, 2.5), (6, 3), (7, 4), (8, 6), (9, 10).
7. b. Plot: (–1, –9), (0, –5), (1, –3), (2, –2), (3, –1.5).
c. Plot: (2, 18), (0, 10), (2, –6), (4, –4), (6, –3).

Section 10.2

1. (3) **2.** (1) **3.** (1) **4.** (3) **5.** (4) **6.** (3)
7. (4) **8.** (3) **9.** (2) **10.** (3) **11.** (2)
12. b. $r_{y=x}$ **13.** b. 5 **14.** c. $y = 2^{-x}$ or $y = \left(\frac{1}{2}\right)^x$

Section 10.3

1. (3) **2.** (2) **3.** (2) **4.** (2) **5.** (4) **6.** (2)
7. (1) **8.** (4) **9.** (3) **10.** (2) **11.** a. y^{x+2} b. 4
12. a. $\frac{1}{2}y - x$ b. 9 **13.** a. (1) $p^2 - \frac{1}{2}q$ (2) $\frac{1}{3}(p-q)$ b. 9998
14. –4, 5 **15.** 2.2

Section 10.4

1. (2) **2.** (3) **3.** (1) **4.** (3) **5.** –2, 1
6. b. ±1, 2 **7.** $x = 4, y = -7$ **8.** a. 0, 1 b. –2.42
9. a. $\frac{5}{3}$ b. 2.22 **10.** a. $h = 7(0.75)^n$ b. 7

Section 10.5

1. (2) **2.** (3) **3.** 3.1 **4.** a. 550 b. 2007
5. a. $a = 5000, b = 1.14$ b. 12, 511 c. 6.7
6. a. 80 b. 61.6% c. 164.2 **7.** 8 yr, 9 months
8. a. 21% b. 2004 **9.** a. 0.34 b. 32% c. 10
10. 7 hr 21 min **11.** 12.2%

CHAPTER 11

Section 11.1

1. $y = 300 + 0.15(x - 1)$ **2.** a. $y = -3.5x + 17.5$ b. 17.5 c. 5:00 P.M.
3. 140 **4.** not linearly related **5.** linearly related
6. linearly related **7.** not linearly related
8. a. $A: y = 0.21x + 11$, $B: y = 0.10x + 20$ b. (1)
9. a. $y = (x - 5)r + C$ b. \$6.85

Section 11.2

1. (3) **2.** (4) **3.** (1) **4.** (1) **5.** (3)
6. a. $w = 8.22L - 77.11$ c. 25.6 d. 17.7
7. a. $a = 0.88$, $b = 49.94$ b. $\overline{x} = 631$, $\overline{y} = 606$ c. 605
8. b. $y = -3.35x + 99.78$ c. 87.5
9. b. $y = 0.72x - 0.13$ c. 2.0 d. 4.3
10. a. $y = -11.28x + 114.40$ b. 46.7 c. 10 hr 8 min
11. a. $y = -6.2x + 12{,}451.2$ b. 20.2 thousand c. 2008

Section 11.3

1. $a = -1, b = 3, c = -8$ **2.** a. $a = 0.045, b = 1.05, c = 0$ b. 194
3. a. $f(x) = -x^2 + 12x + 55$ b. 2 hr 41 min c. 91

Section 11.4

1. a. $y = 77.322(1.013)^x$ b. 364.3 c. 2027
2. a. $y = 61.8(1.2)^x$ b. 62 c. 6 days 17 hr
3. b. $T = 0.63(1.02)^L$, $T = 0.18L^{0.53}$, $T = -0.73 + 0.56 \ln L$ c. power
4. a. $d = 11.37s^{1.05}$, $d = 0.45s^{1.51}$, $d = -265.31 + 111.24 \ln s$
b. power c. 191
5. b. $V = 22.5P^{-1}$ **6.** a. $y = 210.53(0.94)^x$ b. 2014
7. a. $y = 0.2(1.37)^x$ b. 2005
c. exponential function: $r = 0.99$, power function: $r = 0.92$
8. a. $y = -1.4x + 11.6$ b. 4.1 c. $y = 20(0.7)^x$ d. 10 hrs 21 min
e. exponential function, because $r \approx -0.99$ vs. $r \approx -0.92$ for a linear function

CHAPTER 12

Section 12.1

1. (2) **2.** (3) **3.** (3) **4.** (4) **5.** (1) **6.** (3)
7. $\frac{41}{59{,}049}$ **8.** $\frac{10}{64}$ **9.** $\frac{16}{1024}$ **10.** a. $\frac{1024}{16{,}807}$ b. $\frac{6183}{16{,}807}$
11. $\frac{9018}{32{,}768}$ **12.** a. 0.2401 b. 0.3773

Section 12.2

1. (2) **2.** (3) **3.** (1) **4.** (4) **5.** (3) **6.** (1)
7. (3)

Section 12.3

1. 56 **2.** $\frac{289}{9}$ **3.** 98 **4.** 2 **5.** 3 **6.** 15

7. 55 **8.** $\frac{3}{4}$ **9.** $4+5\sqrt{2}$

10. a. (1) 4.24 (2) 0.21 b. $y=5.111x^{0.065}$ c. 4.61

11. a. (1) 79.4 (2) 8.4 b. 66.7% c. $\frac{8}{36}$

12. a. (1) 54.3 (2)5.4 b. 20% c. $\frac{14}{20}$

13. a. (1) 79 (2) 10.1 b. $\frac{84}{435}$ **14.** a. (1) 68.3 (2) 2.3 b. $\frac{2254}{8000}$

Section 12.4

1. (1) **2.** (2) **3.** (4) **4.** (2) **5.** (1) **6.** (3)
7. (3) **8.** (2) **9.** (3) **10.** (1) **11.** 16 **12.** 38

13. $\sigma = 8.7$ and $\bar{x} = 79.7$, so 70% of the scores are within one standard deviation of the mean. To help decide whether the results of the examination approximate a normal distribution, compare the percents of scores that fall within 1, 1.5, and 2 standard deviations, as shown in the accompanying table.

Interval	Percent in the Set of 20 scores	Percent in a Normal Distribution
$\pm 1\sigma$	70%	68.2%
$\pm 1.5\sigma$	90%	86.6%
$\pm 2\sigma$	95%	95.4%

The table indicates that the results of the examination *approximate* a normal distribution.

14. a. 0.624 b. 2.3%

CHAPTER 13

Section 13.1

1. (2) **2.** (2) **3.** (2) **4.** (3) **5.** (1)
6. a. 13.8 b. 690 **7.** 4.1 **8.** a. 12.7 b. 54.7

Section 13.2

1. (4) **2.** (2) **3.** (2) **4.** (4) **5.** (2) **6.** (4)
7. (1) **8.** 9.42 **9.** 2240π **10.** 33π

Section 13.3

1. (4) **2.** (4) **3.** (3) **4.** (3) **5.** (3) **6.** (3)
7. (4) **8.** (4) **9.** (2) **10.** (3) **11.** (3)
12. $\frac{12}{13}$ **13.** $\frac{1}{2}$ **14.** $-\frac{17}{15}$ **15.** $-\frac{17}{15}$

Section 13.4

1. $-\cos 70°$
2. $-\tan 30°$
3. $-\sin 30°$
4. $\cos 20°$
5. $\tan 45°$
6. $\csc 30°$
7. $-\sec 50°$
8. $-\cot 70°$
9. $-\cos\frac{\pi}{3}$
10. $-\sin\frac{\pi}{6}$
11. $-\tan\frac{5\pi}{12}$
12. $-\csc\frac{\pi}{9}$
13. $-\cot 20°$
14. $-\sec 15°$
15. $-\sin 15°$
16. $3\tan x(\tan x+8)(\tan x-8)$
17. $2(\sin\theta+3\cos\theta)(\sin\theta-3\cos\theta)$
18. $3(2\sin x-1)(\sin x+1)$
19. $\frac{1}{\sin y-2}$
20. $\frac{-1}{\tan x+1}$
21. $\frac{1}{\cos x-3}$
22. $\sin x+\cos x$
23. $-\frac{2}{3}$
24. $-\frac{1}{4}, -1$
25. $-\frac{5}{4}, 6$

Section 13.5

1. (4) **2.** (2) **3.** (3) **4.** (1) **5.** (1) **6.** (3)
7. (2) **8.** (1) **9.** (3) **10.** (4) **11.** (1) **12.** (4)
13. (1) **14.** (2) **15.** (4) **16.** (4) **17.** (1)
18. $0, \frac{\pi}{2}, \pi$

Section 13.6

1. (2) **2.** (3) **3.** (4) **4.** (2) **5.** (1) **6.** (2)
7. (1) **8.** (2) **9.** (3) **10.** (1) **11.** $A=0.6, B=0.4\pi$
12. a. July 15 b. 6600

Section 13.7

1. a. See graph b. (3)

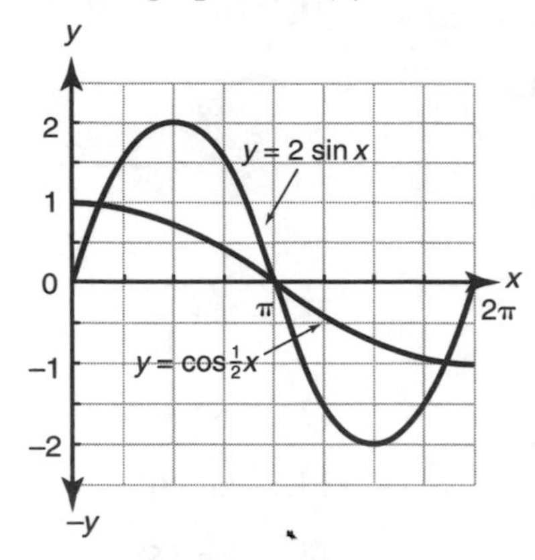

2. a. See graph b. π

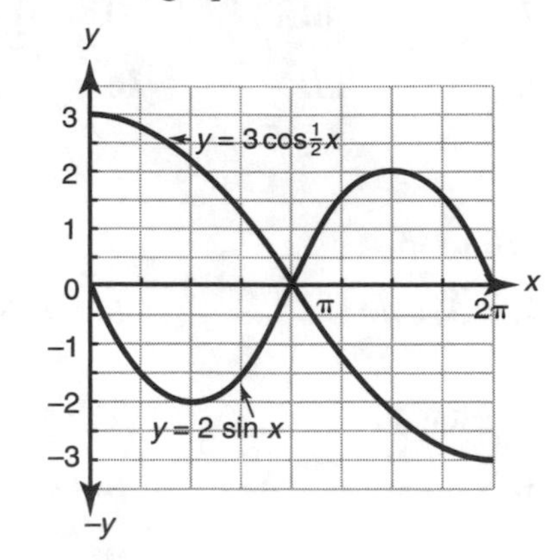

3. a. See graph b. $\pm\frac{\pi}{2}$

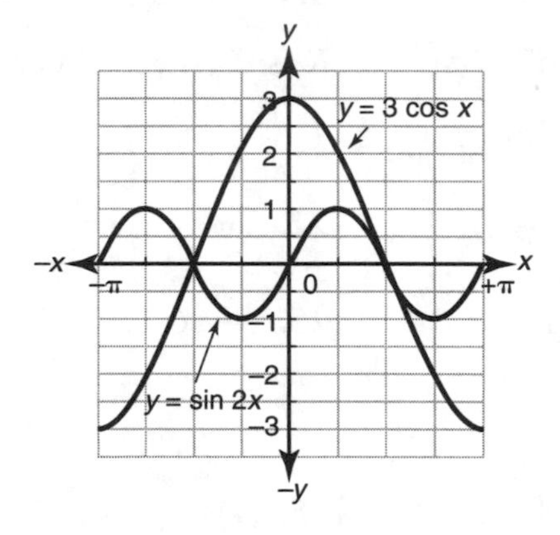

4. a. 6.3 b. 9 c. See graph

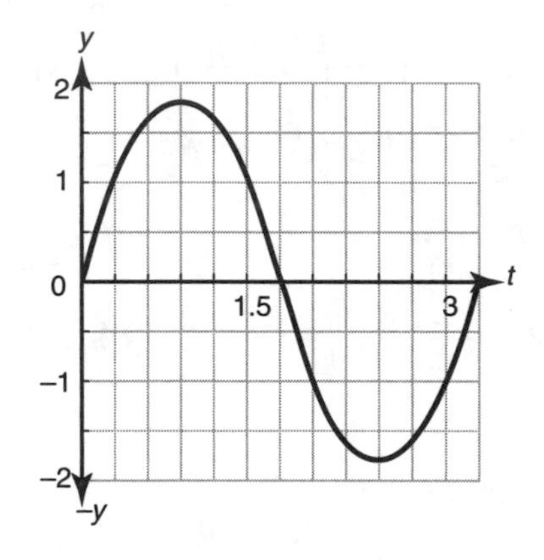

5. a. 40 b. 2 c. See graph

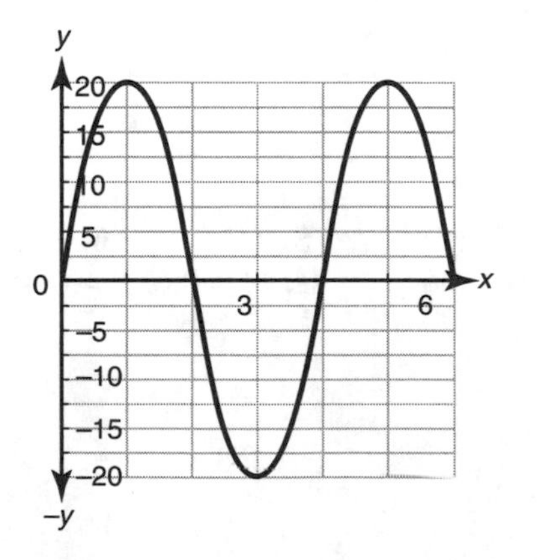

6. a. $d = 8 \sin \frac{\pi}{3} t$ b. See graph

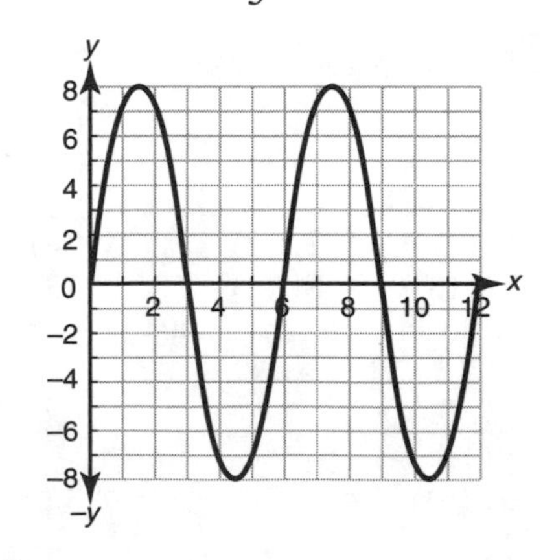

Section 13.8

1. (2) **2.** (3) **3.** (4)
4. (2) **5.** $y = 2 \sin x + 3$
6. $y = 3 \sin x - 4$
7. $y = 3 \cos x - 2$ **8.** $y = 5 \cos x + 4$
9. a. 2 b. $a = \frac{\pi}{3}, b = \frac{7\pi}{3}$
c. See graph.
10. $A = 2000, B = \frac{\pi}{3}(\approx 1.05), D = 4000$
11. $A = 1.5, B = \frac{\pi}{6}(\approx 0.5), D = 6.5$

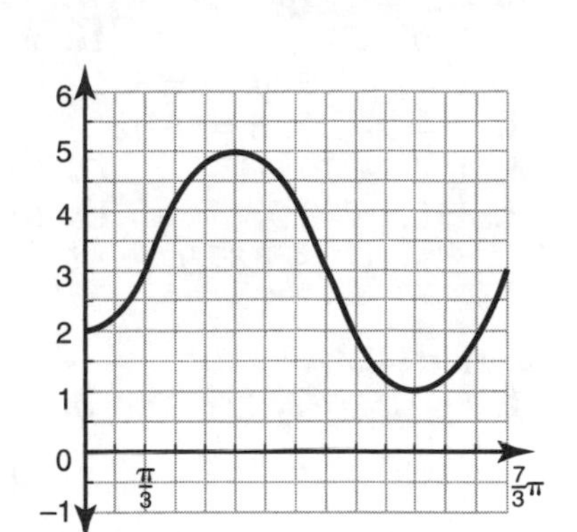

Section 13.9

1. (3) **2.** (4) **3.** (1)
4. a. See graph b. 1

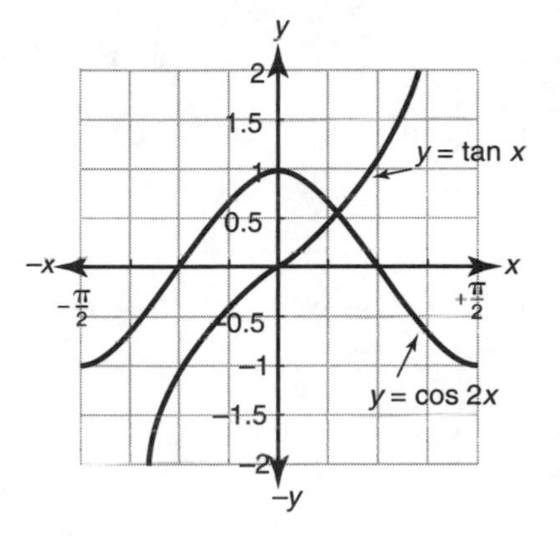

Section 13.10

1. (2) **2.** (3) **3.** (3) **4.** (2) **5.** (2)
6. (1) **7.** (1) **8.** (4) **9.** (4) **10.** (3)

Chapter 14

Section 14.1

1. (1) **2.** (3) **3.** (3) **4.** (4) **5.** (4) **6.** (3)
7. (2) **8.** $\tan^2 A$ **9.** $\sec^2 A$

Section 14.2

1. (3) **2.** (2) **3.** (3) **4.** (1) **5.** (3)
6. 45°, 135°, 225°, 315° **7.** 30°, 150°, 210°, 330°

8. 0°, 120°, 180°, 300° **9.** 90°, 135°, 270°, 315°
10. 120°, 240° **11.** π
12. $x = 240°, 300°; y = 90°$ **13.** 102.68
14. $A = 143.13, B = 66.42$ **15.** 228.6, 311.4
16. 194.5, 270, 345.5 **17.** 45, 108.4, 225, 288.4
18. 111.5, 248.5 **19.** 61.3, 140.5, 241.3, 320.5
20. 53.6, 126.4, 187.9, 352.1

Section 14.3
1. (1) **2.** (2) **3.** (4) **4.** (2) **5.** (3) **6.** (1)
7. (4) **8.** (3) **9.** (4) **10.** $\frac{13}{85}$ **11.** $\frac{36}{85}$
12. $\frac{4-3\sqrt{3}}{3-4\sqrt{3}}$ **13.** $\frac{10}{\sqrt{2}}$ **14.** 0°, 120°, 180°, 240°
15. a. 0.5 b. 66.25

Section 14.4
1. (1) **2.** (4) **3.** (3) **4.** (1) **5.** (2) **6.** $\tan^2 A$
7. 0°, 60°, 180°, 300° **8.** 90, 194.5, 270, 345.5
9. 41.4, 180, 318.6 **10.** 41.8, 138.2, 210, 330

Section 14.5
1. (2) **2.** (3) **3.** (1) **4.** (4) **5.** $-\frac{3}{5}$ **6.** $\frac{1}{\sqrt{15}}$
7. $\frac{10}{9}$ **8.** $\frac{20}{29}$

Chapter 15

Section 15.1
1. (3) **2.** (2) **3.** (2) **4.** (1) **5.** (3) **6.** 150
7. 25 **8.** 49.8, 65.1, 65.1 **9.** 15.1 **10.** 81.3

Section 15.2
1. (3) **2.** (4) **3.** (4) **4.** a. 6.75 b. 16.71
5. a. 41.1 b. 63.7 **6.** 23 ft 5 in. **7.** 796
8. (a) 80 (b) 30° (c) 12.9 **9.** (a) 18 (b) 48.6 **10.** 585.9
11. 254.7

Section 15.3
1. (4) **2.** (3) **3.** (4) **4.** (2) **5.** (3) **6.** (1)
7. (4) **8.** (2) **9.** (1) **10.** (2) **11.** (2) **12.** (3)

Section 15.4

1. (4) **2.** (3) **3.** (2) **4.** (3) **5.** (4) **6.** 54.7
7. 109.3 **8.** a. 100 b. 67 **9.** 10.6 **10.** 42.6
11. a. 5.7 b. 1.1 **12.** $-\frac{2}{7}$ **13.** 16 **14.** a. 9 b. 31.6
15. 955

Section 15.5

1. 66.6 **2.** 105.6 **3.** 60.8 **4.** 451.2 **5.** 240.9
6. a. 26.1 b. 29

GLOSSARY OF MATH B TERMS

abscissa The x-coordinate of a point in the coordinate plane. The abscissa of the point (2, 3) is 2.

absolute value The absolute value of a number x, denoted by $|x|$, is its distance from zero on the number line. Thus, $|x|$ always represents a nonnegative number.

acute angle An angle whose measure is less than 90 and greater than 0.

acute triangle A triangle that contains three acute angles.

additive inverse The opposite of a number. The additive inverse of a number x is $-x$ since $x + (-x) = 0$. The additive inverse of $2 - 3i$ is $-2 + 3i$.

altitude A segment that is perpendicular to the side to which it is drawn.

ambiguous case The situation in which the measures of two sides and an angle that is not included between the two sides are given. These measures may determine one triangle, two triangles, or no triangles.

amplitude One-half of the positive difference between the maximum and minimum values of a periodic function. The amplitude of a sine or a cosine function of the form $y = a \sin bx$ or $y = a \cos bx$ is $|a|$, which is the maximum height of the graphs of these functions.

angle The union of two rays that have the same endpoint.

angle of depression The angle formed by a horizontal line of vision and the line of sight when viewing an object beneath the horizontal line of vision.

angle of elevation The angle formed by a horizontal line of vision and the line of sight when viewing an object above the horizontal line of vision.

antilogarithm The number whose logarithm is given.

arc A curved part of a circle. If the degree measure of the arc is less than 180, the arc is a **minor arc**. If the degree measure of the arc is greater than 180, the arc is a **major arc**. A **semicircle** is an arc whose degree measure is 180.

arccos x The angle A such that $\cos A = x$ and $0° \le A \le 180°$.

arcsin x The angle A such that $\sin A = x$ and $-90° \le A \le 90°$.

arctan x The angle A such that $\tan A = x$ and $-90° < A < 90°$.

area of a triangle One-half the product of the lengths of any two sides of a triangle and the sine of the included angle.

associative property The mathematical law that states that the order in which three numbers are grouped so they can be added or multiplied does *not* matter. For instance, $(2 \times 3) \times 4 = 2 \times (3 \times 4)$.

asymptote A line that a graph approaches but does not intersect as x increases or decreases without bound. The graph of $y = 2^x$ has the negative x-axis as an asymptote. The line $x = \frac{\pi}{2}$ is an asymptote of the graph of $y = \tan x$.

axis of symmetry For the parabola $y = ax^2 + bx + c$, the vertical line $x = \frac{b}{2a}$.

Bernoulli experiment A probability experiment in which there are exactly two possible outcomes. If one of the possible outcomes is considered a "success" with a probability of p, then the remaining outcome is a "failure" with a probability of $1 - p$. The probability of k successes out of n trials is given by the expression ${}_nC_k p^k(1 - p)^{n-k}$.

binomial A polynomial with two unlike terms, as in $2x + y$.

binomial theorem A formula that tells how to expand a binomial of the form $(a + b)^n$, where n is a positive integer, without performing repeated multiplications. For example, $(a + b)^3 = {}_3C_0a^3 + {}_3C_1a^2b^1 + {}_3C_2a^1b^2 + {}_3C_3b^3$.

bisect To divide into two congruent parts.

central angle The angle whose vertex is at the center of a circle and whose sides are radii.

change of base formula The formula $log_b N = \dfrac{\log N}{\log b}$ used to change from a logarithm with base b to an equivalent base 10 logarithm expression.

chord A line segment whose endpoints are points on a circle.

circle The set of points (x, y) in the plane that are a fixed distance r from a given point (h, k) called the *center*. Thus, an equation of a circle is $(x - h)^2 + (y - k)^2 = r^2$.

coefficient of linear correlation A number from −1 to +1, denoted by r, that represents the magnitude and direction of a linear relationship, if any, between two sets of data. If a set of data points are closely clustered about a line, then $|r| \approx 1$. The sign of r depends on the sign of the slope of the line about which the data points are clustered.

collinear points Points that lie on the same line.

combination A selection of objects in which the order of the individual objects is not considered. For example, in selecting a committee of three students from {Alice, Bob, Carol, Dave, Kira}, the combination Alice-Bob-Kira represents the same selection as Bob-Kira-Alice.

combination formula The combination of n objects taken r at a time, denoted by ${}_nC_r$, is given by the formula ${}_nC_r = \dfrac{n!}{r!(n-r)!}$. For example, the number of different 3-member committees that can be selected from a group of 5 students is ${}_5C_3 = \dfrac{5!}{3!(5-3)!} = \dfrac{5 \times 4 \times \cancel{3!}}{\cancel{3!} \times 2 \times 1} = 10$.

common logarithm A logarithm whose base is 10.

commutative property The mathematical law that states that the order in which two numbers are added or multiplied does *not* matter. For example, 2 + 3 = 3 + 2.

complementary angles Two angles whose degree measures add up to 90.

complex fraction A fraction comprised of other fractions, as in $\dfrac{\frac{1}{2}+1}{\frac{3}{4}}$.

complex number A number that can be written in the form $a + bi$, where $i = \sqrt{-1}$ and a and b are real numbers. The set of real numbers is a subset of the set of complex numbers.

composition of functions A function formed by using the output of one function as the input to a second function. The composition of function f followed by function g is the composite function denoted by $g \circ f$, consisting of the set of function values $g(f(x))$, provided $f(x)$ is in the domain of function g.

composition of transformations A sequence of transformations in which a transformation is applied to the image of another transformation.

congruent angles (or sides) Angles (or sides) that have the same measure. The symbol for congruence is ≅.

congruent circles Circles with congruent radii.

congruent parts Pairs of angles or sides that are equal in measure.

congruent polygons Two polygons with the same number of sides are congruent if their vertices can be paired so that all corresponding sides have the same length and all corresponding angles have the same degree measures.

congruent triangles Two triangles are congruent if any one of the following conditions is true: (1) the three sides of one triangle are congruent to the corresponding sides of the other triangle (SSS $\cong$ SSS); two sides and the included angle of one triangle are congruent to the corresponding parts of the other triangle (SAS $\cong$ SAS); two angles and the included side of one triangle are congruent to the corresponding parts of the other triangle (ASA $\cong$ ASA); two angles and the side opposite one of these angles is congruent to the corresponding parts of the other triangle (AAS $\cong$ AAS).

conjugate pair The sum and difference of the same two terms, as in $a + b$ and $a - b$.

constant A quantity that is fixed in value. In the equation $y = x + 3$, x and y are variables and 3 is a constant.

coordinate The real number that corresponds to the position of a point on the number line.

coordinate plane The region formed by a horizontal number line and a vertical number line intersecting at their zero points, called the *origin*.

cosecant The reciprocal of the sine function.

cosine ratio In a right triangle, the ratio of the length of the leg adjacent to a given acute angle to the length of the hypotenuse. If an angle θ is in standard position, then $\cos\theta = \frac{x}{r}$, where $P(x, y)$ is any point on the terminal side of angle θ and $r = \sqrt{x^2 + y^2}$.

cotangent The reciprocal of the tangent function.

coterminal angles Angles in standard position whose terminal sides coincide.

degree A unit of angle measurement defined as 1/360 of one complete rotation of a ray about its vertex.

degree of a monomial The sum of the exponents of its variable factors. For example, the degree of $3x^4$ is 4, and the degree of $-5xy^2$ is 3 since 1 (the power of x) plus 2 (the power of y) equals 3.

degree of a polynomial The greatest degree of its monomial terms. For example, the degree of $x^2 - 4x + 5$ is 2.

dependent variable For a function of the form $y = f(x)$, y is the dependent variable.

diameter A chord of a circle that contains the center of the circle.

dilation A transformation in which a figure is enlarged or reduced in size according to a given scale factor.

direct isometry An isometry that preserves orientation.

discriminant In the quadratic formula $x = \frac{-b \pm \sqrt{b^2 - 4ac}}{2a}$, the discriminant is the quantity underneath the radical sign, $b^2 - 4ac$. If the discriminant is positive, the two roots are real; if the discriminant is 0, the two roots are equal; and if the discriminant is negative, the two roots are imaginary.

distributive property of multiplication over addition For any real numbers a, b, and c, $a(a + c) = ab + ac$ and $(b + c)a = ba + ca$.

domain of a relation The set of all possible first members of the ordered pairs that comprise a relation.

domain of a variable The set of all possible replacements for a variable. Unless otherwise indicated, the domain of a variable is the largest possible set of real numbers.

ellipse An oval-shaped curve, an equation of which is $ax^2 + by^2 = c$, where a, b, and c have the same sign.

equation A statement that two quantities have the same value.

equilateral triangle A triangle in which the three sides have the same length.

equivalent equations Two equations that have the same solution sets. The equations $2x = 6$ and $x + 1 = 4$ are equivalent since they have the same solution, $x = 3$.

event A subset of the set of all possible outcomes of a probability experiment. In flipping a coin the set of all possible outcomes is {head, tail}. One possible event is flipping a head, and another possible event is flipping a tail.

exponent In x^n, the number n is the exponent and indicates the number of times the base x is used as a factor in a product. Thus, $x^3 = x \cdot x \cdot x$.

exponential equation An equation in which the variable appears in an exponent, as in $2^{x+1} = 16$.

exponential function A function of the form $y = b^x$, where b is a positive number other than 1.

exponential regression model See *regression model.*

extrapolation The process of estimating a y-value from a table, graph, or equation using a value of x that falls *outside* the range of observed x-values.

extremes In the proportion $\frac{a}{b} = \frac{c}{d}$, the terms a and d are the extremes. The extremes are the first and fourth terms of the proportion.

factor A number or variable that is being multiplied in a product. A number or variable is a factor of a given product if it divides that product without a nonzero remainder.

factoring The process by which a number of polynomial is written as the product of two or more terms.

factoring completely Factoring a number or polynomial into factors that cannot be factored further.

factorial *n* Denoted by $n!$ and defined for any positive integer n as the product of consecutive integers from n to 1. Thus, $5! = 5 \cdot 4 \cdot 3 \cdot 2 \cdot 1 = 120$.

FOIL The rule for multiplying two binomials horizontally by forming the sum of the products of the first terms (F), the outer terms (O), the inner terms (I), and the last terms (L) of each binomial.

friendly window The viewing rectangle of a graphing calculator sized so that the cursor moves in "friendly" steps of 0.1, or in multiples of 0.1.

function A relation in which no two ordered pairs have the same first member and different second members. A function can be represented by an equation, graph, or table.

fundamental principle of counting If event A can occur in m ways and event B can occur in n ways, then both events can occur in m times n ways.

glide reflection The composite of a line reflection and a translation whose direction is parallel to the reflecting line.

greatest common factor (GCF) The GCF of two or more monomials is the monomial with the greatest coefficient and the variable factors of the greatest degree that are common to all the given monomials. The GCF of $8a^2b$ and $20ab^2$ is $4ab$.

horizontal line test If a horizontal line intersects a graph of a function in, at most, one point, then the graph represents a one-to-one function. If a function is one-to-one, then it has an *inverse* function.

hyperbola A curve that consists of two branches, an equation of which is $ax^2 + by^2 = c$, where a and b have opposite signs and $c \neq 0$. A special type of hyperbola, called a rectangular or equilateral hyperbola, has the equation $xy = k$ ($k \neq 0$) and a graph that is asymptotic to the coordinate axes. A rectangular hyperbola is the graph of an *inverse* variation.

hypotenuse The side of a right triangle that is opposite the right angle.

identity An equation that is true for all possible replacements of the variable, as in $2x + 3 = 8 - (5 - 2x)$.

image In a geometric transformation, the point or figure that corresponds to the original point or figure.

imaginary number A number of the form bi, where b is a real number and i is the imaginary unit.

imaginary unit The number denoted by i, where $i = \sqrt{-1}$.

independent variable For a function of the form $y = f(x)$, x is the independent variable.

index The number k in the expression $\sqrt[k]{x}$ that tells what root of x is to be taken. In a square root radical the index is omitted and is understood to be 2.

indirect proof A mathematical proof that shows a statement is true by assuming its opposite is true and then proving the assumption leads to contradiction of a known fact.

inequality A sentence that compares two quantities using an inequality relation: $<$ (is less than), $\leq$ (is less than or equal to), $>$ (is greater than), $\geq$ (is greater than or equal to), or $\neq$ (is not equal to).

inscribed angle An angle whose vertex is a point on a circle and whose sides are chords.

integer A number from the set $\{\ldots, -3, -2, -1, 0, 1, 2, 3, \ldots\}$.

interpolation The process of estimating a y-value from a table, graph, or equation using a value of x that falls *within* the range of observed x-values.

inverse function The function obtained by interchanging x and y in a one-to-one function and then solving for y. The graphs of a function and its inverse are symmetric to the line $y = x$.

inverse relation The relation obtained by interchanging the first and second members of each ordered pair of a relation.

inverse variation A set of ordered pairs in which the product of the first and second members of each ordered pair is the same nonzero number. Thus, if y varies inversely as x, then $xy = k$ or, equivalently, $y = \frac{k}{x}$, where $k \neq 0$.

irrational number A number that cannot be expressed as the quotient of two integers.

isometry A transformation that preserves distance. Reflections, translations, and rotations are isometries. A dilation is not an isometry.

isosceles triangle A triangle in which two sides have the same length.

Law of Cosines A relationship between the cosine of an angle of a triangle and the lengths of the three sides of the triangle. In ΔABC,

$$a^2 = b^2 + c^2 - 2bc\cos A$$
$$b^2 = a^2 + c^2 - 2ac\cos B$$
$$c^2 = a^2 + b^2 - 2ab\cos C$$

Law of Sines A relationship between two sides of a triangle and the angles opposite these sides. In ΔABC,

$$\frac{a}{\sin A} = \frac{b}{\sin B} = \frac{c}{\sin C}$$

least squares regression A statistical calculation that finds an equation of a function that "best fits" a set of measurement by minimizing the sum of the squares of the vertical distances between the plotted measurements and the function. A graphing calculator has a built-in regression feature that performs the required calculations. See also *regression model*.

leg of a right triangle Either of the two sides of a right triangle that include the right angle.

linear equation An equation in which the greatest exponent of a variables is 1.

linear function A linear equation whose graph is a straight line.

line reflection A transformation in which each point P that is not on a line m is paired with a point P' on the opposite side of line m so that line m is the perpendicular bisector of $\overline{PP'}$. If P is on line m, then P' coincides with P.

linear regression model See *regression model*.

line symmetry A figure has line symmetry when a line m divides it into two parts such that each part is the reflection of the other part in line m.

logarithm of x An exponent that represents the power to which a given base must be raised to produce a positive number x. For example, $\log_2 8 = 3$ because $2^3 = 8$.

logarithmic function The function $y = \log_b x$, which is the inverse of the exponential function $y = b^x$, where b is positive and unequal to 1.

logarithmic regression model See *regression model*.

major arc An arc whose degree measure is greater than 180 and less than 360.

mapping A relation in which each member of one set is paired with exactly one member of a second set.

mean The mean or average of a set of n data values is the sum of the data values divided by n.

origin The zero point on a number line.

parabola The U-shaped graph of a quadratic equation in two variables in which either x or y is squared, but not both. A parabola in which x is squared has a vertical axis of symmetry. A parabola in which y is squared has a horizontal axis of symmetry.

parallelogram A quadrilateral that has two pairs of parallel sides. In a parallelogram, opposite sides are congruent, consecutive angles are supplementary, and diagonals bisect each other.

perfect square A rational number whose square root is also rational.

period The length of the smallest interval of x needed for the graph of a cyclic function to repeat itself. The period of a sine or cosine function of the form $y = a \sin bx$ or $y = a \cos bx$ is $\frac{2\pi}{|b|}$.

periodic function A function whose y-values repeat over equal intervals of x.

permutation An ordered arrangement of objects.

perpendicular lines Two lines that intersect to form right angles.

point symmetry A figure has point symmetry if after it is rotated 180° (a half-turn) the image coincides with the original figure.

polynomial A monomial or the sum or difference of two or more monomials.

positive angle An angle in standard position whose terminal side rotates in a counterclockwise direction.

power regression model See *regression model*.

preimage If under a given transformation, A′ is the image of A ($A \rightarrow A'$), then A is the preimage of A'.

prime factorization The factorization of a polynomial into factors each of which is divisible only by itself or by 1 (or −1).

probability of an event A number from 0 to 1 that represents the likelihood that an event will occur. To find the probability of an event, divide the number of equally likely ways the event can occur by the total number of possible outcomes.

proportion An equation that states that two ratios are equal. In the proportion $\frac{a}{b} = \frac{c}{d}$, a and d are called the *extremes* and b and c are called the *means*. In a proportion, the product of the means is equal to the product of the extremes. Thus, $a \times d = b \times c$.

Pythagorean theorem The square of the length of the hypotenuse of a right triangle is equal to the sum of the squares of the lengths of the two legs of the right triangle.

quadrant One of the four rectangular regions into which the coordinate plane is divided.

quadrantal angle An angle in standard position whose terminal side coincides with a coordinate axis. Quadrantal angles include angles whose measures are integer multiples of 90°.

quadratic equation An equation that can be put into the standard form $ax^2 + bx + c = 0$, provided $a \neq 0$.

quadratic formula The roots of the quadratic equation $ax^2 + bx + c = 0$ are given by the formula $x = \frac{-b \pm \sqrt{b^2 - 4ac}}{2a}$, provided $a \neq 0$.

quadratic function A function that has the form $y = ax^2 + bx + c$, provided $a \neq 0$.

radian The measure of a central angle of a circle that intercepts an arc whose length equals the radius of the circle. To change from degrees to radians, multiply the number of degrees by $\frac{\pi}{180°}$. To change from radians to degrees, multiply the number of radians by $\frac{180°}{\pi}$.

radical equation An equation in which the variable appears underneath a radical sign as part of the radicand.

radicand The expression that appears underneath a radical sign.

range The set of all possible second members of the set of ordered pairs that comprise a relation.

range in data values The positive difference between the greatest and the smallest data values.

rational number A number that can be written in the form $\frac{a}{b}$, where a and b are integers with $b \neq 0$. Decimals in which a set of digits endlessly repeat, like $2500 \ldots \left(= \frac{1}{4}\right)$ and $0.3333 \ldots \left(= \frac{1}{3}\right)$, represent rational numbers.

real number A number that is a member of the set that consists of all rational and irrational numbers.

rectangle A parallelogram with four right angles.

reference angle When an angle is placed in standard position, the acute angle formed by the terminal side and the x-axis.

reflection Reflections of points in the coordinate axes are given by the following rules: $r_{x\text{-axis}}(x, y) = (x, -y)$, $r_{y\text{-axis}}(x, y) = (-x, y)$, and $r_{y=x}(x, y) = (y, x)$. To reflect a point in the origin, use the rule $r_{\text{origin}}(x, y) = (-x, -y)$. See also *line reflection*.

regression model An equation of the function whose graph is fitted to a set of data by a statistical calculation. A *linear regression* model has the form $y = ax + b$, an *exponential regression* model has the form $y = ab^x$, a *logarithmic regression* model has the form $y = a \ln x + b$, and a *power regression* model has the form $y = ax^b$. The regression feature of a graphing calculator allows you to chose the type of regression model and then calculates the constants a and b for the regression model selected. See also *least squares regression*.

relation A set of ordered pairs.

replacement set The set of values that a variable can have.

rhombus A parallelogram with four congruent sides.

right angle An angle whose degree measure is 90. Perpendicular lines intersect at right angles.

right triangle A triangle that contains a right angle.

rotation A transformation in which a point or figure is turned about a fixed point a given number of degrees. The images of points rotated about the origin through angles that are multiples of 90° are given by the following rules: $R_{90°}(x, y) = (-y, x)$, $R_{180°}(x, y) = (-x, -y)$, and $R_{270°}(x, y) = (y, -x)$.

rotational symmetry A figure has rotational symmetry if in less than a full rotation of 360° the image coincides with the original figure.

scalene triangle A triangle in which the three sides have different lengths.

scatterplot A graph obtained by representing two sets of data values as a set of ordered pairs and then plotting the ordered pairs in the coordinate plane.

secant The reciprocal of the cosine function.

secant line A line that intersects a circle in two different points.

semicircle An arc whose degree measure is 180.

sigma (σ) The lowercase Greek letter σ represents standard deviation.

sigma (Σ) The uppercase Greek letter Σ represents the successive summation of terms, as in $\sum_{i=1}^{3} 2^i = 2^1 + 2^2 + 2^3$.

similar figures Figures that have the same shape. Two triangles are similar if two pairs of corresponding angles have the same degree measure. When two triangles are similar, the lengths of corresponding sides are proportional.

sine ratio In a right triangle, the ratio of the length of the leg that is opposite a given acute angle to the length of the hypotenuse. If an angle θ is in standard position, then $\sin\theta = \frac{y}{r}$, where $P(x, y)$ is any point on the terminal side of angle θ and $r = \sqrt{x^2 + y^2}$.

square A parallelogram with four right angles and four congruent sides.

standard deviation A statistic that measures how spread out numerical data are from the mean.

standard position The position of an angle whose vertex is fixed at the origin and whose initial side coincides with the positive x-axis. The side of the angle that rotates is called the *terminal side*. A counterclockwise rotation of the terminal side of an angle represents a positive angle, and a clockwise rotation of the terminal side of an angle produces a negative angle. If an angle θ is in standard position and $P(x, y)$ is any point on the terminal side of angle θ, where $r = \sqrt{x^2 + y^2}$, then for all values of θ for which the trigonometric function are defined:

$$\sin\theta = \frac{y}{r} \qquad \csc\theta = \frac{r}{y}$$

$$\cos\theta = \frac{x}{r} \qquad \sec\theta = \frac{r}{x}$$

$$\tan\theta = \frac{y}{x} \qquad \cot\theta = \frac{x}{y}$$

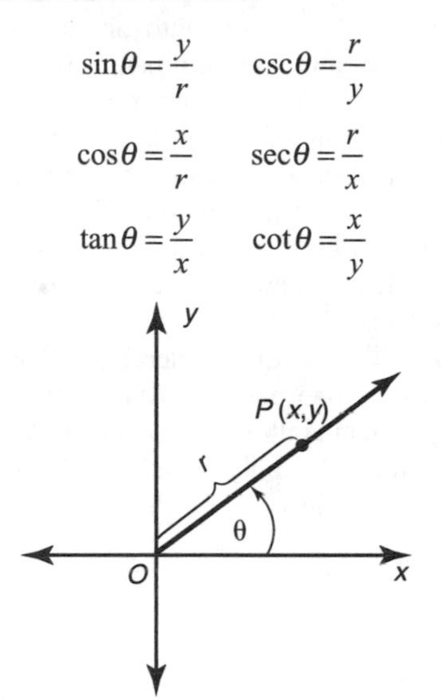

The algebraic signs of the trigonometric functions depend on the signs of x and y in the particular quadrant in which the terminal side of angle θ is located.

standard window The viewing rectangle of a graphing calculator sized so that each coordinate axis has 10 tick marks on either side of the origin.

supplementary angles Two angles whose degree measures add up to 180.

tangent line A line that intersects a circle in exactly one point.

tangent ratio In a right triangle, the ratio of the length of the leg that is opposite a given acute angle to the length of the leg adjacent to that angle. If an angle θ is in standard position, then $\tan\theta = \frac{y}{r}$, where $P(x, y)$ is any point on the terminal side of angle θ.

terminal side The side of an angle in standard position that rotates about the origin while the initial side of the angle remains fixed on the positive x-axis.

theorem A mathematical generalization that can be proved.

transformation A change made to a figure by using a given rule to map each of its points onto one and only one point in a set of image points.

translation A transformation in which each point of a figure is moved the same distance and in the same direction. The transformation $T_{h,k}$ slides a point h units horizontally and k units vertically. Thus, $T_{h,k}(x, y) = (x + h, y + k)$.

trapezoid A quadrilateral in which exactly one pair of sides are parallel. The nonparallel sides are called *legs*.

trinomial A polynomial with three terms, as in $x^2 - 3x + 7$.

unit circle A circle with a radius of 1.

vertex The highest or lowest point of a parabola. The x-coordinate of the vertex of the parabola $y = ax^2 + bx + c$ is $x = -\frac{b}{2a}$. In the vertex form of the equation of the parabola $y = a(x - h)^2 + k$, the vertex is (h, k).

vertical angles Opposite pairs of congruent angles formed when two lines intersect.

vertical line test If no vertical line intersects a graph in more than one point, the graph represents a function.

zero of a function Any value of the variable for which the function evaluates to 0. Each x-intercept of the graph of a function, if any, represents a zero of the function.

The Mathematics B Regents Examination

The Mathematics B Regents Exam is a 3-hour test that is divided into four parts with a total of 34 questions. No choice is permitted in any of the four parts. Part I consists of 20 standard multiple-choice questions. You must record the answers to these 20 questions on a detachable answer sheet located at the end of the question booklet. The answers and the accompanying work for the questions in Parts II, III, and IV must be written directly in the question booklet. Since scrap paper is not permitted for any part of the exam, you must use the blank spaces in the question booklet as scrap paper. Graph paper is included in the question booklet.

How Is the Mathematics B Regents Exam Scored?

Each of your answers to the 20 multiple-choice questions in Part I is scored as either right or wrong. Solutions to questions in Parts II, III, and IV that are not completely correct may receive partial credit according to a special rating guide provided by the New York State Education Department. The accompanying table shows how the Math B Regents Exam breaks down.

Part	Number of Questions	Point Value	Maximum Points
I	20 multiple-choice	2 each	40
II	6 open-ended	2 each	12
III	6 open-ended	4 each	24
IV	2 open-ended	6 each	12
	34 questions	Total rawscore: 88 points	

To receive full credit for a correct answer to a question in Part II, III, or IV, you must show or explain how you arrived at your answer by indicating the necessary steps you took, including appropriate formula substitutions, diagrams, graphs, and charts. A correct numerical answer with no work shown will receive only 1 credit.

Although a graphing calculator is required for the exam, not all graphing questions that appear on the Mathematics B Regents Exam require a graphing calculator. For example, graphing questions involving transformations of points or curves may be answered by plotting points using graph paper. When it is appropriate to answer a question with the help of a graphing calculator:

- Sketch and label graphs in an appropriate viewing window. Include intercepts, coordinates of points of intersection, and tables of values when they are needed for the solution.

- Indicate the number of scores and the mean when calculating the standard deviation.
- Write the regression equation and the coefficient of linear correlation when finding a function that best fits given data. For any interpolations or extrapolations, show the substitutions in the regression equation.

How Is Your Grade Determined?

The raw scores you receive for the four parts of the examination are added together. The maximum total raw score for the Mathematics B Regents Exam is 88 points. A special table will be used to convert your total raw score into a Mathematics B scaled score that falls within the usual 0–100 scale.

NYS Regents Updates

- Unless otherwise directed by the question, use *all* of the digits in the calculator display window. Do not round intermediate values. Rounding, if required, should be done only after the *final* answer is reached.
- For statistics calculations:
  1. Obtain normal curve probabilities using the Normal Curve provided on the formula sheet in the test booklet rather than using a graphing calculator.
  2. To calculate the standard deviation for an entire population of scores, use σ_x. Use S_x, the sample standard deviation, to calculate the standard deviation of a randomly selected subset of the population.

Examination June 2004

Math B

FORMULAS

Area of Triangle

$$K = \frac{1}{2}ab\sin C$$

Function of the Sum of Two Angles

$$\sin (A + B) = \sin A \cos B + \cos A \sin B$$
$$\cos (A + B) = \cos A \cos B - \sin A \sin B$$

Function of the Difference of Two Angles

$$\sin (A - B) = \sin A \cos B - \cos A \sin B$$
$$\cos (A - B) = \cos A \cos B + \sin A \sin B$$

Law of Sines

$$\frac{a}{\sin A} = \frac{b}{\sin B} = \frac{c}{\sin C}$$

Law of Cosines

$$a^2 = b^2 + c^2 - 2bc \cos A$$

Functions of the Double Angle

$$\sin 2A = 2 \sin A \cos A$$
$$\cos 2A = \cos^2 A - \sin^2 A$$
$$\cos 2A = 2 \cos^2 A - 1$$
$$\cos 2A = 1 - 2 \sin^2 A$$

Functions of the Half Angle

$$\sin\frac{1}{2}A = \pm\sqrt{\frac{1-\cos A}{2}}$$

$$\cos\frac{1}{2}A = \pm\sqrt{\frac{1+\cos A}{2}}$$

Normal Curve
Standard Deviation

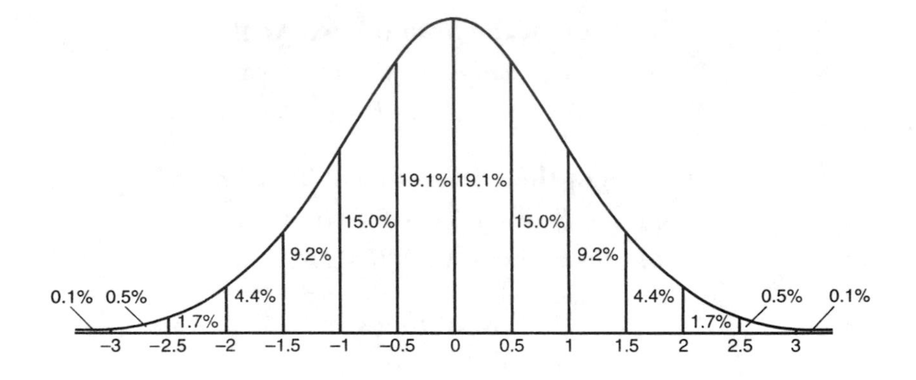

PART I

Answer all questions in this part. Each correct answer will receive 2 credits. No partial credit will be allowed. For each question, write in the spaces provided the numeral preceding the word or expression that best completes the statement or answers the question. [40]

1 What is the sum of $2 - \sqrt{-4}$ and $-3 + \sqrt{-16}$ expressed in $a + bi$ form?

(1) $-1 + 2i$ (3) $-1 + 12i$

(2) $-1 + i\sqrt{20}$ (4) $-14 + i$

1 ____

2 The Hiking Club plans to go camping in a State park where the probability of rain on any give day is 0.7. Which expression can be used to find the probability that it will rain on *exactly* three of the seven days they are there?

(1) ${}_7C_3(0.7)^3(0.3)^4$ (3) ${}_4C_3(0.7)^3(0.7)^4$

(2) ${}_7C_3(0.3)^3(0.7)^4$ (4) ${}_4C_3(0.4)^4(0.3)^3$

2 ____

3 What is the amplitude of the function $y = \frac{2}{3} \sin 4x$?

(1) $\frac{\pi}{2}$ (3) 3π

(2) $\frac{2}{3}$ (4) 4

3 ____

4 Which quadratic function is shown in the accompanying graph?

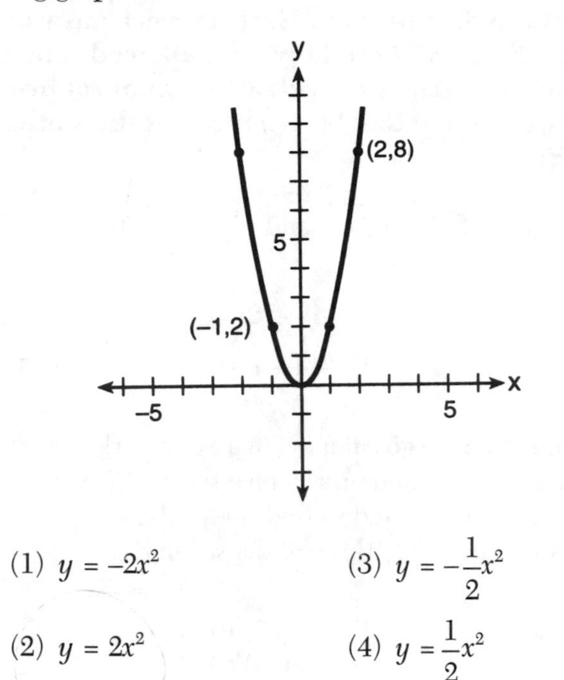

(1) $y = -2x^2$

(2) $y = 2x^2$

(3) $y = -\frac{1}{2}x^2$

(4) $y = \frac{1}{2}x^2$

4 2

y=

5 In the accompanying graph, the shaded region represents set A of all points (x,y) such that $x^2 + y^2 \leq 1$. The transformation T maps point (x,y) to point $(2x,4y)$.

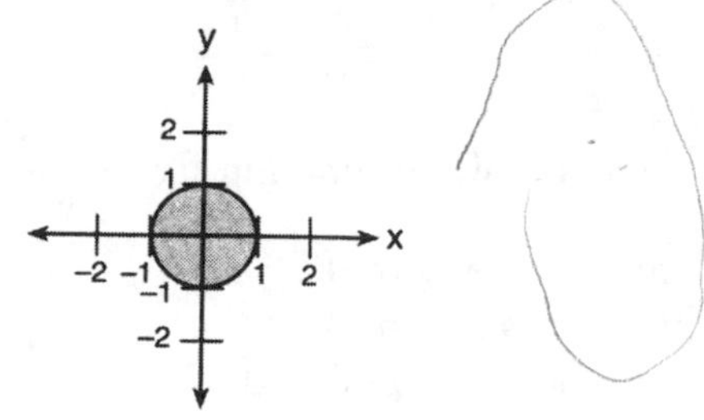

Which graph shows the mapping of set A by the transformation T?

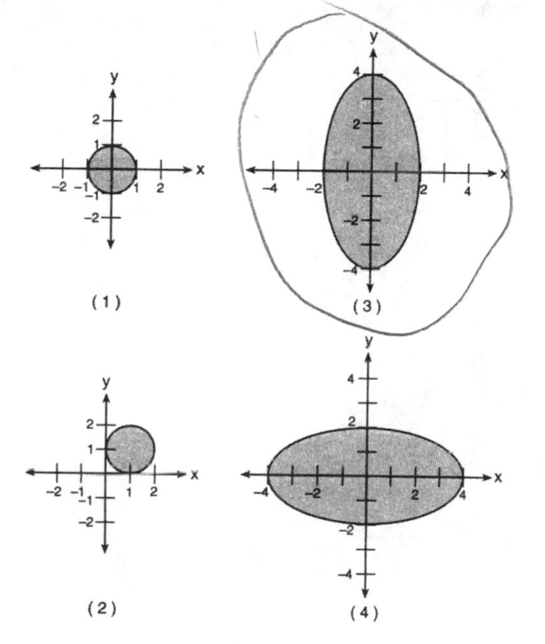

5 ______

6 If $f(x) = 4x^0 + (4x)^{-1}$, what is the value of f(4)?

(1) –12 (3) $1\frac{1}{16}$

(2) 0 (4) $4\frac{1}{16}$

6 ____

7 What is the domain of the function $f(x) = \frac{2x^2}{x^2 - 9}$?

(1) all real numbers except 0
(2) all real numbers except 3
(3) all real numbers except 3 and –3
(4) all real numbers

7 ____

8 Which graph represents an inverse variation between stream velocity and the distance from the center of the stream?

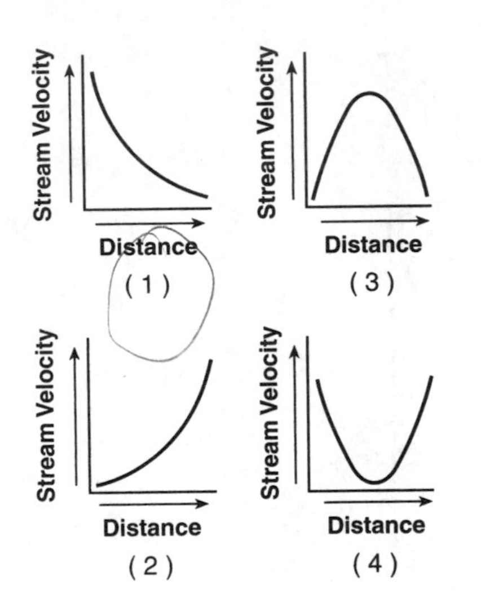

8 ____

9 If $\log_b x = y$, then x equals

(1) $y \bullet b$ (3) y^b

(2) $\frac{y}{b}$ (4) b^y 9 ____

10 The expression $i^0 \bullet i^1 \bullet i^2 \bullet i^3 \bullet i^4$ is equal to

(1) 1 (3) i

(2) -1 (4) $-i$ 10 ____

11 Which equation models the data in the accompanying table?

Time in hours, *x*	0	1	2	3	4	5	6
Population, *y*	5	10	20	40	80	160	320

(1) $y = 2x + 5$ (3) $y = 2x$

(2) $y = 2^x$ (4) $y = 5(2^x)$ 11 ____

12 The amount of juice dispensed from a machine is normally distributed with a mean of 10.50 ounces and a standard deviation of 0.75 ounce. Which interval represents the amount of juice dispensed about 68.2% of the time?

(1) 9.00–12.00 (3) 9.75–11.25

(2) 9.75–10.50 (4) 10.50–11.25 12 ____

13 If θ is an acute angle such that $\sin \theta = \frac{5}{13}$, what is the value of $\sin 2\theta$?

(1) $\frac{12}{13}$ (3) $\frac{60}{169}$

(2) $\frac{10}{26}$ (4) $\frac{120}{169}$ 13 ____

14 Which function is symmetrical with respect to the origin?

(1) $y = \sqrt{x+5}$ (3) $y = -\frac{5}{x}$

(2) $y = |5 - x|$ (4) $y = 5^x$ 14 ____

15 The expression $\dfrac{\frac{1}{x} + \frac{1}{y}}{\frac{1}{x^2} - \frac{1}{y^2}}$ is equivalent to

(1) $\frac{xy}{y-x}$ (3) $\frac{y-x}{xy}$

(2) $\frac{xy}{x-y}$ (4) $y - x$ 15 ____

16 Sam is designing a triangular piece for a metal sculpture. He tells Martha that two of the sides of the piece are 40 inches and 15 inches, and the angle opposite the 40-inch side measures 120°. Martha decides to sketch the piece that Sam described. How many different triangles can she sketch that match Sam's description?

(1) 1 (3) 3
(2) 2 (4) 0 16 ____

17 If $f(x) = x + 1$ and $g(x) = x^2 - 1$, the expression $(g \circ f)(x)$ equals 0 when x is equal to

(1) 1 and –1 (3) –2, only
(2) 0, only (4) 0 and –2 17 ____

18 If θ is a positive acute angle and sin θ = a, which expression represents cos θ in terms of a?

(1) $\sqrt{a}$ (3) $\frac{1}{\sqrt{a}}$

(2) $\sqrt{1-a^2}$ (4) $\frac{1}{\sqrt{1-a^2}}$ 18 ____

19 The expression $\sqrt[4]{16a^6b^4}$ is equivalent to

(1) $2a^2b$ (3) $4a^2b$

(2) $2a^{\frac{3}{2}}b$ (4) $4a^{\frac{3}{2}}b$ 19 ____

20 In the accompanying diagram, $\overline{HK}$ bisects $\overline{IL}$ and $\angle H \cong \angle K$.

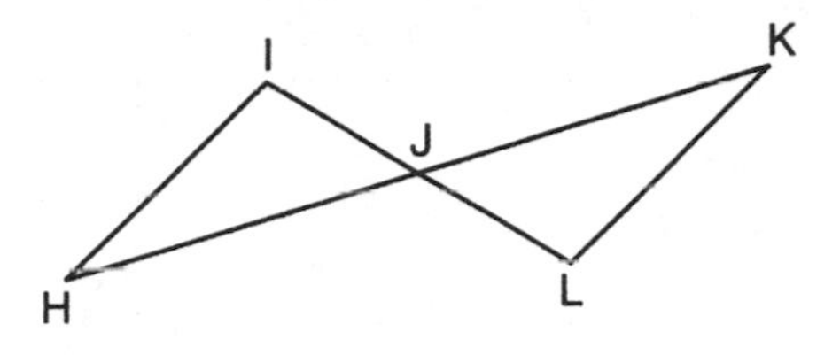

What is the most direct method of proof that could be used to prove $\triangle HIJ \cong \triangle KLJ$?

(1) HL ≅ HL (3) AAS ≅ AAS
(2) SAS ≅ SAS (4) ASA ≅ ASA 20 ____

PART II

Answer all questions in this part. Each correct answer will receive 2 credits. Clearly indicate the necessary steps, including appropriate formula substitutions, diagrams, graphs, charts, etc. For all questions in this part, a correct numerical answer with no work shown will receive only 1 credit. [12]

21 The projected total annual profits, in dollars, for the Nutyme Clothing Company from 2002 to 2004 can be approximated by the model $\sum_{n=0}^{2}(13,567n + 294)$, where n is the year and $n = 0$ represents 2002. Use this model to find the company's projected total annual profits, in dollars, for the period 2002 to 2004.

22 Solve algebraically for x: $27^{2x+1} = 9^{4x}$

23 Find all values of k such that the equation $3x^2 - 2x + k = 0$ has imaginary roots.

24 In the accompanying diagram of square $ABCD$, F is the midpoint of $\overline{AB}$, G is the midpoint of $\overline{BC}$, H is the midpoint of $\overline{CD}$, and E is the midpoint of $\overline{DA}$.

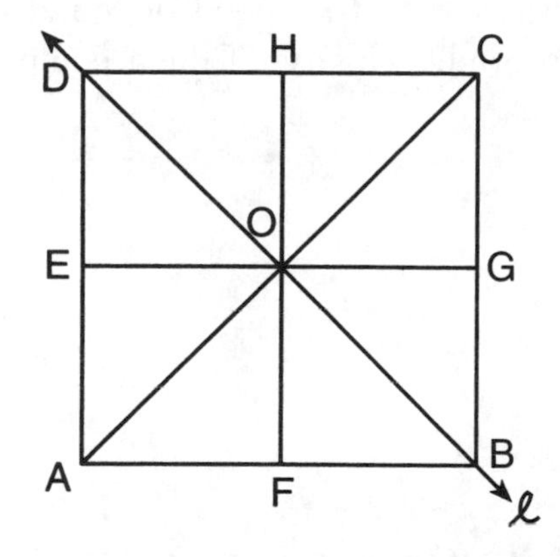

Find the image of $\triangle EOA$ after it is reflected in line ℓ. Is this isometry direct or opposite? Explain your answer.

25 Given: $\triangle ABT$, $CBTD$, and $AB \perp CD$

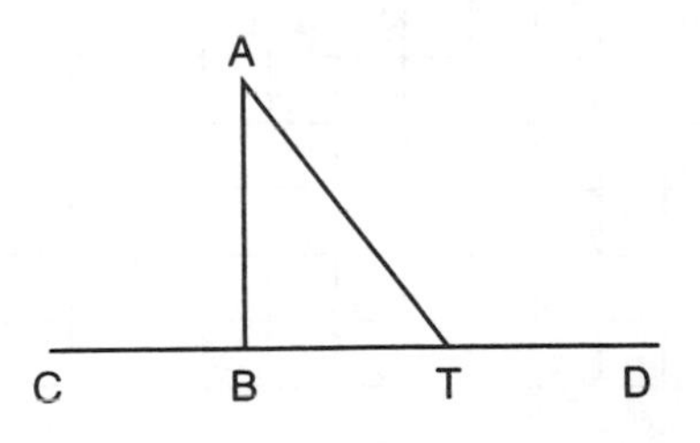

Write an indirect proof to show that $\overline{AT}$ is *not* perpendicular to $\overline{CD}$.

26 The equation $V = 20\sqrt{C+273}$ relates speed of sound, V, in meters per second, to air temperature, C, in degrees Celsius. What is the temperature, in degrees Celsius, when the speed of sound is 320 meters per second? [The use of the accompanying grid is optional.]

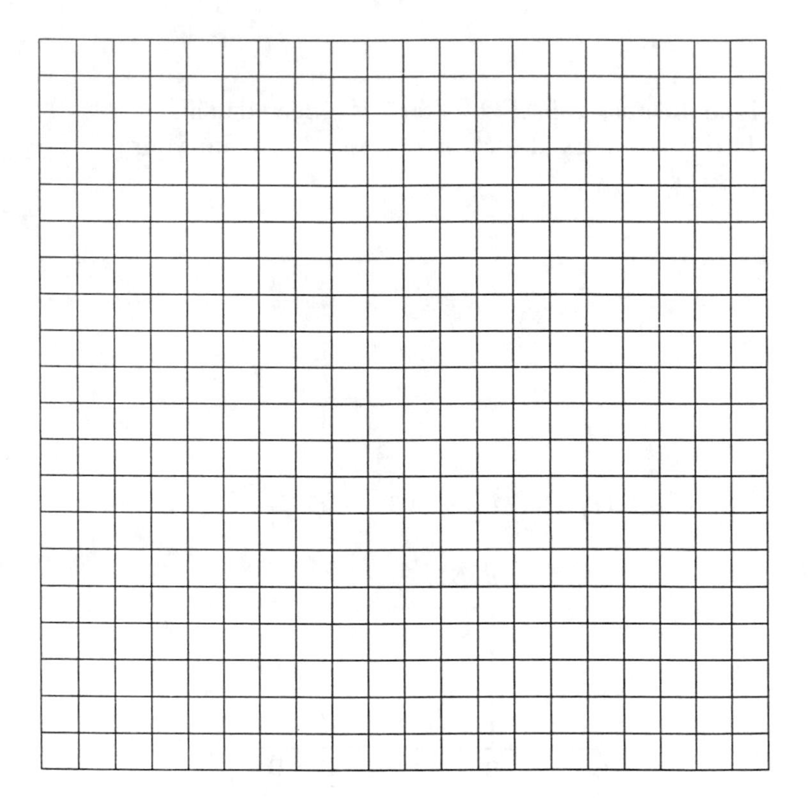

PART III

Answer all questions in this part. Each correct answer will receive 4 credits. Clearly indicate the necessary steps, including appropriate formula substitutions, diagrams, graphs, charts, etc. For all questions in this part, a correct numerical answer with no work shown will receive only 1 credit. [24]

27 Navigators aboard ships and airplanes use nautical miles to measure distance. The length of a nautical mile varies with latitude. The length of a nautical mile, L, in feet, on the latitude line θ is given by the formula $L = 6{,}077 - 31 \cos 2\theta$.

Find, to the *nearest degree*, the angle θ, $0° \leq \theta \leq 90°$, at which the length of a nautical mile is approximately 6,076 feet.

28 Two equal forces act on a body at an angle of 80°. If the resultant force is 100 newtons, find the value of one of the two equal forces, to the *nearest hundredth of a newton.*

29 Solve for x and express your answer in simplest radical form:

$$\frac{4}{x} - \frac{3}{x+1} = 7$$

30 A baseball player throws a ball from the outfield toward home plate. The ball's height above the ground is modeled by the equation $y = -16x^2 + 48x + 6$, where y represents height, in feet, and x represents time, in seconds. The ball is initially thrown from a height of 6 feet.

How many seconds after the ball is thrown will it again be 6 feet above the ground?

What is the maximum height, in feet, that the ball reaches? [The use of the accompanying grid is optional.]

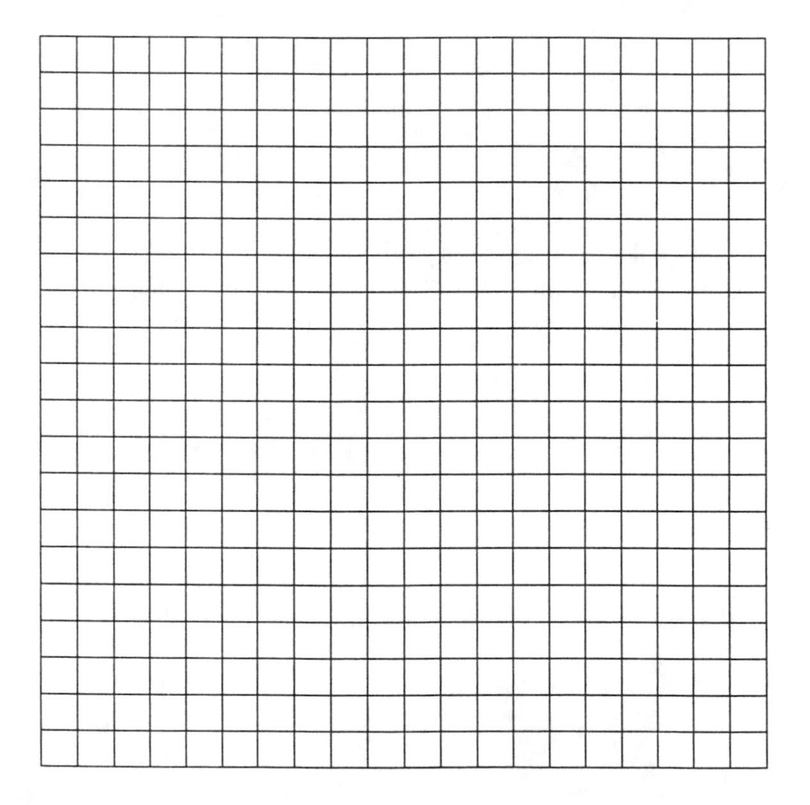

31 An archaeologist can determine the approximate age of certain ancient specimens by measuring the amount of carbon-14, a radioactive substance, contained in the specimen. The formula used to determine the age of a specimen is $A = A_0 2^{\frac{-t}{5760}}$, where A is the amount of carbon-14 that a specimen contains, A_0 is the original amount of carbon-14, t is time, in years, and 5760 is the half-life of carbon-14.

A specimen that originally contained 120 milligrams of carbon-14 now contains 100 milligrams of this substance. What is the age of the specimen, to the *nearest hundred years*?

32 Mrs. Ramírez is a real estate broker. Last month, the sale prices of homes in her area approximated a normal distribution with a mean of \$150,000 and a standard deviation of \$25,000.

A house had a sale price of \$175,000. What is the percentile rank of its sale price, to the *nearest whole number*? Explain what that percentile means.

Mrs. Ramírez told a customer that most of the houses sold last month had selling prices between \$125,000 and \$175,000. Explain why she is correct.

PART IV

Answer all questions in this part. Each correct answer will receive 6 credits. Clearly indicate the necessary steps, including appropriate formula substitutions, diagrams, graphs, charts, etc. For all questions in this part, a correct numerical answer with no work shown will receive only 1 credit. [12]

33 The accompanying diagram shows a circular machine part that has rods $\overline{PT}$ and $\overline{PAR}$ attached at points, T, A, and R, which are located on the circle; $m\overset{\frown}{TA} : m\overset{\frown}{AR} : m\overset{\frown}{RT} = 1{:}3{:}5$; $RA = 12$ centimeters; and $PA = 5$ centimeters.

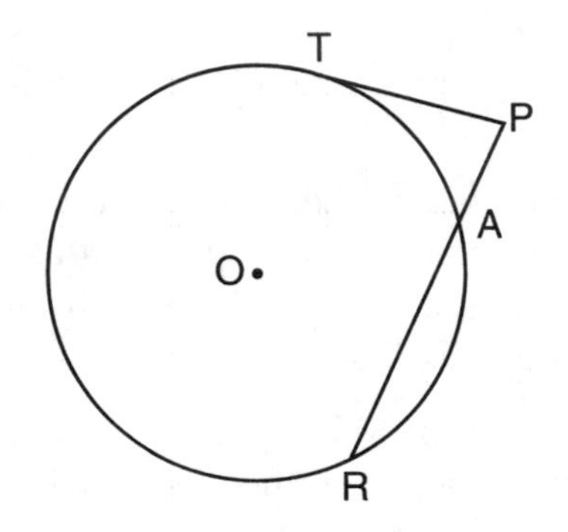

Find the measure of $\angle P$, in degrees, and find the length of rod $\overline{PT}$, to the *nearest tenth of a centimeter.*

34 A surveyor is mapping a triangular plot of land. He measures two of the sides and the angle formed by these two sides and finds that the lengths are 400 yards and 200 yards and the included angle is 50°.

What is the measure of the third side of the plot of land, to the *nearest yard*?

What is the area of this plot of land, to the *nearest square yard*?

Examination August 2004

Math B

FORMULAS

Area of Triangle

$$K = \frac{1}{2}ab\sin C$$

Function of the Sum of Two Angles

$$\sin (A + B) = \sin A \cos B + \cos A \sin B$$

$$\cos (A + B) = \cos A \cos B - \sin A \sin B$$

Function of the Difference of Two Angles

$$\sin (A - B) = \sin A \cos B - \cos A \sin B$$

$$\cos (A - B) = \cos A \cos B + \sin A \sin B$$

Law of Sines

$$\frac{a}{\sin A} = \frac{b}{\sin B} = \frac{c}{\sin C}$$

Law of Cosines

$$a^2 = b^2 + c^2 - 2bc \cos A$$

Functions of the Double Angle

$$\sin 2A = 2 \sin A \cos A$$

$$\cos 2A = \cos^2 A - \sin^2 A$$

$$\cos 2A = 2 \cos^2 A - 1$$

$$\cos 2A = 1 - 2 \sin^2 A$$

Functions of the Half Angle

$$\sin\frac{1}{2}A = \pm\sqrt{\frac{1-\cos A}{2}}$$

$$\cos\frac{1}{2}A = \pm\sqrt{\frac{1+\cos A}{2}}$$

Normal Curve Standard Deviation

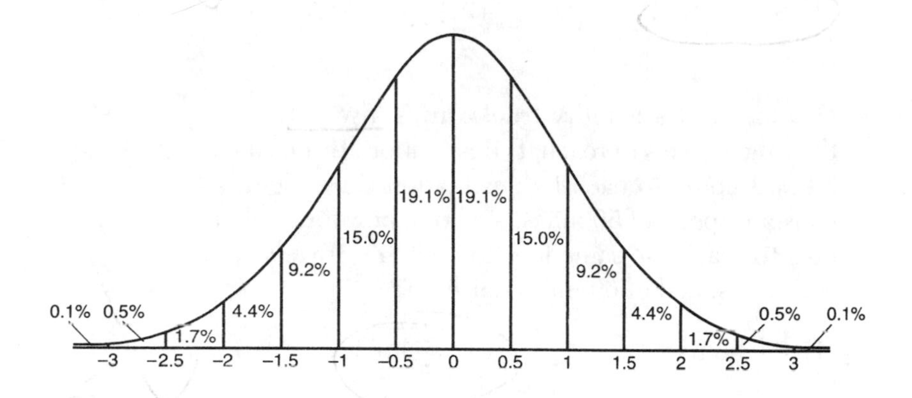

PART I

Answer all questions in this part. Each correct answer will receive 2 credits. No partial credit will be allowed. For each question, write in the space provided the numeral preceding the word or expression that best completes the statement or answers the question. [40]

1 Which condition does *not* prove that two triangles are congruent?

(1) SSS ≅ SSS (3) SAS ≅ SAS
(2) SSA ≅ SSA (4) ASA ≅ ASA

1 ____

2 The speed of a laundry truck varies inversely with the time it takes to reach its destination. If the truck takes 3 hours to each its destination traveling at a constant speed of 50 miles per hour, how long will it take to reach the same location when it travels at a constant speed of 60 miles per hour?

(1) $2\frac{1}{3}$ hours (3) $2\frac{1}{2}$ hours

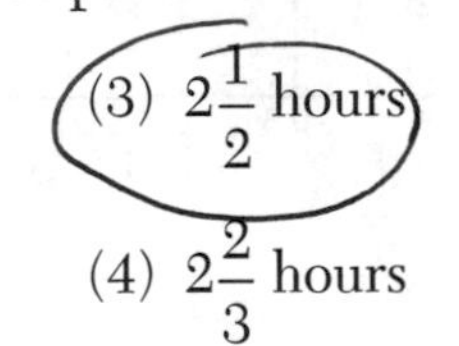

(2) 2 hours (4) $2\frac{2}{3}$ hours

2 ____

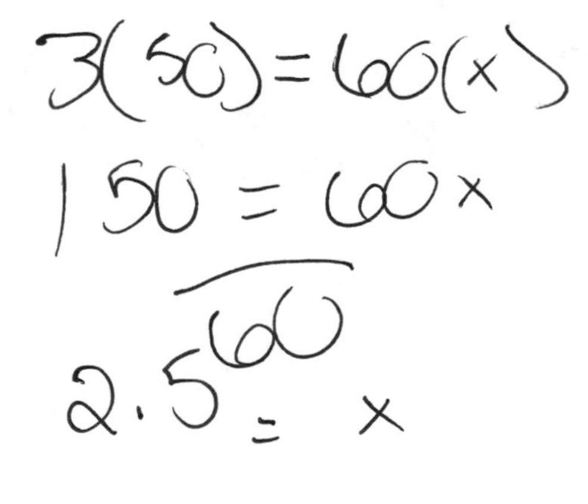

3 Which set of ordered pairs is *not* a function?

(1) {(3,1), (2,1), (1,2), (3,2)}
(2) {(4,1), (5,1), (6,1), (7,1)}
(3) {(1,2), (3,4), (4,5), (5,6)}
(4) {(0,0), (1,1), (2,2), (3,3)}

3 _____

4 A circle has the equation $(x + 1)^2 + (y - 3)^2 = 16$. What are the coordinates of its center and the length of its radius?

(1) (−1,3) and 4
(2) (1,−3) and 4
(3) (−1,3) and 16
(4) (1,−3) and 16

4 _____

5 The mean of a normally distributed set of data is 56, and the standard deviation is 5. In which interval do approximately 95.4% of all cases lie?

(1) 46–56
(2) 46–66
(3) 51–61
(4) 56–71

5 _____

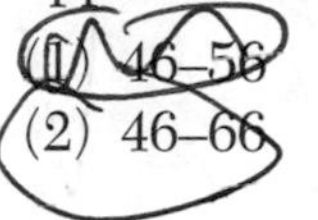

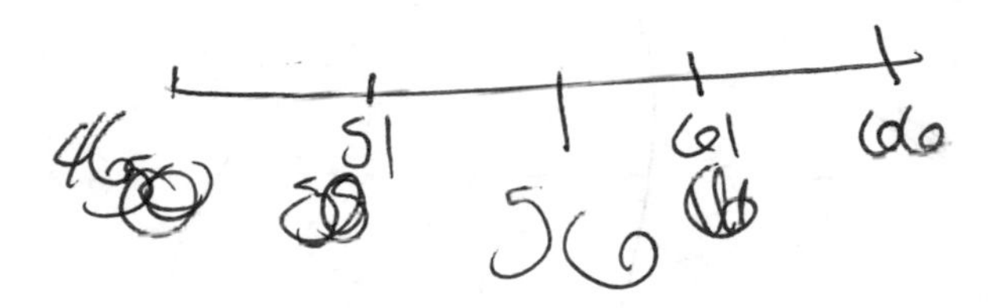

6 The graph below represents $f(x)$.

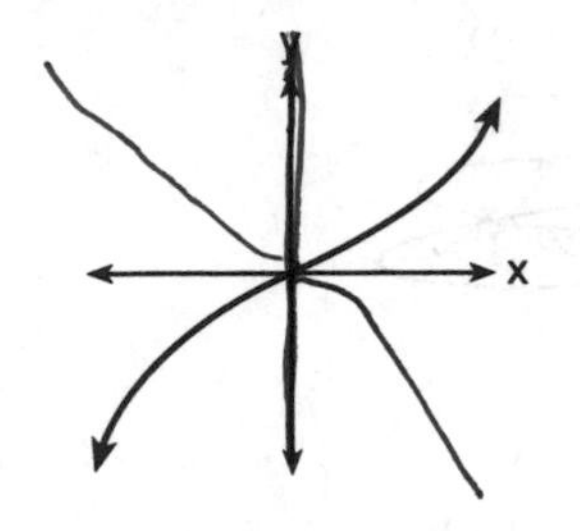

Which graph best represents $f(-x)$?

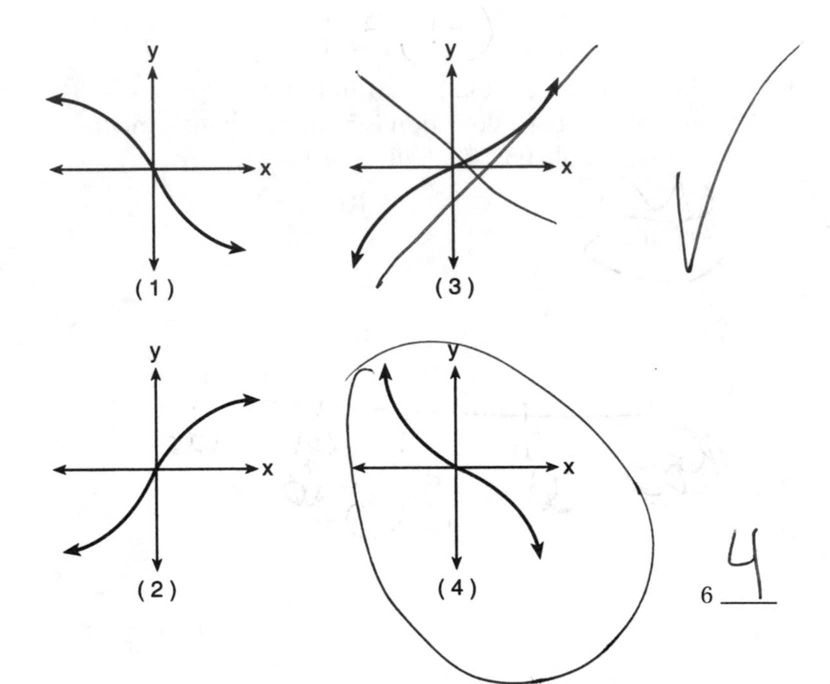

6 4

7 When simplified, $i^{27} + i^{34}$ is equal to

(1) i
(2) i^{61}
(3) $-i - 1$
(4) $i - 1$

7 ____

8 The accompanying diagram shows a child's spin toy that is constructed from two chords intersecting in a circle. The curved edge of the larger shaded section is one-quarter of the circumference of the circle, and the curved edge of the smaller shaded section is one-fifth of the circumference of the circle.

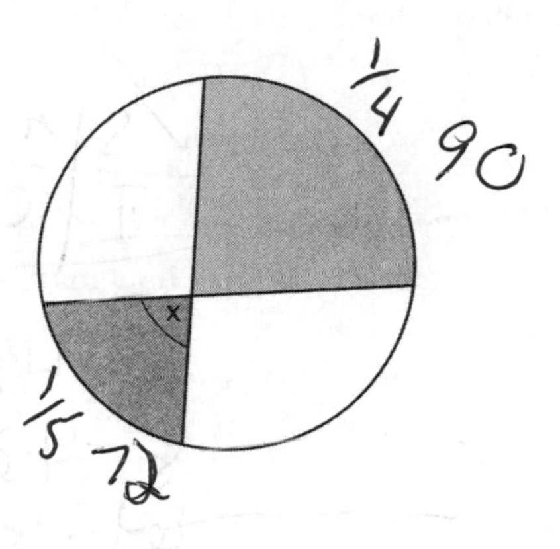

What is the measure of angle x?

(1) 40°
(2) 72°
(3) 81°
(4) 108°

8 ____

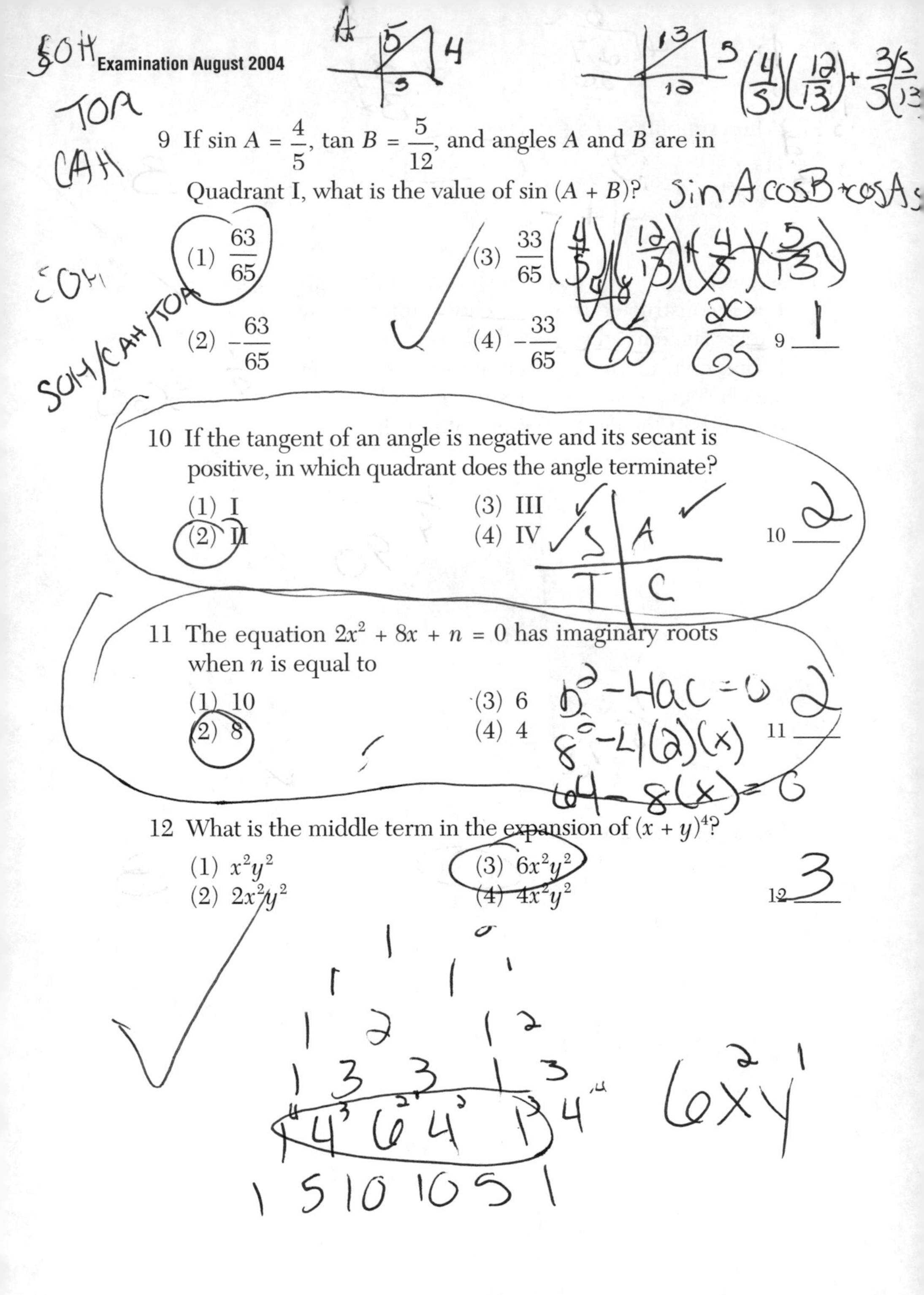

9 If $\sin A = \frac{4}{5}$, $\tan B = \frac{5}{12}$, and angles A and B are in Quadrant I, what is the value of $\sin(A + B)$?

(1) $\frac{63}{65}$
(2) $-\frac{63}{65}$
(3) $\frac{33}{65}$
(4) $-\frac{33}{65}$

9 ____

10 If the tangent of an angle is negative and its secant is positive, in which quadrant does the angle terminate?

(1) I
(2) II
(3) III
(4) IV

10 ____

11 The equation $2x^2 + 8x + n = 0$ has imaginary roots when n is equal to

(1) 10
(2) 8
(3) 6
(4) 4

11 ____

12 What is the middle term in the expansion of $(x + y)^4$?

(1) x^2y^2
(2) $2x^2y^2$
(3) $6x^2y^2$
(4) $4x^2y^2$

12 ____

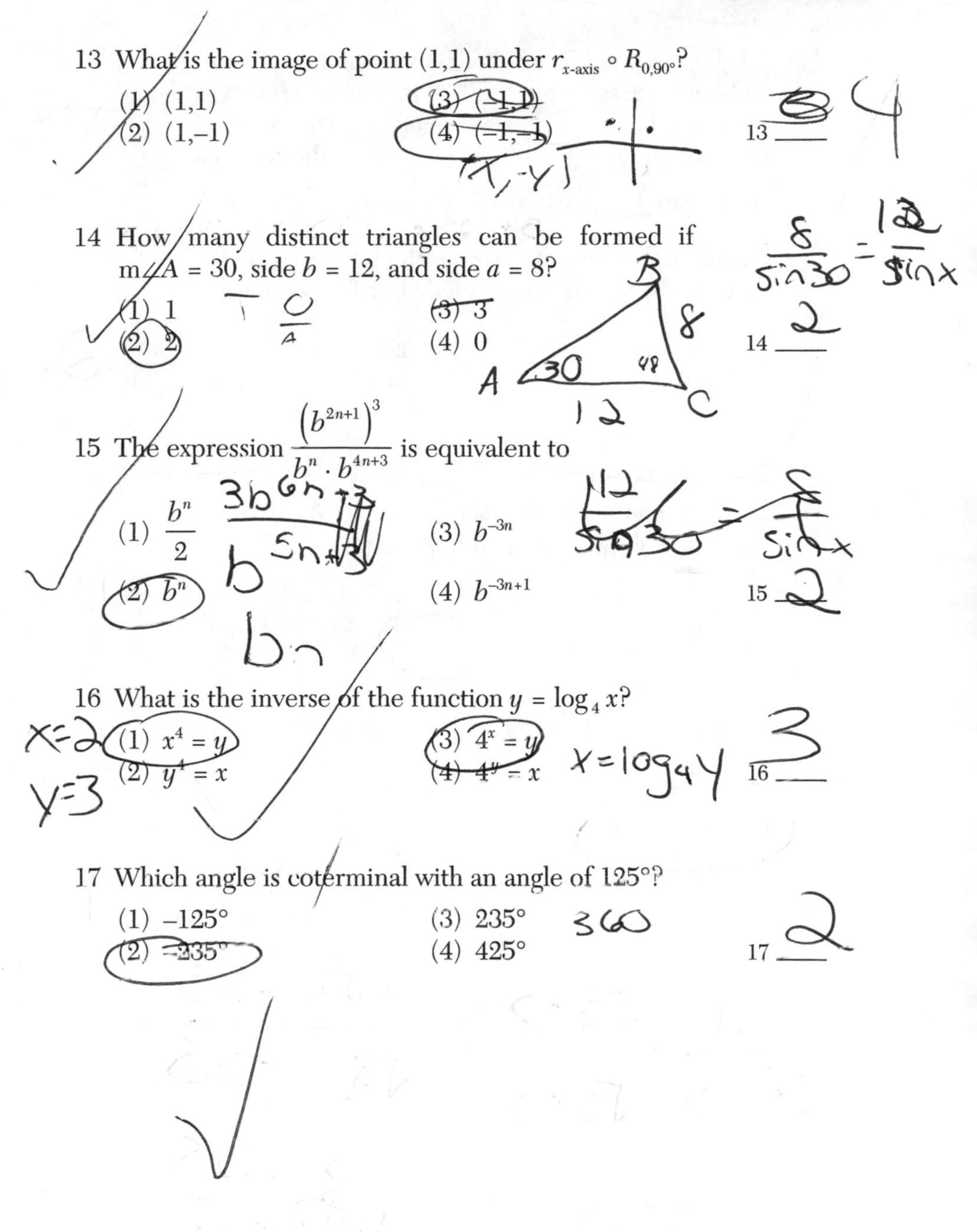

13 What is the image of point (1,1) under $r_{x\text{-axis}} \circ R_{0,90°}$?

(1) (1,1)
(2) (1,–1)
(3) (–1,1)
(4) (–1,–1)

13 ____

14 How many distinct triangles can be formed if m∠A = 30, side b = 12, and side a = 8?

(1) 1
(2) 2
(3) 3
(4) 0

14 ____

15 The expression $\frac{\left(b^{2n+1}\right)^3}{b^n \cdot b^{4n+3}}$ is equivalent to

(1) $\frac{b^n}{2}$
(2) b^n
(3) b^{-3n}
(4) b^{-3n+1}

15 ____

16 What is the inverse of the function $y = \log_4 x$?

(1) $x^4 = y$
(2) $y^4 = x$
(3) $4^x = y$
(4) $4^y = x$

16 ____

17 Which angle is coterminal with an angle of 125°?

(1) –125°
(2) –235°
(3) 235°
(4) 425°

17 ____

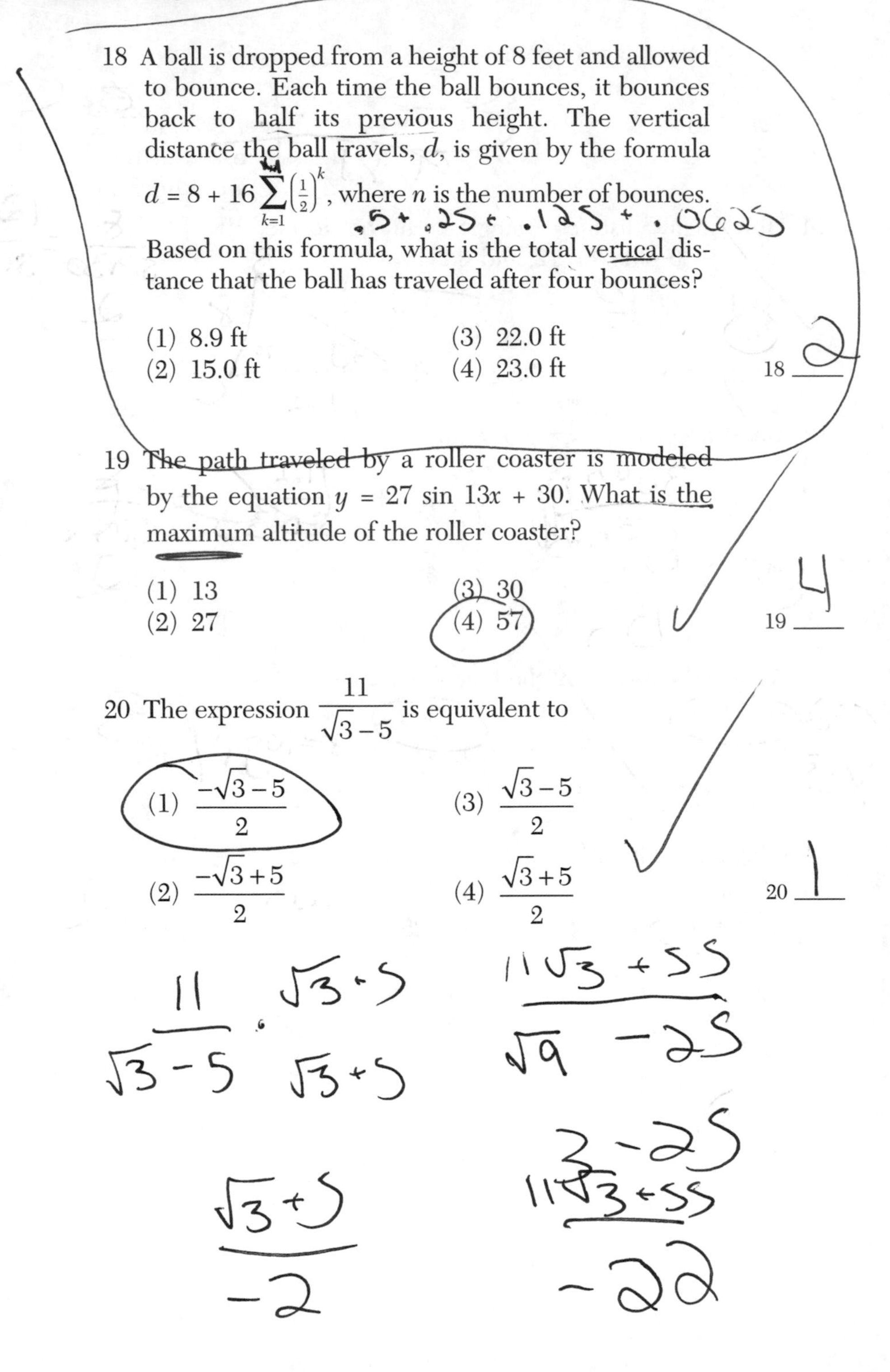

18 A ball is dropped from a height of 8 feet and allowed to bounce. Each time the ball bounces, it bounces back to half its previous height. The vertical distance the ball travels, d, is given by the formula $d = 8 + 16\sum_{k=1}^{n}\left(\frac{1}{2}\right)^{k}$, where n is the number of bounces. Based on this formula, what is the total vertical distance that the ball has traveled after four bounces?

(1) 8.9 ft
(2) 15.0 ft
(3) 22.0 ft
(4) 23.0 ft

18 ____

19 The path traveled by a roller coaster is modeled by the equation $y = 27 \sin 13x + 30$. What is the maximum altitude of the roller coaster?

(1) 13
(2) 27
(3) 30
(4) 57

19 ____

20 The expression $\frac{11}{\sqrt{3}-5}$ is equivalent to

(1) $\frac{-\sqrt{3}-5}{2}$
(2) $\frac{-\sqrt{3}+5}{2}$
(3) $\frac{\sqrt{3}-5}{2}$
(4) $\frac{\sqrt{3}+5}{2}$

20 ____

PART II

Answer all questions in this part. Each correct answer will receive 2 credits. Clearly indicate the necessary steps, including appropriate formula substitutions, diagrams, graphs, charts, etc. For all questions in this part, a correct numerical answer with no work shown will receive only 1 credit. [12]

21 A ski lift begins at ground level 0.75 mile from the base of a mountain whose face has a 50° angle of elevation, as shown in the accompanying diagram. The ski lift ascends in a straight line at an angle of 20°. Find the length of the ski lift from the beginning of the ski lift to the top of the mountain, to the *nearest hundredth of a mile.*

Top of mountain

x 30

20° 130 50°

Beginning of ski lift .75 Base of mountain

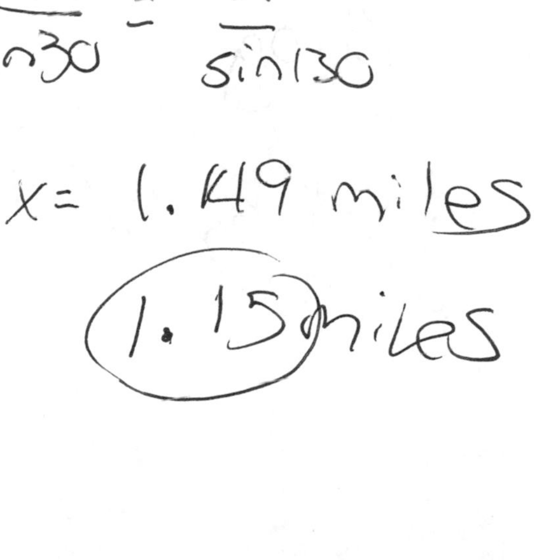

22 Express $\sqrt{-48} + 3.5 + \sqrt{25} + \sqrt{-27}$ in simplest $a + bi$ form.

23 Solve for x: $x^{-3} = \frac{27}{64}$

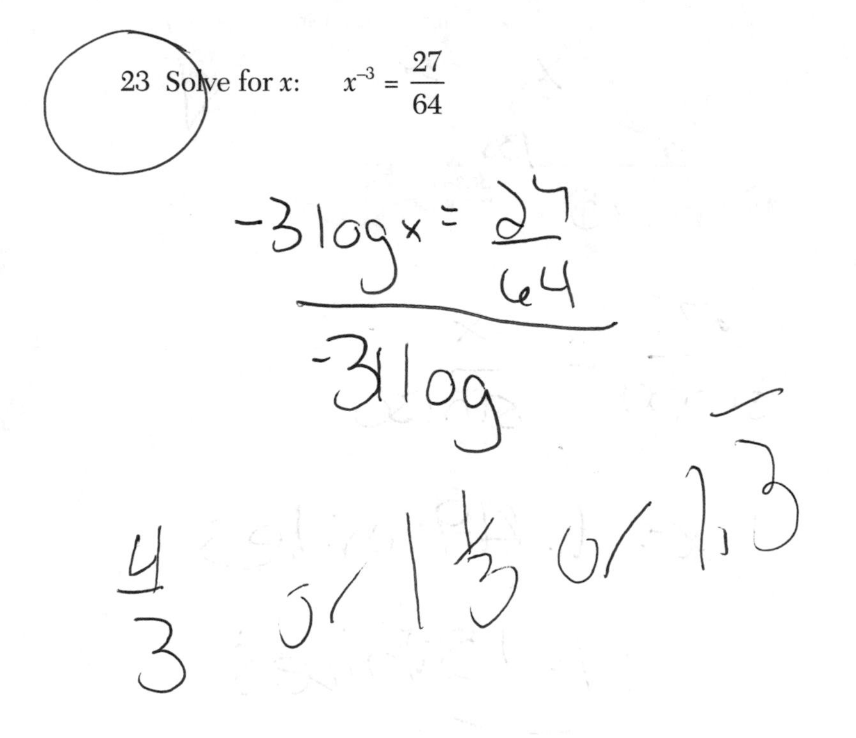

24 The profit a coat manufacturer makes each day is modeled by the equation $P(x) = -x^2 + 120x - 2{,}000$, where P is the profit and x is the price for each coat sold. For what values of x does the company make a profit? [The use of the accompanying grid is optional.]

x = 21 - 100

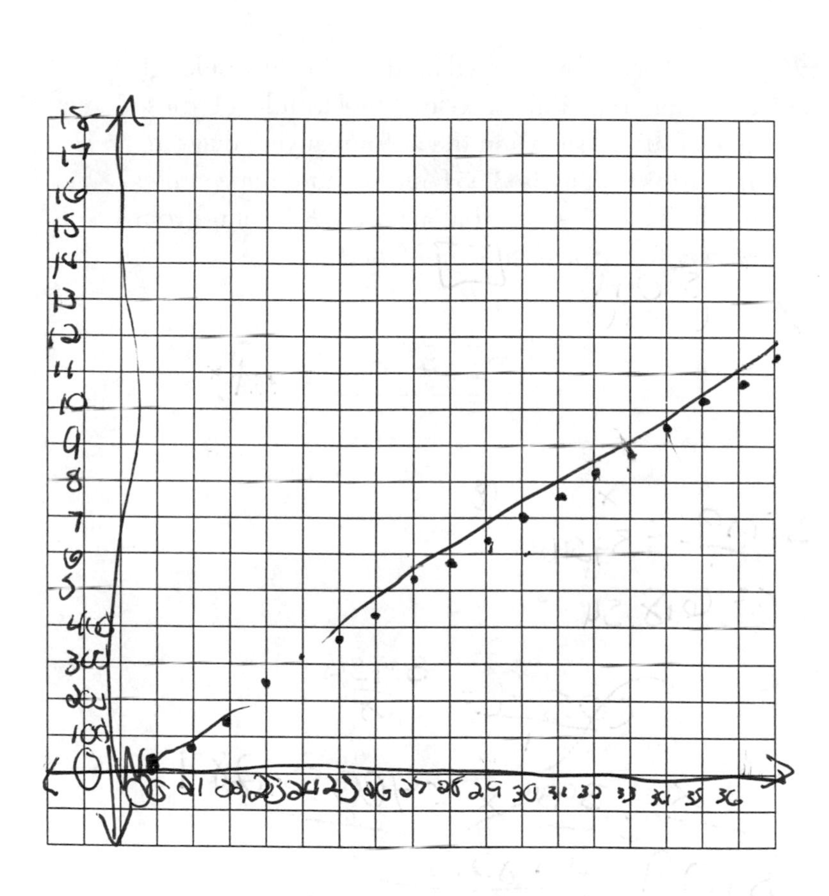

25 Express in simplest form: $\dfrac{\dfrac{1}{r}-\dfrac{1}{s}}{\dfrac{r^2}{s^2}-1}$

26 Cities H and K are located on the same line of longitude and the difference in the latitude of these cities is 9°, as shown in the accompanying diagram. If Earth's radius is 3,954 miles, how many miles north of city K is city H along arc HK? Round your answer to the *nearest tenth of a mile.*

3,954 mi
H
A
9°
K

(Not drawn to scale)

PART III

Answer all questions in this part. Each correct answer will receive 4 credits. Clearly indicate the necessary steps, including appropriate formula substitutions, diagrams, graphs, charts, etc. For all questions in this part, a correct numerical answer with no work shown will receive only 1 credit. [24]

27 A depth finder shows that the water in a certain place is 620 feet deep. The difference between d, the actual depth of the water, and the reading is $|d - 620|$ and must be less than or equal to $0.05d$. Find the minimum and maximum values of d, to the *nearest tenth of a foot*.

$|d - 620| \leq 0.05d$

$d - 620 \leq 0.05d$

621.1

590.5 + 652.6

28 An amount of P dollars is deposited in an account paying an annual interest rate r (as a decimal) compounded n times per year. After t years, the amount of money in the account, in dollars, is given by the equation $A = P\left(1+\frac{r}{n}\right)^{nt}$.

Rachel deposited $1,000 at 2.8% annual interest, compounded monthly. In how many years, to the *nearest tenth of a year,* will she have $2,500 in the account? [The use of the accompanying grid is optional.]

$$\frac{2500}{1000} = \frac{(1000)\left(1+\frac{.028}{12}\right)^{(12)(t)}}{1000}$$

$$2.5 = \left(1+\frac{.028}{12}\right)^{12t}$$

$$2.5 = 1.002^{12t}$$

$$12t \log 1.002$$

$$\frac{.397}{8.667} = \frac{12t(8.677)}{}$$

32.8 years

29 A box containing 1,000 coins is shaken, and the coins are emptied onto a table. Only the coins that land heads up are returned to the box, and then the process is repeated. The accompanying table shows the number of trials and the number of coins returned to the box after each trial.

Trial	0	1	3	4	6
Coins Returned	1,000	610	220	132	45

Write an exponential regression equation, rounding the calculated values to the *nearest ten-thousandth.*

−.9999

Use the equation to predict how many coins would be returned to the box after the eighth trial.

y = 1018.28 • .59685 ^ x

16 coins

30 Tim Parker, a star baseball player, hits one home run for every ten times he is at bat. If Parker goes to bat five times during tonight's game, what is the probability that he will hit *at least* four home runs?

31 A rectangular piece of cardboard is to be formed into an uncovered box. The piece of cardboard is 2 centimeters longer than it is wide. A square that measures 3 centimeters on a side is cut from each corner. When the sides are turned up to form the box, its volume is 765 cubic centimeters. Find the dimensions, in centimeters, of the original piece of cardboard.

32 Solve algebraically for all values of θ in the interval $0° \leq \theta < 360°$ that satisfy the equation $\frac{\sin^2\theta}{1+\cos\theta} = 1$.

PART IV

Answer all questions in this part. Each correct answer will receive 6 credits. Clearly indicate the necessary steps, including appropriate formula substitutions, diagrams, graphs, charts, etc. For all questions in this part, a correct numerical answer with no work shown will receive only 1 credit. [12]

33 The tide at a boat dock can be modeled by the equation $y = -2\cos\left(\frac{\pi}{6}t\right) + 8$, where t is the number of hours past noon and y is the height of the tide, in feet. For how many hours between $t = 0$ and $t = 12$ is the tide at least 7 feet? [The use of the accompanying grid is optional.]

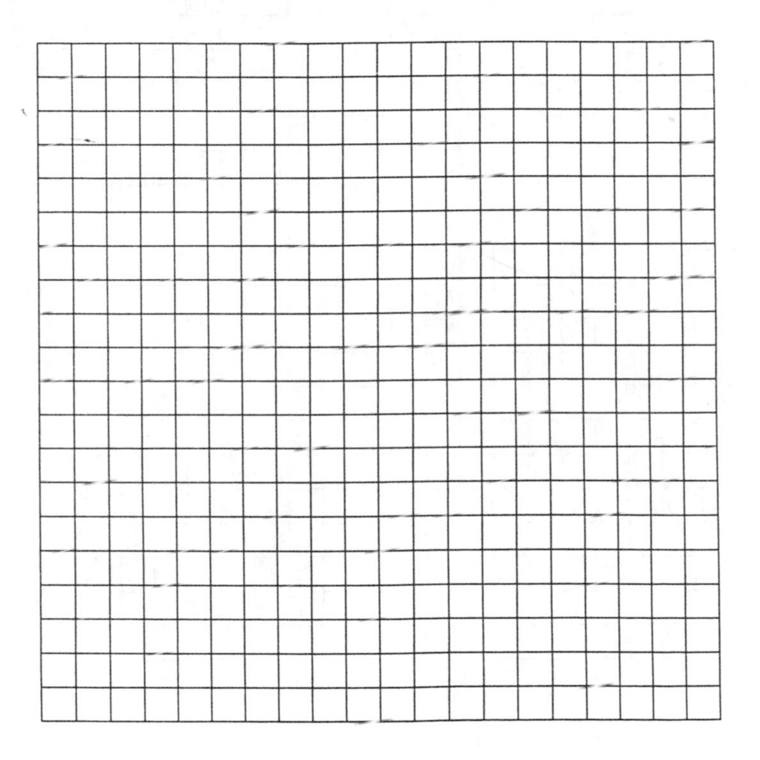

34 The coordinates of quadrilateral *JKLM* are $J(1,-2)$, $K(13,4)$, $L(6,8)$, and $M(-2,4)$. Prove that quadrilateral *JKLM* is a trapezoid but *not* an isosceles trapezoid. [The use of the accompanying grid is optional.]

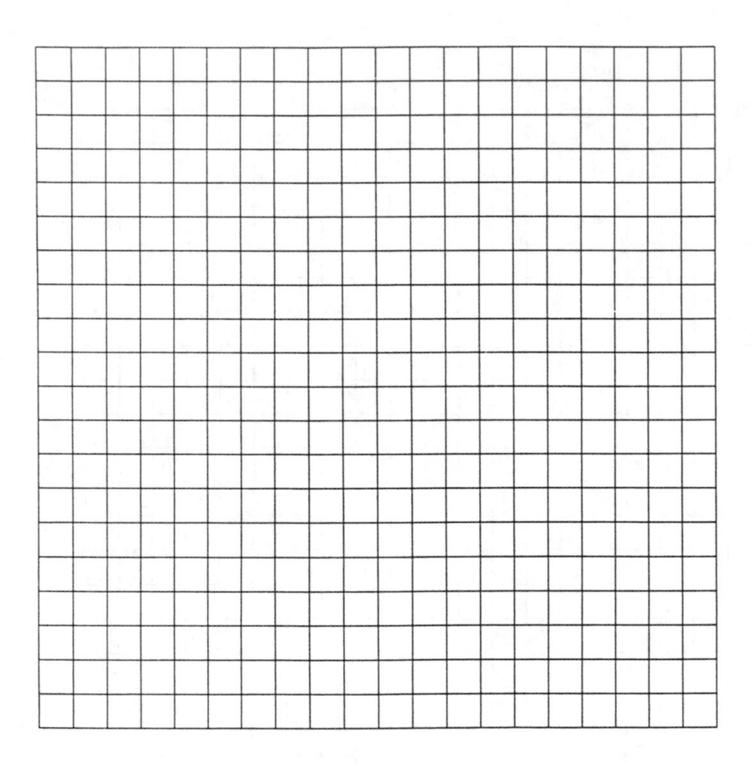

Answers to the Regents Examinations

JUNE 2004

PART I

1. 1	**6.** 4	**11.** 4	**16.** 1
2. 1	**7.** 3	**12.** 3	**17.** 4
3. 2	**8.** 1	**13.** 4	**18.** 2
4. 2	**9.** 4	**14.** 3	**19.** 2
5. 3	**10.** 2	**15.** 1	**20.** 3

PART II

21. 41,583

22. $\frac{3}{2}$

23. $k > \frac{1}{3}$

24. $\triangle HOC$ and opposite

25. Indirect proof showing $\overline{AT} \perp \overline{CD}$

26. –17

PART III

27. 44

28. 65.27

29. $\frac{-3 \pm \sqrt{37}}{7}$

30. 3 and 42

31. 1,500

32. 84

PART IV

33. 80 and 9.2

34. 312 and 30,642

AUGUST 2004

PART I

1. 2	**6.** 4	**11.** 1	**16.** 3
2. 3	**7.** 3	**12.** 3	**17.** 2
3. 1	**8.** 3	**13.** 4	**18.** 4
4. 1	**9.** 1	**14.** 2	**19.** 4
5. 2	**10.** 4	**15.** 2	**20.** 1

PART II

21. 1.15

22. $8.5 + 7i\sqrt{3}$

23. $\frac{4}{3}$ or $1\frac{1}{3}$ or $1.\overline{3}$

24. $20 < x < 100$

25. $-\frac{s}{r(r+s)}$ or $-\frac{s}{r^2+rs}$

26. 621.1

PART III

27. 590.5 and 652.6

28. 32.8

29. $y = 1{,}018.2839(0.5969)^x$ and 16

30. .00046 or $\frac{46}{100{,}000}$

31. 21 by 23

32. 90 and 270

PART IV

33. 8

34. $\overline{JK} \parallel \overline{ML}$, $\overline{MJ} \nparallel \overline{KL}$, and $\overline{MJ} \neq \overline{KL}$

INDEX

NOTES